Green Chemical Synthesis with Microwaves and Ultrasound

Green Chemical Synthesis with Microwaves and Ultrasound

Edited by Dakeshwar Kumar Verma, Chandrabhan Verma, and Paz Otero Fuertes

WILEY VCH

Editors

Dr. Dakeshwar Kumar Verma
Govt. Digvijay Autonomous
Postgraduate College
Department of Chemistry
Rajnandgaon 491441
Chhattisgarh
India

Dr. Chandrabhan Verma
Khalifa University of Science and
Technology
Department of Chemical Engineering
P.O. Box 127788
Abu Dhabi
United Arab Emirates

Dr. Paz Otero Fuertes
University of Vigo
Faculty of Food Science and Technology
Analytical and Food Chemistry
Department
Nutrition and Bromatology Group
Ourense 32004
Spain

Cover Image: © motorolka/Shutterstock

All books published by **WILEY-VCH** are carefully produced. Nevertheless, authors, editors, and publisher do not warrant the information contained in these books, including this book, to be free of errors. Readers are advised to keep in mind that statements, data, illustrations, procedural details or other items may inadvertently be inaccurate.

Library of Congress Card No.: applied for

British Library Cataloguing-in-Publication Data
A catalogue record for this book is available from the British Library.

Bibliographic information published by the Deutsche Nationalbibliothek
The Deutsche Nationalbibliothek lists this publication in the Deutsche Nationalbibliografie; detailed bibliographic data are available on the Internet at <http://dnb.d-nb.de>.

© 2024 WILEY-VCH GmbH, Boschstraße 12, 69469 Weinheim, Germany

All rights reserved (including those of translation into other languages). No part of this book may be reproduced in any form – by photoprinting, microfilm, or any other means – nor transmitted or translated into a machine language without written permission from the publishers. Registered names, trademarks, etc. used in this book, even when not specifically marked as such, are not to be considered unprotected by law.

Print ISBN: 978-3-527-35297-5
ePDF ISBN: 978-3-527-84447-0
ePub ISBN: 978-3-527-84448-7
oBook ISBN: 978-3-527-84449-4

Typesetting: Straive, Chennai, India
Printing and Binding: CPI Group (UK) Ltd, Croydon, CR0 4YY

Contents

About the Editors *xiii*
Preface *xv*

1 Ultrasound Irradiation: Fundamental Theory, Electromagnetic Spectrum, Important Properties, and Physical Principles *1*
Sumit Kumar, Amrutlal Prajapat, Sumit K. Panja, and Madhulata Shukla

1.1 Introduction *1*
1.2 Cavitation History *3*
1.2.1 Basics of Cavitation *3*
1.2.2 Types of Cavitation *5*
1.3 Application of Ultrasound Irradiation *7*
1.3.1 Sonoluminescence and Sonophotocatalysis *9*
1.3.2 Industrial Cleaning *10*
1.3.3 Material Processing *11*
1.3.4 Chemical and Biological Reactions *12*
1.4 Conclusion *14*
Acknowledgments *15*
References *15*

2 Fundamental Theory of Electromagnetic Spectrum, Dielectric and Magnetic Properties, Molecular Rotation, and the Green Chemistry of Microwave Heating Equipment *21*
Raghvendra K. Mishra, Akshita Yadav, Vinayak Mishra, Satya N. Mishra, Deepa S. Singh, and Dakeshwar Kumar Verma

2.1 Introduction *21*
2.1.1 Historical Background *25*
2.1.2 Green Chemistry Principles for Sustainable System *28*
2.2 Fundamental Concepts of the Electromagnetic Spectrum Theory *35*
2.3 Electrical, Dielectric, and Magnetic Properties in Microwave Irradiation *38*
2.4 Microwave Irradiation Molecular Rotation *41*
2.5 Fundamentals of Electromagnetic Theory in Microwave Irradiation *42*
2.5.1 Electromagnetic Radiations and Microwave *43*

2.5.2	Heating Mechanism of Microwave: Conventional Versus Microwave Heating *44*	
2.6	Physical Principles of Microwave Heating and Equipment *46*	
2.7	Green Chemistry Through Microwave Heating: Applications and Benefits *53*	
2.8	Conclusion *57*	
	References *57*	
3	**Conventional Versus Green Chemical Transformation: MCRs, Solid Phase Reaction, Green Solvents, Microwave, and Ultrasound Irradiation** *69*	
	Shailendra Yadav, Dheeraj S. Chauhan, and Mumtaz A. Quraishi	
3.1	Introduction *69*	
3.2	A Brief Overview of Green Chemistry *69*	
3.2.1	Definition and Historical Background *69*	
3.2.2	Significance *70*	
3.3	Multicomponent Reactions *71*	
3.4	Solid Phase Reactions *73*	
3.5	Microwave Induced Synthesis *74*	
3.6	Ultrasound Induced Synthesis *75*	
3.7	Green Chemicals and Solvents *77*	
3.8	Conclusions and Outlook *78*	
	References *79*	
4	**Metal-Catalyzed Reactions Under Microwave and Ultrasound Irradiation** *83*	
	Suresh Maddila, Immandhi S.S. Anantha, Pamerla Mulralidhar, Nagaraju Kerru, and Sudhakar Chintakula	
4.1	Ultrasonic Irradiation *83*	
4.1.1	Iron-Based Catalysts *86*	
4.1.2	Copper-Based Catalysts *89*	
4.1.2.1	Dihydropyrimidinones by Cu-Based Catalysts *91*	
4.1.2.2	Dihydroquinazolinones by Cu-Based Catalysts *92*	
4.1.3	Misalliances Metal-Based Catalysts *94*	
4.2	Microwave-Assisted Reactions *97*	
4.2.1	Solid Acid and Base Catalysts *98*	
4.2.1.1	Condensation Reactions *98*	
4.2.1.2	Cyclization Reactions *100*	
4.2.1.3	Multi-component Reactions *104*	
4.2.1.4	Friedel–Crafts Reactions *106*	
4.2.1.5	Reaction Involving Catalysts of Biological Origin *107*	
4.2.1.6	Reduction *109*	
4.2.1.7	Oxidation *110*	
4.2.1.8	Coupling Reactions *113*	
4.2.1.9	Micelliances Reactions *121*	
4.2.1.10	Click Chemistry *125*	

4.3	Conclusion *127*
	Acknowledgments *128*
	References *128*

5   Microwave- and Ultrasonic-Assisted Coupling Reactions *133*
Sandeep Yadav, Anirudh P.S. Raman, Kashmiri Lal, Pallavi Jain, and Prashant Singh

5.1	Introduction *133*
5.2	Microwave *134*
5.2.1	Microwave-Assisted Coupling Reactions *135*
5.2.2	Ultrasound-Assisted Coupling Reactions *142*
5.3	Conclusion *150*
	References *151*

6   Synthesis of Heterocyclic Compounds Under Microwave Irradiation Using Name Reactions *157*
Sheryn Wong and Anton V. Dolzhenko

6.1	Introduction *157*
6.2	Classical Methods for Heterocyclic Synthesis Under Microwave Irradiation *158*
6.2.1	Piloty–Robinson Pyrrole Synthesis *158*
6.2.2	Clauson–Kaas Pyrrole Synthesis *158*
6.2.3	Paal–Knorr Pyrrole Synthesis *159*
6.2.4	Paal–Knorr Furan Synthesis *160*
6.2.5	Paal–Knorr Thiophene Synthesis *160*
6.2.6	Gewald Reaction *161*
6.2.7	Fischer Indole Synthesis *162*
6.2.8	Bischler–Möhlau Indole Synthesis *162*
6.2.9	Hemetsberger–Knittel Indole Synthesis *163*
6.2.10	Leimgruber–Batcho Indole Synthesis *163*
6.2.11	Cadogan–Sundberg Indole Synthesis *163*
6.2.12	Pechmann Pyrazole Synthesis *164*
6.2.13	Debus–Radziszewski Reaction *164*
6.2.14	van Leusen Imidazole Synthesis *166*
6.2.15	van Leusen Oxazole Synthesis *166*
6.2.16	Robinson–Gabriel Reaction *167*
6.2.17	Hantzsch Thiazole Synthesis *167*
6.2.18	Einhorn–Brunner Reaction *168*
6.2.19	Pellizzari Reaction *169*
6.2.20	Huisgen Reaction *169*
6.2.21	Finnegan Tetrazole Synthesis *171*
6.2.22	Four-component Ugi-azide Reaction *172*
6.2.23	Kröhnke Pyridine Synthesis *172*
6.2.24	Bohlmann–Rahtz Pyridine Synthesis *173*
6.2.25	Boger Reaction *174*

6.2.26	Skraup Reaction	*174*
6.2.27	Gould–Jacobs Reaction	*175*
6.2.28	Friedländer Quinoline Synthesis	*176*
6.2.29	Povarov Reaction	*176*
6.3	Conclusion	*177*
	Acknowledgments	*177*
	References	*177*

7 Microwave- and Ultrasound-Assisted Enzymatic Reactions *185*
Nafseen Ahmed, Chandan K. Mandal, Varun Rai, Abbul Bashar Khan, and Kamalakanta Behera

7.1	Introduction	*185*
7.2	Influence Microwave Radiation on the Stability and Activity of Enzymes	*186*
7.3	Principle of Ultrasonic-Assisted Enzymolysis	*190*
7.4	Applications of Ultrasonic-Assisted Enzymolysis	*192*
7.4.1	Proteins and Other Plant Components Can Be Transformed and Extracted	*192*
7.4.2	Modification of Protein Functionality	*193*
7.4.3	Enhancement of Biological Activity	*194*
7.4.4	Ultrasonic-Assisted Acceleration of Hydrolysis Time	*195*
7.5	Enzymatic Reactions Supported by Ultrasound	*196*
7.5.1	Lipase	*196*
7.5.2	Protease	*196*
7.5.3	Polysaccharide Enzymes	*198*
7.6	Biodiesel Production via Ultrasound-Supported Transesterification	*198*
7.6.1	Homogenous Acid-Catalyzed Ultrasound-Assisted Transesterification	*199*
7.6.2	Transesterification with Ultrasound Assistance and Homogenous Base Catalysis	*199*
7.6.3	Heterogeneous Acid-Catalyzed Ultrasound-Assisted Transesterification	*201*
7.6.4	Heterogeneous Base-Catalyzed Ultrasound-Assisted Transesterification	*205*
7.6.5	Enzyme-Catalyzed Ultrasound-Assisted Transesterification	*207*
7.7	Conclusions	*207*
	Acknowledgments	*209*
	References	*209*

8 Microwave- and Ultrasound-Assisted Synthesis of Polymers *219*
Anupama Singh, Sushil K. Sharma, and Shobhana Sharma

8.1	Introduction	*219*

8.2	Microwave-Assisted Synthesis of Polymers	*220*
8.3	Ultrasound-Assisted Synthesis of Polymers	*223*
8.4	Conclusion	*228*
	References	*229*

9 Synthesis of Nanomaterials Under Microwave and Ultrasound Irradiation *235*

Ahmed A. Mohamed

9.1	Introduction	*235*
9.2	Synthesis of Metal Nanoparticles	*236*
9.3	Synthesis of Carbon Dots	*239*
9.4	Synthesis of Metal Oxides	*240*
9.5	Synthesis of Silicon Dioxide	*243*
9.6	Conclusion	*243*
	References	*244*

10 Microwave- and Ultrasound-Assisted Synthesis of Metal-Organic Frameworks (MOF) and Covalent Organic Frameworks (COF) *249*

Sanjit Gaikwad and Sangil Han

10.1	Introduction	*249*
10.2	Principles	*250*
10.2.1	Principles of Microwave Heating	*250*
10.2.2	Principle of Ultrasound-Assisted Techniques	*250*
10.2.3	Advantages and Disadvantages of Microwave- and Ultrasound-Assisted Techniques	*252*
10.3	MOF Synthesis by Microwave and Ultrasound Method	*252*
10.3.1	Microwave-Assisted Synthesis of MOF	*253*
10.3.2	Ultrasound-Assisted Synthesis of MOFs	*256*
10.4	Factors That Affect MOF Synthesis	*257*
10.4.1	Solvent	*257*
10.4.2	Temperature and pH	*258*
10.5	Application of MOF	*260*
10.6	COF Synthesis by Microwave and Ultrasound Method	*262*
10.6.1	Ultrasound-Assisted Synthesis of COFs	*262*
10.6.2	Microwave-Assisted Synthesis of COF	*262*
10.6.3	Structure of COF (2D and 3D)	*263*
10.7	Factors Affecting the COF Synthesis	*266*
10.8	Applications of COFs	*267*
10.9	Future Predictions	*269*
10.10	Summary	*269*
	Acknowledgments	*269*
	References	*270*

11	**Solid Phase Synthesis Catalyzed by Microwave and Ultrasound Irradiation** *283*	
	R.M. Abdel Hameed, Amal Amr, Amina Emad, Fatma Yasser, Haneen Abdullah, Mariam Nabil, Nada Hazem, Sara Saad, and Yousef Mohamed	
11.1	Introduction *283*	
11.2	Wastewater Treatment *284*	
11.3	Biodiesel Production *289*	
11.4	Oxygen Reduction Reaction *297*	
11.5	Alcoholic Fuel Cells *306*	
11.6	Conclusion and Future Plans *313*	
	References *313*	
12	**Comparative Studies on Thermal, Microwave-Assisted, and Ultrasound-Promoted Preparations** *337*	
	Tri P. Adhi, Aqsha Aqsha, and Antonius Indarto	
12.1	Introduction *337*	
12.1.1	Background on Preparative Techniques in Chemistry *337*	
12.1.2	Overview of Thermal, Microwave-Assisted, and Ultrasound-Promoted Preparations *338*	
12.1.3	Significance of Comparative Studies in Enhancing Synthetic Methodologies *341*	
12.1.3.1	Optimization of Conditions *341*	
12.1.3.2	Efficiency Improvement *342*	
12.1.3.3	Methodological Advances *343*	
12.1.3.4	Sustainability and Green Chemistry *343*	
12.2	Fundamentals of Thermal, Microwave-Assisted, and Ultrasound-Assisted Reactions *345*	
12.2.1	Explanation of Thermal Reactions and Their Advantages and Limitations *345*	
12.2.2	Introduction to Microwave-Assisted Reactions and How They Differ from Traditional Method *346*	
12.2.3	Understanding the Principles and Mechanisms of Ultrasound-Promoted Reactions *347*	
12.3	Case Studies in Organic Synthesis *349*	
12.3.1	Examining Examples of Organic Reactions Performed Under Thermal Conditions *349*	
12.3.1.1	Esterification Reaction Under Thermal Conditions *349*	
12.3.1.2	Dehydration of Alcohols *349*	
12.3.1.3	Oxidation of Aldehydes to Carboxylic Acids Using Water *350*	
12.3.2	Case Studies Showcasing the Application of Microwave-Assisted Reactions *350*	
12.3.2.1	Microwave-Assisted C—C Bond Formation *351*	
12.3.2.2	Microwave-Assisted Cyclization *352*	
12.3.2.3	Microwave-Assisted Dehydrogenation Reactions *353*	
12.3.2.4	Microwave-Assisted Organic Synthesis *353*	

12.3.3	Highlighting Successful Instances of Ultrasound-Promoted Organic Synthesis *353*	
12.3.3.1	Ultrasound-Promoted in Organic Synthesis *354*	
12.3.3.2	Ultrasound-Promoted Oxidations *354*	
12.3.3.3	Ultrasound-Promoted Esterification *354*	
12.3.3.4	Ultrasound-Promoted Cyclization *354*	
12.4	Scope and Limitations *355*	
12.4.1	Discussing the Applicability of Each Method to Different Reaction Types *355*	
12.4.2	Identifying the Limitations and Challenges Faced by Each Technique *357*	
12.4.3	Opportunities for Combining Approaches to Overcome Specific Limitations *358*	
12.5	Future Directions and Emerging Trends *359*	
12.5.1	Overview of Recent Advancements and Ongoing Research in Thermal, Microwave, and Ultrasound-Assisted Preparations *359*	
12.5.1.1	Food Processing Technologies *360*	
12.5.1.2	Chemical Routes to Materials: Thermal Oxidation of Graphite for Graphene Preparation *360*	
12.5.1.3	Environmental and Sustainable Applications: Waste to Energy *361*	
12.5.2	Recent Findings in Microwave-Assisted Preparation *361*	
12.5.2.1	Catalyst *361*	
12.5.2.2	Nanotechnology *362*	
12.5.3	Food Processing Technologies *362*	
12.5.4	Ultrasound-Assisted Preparations *363*	
12.5.4.1	Biomedical *363*	
12.5.4.2	Artificial Intelligence (AI) *363*	
12.6	Identification of Potential Areas for Further Exploration and Improvement *363*	
12.6.1	Reaction Mechanisms and Kinetics *363*	
12.6.2	Synergistic Effects *364*	
12.6.3	Green Chemistry and Sustainability *366*	
12.6.4	Scale-Up and Industrial Application *366*	
12.6.5	Catalysis and Selectivity *367*	
12.6.6	In Situ Monitoring and Control *367*	
12.6.7	Mechanistic Studies *368*	
12.6.8	Temperature and Energy Management *368*	
12.6.9	Materials Processing *369*	
12.6.10	Biomedical Applications *370*	
12.7	The Role of Artificial Intelligence and Computational Approaches in Optimizing Preparative Techniques *370*	
	References *372*	

Index *381*

About the Editors

Dr. Dakeshwar Kumar Verma holds a PhD in chemistry and serves as Assistant Professor of Chemistry at Government of Digvijay Autonomous Postgraduate College in Rajnandgaon, Chhattisgarh, India. He is driven by a profound passion for scientific exploration, and his research focuses primarily on the preparation and design of organic compounds for diverse applications. With an impressive track record, Dr. Verma has authored more than 100 research articles, review articles, and book chapters that have found their places in esteemed peer-reviewed international journals, including those published in *Nature* and by ACS, RSC, Wiley, Elsevier, Springer, and Taylor & Francis, among others. This extensive body of work showcases his dedication to contributing to the collective knowledge of the scientific community. Beyond his role as an author, Dr. Verma is an active and valued participant in the peer-review process. As a testament to his academic influence, Dr. Verma has taken on editorial/authored responsibilities for various published and upcoming books, slated to be published by renowned publishers such as ACS, RSC, Wiley, Springer, Taylor & Francis, Elsevier, and De Gruyter. This role highlights his dedication to shaping and disseminating knowledge in various scientific domains. With a cumulative citation count of more than 1850, an h-index of 26, and an i-10 index of 34, Dr. Verma's impact is evident through the recognition and relevance of his work in the scientific community. His commitment to fostering the growth of future scholars is also evident in his supervision of two full-time PhD research scholars. Dr. Verma's dedication to research excellence has been acknowledged through prestigious awards such as the Council of Scientific and Industrial Research Junior Research Fellowship award in 2013. During his PhD journey in 2013, he also achieved the MHRD National fellowship, further solidifying his commitment to academic growth and advancement. His multifaceted role as a researcher, reviewer, editor, invited speaker, resource person, and mentor underscores his substantial impact on the academic landscape.

Chandrabhan Verma works at the Department of Chemical Engineering, Khalifa University of Science and Technology, Abu Dhabi, United Arab Emirates. Dr. Verma obtained BSc and MSc degrees from Udai Pratap Autonomous College, Varanasi (UP), India. He received his PhD from the Department of Chemistry, Indian Institute of Technology (Banaras Hindu University) Varanasi, under the supervision of Prof. Mumtaz A. Quraishi in Corrosion Science and Engineering. He is a member of the American Chemical Society (ACS) and a lifetime member of the World Research Council (WRC). He is a reviewer and editorial board member of various internationally recognized journals from ACS, RSC, Elsevier, Wiley, and Springer. Dr. Verma has published numerous research and review articles with ACS, Elsevier, RSC, Wiley and Springer, etc., in different areas of science and engineering. His current research focuses on designing and developing industrially applicable corrosion inhibitors. Dr. Verma has edited a few books for the ACS, Elsevier, RSC, Springer, and Wiley. He has received several awards for his academic achievements, including a gold medal in MSc (Organic Chemistry; 2010) and Best Publication awards from the Global Alumni Association of IIT-BHU (Second Prize 2013).

Dr. Paz Otero Fuertes received her bachelor's degree in food science from the University of Santiago de Compostela (USC), Spain. After that, she completed her PhD at the Pharmacology Department of the Veterinary Faculty in the same University and was awarded with the distinction Cum Laude and the Special PhD Award. Dr. Paz Otero has held postdoctoral research positions in Limerick Institute of Technology, Ireland (2014–2017), and The Institute of Food Science Research, CIAL (2017–2018) at the Autonomous University of Madrid. She has published extensively in the field of food chemistry, toxicology, analytical chemistry, pharmacology, nutrition, and phycotoxin biology with more than 56 authored research articles, 70 contributions to international congress, and 12 book chapters to date. Her h-index is 28, and her total citation is 1550 in Google Scholar database. Dr. Paz Otero Fuertes also serves as invited reviewer for several research journals, editorial board member for online free-access journals, and guest editor for special issues.

Preface

Chemical transformations mediated by ultrasound (US) and microwaves (MW) benefit different chemical processes. The following are some of the main advantages of each approach: Green chemistry principles are often employed in assessing the environmental impact of chemical reactions, particularly those facilitated by ultrasound and microwave radiation. Green chemistry aims to create and construct procedures that use less energy, produce fewer dangerous compounds, and are as efficient as possible. Reaction mixtures can be heated quickly and precisely using microwave heating, as is well known. Comparing this to traditional heating techniques can result in quicker reaction times and lower energy usage. Through the acceleration of chemical reactions and the promotion of effective mass transfer, ultrasonic waves can also increase reaction rates. This may lead to lower energy needs and more general efficiency. The use of large volumes of solvents is frequently reduced or eliminated when reactions may be conducted under milder circumstances, thanks to the capabilities of both microwave and ultrasonic technologies. This is consistent with the green chemistry idea of reducing the amount of hazardous or environmentally damaging solvents used. Specific reactions can be more selective than others due to the cavitation effects of ultrasound and the selective heating created by microwaves. Doing so can decrease waste and byproduct production, making the process more environmentally friendly. Reaction times are frequently shortened by the faster reaction rates associated with ultrasound and microwave techniques. This can save time and energy in producing a given amount of goods, positively affecting the environment and the economy. By reducing the possibility of thermal runaway or adverse reactions, the mechanical impacts of ultrasonic waves and the regulated and targeted heating offered by microwaves can help provide safer reaction conditions. Specific reactions mediated by microwaves and ultrasonography might be readily scaled up, enabling more significant, environmentally friendly operations.

This book explores the fundamentals, contemporary trends, obstacles, and potential future applications of microwave- and ultrasound-assisted chemical transformation irradiations, demonstrating their worth and range. Each of the 12 chapters in this book covers a distinct facet of nonconventional chemical reactions. The fundamental theories and principles of ultrasound-mediated reactions are covered in Chapter 1, along with the opportunities and problems that exist today.

The theory and fundamentals of microwave-mediated reactions are covered in Chapter 2. Chapter 3 compares the challenges and prospects of conventional and MW-/US-mediated chemical transformation. Metal-catalyzed and coupling processes under MW and US irradiation are covered in Chapters 4 and 5, respectively. The synthesis of bioactive heterocycles, enzymatic processes, polymers, and nanomaterials under MW and US irradiation are covered in Chapters 6, 7, 8, and 9, respectively. The synthesis of covalent organic frameworks (COFs) and metal–organic frameworks (MOFs) mediated by MW and US is covered in Chapter 10. The benefits of MW and US irradiation in solid-phase syntheses are discussed in Chapter 11. Chapter 12 concludes with a comparison of chemical changes facilitated by thermal, microwave, and ultrasonic heating.

We are very thankful to the authors of all chapters for their outstanding and passionate efforts in making this book. Special thanks to the Wiley staff Dr. Sakeena Quraishi (Commissioning Editor), Judit Anbu Hena (Content Refinement Specialist), Shwathi Srinivasan (Managing Editor, Advanced Chemistry and Chemical Engineering), and Tanya Domeier for their dedicated support and help during this project. In the end, all thanks to Wiley for publishing the book.

March 2024

Dakeshwar Kumar Verma
Chhattisgarh, India
Chandrabhan Verma
Abu Dhabi, United Arab Emirates
Paz Otero Fuertes
Ourense, Spain

1

Ultrasound Irradiation: Fundamental Theory, Electromagnetic Spectrum, Important Properties, and Physical Principles

Sumit Kumar[1], Amrutlal Prajapat[2], Sumit K. Panja[2], and Madhulata Shukla[3]

[1]Magadh University, Department of Chemistry, Bodh Gaya 824234, Bihar, India
[2]Uka Tarsadia University, Tarsadia Institute of Chemical Science, Maliba Campus, Gopal Vidyanagar, Bardoli, Mahuva Road, Surat 394350, Gujarat, India
[3]Veer Kunwar Singh University, Gram Bharti College, Department of Chemistry, Ramgarh, Kaimur 821110, Bihar, India

1.1 Introduction

US, also referred to as ultrasonic treatment or sonication, employs high frequency sound waves to agitate particles in a liquid or solid medium [1]. This process relies on the phenomenon of cavitation, which happens when high-intensity sound waves create small bubbles in a liquid. These bubbles rapidly expand and collapse, producing pressure and temperature gradients that can break down particles and disrupt chemical bonds. This is known as acoustic cavitation, and it can be utilized for various purposes, including emulsification, dispersion, mixing, and extraction. Additionally, US can increase the surface area of reactants and enhance chemical reactions by promoting mass transfer between phases. It can also induce the formation of free radicals, which can react with target compounds and break them down. US is widely used in a range of fields, such as wastewater treatment, food processing, pharmaceuticals, and materials science [2–4]. The effectiveness of US depends on several factors, such as the frequency and intensity of the sound waves, the duration of exposure, and the characteristics of the medium being treated. Cavitation can be generated either by passing ultrasonic energy in the liquid medium or by utilizing alterations in the velocity/pressure in hydraulic systems. The intensity of cavitation, and hence the net chemical/physical effects, relies heavily on the operating and design parameters, including reaction temperature, hydrostatic pressure, irradiation frequency, acoustic power, and ultrasonic intensity. To increase the extent or rate of reaction, cavitation can be combined with one or more irradiations or some additives can be utilized, which can be solids or gases and can sometimes have catalytic effects. The free radicals generated during the oxidation process consist of hydroxyl ($^{\bullet}OH$), hydrogen ($^{\bullet}H$), and hydroperoxyl ($HO_2^{\bullet}$) radicals. Overall, the theory behind US is based on the principles of acoustic cavitation, which can be harnessed to achieve a variety of physical, chemical, and biological effects.

Green Chemical Synthesis with Microwaves and Ultrasound, First Edition.
Edited by Dakeshwar Kumar Verma, Chandrabhan Verma, and Paz Otero Fuertes.
© 2024 WILEY-VCH GmbH. Published 2024 by WILEY-VCH GmbH.

US refers to the application of high-frequency sound waves to a target material or medium. Here are some properties of US:

Frequency: Ultrasound waves have frequencies above the upper limit of human hearing, typically between 20 kHz and several MHz (megahertz). The frequency determines the energy and penetration depth of the ultrasound waves.

Wavelength: The wavelength of ultrasound waves is inversely proportional to the frequency. Higher frequencies result in shorter wavelengths. This property allows ultrasound waves to interact with small-scale structures and particles.

Intensity: Ultrasound intensity refers to the amount of energy carried by the sound waves per unit area. It determines the strength of the ultrasound waves and their effect on the target material. Ultrasound intensity is typically measured in units of watts per square centimeter (W/cm^2).

Propagation: Ultrasound waves propagate through materials as longitudinal waves, causing the particles of the medium to vibrate in the direction of wave propagation. This enables the transmission of energy and information through the medium.

Absorption: Ultrasound waves can be absorbed by materials they pass through. The extent of absorption depends on the properties of the material, such as its density, viscosity, and composition. Absorption leads to the conversion of ultrasound energy into heat, which can be utilized in various applications.

Reflection and refraction: When ultrasound waves encounter an interface between two different media, such as air and a solid object, some of the waves are reflected back and some are transmitted into the new medium. The angles of reflection and refraction obey the laws of physics similar to those governing light waves.

Cavitation: US can induce a phenomenon known as cavitation, where the rapid changes in pressure cause the formation and implosion of tiny bubbles in a liquid medium. Cavitation can generate localized high temperatures and pressures, which can be utilized in processes like sonochemistry and ultrasonic cleaning.

Noninvasiveness: Ultrasound waves can be transmitted through the body noninvasively, making them useful in medical imaging techniques like ultrasound scans and sonograms. They provide real-time visualization of internal organs, tissues, and structures without the need for surgery or ionizing radiation.

Doppler effect: The Doppler effect occurs when there is relative motion between the source of ultrasound waves and the target. This effect causes a shift in the frequency of the reflected waves, enabling the measurement of blood flow, velocity, and direction in medical applications like Doppler ultrasound [5, 6].

Safety: US is generally considered safe for medical and industrial applications, as it does not involve ionizing radiation like X-rays or gamma rays. However, high-intensity ultrasound can cause thermal effects, and prolonged exposure to certain intensities may have biological effects. Safety guidelines and standards are in place to ensure the safe use of ultrasound in different applications.

1.2 Cavitation History

The phenomenon of cavitation was first observed by Thornycroft and Barnaby in 1895 when the propeller of their submarine became pitted and eroded over a short operating period. This was due to collapsing bubbles caused by hydrodynamic cavitation, which generated intense pressure and temperature gradients in the surrounding area [7]. In 1917, Rayleigh published the first mathematical model describing a cavitation event in an incompressible fluid [8]. It was not until 1927, when Loomis reported the first chemical and biological effects of ultrasound, that researchers realized the potential of cavitation as a useful tool in chemical reaction processes [9]. One of the earliest applications of ultrasound-induced cavitation was the degradation of a biological polymer [10]. Since then, the use of acoustic cavitation has become increasingly popular, particularly as a novel alternative to traditional methods for polymer production, enhancing chemical reactions, emulsifying oils, and degrading chemical or biological pollutants [11]. The advantage of utilizing acoustic cavitation for these applications is that it allows for much milder operating conditions compared to conventional techniques, and many reactions that may require toxic reagents or solvents are not necessary.

1.2.1 Basics of Cavitation

Ultrasound is a type of sound wave with a frequency above 20 kHz, and when it propagates through a liquid medium, it can create conditions for cavitation. Ultrasound has been extensively used as an intensifying approach in various fields, including chemical synthesis, electrochemistry, food technology, environmental engineering, materials, and nanomaterial science, biomedical engineering, biotechnology, sonocrystallization, and atomization [2, 12–21]. The use of ultrasound can lead to greener intensified processing with significant economic savings [22, 23]. Ultrasound-induced cavitation, also known as acoustic cavitation, is mainly due to the alternate compression and rarefaction cycles that drive the various stages of cavity inception, growth, and final collapse, as shown in Figure 1.1 [12].

When cavities collapse, a significant amount of energy is released, leading to the formation of acoustic streaming associated with turbulence resulting from the continuous generation and collapse of cavities in the system. Moreover, chemical effects, such as the occurrence of local hotspots in the interfacial region between the bubble and adjacent liquid, can generate free radicals [24]. The primary reactions that occur during sonication can be considered the initiator of a series of radical reactions depending on the species:

$$H_2O \leftrightarrow {}^{\bullet}OH + {}^{\bullet}H \tag{1.1}$$

$$H^{\bullet} + H^{\bullet} \leftrightarrow H_2 \tag{1.2}$$

$${}^{\bullet}OH + {}^{\bullet}OH \leftrightarrow H_2O_2 \tag{1.3}$$

$$H^{\bullet} + O_2 \leftrightarrow HO_2^{\bullet} \tag{1.4}$$

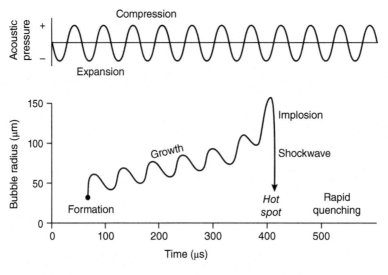

Figure 1.1 Schematic representation of the mechanism of generation of acoustic cavitation. Source: Reproduced from Gogate et al. [12]/John Wiley & Sons.

$$H^\bullet + HO_2^\bullet \leftrightarrow H_2O_2 \tag{1.5}$$

$$HO_2^\bullet + HO_2^\bullet \leftrightarrow H_2O_2 + O_2 \tag{1.6}$$

When ultrasound is applied to water, it causes the generation of $^\bullet OH$ and $H^\bullet$ radicals, which subsequently leads to the production of hydrogen peroxide (H_2O_2). Both of these agents are strong oxidizing agents. As the cavitation bubble collapses, it generates tremendous local pressure gradients, temperature, and microjets in the liquid at the collapse point [25]. The release of the accumulated energy during bubble collapse in the form of shock waves and hot spots can significantly enhance the reaction rate [26]. In large-scale sonochemical reactors, the two most important features of cavity dynamics are the maximum size reached before the violent collapse and the intensity of the collapse. Maximizing both of these effects in large-scale designs is necessary to achieve the desired processing efficacy.

The chemical changes associated with cavitation induced by the passage of sound waves are referred to as sonochemistry [1]. Ultrasound's chemical effects do not arise from direct interaction with molecular species but rather from acoustic cavitation, which involves the formation, growth, and implosive collapse of bubbles in a liquid, resulting in very high energy densities of $1–1018\,kW/m^3$ [1, 27]. Figure 1.2 depicts the mechanism of cavitation growth and collapse in liquid. The collapse takes place in microseconds and can be considered adiabatic. Cavitation can occur at millions of locations in a reactor simultaneously and generate conditions of very high temperatures and pressures (a few thousand atmospheres of pressure and a few thousand Kelvin of temperature) locally, while the overall environment remains at ambient conditions. As a result, chemical reactions that require stringent conditions can be effectively carried out using cavitation at ambient conditions.

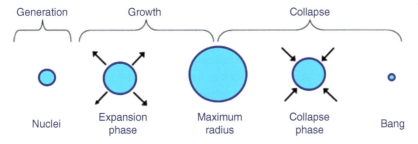

Figure 1.2 Mechanism of cavitation growth and collapse in liquid medium.

Acoustic cavitation is the process of nucleus growth and collapse of micro-gas bubbles or cavities in a liquid. This occurs rapidly, releasing large amounts of energy over a small area and creating extreme temperature and pressure gradients [23, 28, 29]. Cavitation generates high temperatures (between 1000 and 15 000 K) and pressures (between 500 and 5000 bar) locally and can occur at millions of locations within the reactor. Additionally, cavitation leads to acoustic streaming, intense shear stress near the collapsing bubble, and the formation of micro-jets. The local effects of cavitation are advantageous for reactions, including the generation of free radicals due to the dissociation of vapors trapped in the cavitating bubbles, which can intensify chemical reactions or cause unexpected reactions. The collapse of cavities also creates acoustic streaming and turbulence, promoting reaction rates [1, 13, 30]. Therefore, cavitation is useful for generating local turbulence and liquid micro-circulation and enhancing transport processes.

1.2.2 Types of Cavitation

Cavitation is a physical process that can happen when ultrasound is applied to a liquid medium, causing the formation, growth, and subsequent collapse of bubbles or voids in the liquid. The effects of ultrasound on the liquid medium can either be beneficial or detrimental, depending on the type of cavitation. Ultrasound radiation can stimulate various types of cavitation, such as stable cavitation, transient cavitation, inertial cavitation, and acoustic cavitation, depending on the properties of the liquid medium and the intensity and frequency of the ultrasound. To optimize ultrasound-based processes and minimize potential harmful effects, it is crucial to understand the different types of cavitation that can occur during US. The following are the various types of cavitation that can occur during US:

Stable cavitation: Stable cavitation occurs when bubbles are formed and oscillate in a liquid medium under the influence of ultrasound. Unlike other types of cavitation, the bubbles in stable cavitation do not collapse completely but rather oscillate at a specific frequency. The oscillation of these bubbles can generate acoustic streaming and microstreaming, which can enhance the mixing and mass transfer of the liquid medium. Stable cavitation has been utilized in several applications such as ultrasound-assisted emulsification, sonochemistry, and ultrasound-assisted extraction [31, 32].

Transient cavitation: Transient cavitation occurs when bubbles are formed, grow, and rapidly collapse in a liquid medium under the influence of ultrasound [33]. The collapse of these bubbles can produce high-pressure waves and shock waves, which can cause mechanical damage to cells and tissues. Although transient cavitation can be useful in applications such as sonoporation, which involves the temporary formation of pores in cell membranes to enhance drug delivery, excessive or prolonged exposure to it can result in tissue damage and cell death.

Inertial cavitation: Inertial cavitation occurs when bubbles in a liquid medium grow and collapse violently due to ultrasound exposure. The collapse of the bubbles produces high-pressure waves and shock waves that may result in mechanical damage to cells and tissues. Inertial cavitation can also create high temperatures and pressures that can trigger chemical reactions in the liquid medium [34]. This type of cavitation is usually unwanted in many applications due to the risk of tissue damage and chemical degradation.

Acoustic cavitation: Acoustic cavitation is a physical phenomenon that involves the formation and collapse of bubbles in a liquid medium under the influence of ultrasound. The type of cavitation can either be stable or transient, depending on the intensity of the ultrasound. Acoustic cavitation can produce high temperatures and pressures that can induce chemical reactions in the liquid medium, as well as generate free radicals and other reactive species that can cause chemical degradation.

Furthermore, cavitation can be categorized into four principal types, which are acoustic, hydrodynamic, optic, and particle cavitation, as illustrated in Figure 1.3. Acoustic and hydrodynamic cavitation is the result of tensions that exist in a liquid, while optic and particle cavitation arise from the local deposition of energy. The classification of cavitation based on the method of technique used and the process of cavity generation is important for understanding the effects of ultrasound on a liquid medium and for optimizing ultrasound-based processes.

Acoustic cavitation: Acoustic cavitation is the process of forming and collapsing bubbles in a liquid medium through the use of sound waves, particularly ultrasound with frequencies ranging from 16 kHz to 100 MHz. The phenomenon of chemical changes induced by acoustic cavitation is commonly known as sonochemistry [35]. It involves the combination of ultrasound and chemistry.

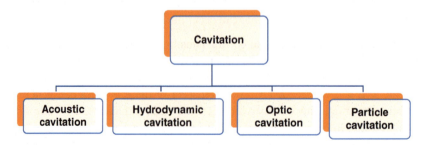

Figure 1.3 Types of cavitation based on technique used.

Hydrodynamic cavitation: Hydrodynamic cavitation is a type of cavitation that is produced by pressure variations created through the geometry of the system, which creates velocity variation. For instance, by leveraging the system's geometry, the interchange of pressure and kinetic energy can be achieved, leading to the formation of cavities, as seen in the case of flow through an orifice, venturi, and other similar systems.

Optic cavitation: Optic cavitation involves the use of high-intensity light, typically from a laser, to create cavitation in a liquid medium. The photons of the light can rupture the liquid continuum and generate bubbles or voids.

Particle cavitation: Particle cavitation is induced by a stream of elementary particles, such as a neutron beam, disrupting a liquid medium. This type of cavitation is commonly observed in devices like bubble chambers.

When it comes to cavitation, two types are frequently employed due to their efficacy in generating the necessary intensities for chemical or physical transformations: acoustic and hydrodynamic cavitation. The extent of cavitational impact hinges on both the turbulence intensity and the number of cavities formed. In essence, ultrasound wave propagation through medium results in acoustic cavitation, whereas hydrodynamic cavitation occurs as the flow's velocity changes due to alterations in the flow path geometry.

1.3 Application of Ultrasound Irradiation

US has a wide range of applications across various fields. Here are some notable applications of US:

Medical sciences: Ultrasound imaging is commonly used in medical diagnostics to visualize internal organs, tissues, and structures in real-time [36]. It is a noninvasive and radiation-free imaging technique that is particularly useful for examining the abdomen, pelvis, heart, blood vessels, and developing fetus during pregnancy (see Figure 1.4). There are some other applications, which are explained below.

Diagnostic imaging: One of the most common uses of ultrasound in medicine is diagnostic imaging. Ultrasound imaging allows noninvasive visualization of internal organs, tissues, and structures in real-time. It is used to examine various body parts, including the abdomen, pelvis, heart, blood vessels, musculoskeletal system, and the developing fetus during pregnancy [38, 39].

Obstetrics and gynecology: Ultrasound is extensively used in obstetrics and gynecology to monitor the progress of pregnancy, assess fetal development, determine the position of the fetus, and detect any abnormalities. It is also used for evaluating the female reproductive system, such as examining the uterus, ovaries, and fallopian tubes.

Cardiology: Ultrasound plays a crucial role in cardiology for evaluating the structure and function of the heart. Echocardiography, a type of ultrasound imaging, allows visualization of the heart's chambers, valves, and blood flow patterns.

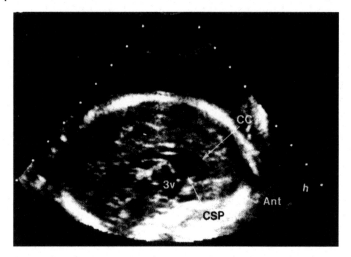

Figure 1.4 The abdominal sonography of the brain of a 21-week-old fetus. Source: Reproduced with permission from Pilu et al. [37]/John Wiley & Sons.

It helps in diagnosing and monitoring various heart conditions, such as heart valve disorders, congenital heart defects, and heart muscle abnormalities.

Vascular imaging: Ultrasound is used to examine blood vessels and assess blood flow patterns. Doppler ultrasound is particularly valuable in measuring the velocity and direction of blood flow, detecting blockages, or narrowing of vessels (such as in cases of deep vein thrombosis or arterial stenosis), and evaluating vascular abnormalities.

Interventional procedures: Ultrasound guidance is employed during certain minimally invasive procedures to enhance accuracy and safety. For example, ultrasound can be used to guide the insertion of needles for biopsies, aspirations, or injections. It helps in precisely targeting the intended area and avoiding damage to surrounding structures.

Sonography-guided therapies: Ultrasound is utilized in various therapeutic procedures. High-intensity focused ultrasound (HIFU) is used to precisely deliver focused energy to treat tumors or ablate abnormal tissues, such as uterine fibroids or prostate tumors, without the need for surgery. Additionally, ultrasound can be used for targeted drug delivery or gene therapy by utilizing microbubbles that enhance the permeability of cell membranes.

Guidance for minimally invasive surgeries: During minimally invasive surgeries, such as laparoscopic or robotic procedures, ultrasound can be used to provide real-time imaging guidance. It helps surgeons visualize and navigate internal structures, locate tumors or lesions, and ensure precise surgical instrument placement.

Therapeutic treatments: HIFU is utilized for therapeutic purposes. It involves focusing ultrasound waves on specific target tissues to generate heat or mechanical effects, leading to tissue ablation, tumor destruction, and targeted drug delivery. HIFU is used in the treatment of various conditions, including uterine fibroids, prostate cancer, liver tumors, and pain management.

Physiotherapy and rehabilitation: Ultrasound therapy is used in physiotherapy to provide deep tissue heating and promote healing. It is employed to treat conditions like muscle strains, sprains, joint inflammation, and sports injuries. The thermal effects of ultrasound can increase blood flow, relax muscles, and alleviate pain.

Dental applications: Ultrasound is utilized in dentistry for various procedures. It is commonly used for dental imaging, such as imaging the teeth and supporting structures. Ultrasonic scalers are also employed for dental cleanings and the removal of plaque and tartar from teeth.

Scaling and root planning, endodontic treatment, periodontal treatment, implantology, restorative dentistry, and dental prosthetics are important procedures to employ the ultrasonic iterations. Ultrasonic scalers are commonly used in dental hygiene for scaling and root planning procedures. These devices use ultrasonic vibrations to remove tartar, plaque, and bacterial deposits from the teeth and gums. The high-frequency vibrations generated by the ultrasonic scaler help to break down and dislodge the deposits, making the cleaning process more efficient and comfortable for the patient.

Ultrasonic instruments are utilized in endodontics, which involves the treatment of the tooth's pulp and root canal system. Ultrasonic tips, such as ultrasonic files or ultrasonic irrigators, are employed to remove infected or necrotic pulp tissue, clean and shape the root canals, and facilitate the irrigation of disinfectants or irrigation solutions. Ultrasonic vibrations aid in the removal of debris, disinfection of the canals, and better penetration of irrigants into complex root canal anatomy.

Ultrasonic devices are utilized in periodontal therapy to treat gum diseases and perform various procedures. Ultrasonic scalers and tips are used for subgingival debridement, which involves removing calculus and bacteria from below the gum line. The ultrasonic vibrations help to disrupt and remove the biofilm and tartar from periodontal pockets, promoting better healing and reduced pocket depths. Ultrasonic instruments are also employed in implant dentistry for the placement and maintenance of dental implants. During implant surgery, ultrasonic tips can be used for site preparation, osteotomy, and socket cleaning. Ultrasonic instruments are also useful for implant maintenance and cleaning around implant surfaces, removing plaque and calculus without damaging the implant or surrounding tissues.

The applications of ultrasonic irradiation in restorative dentistry procedures are as well. Ultrasonic instruments can be used for the removal of old restorative materials, such as amalgam or composite fillings, by gently vibrating and loosening the material for easier removal. Ultrasonic tips can also aid in the cleaning and preparation of the tooth structure before placing restorations like dental crowns or veneers.

1.3.1 Sonoluminescence and Sonophotocatalysis

Sonoluminescence refers to the emission of light from collapsing bubbles in a liquid medium under the influence of ultrasound. It is a fascinating phenomenon with potential applications in fields such as chemistry, physics, and materials

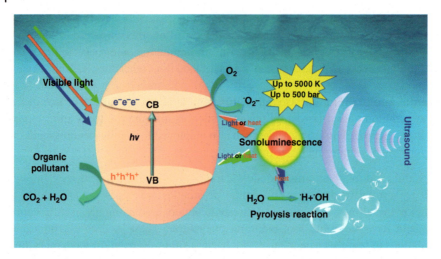

Figure 1.5 Schematic illustration for sonophotocatalytic mechanism. Source: Reproduced with permission from Wang and Cheng [40]/MDPI/Licensed under CC BY 4.0.

science. Sonophotocatalysis (see Figure 1.5) involves combining ultrasound with photocatalytic reactions to enhance the efficiency of photocatalysts for water treatment, pollution remediation, and energy production.

When used in conjunction with light and a photocatalyst, the sonophotocatalytic process can have a synergistic impact that speeds up the breakdown of organic contaminants in wastewater. The increased generation of reactive free radicals as well as the enhanced mass transfer of the contaminants to the photocatalyst surface are two reasons for the synergistic impact [40]. The enhanced creation of reactive radicals like ˙OH (see Figure 1.5), which are particularly effective at destroying organic pollutants, is one of the main benefits of sonophotocatalysis. Ultrasonic waves have the ability to cause cavitation, which produces high-energy bubbles that burst and emit shockwaves and heat, leading reactive radicals to develop.

Yun et al. [41] have developed an efficient catalyst that can produce H_2O_2 and destroy refractory pollutants. This study uses an in situ precipitation technique to rationally construct a number of new $Ag_6Si_2O_7/SmFeO_3$ (ASF) heterojunction catalysts. Several characterization procedures were used to confirm the characteristics of the manufactured ASF nanocomposites. With an adequate concentration of ciprofloxacin (CIP) of 10 mg/l at 400 W US power, 0.6 g/l catalyst dosage, pH of 5.0, as well as 40 kHz US frequency during irradiation time of 60 minutes, the ASF-1.5 sample in particular displays high efficiency (94.9%) of sonophotocatalytic.

1.3.2 Industrial Cleaning

US is applied in industrial cleaning processes such as ultrasonic cleaning [42, 43]. It involves immersing objects in a cleaning solution and subjecting them to high-frequency sound waves. The cavitation effect generated by ultrasound helps remove dirt, contaminants, and deposits from the surfaces of objects, making it useful for cleaning delicate or intricate items.

Ultrasonic cleaning systems consist of a cleaning tank filled with a suitable cleaning solution or solvent. The object to be cleaned is immersed in the liquid, and ultrasonic transducers located in the tank generate high-frequency sound waves. These sound waves create alternating high- and low-pressure zones in the liquid, leading to the formation and collapse of cavitation bubbles near the object's surface. The collapse of these bubbles generates intense local energy, effectively scrubbing away contaminants.

Ultrasonic cleaning is highly effective in removing a wide range of contaminants, including oils, grease, dirt, rust, scale, and other residues [43]. The cavitation action reaches into complex geometries and crevices that are difficult to access using other cleaning methods. This makes it particularly useful for cleaning intricate parts, such as machine components, automotive parts, electronics, jewelry, medical instruments, and precision equipment.

One typical aspect of dairy processing is the ultrafiltration of whey solutions. The economics of such a process are, however, greatly impacted by the regular fouling of ultrafiltration membranes and the following cleaning cycle. In this study performed by Muthukumaran et al. [44], it is monitored into how ultrasonics affect the cleansing of whey-fouled membranes and what factors affect this result. A tiny single-sheet membrane unit that was completely submerged in an ultrasonic bath was used for the experiments.

An earlier solution to the problems produced by the acidic ammonium salt crystallization of vanadium was the ultrasound crystallization (UC) technique [45]. This study looked closely at how several parameters affected the properties of vanadium crystallization [45]. The results demonstrated that using ultrasonic power of 600 W, a baseline pH value of 2.0, ambient temperature of 95 °C, ammonium salt addition coefficient of 0.5, period of five minutes, excessive vanadium precipitation ratio (99.67%), and vanadium level of 20 g/l, along with V_2O_5 purity (99.50%) of the outcomes of the reaction can be achieved.

1.3.3 Material Processing

Ultrasound is used in various material processing applications. It can be employed for emulsification, dispersion, and homogenization of liquids, as well as particle size reduction. Ultrasonic devices are also used for degassing, degreasing, and defoaming processes in industries like food and beverage, pharmaceuticals, and cosmetics.

Ji et al. have studied the crystalline structures of Sn–Ag–Cu alloy ingots formed through ultrasound-assisted solidification, with an emphasis on the restrictions on ultrasonic processing depth and time imposed by the melt solidification's cooling rate [46]. Raising the ultrasonic power during cooling by air caused the –Sn phase to split from a dendritic structure into a circle-like equiaxed shape by lowering the undercooling temperature and lengthening the process of the solidification period. The grain size was reduced from 300 to 20 mm.

Using Y_2O_3, $CuCl_2$ as well as $BaCl_2$ as the starting components for the co-precipitation process, Jian-feng et al. [47] have produced Y_2BaCuO_5 nanocrystallites with the aid of ultrasonic irradiation. Transmission electron microscopy (TEM) and X-ray diffraction (XRD) were used to characterize the crystallization and morphologies of nanoparticles as prepared. Results demonstrate that using a mixture of NaOH and Na_2CO_3 as a precipitator, Y_2BaCuO_5 monophase can be produced at calcining temperatures up to 900 °C. With a rise in sonicating power, Y_2BaCuO_5 crystallites' particle size reduces. When the sonicating power is increased to 300 W, it is possible to produce Y_2BaCuO_5 crystallites that are around 30 nm in size.

In order to establish an affordable method for producing bioethanol, the effort focuses on intensifying delignification and subsequent enzymatic-hydrolysis of sustainable biomass such as coconut coir groundnut shells, and pistachio shells utilizing an ultrasound-aided methodology [48]. The obtained results for delignification of biomass showed that the extent of delignification for groundnut shells, coconut coir, and pistachio shells under conventional alkaline treatment was 41.8%, 45.9%, and 38%, respectively, while it raised to 71.1%, 89.5%, and 78.9%, providing a nearly 80–100% boost under the ultrasound supported technique. The traditional technique produced reducing sugar yields of 10.2, 12.1, and 8.1 g/l for groundnut shells, coconut coir, and pistachio shells, respectively, under optimal conditions. In contrast, the yields from ultrasound-assisted enzymatic hydrolysis were 21.3, 23.9, and 18.4 g/l in the identical amount of biomass.

1.3.4 Chemical and Biological Reactions

US is employed in chemical and biological reactions to enhance reaction rates, promote mixing, and improve mass transfer. It is used for various processes such as the synthesis of nanoparticles, extraction of bioactive compounds from plants, sonochemistry, and sono-organic reactions.

Although the use of ultrasound in biotechnology is still relatively recent, it has been found to trigger a number of mechanisms that happen when cells or enzymes are present [49]. The enzymes are denaturized, and the cells are broken by intense ultrasonic vibrations. Low-intensity ultrasonic waves have the ability to alter cellular metabolism or enhance the mass transfer of substances via the boundary layer, cellular membrane, or wall. The most significant aspect in the case of enzymes appears to be an increase in the mass transfer rate of the reagents to the active site. Native enzymes are more susceptible to the heat deactivation caused by ultrasound than immobilized enzymes. Enzymes can perform synthesis using reverse micelles. The use of ultrasound in biotechnology is considered in a number of applications. Molecular complexes stabilized by hydrogen bonding and dispersion interaction can also be studied [50–52]. Ngoc and colleagues conducted a study on the impact of ultrasound stimulation on hydrogen bonding within a composite slurry comprising networked alumina and polyacrylic acid.

Underwater communication and sensing: Ultrasound waves can travel long distances in water with minimal attenuation. This property makes ultrasound suitable for underwater communication and sensing applications [53, 54]. It is used for underwater navigation, fish finding, underwater imaging, and marine research. As a result of the Internet of Underwater Things (IoUT), new maritime commercial, scientific, and military applications will be possible. Since these systems are often powered by conventional batteries, powering electrical equipment in deep water still poses a significant difficulty. The first underwater sensor node without batteries has been designed by Guida et al. [53], and it can wirelessly recharge using ultrasonic waves over greater distances than are now possible. An undersea platform's architectural design that can draw electrical energy from ultrasonic waves is presented.

In a novel underwater self-powered all-optical wireless ultrasonic sensor (SAWS) that utilizes triboelectrification-induced electroluminescence (TIEL), Tian et al. [54] have demonstrated that accurate and persistent TIEL can be created under the excitation of ultrasonic waves. The SAWS has been shown to be an attractive option for real-time optical signal exchange because it has an ultrafast response time of below 50 ms and an ultrahigh signal-to-noise ratio (SNR) of 26.02 dB. It has also been demonstrated to precisely identify the exact position of the ultrasonic source with a margin of error of less than 4.6%.

Non-destructive testing: Ultrasound is employed in non-destructive testing (NDT) techniques to evaluate the integrity of materials and structures without causing damage [55, 56]. Ultrasound-based methods, such as ultrasonic testing (UT), are used for flaw detection, thickness measurement, and characterization of materials in industries like aerospace, automotive, and construction. Wire and Arc Additive Manufacturing (WAAM) parts are characterized using NDT techniques for materials techniques and inspection [56]. Electrical and ultrasonic tests were performed on the samples used to compare the techniques' capabilities. Metallographic, hardness, and electrical conductivity analysis were also applied to the same samples for product characterization.

Application of ultrasonic irradiation in synthesis: Nowadays, in the field of organic chemistry synthesis, ultrasonic irradiation techniques have been widely used, and are the most frequently used techniques among all the synthesis procedures. This is because of the fact that reaction rates can be improved, and one can adjust the selectivity performance of the reaction [57]. Ultrasonic irradiation techniques are considered to be the green synthetic process in chemistry, as this technique reduces the waste product, and less energy is required to carry out a reaction [58]. Microwaves (MW) and ultrasonic irradiation techniques are existing methods for process amplification and are being used both in industry and academy these days. Their mutual usage in the identical reactor led to significant consequences [59]. Ultrasonication permits the quick distribution of solids, the establishment of porous materials and nanostructures, and the decay of organics containing biological components [58]. Ultrasonic irradiation leads to the formation of particular reactive chemical species that cannot be produced from conventional conditions due to the physical and chemical effects of ultrasound. There are several advantages of this ultrasound irradiation technique over the conventional process for synthesizing organic compounds. The most important are higher percentage yield, selective product formation, easy generation of reactive species, and the usage of hazardous reagents. Ultrasonic irradiation techniques are used mainly for those organic synthesis reactions where traditional methods require very long reaction time and drastic conditions are required for initiating a reaction [60–62]. US procedure pathway is reported to have a strong influence on reaction rates, purity, as well as high yields in comparison to the conventional heating process [63]. Unique and proficient synthetic methods have been reported by Karmakar et al. for the formation of N/O-containing five- to seven-membered heterocycles and their fused analogs. Studies on chiral polymers have provoked excessive consideration between chiral supramolecular materials depending on their properties. The synthesis of chiral polymeric composites (CMNPs/1,4-Zbtb and 1,3-Zbtb) has been reported by Sharifzadeh et al. using solvothermal, mechanical stirring, and ultrasonic irradiation, three different methods. It has been found that synthesis carried out using ultrasound is much more proficient and also economically cost-effective than the other two methods. This process is also beneficial from an energy-saving and time-consuming point of view [64]. US techniques are also applied to minimize the waste going into the environment. Industrial solid waste (slag) obtained from the synthesis of catalysts contains SiO_2, Fe_2O_3, $Ca(OH)_2$, Al_2O_3, $CaCO_3$, and TiO_2 which can be dissolve using US [65]. Defective graphene nanosheets (dGN4V) having 5-9, 5-8-5, and point defects were synthesized using a sonoelectrochemical technique with an applied potential of 4 V (vs. Ag/AgCl) to initiate the rapid intercalation of phosphate ions amongst the layers of the graphite foil as a working electrode [66]. Synthesis of 1,5-dinitroaryl-1,4-pentadien-3-ones has been reported using US in the presence of K_2CO_3 as a catalyst [62]. For the green synthesis process, the ionic liquid is considered to be the best medium to carry out reactions in the microwave or ultrasonic irradiation [67]. Different compounds of 4,6-disubstituted-1,3,5-triazines series comprising hydrazone derivatives were synthesized using both ultrasonic irradiation and conventional heating methods. It has been observed that ultrasonication produces the required products of more yields, and of greater purity, in less time as compared to the normal conventional technique.

1.4 Conclusion

High frequency sound waves are used in the process of US to change the physical and chemical properties of materials. Applications for ultrasound can be found in a variety of disciplines, such as chemistry, material science, medicine, and food processing. The use of ultrasonic technology might theoretically result in greener and more intensified processing, which also saves a lot of money. US is a successful method to accelerate chemical reactions, improve yields, and shorten reaction times in chemistry and the chemical industries. Additionally, this method can be used to create composites, polymers, and nanoparticles, among other materials. US is used to process and modify materials in the field of material science, including surface cleaning, surface modification, and particle dispersion. Through the use of this technology, advanced materials with enhanced characteristics and performance have been created. US is frequently employed in medicine for therapeutic purposes such as tissue ablation and targeted medication distribution. US is useful in the food business for food processing and preservation, including compound extraction, food quality preservation, and texture enhancement. Overall, it has been demonstrated that cavitation has emerged as a key tool in a variety of industries with significant potential for environmentally friendly and sustainable processing. It is a promising instrument for the future because of its scope for improvement and development.

Acknowledgments

Sumit Kumar would like to thank Magadh University, Bodh Gaya, Bihar, India for providing infrastructure as well as instrument facilities, and SERB, DST, India (Grant No. SRG/2019/002284) for financial support.

References

1 Gogate, P.R., Tayal, R.K., and Pandit, A.B. (2006). Cavitation: a technology on the horizon. *Curr. Sci.* 91: 35–46.
2 Gogate, P.R. and Prajapat, A.L. (2015). Depolymerization using sonochemical reactors: a critical review. *Ultrason. Sonochem.* 27: 480–494. https://doi.org/10.1016/j.ultsonch.2015.06.019.
3 Katoch, G. (2023). Sol-gel auto-combustion developed Nd and Dy co-doped Mg nanoferrites for photocatalytic water treatment, electrocatalytic water splitting and biological applications. *J. Water Process Eng.* 53: 103726.
4 Kumar, A., Kumar, S., Pathak, A.K. et al. (2023). Recent progress in nanocomposite-oriented triboelectric and piezoelectric energy generators: an overview. *Nano-Struct. Nano-Objects* 36: 101046. https://doi.org/10.1016/j.nanoso.2023.101046.
5 Wang, L.V. (2008). Prospects of photoacoustic tomography. *Med. Phys.* 35: 5758–5767.
6 Serr, D.M., Padeh, B., Zakut, H. et al. (1971). Studies on the effects of ultrasonic waves on the fetus. In: *Perinatal Medicine: 2nd European Congress, London, April 1970* (ed. P.J. Huntingford, R.W. Beard, F.E. Hytten, and J.W. Scopes), 302. Karger.
7 Thorneycroft, J. and Barnaby, S.W. (1895). Torpedo-boat destroyers. *Inst. Civil Eng.* 122: 51–69.
8 Rayleigh, L. (1917). On the pressure developed in a liquid during the collapse of a spherical cavity. *Lond. Edinb. Dublin Philos. Mag.* 34 (199–04): 94–98.
9 Richards, W.T. and Loomis, A.L. (1927). The chemical effects of high frequency sound waves I. A preliminary survey. *J. Am. Chem. Soc.* 49: 3086–3100.
10 Brohult, S. (1937). Splitting of the haemocyanin molecule by ultrasonic waves. *Nature* 140: 805.

11 Leong, T., Ashokkumar, M., and Kentish, S. (2011). The fundamentals of power ultrasound – a review. *Acoust. Aust.* 39: 54–63.

12 Gogate, P.R., Mujumdar, S., and Pandit, A.B. (2003). Large-scale sonochemical reactors for process intensification: design and experimental validation. *J. Chem. Technol. Biotechnol.* 78: 685–693. https://doi.org/10.1002/jctb.697.

13 Gogate, P.R. and Pandit, A.B. (2005). A review and assessment of hydrodynamic cavitation as a technology for the future. *Ultrason. Sonochem.* 12: 21–27.

14 Gogate, P.R. and Pandit, A.B. (2004). A review of imperative technologies for wastewater treatment I: oxidation technologies at ambient conditions. *Adv. Environ. Res.* 8: 501–551.

15 Mason, T.J., Lorimer, J.P., and Walton, D.J. (1990). Sonoelectrochemistry. *Ultrasonics* 28: 333–337.

16 Prajapat, A.L. and Gogate, P.R. (2015). Depolymerization of guar gum solution using different approaches based on ultrasound and microwave irradiations. *Chem. Eng. Process. Process Intensif.* 88: 1–9.

17 Theerthagiri, J., Madhavan, J., Lee, S.J. et al. (2020). Sonoelectrochemistry for energy and environmental applications. *Ultrason. Sonochem.* 63: 104960. https://doi.org/10.1016/j.ultsonch.2020.104960.

18 Wu, P., Wang, X., Lin, W., and Bai, L. (2022). Acoustic characterization of cavitation intensity: a review. *Ultrason. Sonochem.* 82: 105878. https://doi.org/10.1016/j.ultsonch.2021.105878.

19 Wu, X. and Mason, T.J. (2017). Evaluation of power ultrasonic effects on algae cells at a small pilot scale. *Water (Switzerland)* 9: 1–8. https://doi.org/10.3390/w9070470.

20 Kruus, P., Neill, M.O., and Robertson, D. (1990). Ultrasonic initiation of polymerization. *Ultrasonics* 28: 304–309.

21 Sada, P.K., Bar, A., Jassal, A.K. et al. (2023). A novel rhodamine probe acting as chemosensor for selective recognition of Cu^{2+} and Hg^{2+} ions: an experimental and first principle studies. *J. Fluoresc.* https://doi.org/10.1007/s10895-023-03412-y.

22 Mason, T.J. (1992). *Practical Sonochemistry: Users Guide in Chemistry and Chemical Engineering*, Ellis Horwood Series in Organic Chemistry. Chichester, UK: Ellis Horwood Publishers.

23 Gogate, P.R. (2008). Cavitational reactors for process intensification of chemical processing applications: a critical review. *Chem. Eng. Process. Process Intensif.* 47: 515–527.

24 Czechowska-biskup, R., Rokita, B., Lotfy, S. et al. (2005). Degradation of chitosan and starch by 360-kHz ultrasound. *Carbohydr. Polym.* 60: 175–184. https://doi.org/10.1016/j.carbpol.2004.12.001.

25 Brenner, C.E. (1995). *Cavitation & Bubble Dynamics*. New York: Oxford University Press Inc.

26 Mason, T.J. and Lorimer, J.P. (2002). *Applied Sonochemistry: Use of Power Ultrasound in Chemistry and processing*. Weinheim: Wiley-VCH Verlag Gmbh.

27 Suslick, K.S. (1979). Sonochemistry. *Science* 247 (1990): 1441–1445. https://doi.org/10.2307/j.ctv1zckxc3.15.

28 Leighton, T.G. (1994). *The Acoustic Bubble*. London: Academic Press.

29 Young, F.R. (1989). *Cavitation*. London: McGraw-Hill.

30 Crum, L.A. (1982). Acoustic cavitation. *Ultrasonic Symposium*. 1–11.

31 Mason, T.J. (1990). *Sonochemistry: Theory, Applications, and Uses of Ultrasound in Chemistry*. Ellis Horwood.

32 Gogate, P.N. and Patil, P.R. (n.d.). *Cavitation: A Novel Technology for Intensification of Chemical and Biological Processes*. Springer.

33 Ashokkumar, M. (2011). The characterization of acoustic cavitation bubbles – an overview. *Ultrason. Sonochem.* 18: 864–872. https://doi.org/10.1016/j.ultsonch.2010.11.016.

34 Fabiilli, M.L., Haworth, K.J., Fakhri, N.H. et al. (2009). The role of inertial cavitation in acoustic droplet vaporization. *IEEE Trans. Ultrason. Ferroelectr. Freq. Control* 56: 1006–1017. https://doi.org/10.1109/TUFFC.2009.1132.

35 Gadipelly, C., Pérez-González, A., Yadav, G.D. et al. (2014). Pharmaceutical industry wastewater: review of the technologies for water treatment and reuse. *Ind. Eng. Chem. Res.* 53: 11571–11592. https://doi.org/10.1021/ie501210j.

36 Hill, C.R. (1970). Calibration of ultrasonic beams for bio-medical applications. *Phys. Med. Biol.* 15: 241.

37 Pilu, G., Sandri, F., Perolo, A. et al. (1993). Sonography of fetal agenesis of the corpus callosum: a survey of 35 cases. *Ultrasound Obstet. Gynecol.* 3: 318–329. https://doi.org/10.1046/j.1469-0705.1993.03050318.x.

38 Nguyen, C.P. and Goodman, L.H. (2012). Fetal risk in diagnostic radiology. In: *Seminars in Ultrasound, CT and MRI*, 4–10. Elsevier.

39 Pelsang, R.E. (1998). Diagnostic imaging modalities during pregnancy. *Obstet. Gynecol. Clin. North Am.* 25: 287–300.

40 Wang, G. and Cheng, H. (2023). Application of photocatalysis and sonocatalysis for treatment of organic dye wastewater and the synergistic effect of ultrasound and light. *Molecules* 28. https://doi.org/10.3390/molecules28093706.

41 Yun, K., Saravanakumar, K., Jagan, G. et al. (2023). Fabrication of highly effective $Ag_6Si_2O_7/SmFeO_3$ heterojunction with synergistically enhanced sonophotocatalytic degradation of ciprofloxacin and production of H_2O_2: influencing factors and degradation mechanism. *Chem. Eng. J.* 468: 143491. https://doi.org/10.1016/j.cej.2023.143491.

42 Mason, T.J. (2016). Ultrasonic cleaning: an historical perspective. *Ultrason. Sonochem.* 29: 519–523.

43 Leonelli, C. and Mason, T.J. (2010). Microwave and ultrasonic processing: now a realistic option for industry. *Chem. Eng. Process. Process Intensif.* 49: 885–900.

44 Muthukumaran, S., Yang, K., Seuren, A. et al. (2004). The use of ultrasonic cleaning for ultrafiltration membranes in the dairy industry. *Sep. Purif. Technol.* 39: 99–107. https://doi.org/10.1016/j.seppur.2003.12.013.

45 Chen, B., Bao, S., Zhang, Y., and Li, C. (2023). Reactive crystallization of ammonium polyvanadate from vanadium-bearing solution assisted by efficient ultrasound irradiation: crystallization characteristics and growth process. *J. Mater. Res. Technol.* https://doi.org/10.1016/j.jmrt.2023.05.250.

46 Ji, H., Wang, Q., Li, M., and Wang, C. (2014). Effects of ultrasonic irradiation and cooling rate on the solidification microstructure of Sn–3.0Ag–0.5Cu alloy.

J. Mater. Process. Technol. 214: 13–20. https://doi.org/10.1016/j.jmatprotec.2013.07.013.

47 Jian-feng, H., Xie-rong, Z., Li-yun, C., and Xin-bo, X. (2009). Preparation of Y_2BaCuO_5 nanoparticles by a co-precipitation process with the aid of ultrasonic irradiation. *J. Mater. Process. Technol.* 209: 2963–2966. https://doi.org/10.1016/j.jmatprotec.2008.07.001.

48 Subhedar, P.B., Ray, P., and Gogate, P.R. (2018). Intensification of delignification and subsequent hydrolysis for the fermentable sugar production from lignocellulosic biomass using ultrasonic irradiation. *Ultrason. Sonochem.* 40: 140–150. https://doi.org/10.1016/j.ultsonch.2017.01.030.

49 Sinisterra, J.V. (1992). Application of ultrasound to biotechnology: an overview. *Ultrasonics* 30: 180–185. https://doi.org/10.1016/0041-624X(92)90070-3.

50 Panja, S.K., Dwivedi, N., Noothalapati, H. et al. (2015). Significance of weak interactions in imidazolium picrate ionic liquids: spectroscopic and theoretical studies for molecular level understanding. *Phys. Chem. Chem. Phys.* 17: 18167–18177. https://doi.org/10.1039/C5CP01944C.

51 Kumar Panja, S., Kumar, S., Fazal, A.D., and Bera, S. (2023). Molecular aggregation kinetics of heteropolyene: an experimental, topological and solvation dynamics studies. *J. Photochem. Photobiol., A* 445: 115084. https://doi.org/10.1016/j.jphotochem.2023.115084.

52 Kumar, S., Singh, S.K., Calabrese, C. et al. (2014). Structure of saligenin: microwave, UV and IR spectroscopy studies in a supersonic jet combined with quantum chemistry calculations. *Phys. Chem. Chem. Phys.* 16: 17163. https://doi.org/10.1039/C4CP01693A.

53 Guida, R., Demirors, E., Dave, N. et al. (2018). An acoustically powered battery-less internet of underwater things platform. *2018 Fourth Underwater Communications and Networking Conference (UComms)*, IEEE. 1–5.

54 Tian, Z., Su, L., Wang, H. et al. (2022). Underwater self-powered all-optical wireless ultrasonic sensing, positioning and communication with ultrafast response time and ultrahigh sensitivity. *Adv. Opt. Mater.* 10: 2102091.

55 Umar, M.Z., Vavilov, V., Abdullah, H., and Ariffin, A.K. (2016). Ultrasonic infrared thermography in non-destructive testing: a review. *Russ. J. Nondestr. Test.* 52: 212–219.

56 Lopez, A., Bacelar, R., Pires, I. et al. (2018). Non-destructive testing application of radiography and ultrasound for wire and arc additive manufacturing. *Addit. Manuf.* 21: 298–306.

57 Li, J.-T., Yang, W.-Z., Wang, S.-X. et al. (2002). Improved synthesis of chalcones under ultrasound irradiation. *Ultrason. Sonochem.* 9: 237–239. https://doi.org/10.1016/S1350-4177(02)00079-2.

58 Cintas, P. and Luche, J.-L. (1999). Green chemistry. The sonochemical approach. *Green Chem.* 1: 115–125. https://doi.org/10.1039/A900593E.

59 Călinescu, I., Vinatoru, M., Ghimpețeanu, D. et al. (2021). A new reactor for process intensification involving the simultaneous application of adjustable ultrasound and microwave radiation. *Ultrason. Sonochem.* 77: 105701. https://doi.org/10.1016/j.ultsonch.2021.105701.

60 Mady, M.F., El-Kateb, A.A., Zeid, I.F., and Jørgensen, K.B. (2013). Comparative studies on conventional and ultrasound-assisted synthesis of novel homoallylic alcohol derivatives linked to sulfonyl dibenzene moiety in aqueous media. *J. Chem.* 2013: 364036. https://doi.org/10.1155/2013/364036.

61 Singla, M. and Sit, N. (2021). Application of ultrasound in combination with other technologies in food processing: a review. *Ultrason. Sonochem.* 73: 105506.

62 Ding, L., Wang, W., and Zhang, A. (2007). Synthesis of 1,5-dinitroaryl-1,4-pentadien-3-ones under ultrasound irradiation. *Ultrason. Sonochem.* 14: 563–567. https://doi.org/10.1016/j.ultsonch.2006.09.008.

63 Karmakar, R. and Mukhopadhyay, C. (2021). Chapter 15 – Ultrasonication under catalyst-free condition: an advanced synthetic technique toward the green synthesis of bioactive heterocycles. In: *Green Synthetic Approaches for Biologically Relevant Heterocycles*, 2e (ed. E. Brahmachari), 497–562. Elsevier. https://doi.org/10.1016/B978-0-12-820586-0.00014-5.

64 Sharifzadeh, Z., Berijani, K., and Morsali, A. (2021). High performance of ultrasonic-assisted synthesis of two spherical polymers for enantioselective catalysis. *Ultrason. Sonochem.* 73. 105499. https://doi.org/10.1016/j.ultsonch.2021.105499.

65 Kholkina, E., Kumar, N., Eränen, K. et al. (2021). Ultrasound irradiation as an effective tool in synthesis of the slag-based catalysts for carboxymethylation. *Ultrason. Sonochem.* 73. https://doi.org/10.1016/j.ultsonch.2021.105503.

66 Wang, T.P., Lee, C.L., Kuo, C.H., and Kuo, W.C. (2021). Potential-induced sonoelectrochemical graphene nanosheets with vacancies as hydrogen peroxide reduction catalysts and sensors. *Ultrason. Sonochem.* 72: 105444. https://doi.org/10.1016/j.ultsonch.2020.105444.

67 Jasim, S.A., Tanjung, F.A., Sharma, S. et al. (2022). Ultrasound and microwave irradiated sustainable synthesis of 5- and 1-substituted tetrazoles in TAIm[I] ionic liquid. *Res. Chem. Intermed.* 48: 3547–3566. https://doi.org/10.1007/s11164-022-04756-z.

2

Fundamental Theory of Electromagnetic Spectrum, Dielectric and Magnetic Properties, Molecular Rotation, and the Green Chemistry of Microwave Heating Equipment

Raghvendra K. Mishra[1], Akshita Yadav[2], Vinayak Mishra[3], Satya N. Mishra[4], Deepa S. Singh[5], and Dakeshwar Kumar Verma[6]

[1] Cranfield University, School of Aerospace, Transport and Manufacturing, College Rd, Wharley End, Bedford MK43 0AL, United Kingdom
[2] State University of New York at Buffalo, Department of Mechanical and Aerospace Engineering, Buffalo, NY 14260-4400, USA
[3] International Institute of Information Technology, Department of Computer Aided Structural Engineering, CR Roa Road, Telangana 500032, India
[4] Brahmanand PG College, Department of Mathematics, Mall Road, Kanpur, Uttar Pradesh 208011, India
[5] M.J.P. Rohilkhand University, Hindu College, Department of Zoology, Station Road, Moradabad, Uttar Pradesh 244001, India
[6] Govt. Digvijay Autonomous Postgraduate College, Department of Chemistry, Rajnandgaon, Chhattisgarh 491441, India

2.1 Introduction

The fundamental concept revolves around designing hybrid microwave-ultrasonication irradiation systems to enhance the efficiency and selectivity of green chemical reactions. These systems function by combining microwave and ultrasonic radiation to promote reactions while overcoming limitations like nonuniformity. This is significant for sustainable chemistry due to its potential to reduce energy consumption and harmful solvent usage. Applications are observed in various fields, motivating research in disciplines such as chemistry and materials science. Researchers work with materials that can focus and distribute radiation, aiming to optimize system efficiency. Ongoing research explores mechanisms and principles governing this technology, with implications for greener chemical synthesis [1]. The fundamental concept involves the design of a hybrid material for sustainable high-value chemical production in industrial settings. This material system combines to enhance reaction efficiency when interacting with electromagnetic radiation while minimizing its environmental impact. It functions by utilizing metamaterial-based applicators and pulsed radiation to focus and distribute energy effectively. Significantly, it offers scalable, cost-effective, and safe operation, motivating research in fields like materials science and engineering. Key materials include transparent reactors and durable applicators. Ongoing research explores sensor-based control systems for optimized production, marking the forefront of

Green Chemical Synthesis with Microwaves and Ultrasound, First Edition.
Edited by Dakeshwar Kumar Verma, Chandrabhan Verma, and Paz Otero Fuertes.
© 2024 WILEY-VCH GmbH. Published 2024 by WILEY-VCH GmbH.

current research in sustainable chemical synthesis [2]. The fundamental concept involves the utilization of green synthesis, specifically supercritical fluids, and biocatalysis, to enhance the production of high-value products with increased efficiency and sustainability. Supercritical fluids, operating above critical conditions, act as nontoxic and nonflammable solvents, reducing waste impact on surroundings, and finding applications in pharmaceuticals, nanomaterials, and food additives. Biocatalysis employs enzymes for selective reactions in mild conditions, minimizing the use of hazardous chemicals. Both technologies motivate research in materials science and engineering. Their significance lies in reducing cost, waste production, and energy consumption while providing sustainable alternatives to traditional methods [3]. Green chemistry is a scientific concept focused on designing and implementing chemical processes and products to minimize or eliminate hazardous substances, thereby reducing ecological disasters. It functions through principles such as waste prevention, maximizing atom economy, using safer chemicals and solvents, and designing products for easy degradation. Its significance lies in promoting sustainable and environmentally friendly chemical practices. Applications are employed in various industries, driven by the need to reduce environmental harm, improve efficiency, and create more eco-friendly products. This field engages principles of thermodynamics, kinetics, catalysis, and material science to develop greener chemical synthesis routes, offering the potential to transform the chemical industry and benefit the ecosystem [4]. Figure 2.1 visually represents the 12 fundamental principles of green chemistry, offering a comprehensive framework for sustainable chemical practices that prioritize waste reduction, safe solvents, and energy efficiency. These principles are put into practical applications, notably in pharmaceutical synthesis, where eco-friendly reactions align with the goals of green chemistry by minimizing waste. The Nobel Prize recognition underscores the significance of sustainable chemistry, with metathesis applications benefiting various fields by reducing waste and energy consumption. The emergence of green chemistry in the 1990s, coupled with Anastas and Warner's principles, has shifted chemical research toward proactive pollution prevention, leaving a lasting impact on academia. Moreover, extending these principles to encompass concepts like the circular economy, sustainability metrics, and global collaboration further enhances global sustainability efforts within chemical processes.

In 2012, Elevance Renewable Sciences earned the Presidential Green Chemistry Challenge Award for their pioneering use of metathesis to convert natural oils into eco-friendly chemicals, including concentrated cold-water detergents. This approach reduced waste and improved energy saving, showcasing green chemistry's impact across industries. Similarly, Los Alamos National Laboratory's use of supercritical carbon dioxide transformed computer chip manufacturing, reducing resource consumption, and aligning with sustainability principles. These examples illustrate how innovative technologies and green chemistry principles are reshaping industries, promoting eco-conscious solutions in line with academic research and scientific goals [6]. Richard Wool's innovative use of chicken feathers for keratin-based fibers has the potential to revolutionize printed circuit board technology, offering lightweight and durable solutions applicable to chip technology

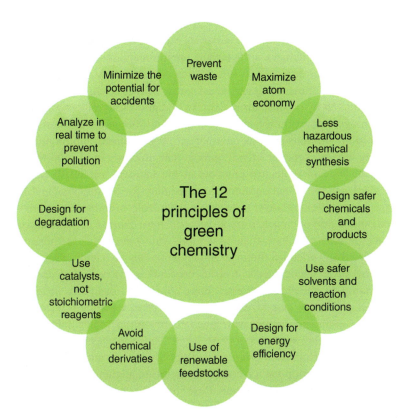

Figure 2.1 Scheme of the 12 principles of green chemistry. Source: Trombino et al. [5]/ MDPI/Public Domain/CC by 4.0.

and biofuel production, in line with sustainability principles. In the pharmaceutical sector, the collaboration between Merck and Codexis has led to a greener synthesis for sitagliptin, reducing waste, improving yield, and safety, and eliminating the need for metal catalysts. This approach promises more sustainable pharmaceutical manufacturing and safer medication products, emphasizing the significance of eco-conscious chemical routes in achieving safer and environmentally friendly outcomes [7]. Green chemistry finds extensive applications across various sectors, focusing on the development of environmentally friendly compounds and eco-conscious solvents, particularly in industries like paint and coatings. These applications prioritize sustainability through energy conservation, waste reduction, and the substitution of hazardous substances with safer alternatives. Green chemistry's alignment with academic research and scientific objectives underscores its vital role in advancing environmentally responsible chemical practices, contributing to a more sustainable and eco-conscious future across diverse domains [8].

Microwave-assisted chemistry represents a transformative approach to chemical synthesis, offering rapid reactions, improved yields, and higher purity while adhering to green chemistry principles. In pharmaceutical synthesis, it embodies biodegradable practices, promoting sustainability and shaping academic research and scientific pursuits. Beyond its current applications, microwave chemistry holds promise in diverse fields such as materials science, environmental science, and food processing, signifying a shift toward environmentally responsible and resource-efficient chemical practices [9]. Microwave chemistry revolutionizes chemical reactions with dramatically shortened reaction times, leading to improved yields and minimizing by-product formation. This aligns with green chemistry's principles of energy conservation and waste reduction. Microwave synthesis has found success in applications such as pharmaceutical manufacturing, promoting faster and more environmentally friendly production processes, and reflecting a shift toward efficient and ecological practices at the industry level [10]. Nevertheless, the challenge of limited equipment accessibility for researchers persists. To overcome this hurdle, initiatives focused on affordable equipment, materials development, and hybrid approaches are necessary. These efforts can democratize the use of microwave technology, making it more widely available, and thereby promoting environment-friendly chemical implementations in scientific research while advancing sustainability goals in academia and industry [11]. Excessive microwave heating in chemical processes does pose risks, including potential harm to tissues and DNA due to radioactive decay and hazardous reactions. However, when used in conjunction with green chemistry principles, microwave exposure offers sustainable synthesis options. The effective management of these risks through safety measures and protocols enables researchers and industries to harness microwave irradiation for eco-conscious exercises, aligning with scientific research goals for a cleaner and safer future. Case studies further highlight the success of this synergy, showcasing advancements in eco-friendly chemistry practices operations [12]. Green chemistry plays a pivotal role by prioritizing environmentally responsible practices and biodegradable products. Microwave and ultrasonication irradiation techniques, based on electromagnetic principles, offer sustainability in various fields, including physics and medicinal chemistry. Their alignment with green chemistry's objectives promotes eco-friendly applications, driving progress in the research domain toward a cleaner and more sustainable future [13]. These technologies optimize chemical transformations, fostering sustainability across various aspects, as exemplified by real-world applications, effectively advancing sustainable goals in academic research as well as scientific exploration [14]. Microwave and vibrational irradiation methods exhibit remarkable versatility in addressing environmental concerns, effectively removing pollutants from soil and water without generating harmful waste products. These eco-friendly applications seamlessly align with green chemistry principles, emphasizing sustainable approaches to remediate polluted areas and restore ecosystems. Their capacity to tackle environmental issues underscores their significance in academic research and scientific endeavors, promoting cleaner and more sustainable practices for a greener future, and reflecting a commitment to eco-conscious chemistry practices with a strong focus on environmental stewardship [15]. Aligned with sustainability and environmental responsibility, these methods have diverse applications across industries, promoting cleaner and more sustainable chemical practices. In academic research and scientific exploration, they represent a pivotal step toward a

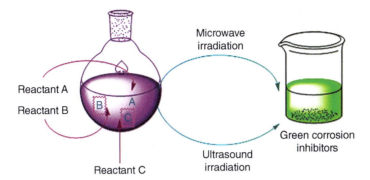

Figure 2.2 Illustration showcasing the transformative impact of microwave and ultrasonication irradiation in green chemistry. Source: Verma et al. [18]/Elsevier.

greener and more innovative future, emphasizing a commitment to eco-conscious chemistry practices and a transformative shift in chemical processes toward a more sustainable and environmentally friendly direction [16]. In each context, they contribute to cleaner and more efficient processes while minimizing the ecological footprint of industrial activities [17]. Figure 2.2 illustrates the transformative impact of microwave and ultrasonication irradiation in green chemistry. These techniques align seamlessly with green chemistry principles, offering benefits such as rapid reaction times, improved product selectivity, and minimized environmental impact. They have become indispensable tools for sustainable chemical processes in various industries. In the following sections, we will delve deeper into the fundamental principles governing microwave and ultrasonication irradiation, highlighting how they optimize chemical reactions. Additionally, we will explore their advantages, including enhanced selectivity and reduced reaction times, and examine their applications in specific industries. Through this detailed examination, we aim to provide a comprehensive understanding of the transformative potential of these technologies in the realm of green chemistry and their pivotal role in shaping a more sustainable future.

2.1.1 Historical Background

The integration of microwave and ultrasonication methods into green chemistry practices is driven by the need to address environmental concerns and promote sustainable chemical processes. These methods offer rapid and efficient reactions, aligning with green chemistry's principles of energy efficiency and waste reduction. They find applications across diverse industries, from pharmaceuticals to materials science, exemplifying a transformative shift toward more sustainable and environmentally friendly chemical processes [19]. The integration of microwave and ultrasonication methods into green chemistry practices is driven by the imperative to uphold sustainability and environmental responsibility. This entails aligning chemical transformations with eco-conscious values and striving for cleaner and more resource-efficient processes. This transformative incorporation, exemplified by the introduction of microwave irradiation in organic synthesis in the 1980s, has since catalyzed ongoing advancements in academia and research, fostering cleaner and more efficient chemical processes across diverse industries [19]. The historical underpinnings of microwave irradiation's integration into organic synthesis have paved the way for the continuous development of eco-friendly chemical transformations. These foundations, aligned with the unwavering pursuit of environmentally conscious and efficient processes, exemplify the field's commitment to innovation and ecological awareness. This evolution represents a persistent drive for progress within the realm of chemical science. Microwave irradiation's consistent expansion in utilization, driven by its effectiveness in reducing reaction times and increasing product yields, aligns seamlessly with academic research and scientific objectives [20]. Microwave irradiation technology accelerates chemical reactions by inducing molecular rotations, effectively enhancing reaction kinetics, and enabling precise control by selectively heating specific portions of the reaction mixture. Conversely, ultrasonication irradiation has a historical presence in the food industry, notably improving plant compound extraction and fat emulsification since the 1950s. Both techniques exemplify the multifaceted utility of innovative methods in optimizing chemical processes. Microwave irradiation finds applications in pharmaceutical synthesis, reducing reaction times and improving yields, aligning with academic research and scientific goals. Ultrasonication plays a pivotal role in food technology, enhancing flavor extraction and emulsion stability. These applications highlight their versatility and significance in academia and research, promoting efficiency and precision in chemical processes [21]. Ultrasonication irradiation gained recognition in the 1990s for its pivotal role in advancing environmentally responsible chemical methodologies. The concept of integrating microwave and ultrasonication irradiation for green chemical transformations emerged in the early 2000s, offering the promise of improved reaction efficiency and selectivity. This innovative approach embodies the ongoing pursuit of more sustainable and efficient chemical processes, aligning with the tenets of green chemistry. It signifies the evolution of techniques aimed at optimizing chemical reactions while reducing environmental impact. The combination of microwave and ultrasonication irradiation works synergistically to enhance desired chemical product yields while simultaneously minimizing the formation of undesired byproducts, meeting the objectives of academic exploration [22]. The strategy of integrating microwave and ultrasonication irradiation represents a pioneering approach dedicated to enhancing chemical processes, harmonizing perfectly with the core principles of green chemistry that prioritize sustainability and environmental conscientiousness. The collaboration between these two irradiation methods marks a noteworthy progression in the quest for more efficient and environmentally responsible chemical transformations. This evolution of green chemical transformation is a testament to the dynamic interplay between scientific principles and technological innovations, in alignment with the overarching ideas of academic and scientific findings, aimed at achieving both precision and environmental obligation in chemical processes [23]. The late 1990s marked a significant juncture in the consolidation of green chemistry principles within the scientific community. During this period, there was a resounding recognition of the imperative for chemical processes to pivot toward renewable and environmental management, and economic viability. This pivotal era symbolized a deliberate commitment to bridging scientific principles with practical solutions, reflecting the ongoing pursuit of more sustainable chemical practices. It underscored the growing awareness of the need for environmentally conscious methodologies to address contemporary environmental and economic complexities, all in alignment with the overarching goals of academic as well as scientific study [3].

During the late 1990s, there was a concerted effort to reduce the use of hazardous chemicals, minimize waste generation, and enhance process efficiency, aligning with the principles of green chemistry. A pivotal development in this context was the introduction of microwave irradiation in the 1980s, revolutionizing organic synthesis. Its unique capability to induce molecular rotations facilitated rapid heating, resulting in significantly shorter reaction times. Furthermore, its precision in selectively heating specific components within a reaction mixture expanded its potential to improve reaction selectivity. This innovation epitomizes the fusion of scientific ingenuity with sustainability objectives, underscoring its profound impact on advancing eco-conscious chemistry practices, as emphasized in systematic exploration [24]. The historical context underscores a steadfast commitment to adopting environmentally conscious and efficient methodologies in chemical processes. Ultrasonication, originally utilized in the food industry in the 1950s for tasks like plant compound extraction and fat emulsification, gradually gained recognition within organic synthesis, particularly in the 1990s. However, a pivotal turning point emerged in the early 2000s with the conceptualization of combining microwave and ultrasonication irradiation methods. This marked a significant leap forward in the pursuit of sustainable and efficient chemical transformations, aligning seamlessly with the goals of academic research and scientific exploration, showcasing the evolving landscape of eco-conscious chemistry practices [25]. The innovative amalgamation of microwave and ultrasonication irradiation techniques represented a significant leap toward advancing the efficiency and sustainability of chemical processes. This collaborative integration harnessed the potent capabilities of both methods to optimize chemical reactions fundamentally. It stemmed from the recognition that their combined application had the potential to enhance product yield while concurrently reducing the generation of undesired by-products. The evolution of these techniques signifies a dynamic stride in the pursuit of more environmentally responsible and efficient chemical practices, illustrating their pivotal role in shaping the landscape of sustainable chemistry [26]. The innovative fusion of microwave and ultrasonication irradiation techniques represents a significant leap toward achieving rapid, efficient, and highly selective chemical transformations, all while eliminating the use of hazardous solvents and reagents. This approach signifies a substantial stride in realizing environmentally responsible and efficient chemical methodologies, underpinned by a historical trajectory reflecting an enduring commitment to advancing eco-conscious chemical practices. This merger in green chemical transformation exemplifies the transition from theoretical principles to practical application, marking a pivotal advancement in developing more sustainable and eco-friendly chemical processes. It underscores the central themes of innovation and ecological responsibility in the field of chemistry, aligning seamlessly with academic research and scientific goals, embodying the transformative potential of eco-conscious chemistry practices [27]. The integration of microwave and ultrasonication irradiation techniques embodies a resolute commitment to bridging the chasm between scientific ideals and practical applications, propelling the evolution of more sustainable and efficient chemical methodologies. This journey epitomizes an unwavering pursuit of ecological responsibility and heightened efficiency in chemical processes, with innovation and adaptability at its core. It underscores the transformative potential of eco-conscious chemistry practices, symbolizing a determined endeavor toward a more environmentally aware and efficient future in the domain of chemistry, a journey of profound significance embraced by academia and scientific exploration. It reflects a commitment to continuously explore and implement advanced methodologies that align with sustainability principles. This drive for innovation is central to addressing environmental challenges and optimizing chemical processes for a greener and more sustainable tomorrow.

2.1.2 Green Chemistry Principles for Sustainable System

Green chemistry is a transformative paradigm emphasizing sustainability and reduced environmental impact in chemical products and processes. It significantly advances scientific knowledge by promoting innovation in ethical and sustainable chemical methodologies. For instance, it can replace hazardous solvents with water, develop greener pharmaceutical synthesis, and catalyze reactions to minimize waste. In research grant peer reviews, projects tackling issues like wastewater treatment with environmentally friendly catalysts exemplify its potential impact in building a cleaner, more sustainable future [28]. In the realm of drug delivery systems, the paramount focus is on selecting solvents with low toxicity, biodegradability, and minimal harm to human health, aligning with green chemistry principles. Ionic liquids (ILs), known for their unique properties and ability to remain in a liquid state at lower temperatures, stand out as attractive solvents. Recycling ILs emerges as a pivotal facet, reducing waste generation and meeting the ongoing demand for these chemicals. This approach underscores a commitment to eco-conscious pharmaceutical research, aligning with academic research and scientific objectives where safety and environmental responsibility are paramount. ILs offer advantages in improving drug solubility, stability, and bioavailability while adhering to sustainable practices, making them a noteworthy choice in advancing drug delivery systems [29]. Green chemistry is a transformative approach that prioritizes environmentally responsible chemical products and processes, aligning with sustainability principles by minimizing the use and production of harmful substances. ILs, chosen for their low toxicity and biodegradability, exemplify eco-conscious solvent selection. Recycling ILs is integral to sustainability, reducing waste, and production demands. This approach mirrors the commitment to eco-conscious pharmaceutical research, emphasizing safety and environmental stewardship. The "cradle-to-cradle" design philosophy underscores the importance of minimizing environmental impact throughout a product's lifecycle. Green chemistry plays a pivotal role in addressing global challenges, offering a path toward a more sustainable and eco-friendly future [30]. Green chemistry principles play a pivotal role in the development of drug delivery systems that offer precise and controlled drug release, thereby reducing waste and potential adverse effects. For instance, biodegradable polymers, such as poly(lactic-*co*-glycolic acid) (PLGA), are used to encapsulate drugs and slowly release them over time, improving patient adherence by eliminating the need for frequent dosing. This aligns with academic research and scientific objectives that prioritize eco-conscious pharmaceutical solutions. By minimizing the use of harmful substances and optimizing drug release, green chemistry not only enhances patient outcomes but also contributes to environmental stewardship. This approach underscores the broader commitment to sustainable healthcare practices, emphasizing safety and effectiveness while minimizing ecological impact [31]. In the realm of green drug delivery systems, material selection plays a pivotal role in ensuring both safety and environmental sustainability. Biocompatible substances like chitosan and hyaluronic acid (HA) are favored choices due to their compatibility with biological systems, minimizing the risk of adverse reactions. This aligns with the core principles of green chemistry, which aims to reduce the generation of harmful by-products in chemical processes. For instance, when considering chitosan-based drug delivery, its natural origin and biodegradability exemplify green chemistry in action. Chitosan's ability to encapsulate and release drugs in a controlled manner not only enhances therapeutic outcomes but also reduces the environmental burden associated with drug manufacturing and disposal [32]. ILs are a promising class of solvents for drug delivery. They have a number of unique properties that make them attractive for this application, including low volatility, nonflammability, and high solvation capabilities. ILs can be used to design greener drug delivery systems by reducing the use of hazardous solvents and reagents and by creating more targeted drug delivery systems. ILs have the potential to make drug delivery safer, more effective, and more environmentally friendly [33]. Chitosan-based hydrogels and nanoparticles are promising materials for drug delivery due to their biocompatibility, tunable properties, and ability to encapsulate and control the release of drugs. These materials are being used to develop new and more effective treatments for a variety of diseases, including diabetes, cancer, and infectious diseases. For example, chitosan-based hydrogels can be implanted under the skin to release insulin in a controlled manner, improving blood sugar control in patients with diabetes. Chitosan-based nanoparticles can be targeted at tumors to deliver cancer drugs more effectively and reduce side effects. The development of chitosan-based hydrogels and nanoparticles for drug delivery is a significant scientific advancement with the potential to revolutionize the field of pharmaceutical science. These materials offer the promise of safer, more effective, and more targeted drug delivery [34]. Click chemistry is a group of chemical reactions that are highly efficient and specific, making them ideal for use in drug delivery. It can be used to functionalize and crosslink drug delivery carriers in a variety of ways. Targeting ligands are molecules that bind to specific cells or tissues, and they can be used to direct drug delivery carriers to specific sites in the body. This can improve the efficacy of treatment and reduce side effects. Crosslinking involves linking two or more drug delivery carriers together to form a network. This can improve the stability of the drug delivery system in the body and extend its release profile. Click chemistry is being used to develop new and more effective drug delivery systems for a variety of diseases, including cancer, infectious diseases, and chronic diseases. For example, click chemistry is being used to develop targeted nanoparticles for the delivery of cancer drugs, crosslinked hydrogels for the delivery of proteins and other biological drugs, and new types of drug delivery systems, such as micelles and vesicles [35]. Polymer grafting and environmentally friendly processes can be combined to design more effective and sustainable drug delivery systems. This approach has the potential to improve the treatment of diseases and make drug delivery more accessible to people around the world. Specific examples of this approach include water-soluble nanoparticles for cancer drug delivery, hydrogels for the sustained release of proteins, and other biological drugs, and new types of drug delivery systems, such as micelles and vesicles [36]. Chitosan-based hydrogels and stabilized metal nanoparticles are promising materials for drug delivery. Chitosan-based hydrogels can encapsulate and release drugs in a controlled manner. Stabilized metal nanoparticles can target drugs to specific cells or tissues. Combining these technologies can lead to more effective and precise drug delivery systems. Examples of applications include controlled delivery of chemotherapy drugs and targeted delivery of cancer drugs. These technologies have the potential to revolutionize drug delivery for a wide range of diseases [37]. HA-based hydrogels, known for their biocompatibility, have garnered attention across diverse biomedical fields, including tissue regeneration, diagnostics, and drug delivery. HA-based hydrogels are promising materials for therapeutic and diagnostic applications. HA-based hydrogels are biocompatible, tunable, and versatile. HA-based hydrogels can be used to deliver drugs to specific cells or tissues. HA-based hydrogels can also be used to deliver proteins and other biological drugs to specific tissues. HA-based hydrogels can also be used to support tissue regeneration. HA-based hydrogels are being developed as biosensors and imaging contrast agents [38]. Targeting ligands and crosslinkers are promising tools for designing more effective and targeted drug delivery systems. Targeting ligands can be used to guide drugs to specific cells or tissues, improving the efficacy of treatment and reducing side effects. Crosslinkers can be used to create stable drug delivery systems for controlled drug release. Targeting ligands and crosslinkers are being used to develop new drug delivery systems for a variety of diseases. These technologies have the potential to revolutionize the field of medicine and make it possible to develop more effective and personalized treatments for a wide range of diseases [39]. IEDDA click reaction-based HA hydrogels are promising materials for targeted drug delivery. These hydrogels are catalyst-free, stimuli-responsive, and can be targeted to specific cells or tissues by attaching targeting ligands. IEDDA click reaction-based HA hydrogels are formed by the reaction of tetrazine and trans-cyclooctene groups, which is a catalyst-free and stimuli-responsive reaction. These hydrogels can be targeted to specific cells or tissues by attaching targeting ligands to the hydrogel surface. IEDDA click reaction-based HA hydrogels are being used to develop new drug delivery systems for a variety of diseases, including cancer, infectious diseases, and cardiovascular diseases. For example, researchers are developing IEDDA click reaction-based HA hydrogels to deliver chemotherapy drugs to cancer cells. The hydrogels are targeted at cancer cells by attaching folic acid to the hydrogel surface. The hydrogels are also designed to be pH-responsive so that the drugs are released in the tumor microenvironment. These hydrogels have the potential to revolutionize the field of medicine and make it possible to develop more effective and personalized treatments for a wide range of diseases [40]. Starch-based materials (SHs) are a promising new class of materials with a wide range of potential applications. They are derived from starch, a renewable and abundant resource. SHs are biodegradable, compostable, nontoxic, and versatile. SHs can be used to develop sustainable materials and solutions for a variety of global challenges, such as climate change, food insecurity, and plastic pollution. For example, SHs can be used to produce biofuels, sustainable packaging materials, biodegradable food packaging materials, edible films and coatings, biodegradable straws, utensils, food packaging materials, and bio-based plastics. By understanding the unique properties of SHs and developing new ways to use them, we can create new and innovative ways to reduce our environmental impact and create a more sustainable future [41]. SHs are a promising class of materials for sustainable development. They are derived from renewable resources, exhibit properties that make them versatile for various sectors and can be used to develop sustainable solutions for global challenges. SHs are aligned with green chemistry principles and can be used to develop a wide range of products, including packaging materials, food additives, textiles, and drug delivery systems. SHs can also be used to develop sustainable solutions for global challenges, such as climate change, food insecurity, and plastic pollution. Some specific examples of SH-based products that are currently available or under development – edible films and coatings, biodegradable straws, utensils, food packaging materials, bio-based plastics, and SH-based drug delivery systems. As research on SHs continues, we can expect to see even more innovative and sustainable products made from these materials in the future [42]. SHs are a promising class of sustainable materials derived from starch, a renewable and abundant resource. SHs are biodegradable, compostable, nontoxic, and versatile. SHs can be used to develop sustainable packaging materials, food additives, textiles, drug delivery systems, and solutions for global challenges, such as climate change, food insecurity, and plastic pollution. It is important to consider the ethical and social implications of using SHs, such as their impact on food security and the environment. By carefully weighing the benefits and drawbacks, we can ensure that SHs are used in a responsible and sustainable manner [43]. Starch's molecular arrangement, defined by amylose and amylopectin, plays a vital role in its properties. Understanding this structure is essential for developing sustainable materials and enhanced functionality in food science, biotechnology, and material engineering. Interdisciplinary research focuses on optimizing amylose and amylopectin components to achieve desired properties [44]. Amylose and amylopectin are the two major components of starch, with amylopectin being the dominant one. The relative composition of these two components varies depending on the source of starch, such as corn, wheat, potato, and rice. This variation in composition also affects the physical properties of starch, such as its amorphous-crystalline balance, digestibility, viscosity, and texture. Interdisciplinary research is ongoing to leverage the diversity of starch and its components to develop tailored materials with functional and sustainable properties. For example, researchers are developing starch-based bioplastics, food additives, textiles, and drug delivery systems with enhanced properties by manipulating the amylose and amylopectin components [45]. By understanding the role of amylose and amylopectin in the crystallinity and biodegradability of starch, researchers can design starch-based bioplastics with tailored properties for specific applications. Starch-based food additives, such as thickeners, emulsifiers, and stabilizers, are widely used in the food industry. Researchers can leverage the unique properties of amylose and amylopectin to develop starch-based food additives with improved functionality and reduced environmental impact. Amylose and amylopectin can be used to engineer starch-based drug delivery systems with controlled release of drugs [46]. Native starch has limitations in dimensional stability, mechanical properties, and processing efficiency. To address these challenges, researchers are developing cutting-edge approaches to modify starch properties using green chemical methods, composite materials, and scalable processing methods. These new materials offer eco-friendly alternatives with tailored functionalities for diverse applications [47]. Researchers employ green chemistry principles to modify starch for diverse applications, driven by sustainability concerns. This approach aims to enhance starch performance while reducing environmental impact. Interdisciplinary collaboration in materials science and chemistry is essential to explore innovative materials and compositions. Ongoing cutting-edge research focuses on sustainable starch-based solutions for global challenges [48]. Radiation-based modifications of starch offer a promising approach to developing sustainable materials with enhanced properties for diverse applications. This green chemistry-inspired method allows precise and eco-friendly alterations to starch's properties, promoting sustainability. Interdisciplinary efforts involving materials science and radiation physics are essential to advance this technique, with ongoing research aimed at optimizing radiation-based modifications for various applications [17]. Green chemistry-inspired radiation-based modification of starch is a promising approach to developing sustainable materials with enhanced properties for diverse applications. This method allows precise and eco-friendly alterations to starch properties while ensuring safety. Interdisciplinary efforts encompassing radiation physics and materials science are essential to advance this technique. Researchers are exploring sustainable techniques to further enhance performance and reduce environmental impact, such as using natural crosslinkers [49]. Starch citrate, a sustainable and versatile material with natural antibacterial properties, can be further enhanced through green chemistry-inspired radiation-based modification. This approach allows precise and eco-friendly alterations to starch citrate's properties, tailoring its performance for specific applications. Interdisciplinary efforts encompassing radiation physics, materials science, and microbiology are essential to advance this technique [50].

Green chemistry-inspired radiation-based modification of starch citrate and lignin is a promising approach to developing sustainable materials with enhanced antimicrobial properties and improved performance for diverse applications. This technique offers precise and eco-friendly alterations to material properties, enabling tailoring to specific needs. Interdisciplinary collaboration and optimization are essential to advance this field. Lignin-based innovations are gaining traction as researchers seek to address environmental challenges and create more sustainable materials. Larraneta et al. [51] introduced an eco-friendly approach to preparing these materials by combining LIG with poly(ethylene glycol) and poly(methyl vinyl ether-co malic acid) through an esterification reaction, significantly accelerating the process using microwave (MW) radiation. Microwave-assisted modification of starch citrate and lignin is a promising green chemistry-inspired approach to developing sustainable materials with enhanced properties and performance for diverse applications. This method allows precise and eco-friendly alterations to material properties, enabling tailoring to specific needs. Interdisciplinary collaboration and optimization are essential to advance this field. Microwave technology offers numerous advantages, such as reduced production time, energy consumption, and costs, making it a particularly valuable tool for industries seeking more efficient and eco-friendly manufacturing [52]. Sustainable drug delivery systems based on starch citrate, lignin, and inulin offer a promising green chemistry-inspired approach to developing innovative and eco-friendly materials for targeted and controlled drug delivery. These systems have the potential to address the limitations of traditional drug delivery systems, such as systemic side effects and poor drug bioavailability. Interdisciplinary collaboration and optimization are essential to advance this field and bring these systems to fruition [53]. Researchers could design a sustainable and scalable green chemistry-inspired drug delivery system based on linseed mucilage for the controlled release of an oral insulin formulation. The pH-responsive hydrogel would release insulin in the acidic environment of the stomach, ensuring direct delivery to the bloodstream. The system would be biocompatible, biodegradable, and adaptable for use with other drugs and delivery routes. This example demonstrates the potential of linseed mucilage-based materials in developing sustainable and innovative drug delivery systems [54].

Green chemistry-inspired synthesis of Ag-NPs is a promising approach to developing sustainable and effective antimicrobial materials. These Ag-NPs possess antioxidant and antimicrobial properties, making them promising candidates for a variety of applications beyond biomedicine, including environmental remediation and food packaging [55]. Gellan gum (GG) stands out as a sustainable biopolymer, boasting biocompatibility, hydrophilicity, and gelling properties. Its applications span agriculture, where GG-based formulations control pesticide and fertilizer release, mitigating environmental concerns. In food technology, GG's gelling prowess elevates textures and prolongs product shelf life, reducing food waste. Moreover, GG's potential in biomedicine encompasses drug delivery and tissue engineering. GG's versatility and eco-friendliness make it a key player in crafting innovative and sustainable solutions across these diverse fields [56]. Electro-stimulated drug release devices (EDRDs) are innovative medical tools designed to respond to electrical signals, offering precise and adaptable controlled substance release. Notably, they hold immense promise in the realm of insulin delivery for diabetes management. These devices are characterized by their precision, ensuring patients receive the correct insulin dose when needed, enhancing safety through biocompatible materials, and adaptability to individual insulin requirements. For instance, researchers envision an EDRD comprising an implantable device equipped with a reservoir for insulin and a microfluidic network. It can be powered by a battery, rechargeable with an external transmitter, and programmed to adjust insulin release based on real time blood sugar monitoring. Such EDRDs, typically constructed from durable materials like silicone and titanium, offer improved blood sugar control, reduced diabetes-related complications, and an enhanced quality of life for individuals with diabetes [57]. Genipin-based crosslinked materials represent a sustainable avenue for creating controlled release systems in agriculture, benefiting both the environment and crop health. Genipin, a natural and biodegradable crosslinking agent, serves as an eco-friendly choice. An example of such a system is the biodegradable capsule, which combines pesticides, nutrients, and genipin to ensure prolonged chemical release near plant roots. By adjusting capsule attributes and genipin concentration, this approach can be tailored to meet specific crop needs, reducing environmental impact, enhancing crop efficiency, and maintaining nutrient consistency [58]. Genipin-crosslinked materials, leveraging the natural and biodegradable properties of genipin, offer a sustainable solution for wastewater treatment. They excel in pollutant removal, including heavy metals, due to their tailored affinity. These materials enhance water quality, reduce environmental impacts, and protect public health, exemplifying their versatile applications in improving wastewater treatment systems [59]. Green chemistry principles are driving the development of sustainable and scalable nanomaterial synthesis methods, particularly for applications in consumer electronics and renewable energy. These approaches prioritize eco-friendliness by minimizing hazardous chemicals and waste production. Examples include using natural materials like cellulose nanofibers, silver nanowire synthesis, and employing aqueous solutions and mild conditions for silicon nanocrystal production. Furthermore, continuous flow reactors facilitate large-scale nanomaterial production. These advancements are crucial for integrating nanotechnology into consumer electronics and renewable energy solutions [60]. Green composites are ushering in a transformation in the transportation sector, particularly in lightweight auto parts and aerospace applications. These composites, crafted from renewable and recyclable materials like plant fibers and bio-based resins, offer substantial environmental advantages compared to petroleum-based counterparts. Notable examples include natural fiber-reinforced polymer (NFRP) composites for lightweight auto parts and carbon fiber-reinforced bioresin (CFRB) composites for aerospace applications. These innovations hold the potential to revolutionize the industry, making vehicles more fuel-efficient, lighter, and eco-friendly [61]. Sustainable food packaging materials are emerging as a pivotal solution to address global food waste issues, enhancing food security and reducing the environmental footprint of the food industry. Two notable examples include edible coatings, which preserve product freshness by reducing moisture loss and inhibiting microbial growth, and active packaging materials, designed to interact with food or the environment to extend shelf life. These innovations hold the potential to revolutionize the food industry by ensuring safer, fresher, and more sustainable food products [62]. Academic sustainable chemistry programs educate future scientists on eco-friendly practices. Research institutions prioritize sustainability, advancing green technologies globally. Sustainable materials and green chemistry foster an environmentally responsible future, addressing global challenges while promoting harmony with the planet.

2.2 Fundamental Concepts of the Electromagnetic Spectrum Theory

The electromagnetic spectrum theory is a foundational concept in physics, describing the behavior of electromagnetic waves characterized by oscillating electric and magnetic fields propagating through space at the speed of light. This theory plays a pivotal role in numerous scientific and technological domains, including communication, imaging, and scientific research, spanning radio waves to gamma rays. Some key principles include the transverse nature of electromagnetic waves, their constant speed in a vacuum, and the inverse relationship between frequency and wavelength. One challenging application is the design of an early stage cancer cell detection system utilizing electromagnetic waves. To accomplish this, researchers would need to determine the most suitable type of electromagnetic waves for precise imaging, establish methods to differentiate between cancerous and healthy cells, enhance sensitivity for early detection, and ensure the system's safety and affordability [63]. The electromagnetic spectrum encompasses a wide range of electromagnetic radiation, from long radio waves to short gamma rays, capable of traveling through various mediums, even vacuum. Frequency, measured in Hertz (Hz), denotes the number of wave cycles within a unit of time, while wavelength represents the spatial length of one complete cycle, inversely related to frequency. The speed of light (c) in a medium is the product of frequency and wavelength, as expressed in the equation $c = f\lambda$, highlighting their fundamental interdependence. For instance, visible light spans a frequency range of 400–700 THz, with corresponding wavelengths of 400–700 nm. In a vacuum, light travels at a constant speed of approximately 300 000 km/s, where the inverse relationship between frequency and wavelength holds true, distinguishing colors like red and blue based on their frequencies and wavelengths. A challenging task related to this topic involves designing an electromagnetic wave-based imaging system to detect cancer cells in their early stages. This challenge demands a profound understanding of electromagnetic theory to select the appropriate waves, devise mechanisms for cancer cell differentiation, enhance sensitivity for early detection, and ensure safety and affordability for patients [64]. In the realm of electromagnetic waves and energy harvesting, it is crucial to understand that the energy of these waves corresponds directly to their frequency. The electromagnetic spectrum covers an extensive range of frequencies, from extremely low to extremely high, each with its unique applications across various fields, including telecommunications, medical imaging, and more. Energy harvesting, on the other hand, involves converting environmental energy into electrical power, and electromagnetic wave-based systems offer the potential to harness energy from any part of the spectrum. An illustrative example introduces a broadband electromagnetic energy harvester capable of converting a wide spectrum of frequencies into electricity [65]. This innovative technology holds promise for powering wireless sensor nodes and small electronic devices, albeit with the ongoing challenge of achieving cost-effective and efficient energy harvesting across a broader range of applications. Potential applications for electromagnetic wave-based energy harvesting are diverse, ranging from powering environmental sensors and small electronics to providing energy solutions in remote or off-grid areas, including disaster zones. Furthermore, this technology can play a role in sustainable energy generation from renewable sources like solar and wind. However, several challenges must be addressed, including refining the efficiency and cost-effectiveness of energy harvesting systems, ensuring they can harness energy across a wide frequency range, enhancing their durability and reliability, and integrating them effectively with existing energy grids [66]. The utilization of electromagnetic waves in the quest to detect and identify exoplanets presents promising opportunities. Different approaches involve designing devices to detect how exoplanets interact with electromagnetic waves, either by absorbing or reflecting them. Additionally, the unique spectral signatures of exoplanets, shaped by their composition and atmosphere, offer another avenue for identification. A recent example in this field highlights the potential of using radio waves to detect exoplanets, particularly smaller and fainter ones, through reflection [67]. Nonetheless, challenges persist, including the need for highly sensitive and affordable devices, as well as the distinction between exoplanets and other celestial objects. Despite these challenges, the development of electromagnetic wave-based devices for exoplanet detection is a vibrant research area with the potential to revolutionize exoplanetary astronomy and expand our search for extraterrestrial life [68]. The wave-particle duality of electromagnetic radiation, inherent in the quantum nature of photons, has profound implications for various scientific and technological applications. Notably, this duality plays a pivotal role in the functionality of devices harnessing electromagnetic waves. Quantum devices, capitalizing on the unique properties of quantum systems, offer capabilities beyond classical counterparts. Quantum computers, for instance, leverage the superposition and entanglement of quantum particles, enabling computations unattainable by classical computers. Similarly, quantum cryptography devices exploit the quantum properties of light to establish unhackable communication channels. A pertinent example from research explores the potential of using photons to construct quantum computers, shedding light on the advantages and challenges of this approach [69]. These quantum devices hold promise in diverse fields, including materials science, secure communication, high-precision sensing, and advanced imaging. However, their development requires meticulous control of quantum systems and addressing sensitivity to external disturbances. Despite these challenges, ongoing research in quantum devices stands poised to usher in transformative technologies with far-reaching implications [70]. Microwaves are a type of electromagnetic radiation with wavelengths ranging from approximately 1 mm to 30 cm. They have a variety of applications, including cooking, communication, and medical imaging. One of the potential applications of microwaves is in the noninvasive detection and identification of tumors. Microwaves can interact with matter in a variety of ways, depending on the dielectric properties of the material. Dielectric properties are the electrical properties of a material, and they can vary depending on the composition of the material. Tumors typically have different dielectric properties than healthy tissue, so it may be possible to use microwaves to detect the presence of tumors. There are a number of different approaches to designing microwave-based devices for detecting and identifying tumors. One approach is to design a device that can detect the difference in dielectric properties between tumors and healthy tissue. Another approach is to design a device that can detect the metabolic activity of tumors. Tumors typically have higher metabolic activity than healthy tissue, so it may be possible to use microwaves to detect the increased metabolic activity of tumors [71]. Microwaves, with their unique interaction capabilities based on material dielectric properties, are at the forefront of material property control. Their applications span multiple domains, from enhancing the conductivity of semiconductors to crafting patterns in materials and guiding crystal growth. Notably, microwave-assisted synthesis of nanomaterials offers significant promise, as highlighted in research on nanomaterial production. Challenges encompass the need for precise, scalable, and cost-effective control devices, as well as deeper insights into fundamental microwave-material interactions. Nevertheless, ongoing research in this dynamic field is poised to revolutionize various industries, leading to improved material properties, more efficient manufacturing processes, and the creation of novel nanomaterials. This burgeoning area of microwave-based material control holds the potential to transform industries and drive innovation across multiple sectors [72]. Microwave cooking, a popular and convenient method, harnesses electromagnetic radiation with wavelengths from 1 mm to 30 cm. It relies on the interaction of microwaves with polar molecules, which possess positive and negative ends. When exposed to microwaves, these molecules rotate and generate heat. The heat generated depends on microwave frequency and the material's dielectric properties, which vary with its composition. This cooking technique exploits the high dielectric properties of water, a key component in most foods. As microwaves interact with water molecules in the food, they induce molecular rotation, resulting in heat generation and food cooking. Beyond culinary applications, microwave heating finds utility in scientific realms like food processing, chemical synthesis, and materials science, facilitating quick and efficient drying, pasteurization, chemical synthesis, and annealing of materials. Understanding the fundamental interaction between microwaves and polar molecules is crucial for mastering microwave cooking and optimizing its application in scientific research. It plays a pivotal role in food preparation, from cooking diverse foods to defrosting, reheating, and even making popcorn. In the scientific domain, microwave technology has enabled advances in food processing, chemical synthesis, and materials science, enhancing efficiency and enabling innovative research [73]. Electromagnetic radiation comprises oscillating electric and magnetic fields traveling at the speed of light, guided by Maxwell's equations. Microwave irradiation's efficacy lies in its selective interaction with molecules like water, inducing their rotation and generating heat. This principle finds applications in various fields, including science and cooking, emphasizing the importance of understanding these interactions [74]. Understanding these principles elucidates microwave heating mechanisms, crucial for efficient cooking as microwaves penetrate and heat food from within. Beyond cooking, microwaves are applied in diverse fields, including chemistry and materials research, due to their selective interaction with specific materials. This fundamental understanding is vital for optimizing microwave applications in various disciplines [75]. Microwave selectivity facilitates precise manipulation and research in scientific and industrial sectors. This versatility highlights its value in diverse contexts, from chemistry to materials research, enhancing efficiency and precision. Understanding this capability is fundamental for optimizing microwave applications across disciplines [76]. A profound understanding of the electromagnetic spectrum and MW principles reveals their wide-ranging applications in daily life, science, and industry. Microwaves, as a part of the spectrum, are harnessed for tasks like cooking and scientific research. Understanding wave-particle duality, wavelength-frequency relationships, and microwave-material interactions is crucial for effective utilization in diverse applications, laying the foundation for comprehending their significance in various fields.

2.3 Electrical, Dielectric, and Magnetic Properties in Microwave Irradiation

Microwave irradiation employs electromagnetic waves within the microwave frequency range (300 MHz to 300 GHz) to induce dielectric and magnetic properties in materials, facilitating various scientific and industrial applications. These interactions are crucial for fields such as chemistry, materials research, and communication technologies. Understanding these principles is fundamental for harnessing microwave irradiation's potential in diverse contexts [77]. Understanding microwave-material interactions, driven by dielectric characteristics and polarization, is essential for the effective utilization of this technology. Its broad applications encompass materials science, chemistry, and communication, underscoring its importance in diverse scientific and practical realms. Researchers explore novel materials and compositions, advancing current understanding and pushing the boundaries of microwave-based research in various disciplines [78]. The dielectric constant, or relative permittivity, characterizes a material's ability to store electrical energy under an electric field, influencing electromagnetic wave propagation. This property is crucial in optimizing microwave-material interactions, particularly in materials science and communication. Materials with higher dielectric constants exhibit greater microwave absorption, impacting various scientific and industrial applications [79]. Dielectric loss (tan δ) quantifies energy dissipation as heat in response to an electric field's polarization phase angle. Materials with higher tan δ

absorb more microwave energy, vital for controlled heating in applications like cooking and scientific research. Understanding and manipulating tan δ in various materials are significant for optimizing microwave-based processes [80]. Magnetic permeability is vital for materials' interaction with magnetic components of electromagnetic waves, especially microwaves. High magnetic permeability enhances this interaction, impacting microwave-based processes. This knowledge is critical in fields like materials science and electromagnetic applications, where precise control of materials' magnetic response is significant for optimizing performance [81]. Understanding these properties is crucial for optimizing material behavior in various applications. Materials with dielectric losses can be heated using microwaves and applied in industrial processes like curing and medical hyperthermia for cancer treatment. This knowledge is relevant in materials science, engineering, and medical disciplines for efficient energy absorption and controlled heating [82]. Microwave irradiation is versatile, used in polymerization, chemical synthesis, and nanoparticle production. Knowledge of dielectric and magnetic properties is crucial for antenna and radar system design, but these properties can vary with material composition and conditions. Microwave technology has diverse applications in both scientific and industrial domains. A comprehensive understanding of material properties is essential for optimizing microwave applications in diverse fields, enabling tailored and enhanced performance in scientific and industrial settings. Conductive polymer composites (CPCs) are integral in applications involving microwave irradiation due to their critical electrical, dielectric, and magnetic properties. These properties are essential for applications like electromagnetic interference (EMI) shielding, capacitors, and integrated circuits when exposed to MW. CPCs offer versatility in materials and compositions, driving ongoing cutting-edge research across multiple disciplines to optimize their performance in various scientific and industrial contexts [83]. In the realm of microwave irradiation, optimizing CPCs demands a balance between achieving favorable electrical and dielectric properties while mitigating challenges related to high nanofiller content. This balance is crucial for applications like EMI shielding and integrated circuits. Interdisciplinary research focuses on reducing the percolation threshold, enhancing electrical conductivity, and addressing processing complexities to improve CPC performance in diverse scientific and industrial contexts [84]. In the context of microwave applications, the "segregated network" strategy involves selectively dispersing nanofillers at composite interfaces and impacting electrical and dielectric properties. This approach is relevant for applications like EMI shielding but may pose challenges related to polymeric molecule diffusion and structural integrity. Achieving the right balance in electrical conductivity is crucial, driving interdisciplinary research in materials science and engineering to optimize CPC performance in diverse contexts [85]. In microwave applications, CPCs with electrical conductivity below $1\,S\,m^{-1}$ are sought for effective EMI shielding, but achieving both low conductivity and microwave transparency is challenging. Commercially, EMI shielding effectiveness targets are often below 20 dB. Carbon-based nanofillers such as multi-walled carbon nanotubes (MWCNTs), graphene nanoplatelets, and carbon nanofibers, known for high aspect ratios and electrical conductivity, are favored for reinforcing CPCs in microwave scenarios, driving interdisciplinary research in materials science and engineering for improved performance [86]. Research on CPCs for microwave applications focuses on unraveling structure-property relationships. This entails investigating how nanofiller distribution, concentration, and type influence CPCs' electrical, dielectric, and magnetic characteristics under microwave irradiation. Achieving the right balance of these properties is vital for optimizing CPCs in applications requiring EMI shielding and microwave transparency, driving interdisciplinary research in materials science and engineering for enhanced CPC performance [87]. Studying dielectric constants in microwave irradiation is essential for optimizing the materials' microwave absorption. This property impacts various scientific and industrial applications, especially in microwave heating. Materials with higher dielectric constants absorb more microwave energy, making them valuable for processes like cooking and materials processing. This research contributes to understanding and harnessing microwave energy for diverse purposes, involving materials science and engineering disciplines [87]. Microwave irradiation can enhance the electrical conductivity of materials by promoting ion mobility. This phenomenon is relevant in material processing and chemical reactions, where increased conductivity can accelerate processes. The interdisciplinary field involves materials science and chemistry, aiming to optimize conductivity for various industrial applications. Understanding these mechanisms contributes to more efficient processes and innovative technologies [88]. The dielectric loss factor (tan δ) signifies a material's ability to convert microwave energy into heat. Materials with higher tan δ values absorb and dissipate microwave energy efficiently. This phenomenon, known as dielectric relaxation, involves rapid dipole oscillation and is crucial for optimizing microwave heating processes in various scientific and industrial applications. Research in this field explores how different materials and compositions impact dielectric behavior, contributing to improved microwave-based technologies [89]. Magnetic permeability characterizes a material's response to a magnetic field, influencing its interaction with electromagnetic waves during microwave irradiation. High magnetic permeability materials can significantly affect heating behavior in microwave applications, making their study important in optimizing such processes. Research in this area explores how material compositions impact magnetic properties, contributing to advancements in microwave technology and applications [90]. Research into materials like iron-based compounds explores their strong magnetic responses to microwaves, enabling applications like induction heating. This research delves into how these properties vary with factors like frequency and temperature, providing insights into energy absorption mechanisms. It is crucial for optimizing microwave-based heating processes in various fields, from metallurgy to materials science [91]. Research in tailoring materials and processes for electrical, dielectric, and magnetic responses to microwave irradiation has wide-ranging applications, spanning from food processing to environmental remediation. It promises enhanced energy efficiency and product quality across industries. This interdisciplinary field involves materials science, engineering, and physics, focusing on optimizing energy absorption mechanisms for diverse industrial and scientific purposes.

2.4 Microwave Irradiation Molecular Rotation

Microwave irradiation employs microwave electromagnetic waves to induce molecular rotation in materials. This technique is vital in scientific research, chemistry, and various industrial applications. Researchers from disciplines such as chemistry, physics, and materials science study this process to optimize microwave-related outcomes, such as heating and material processing, across a spectrum of materials and compositions. Understanding these principles contributes to advancements in energy-efficient technologies and diverse scientific domains [92, 93]. Understanding molecular rotation induced by MW is pivotal for optimizing its applications in scientific research and industrial processes. This phenomenon occurs in polar molecules due to the interaction between their electric dipole moments and microwave electromagnetic fields. Researchers across disciplines such as chemistry, physics, and engineering investigate these processes, contributing to advancements in diverse fields and the development of energy-efficient technologies [94]. Molecular rotation induced by microwave energy is essential for diverse applications in chemistry, materials science, and communication. It involves polar molecules absorbing microwave energy and causing transitions between rotational states. Researchers from various disciplines investigate these processes, advancing scientific knowledge and technological innovations across industries [78]. Microwave ovens utilize molecular rotation of water molecules to efficiently heat food through MW absorption, causing friction and generating heat. This process is fundamental in cooking and has broad applications in scientific contexts. Understanding this heating mechanism is crucial for efficient microwave technology utilization in various fields, involving interdisciplinary research and practical significance [95]. Microwave irradiation's role in molecular rotation is pivotal for accelerating chemical reactions, particularly in organic synthesis, where it offers precision and speed. This technique finds application in diverse scientific and industrial contexts, enhancing reaction efficiency. Understanding these mechanisms is essential for optimizing microwave irradiation across various disciplines, involving material-specific applications and ongoing research into reaction kinetics [96]. Microwave irradiation, operating in the 300 MHz to 300 GHz frequency range, has diverse applications in materials science and medical chemistry. Nonpolar molecules do not undergo molecular rotation in response to MW, distinguishing their behavior from polar molecules. This understanding of microwave characteristics is vital for its application and significance across scientific disciplines and industrial domains [97]. Molecular responses to MW are pivotal in materials science and medical chemistry. Microwaves induce rotations in molecules with a dipole moment, altering their energy levels, impacting diverse applications. This knowledge is pivotal for precise molecular control in scientific and industrial contexts [98]. Microwave technology, utilized in applications like microwave ovens and chemical synthesis, selectively excites water molecules, inducing molecular rotation and heat generation. This accelerates chemical reactions, particularly in chemistry and materials research, enhancing precision and efficiency. Understanding these principles is crucial for advancing scientific and industrial processes reliant on controlled molecular interactions [99]. Microwave irradiation enhances process efficiency and precision in scientific and industrial applications. Tailored utilization of this technology relies on understanding molecular response to microwaves, considering factors like structure and dipole moment. This knowledge informs applications in various disciplines, optimizing processes dependent on controlled molecular behavior [100]. Microwave irradiation, with limitations, is employed for selective heating of polar molecules, particularly in materials like water. Its application is constrained by the need for compatible materials and specific frequency and wavelength ranges. Microwave irradiation, despite limitations, is a versatile tool for scientific and industrial applications. Its impact on molecular rotation informs tailored approaches in various domains, emphasizing precision in controlling molecular behavior. A profound understanding of these principles underpins its valuable real-world applications.

2.5 Fundamentals of Electromagnetic Theory in Microwave Irradiation

Chemical reactions are initiated using diverse energy sources, such as heat, light, pressure, plasma, ultrasound, and microwaves. These methods enable bond breaking and formation, vital for reactions across chemistry, materials science, and industry. Their application varies, optimizing processes in different fields, and ongoing research explores their mechanisms and significance in precision-controlled reactions [101]. Microwave heating, also called dielectric heating, is an efficient method using electromagnetic waves to induce rapid and precise heating in materials, replacing traditional conductive heating. Its significance lies in its speed and precision, finding applications in chemistry, materials synthesis, and industry. Ongoing research explores its mechanisms and potential for controlled heating processes [102]. Microwave heating is often referred to as dielectric heating because it primarily relies on the dielectric properties of materials to generate heat when exposed to MW. Dielectric heating occurs due to the interaction between the alternating electric field of microwaves and polar molecules within the material. The rapid changes in the electric field cause the polar molecules to continuously reorient themselves, generating heat through friction and molecular movement. This dielectric heating mechanism is the fundamental principle behind how microwave ovens and other microwave heating applications work. Microwave heating harnesses the ability of materials, both liquids and solids, to convert MW into internal heat. This method is pivotal in various scientific and industrial processes, offering rapid and precise heating capabilities. Its applications span chemistry, materials synthesis, and product testing, where efficiency and controlled heating are essential for enhancing process outcomes and productivity. Researchers in disciplines like chemistry and materials science continue to explore innovative applications and mechanisms to optimize microwave heating for diverse purposes [103].

2.5.1 Electromagnetic Radiations and Microwave

Researchers are advancing electromagnetic radiation-based devices to finely manipulate atomic and molecular-level properties of materials, offering transformative impacts across industries such as manufacturing, medicine, and electronics. These devices encompass lasers, microwaves, and terahertz radiation, enabling precise control over chemical reactions, material properties, and even molecular rotation. Challenges include controlled manipulation, affordability, and integration with existing processes, making this field of research highly promising for revolutionizing various industries and enhancing material properties, manufacturing efficiency, and medical applications, as shown in Figure 2.3a [104]. Electromagnetic radiation-based devices for detecting and identifying exoplanets are crucial for exploring distant worlds beyond our solar system. They tackle the challenge of detecting faint exoplanets in the presence of bright stars. Two primary devices are transit photometers, which observe the dimming of a star when an exoplanet passes in front of it, revealing the exoplanet's size and orbit, and radial velocity spectrometers, which measure the star's Doppler shift caused by the gravitational pull of an orbiting exoplanet, allowing for the determination of its mass. These devices offer profound implications, from discovering potentially habitable exoplanets to studying their atmospheres for signs of life and furthering our knowledge of exoplanetary system formation and evolution. The interdisciplinary nature of this field involves astronomy, physics, engineering, and materials science, pushing the boundaries of our understanding of exoplanets and the cosmos [105]. Microwave-based energy harvesting devices harness MW to generate electricity, offering a clean and renewable energy source. They can convert MW into electrical power, primarily through devices like rectennas and thermoelectric converters. Microwaves are abundant in the environment, originating from various sources, including the sun and human-made devices. Research in this field aims to develop efficient and affordable devices that can power a wide range of applications, from smartphones to remote sensors, and even utilize waste heat for energy harvesting. Interdisciplinary research spanning engineering, physics, and materials science drives innovation, with a focus on improving efficiency and compatibility with existing energy systems, thus potentially reducing reliance on fossil fuels and advancing sustainable energy solutions, as shown in Figure 2.3b. Microwave-based devices in chemical reactions aim to enhance efficiency and safety. By utilizing MW, these devices heat reactants rapidly and uniformly, as seen in microwave synthesis reactors, leading to shorter reaction times and increased yields. Additionally, microwave plasma reactors can create plasmas to accelerate various chemical reactions. Safety is improved by decomposing hazardous chemicals through microwave heating or radiation exposure with catalysts. Research spans various disciplines, including chemistry and materials science, aiming to design cost-effective devices compatible with existing equipment and scale up reactions from labs to industry. This innovation holds the potential to revolutionize chemical, pharmaceutical,

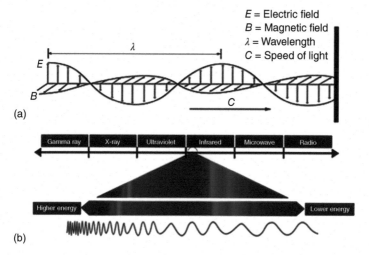

Figure 2.3 (a) Electromagnetic radiation and (b) electromagnetic spectrum.

and materials industries, promoting environmental sustainability and creating enhanced products [106].

2.5.2 Heating Mechanism of Microwave: Conventional Versus Microwave Heating

Microwave-based devices for heating materials offer efficient and controlled heating in industrial and scientific applications. They utilize MW to heat materials rapidly and uniformly, reducing processing times and enhancing product quality. Examples include microwave volumetric heaters and flow-through reactors, which minimize heat loss and support continuous processing. Challenges include designing precise and compatible devices while ensuring affordability and scalability. Advancements in this field have the potential to revolutionize manufacturing, food processing, and medical devices, offering improved efficiency and innovation in various industries [107]. The dominant role of the electric field component of MW in heating polar molecules arises from the tendency of polar molecules to align with the electric field, leading to rapid molecular rotation and heat generation through collisions. This principle is harnessed in various innovative microwave-based devices. This phenomenon is crucial for microwave ovens, microwave plasma reactors, and medical devices that selectively heat cells, with applications spanning food processing, chemical reactions, and medical treatments. Challenges include device efficiency, affordability, and safety [108]. The fundamental concept involves designing microwave-based devices to precisely control microwave-matter interactions at the nanoscale for synthesizing unique nanomaterials. MW offers advantages such as rapid, uniform heating and acceleration of chemical reactions, revolutionizing nanomaterial synthesis. This research addresses the challenge of precise control over nanomaterial properties, aiming to impact industries such as electronics, energy, and healthcare. Interdisciplinary efforts encompass chemistry, materials science, and engineering. The forefront involves combining microwave volumetric heating and plasma synthesis for nanomaterial design, with potential implications for advanced technologies [108]. Microwave-based technologies hold the potential to address major challenges in materials processing, food science, medical devices, and scientific research. By enabling rapid and uniform heating, they facilitate the synthesis of novel materials, energy-efficient food processing, minimally invasive medical procedures, and advanced scientific instrumentation, revolutionizing these fields. Interdisciplinary engagement spanning materials science, engineering, and biomedical sciences is essential to harness microwave-based innovations, ushering in a new era of efficient and impactful solutions [109]. The fundamental concept of microwave-based heating in science and engineering is based on the interaction of MW with polar molecules and ions in materials. Microwaves cause rapid dipolar rotation and ionic conduction, generating heat within the material. This technology is significant for its ability to provide efficient and precise heating, finding applications in materials processing, such as sintering ceramics and drying materials. Its study is motivated by the pursuit of energy-efficient heating methods. Academic disciplines engaged include materials science, electrical engineering, chemistry, and physics. Researchers explore advanced microwave technologies, control systems, and novel materials. The field's forefront involves refining microwave heating methods and understanding complex microwave-matter interactions, with implications spanning improved materials processing and scientific research tools, contributing significantly to advancements in science and engineering. The fundamental concept here is harnessing dipolar rotation and ionic conduction through MW to create new microwave-based devices for efficient and precise material heating. These devices function by applying microwaves to polar molecules and ions in materials, inducing rapid rotation and conduction, thereby generating heat. This concept holds significance in various applications, including materials processing, medical devices, and scientific research, where precise heating is crucial. The process involves controlled microwave-matter interactions for targeted heating. This area of study is motivated by the pursuit of energy-efficient and precise heating methods, driving research toward innovative device designs, control systems, and materials utilization. Dipolar rotation involves the movement of polar molecules aligning with electric fields, generating heat in microwave heating. Ionic conduction is the movement of charged ions responding to electric fields, contributing to heat generation in materials. These mechanisms underlie efficient microwave-based heating in various applications, including materials processing and medical devices [110]. Microwave technology, a field of significant scientific and engineering importance, revolves around the utilization of electromagnetic waves in the microwave frequency range for various applications. Fundamentally, microwaves interact with materials at a molecular level, causing dipolar rotation and ionic conduction, generating heat. This concept is significant as it offers rapid, uniform, and efficient heating, impacting diverse areas such as materials processing, medical devices, and scientific research. Applications include microwave ovens, medical hyperthermia treatments, and materials synthesis. Researchers across science and engineering disciplines collaborate to optimize this technology, exploring novel

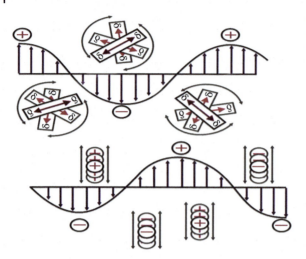

Figure 2.4 Translational motions with changing electric field.

materials, control methods, sensors, and energy harvesting devices. Governing principles involve electromagnetic wave propagation and material interactions, with implications spanning improved industrial processes, medical treatments, and advanced materials development, as shown in Figure 2.4.

2.6 Physical Principles of Microwave Heating and Equipment

The fundamental concept involves using MW to accelerate and control chemical reactions for the synthesis of new materials. MW rapidly heats materials through dipolar rotation and ionic conduction, enabling precise control over reaction conditions. This is significant because it offers a more efficient and uniform method for synthesizing materials with unique properties. Applications are observed in various fields, motivating interdisciplinary research. Materials and compositions can vary widely, from nanomaterials to polymers. Current research focuses on designing microwave-based devices that provide precise control over reaction parameters and scalability for industrial production. Principles of electromagnetic wave interaction with matter govern this technology, with implications spanning materials science, electronics, energy, and catalysis. The pivotal scientific implications include the potential to revolutionize materials synthesis processes across industries [111]. The fundamental concept involves using metamaterials and nano-optics to design a microwave-based device that selectively and precisely heats materials. Metamaterials are engineered materials with specific electromagnetic properties, allowing for the creation of devices that can selectively heat materials or regions of a material. Nano-optics, on the other hand, deals with the precise interaction of light with matter at the nanoscale, enabling the focused application of MW with high precision. This concept is significant because it offers a novel approach to tailored material processing and chemical synthesis, potentially leading to improved product quality, reduced waste, and the development of new materials. Applications are observed in various fields, motivating interdisciplinary research, and the materials used can vary depending on the specific application. Current research focuses on designing effective metamaterial structures and refining nano-optical techniques for precise microwave heating [112]. The fundamental concept involves using metamaterials and nano-optics to design a microwave-based device capable of highly selective and precise material heating. Metamaterials, engineered for specific electromagnetic properties, enable the creation of devices, selectively heating materials or regions within them. Nano-optics, dealing with nanoscale light-matter interactions, provides the means to focus MW with extraordinary precision. Combining these techniques leads to a device with applications in tailored material processing and chemical synthesis. Research challenges include designing metamaterials with desired electromagnetic properties, developing precise nano-optical techniques, and integrating these into existing equipment. The successful development of such microwave-based devices holds the potential to revolutionize various industries, enhancing material properties, and enabling more efficient chemical synthesis methods, ultimately impacting fields like manufacturing and chemistry [113]. Metamaterials are engineered materials designed at the micro or nanoscale with unique electromagnetic properties not found in nature. They can manipulate electromagnetic waves, enabling effects like negative refraction and invisibility cloaking. Their applications span optics, telecommunications, radar, and materials science, with the potential for revolutionary advances in technology. One innovative approach to designing a new microwave-safe container or utensil involves the integration of microwave-transparent and microwave-absorbing materials in its construction. This concept aims to achieve uniform and controlled microwave heating by allowing the container to absorb and evenly distribute MW while shielding the contents from overheating. Another promising avenue is the incorporation of sensors and feedback mechanisms within the container to monitor and regulate the heating process. These sensors can assess the temperature of the material being heated and adjust the microwave power output, accordingly, ensuring precise and even heating. This research area holds significance in various applications, including culinary arts, scientific research, and industrial processes, as it can enhance safety, precision, and convenience in microwave heating procedures, ultimately reducing waste, and improving outcomes across diverse sectors [114]. One innovative approach to enhancing microwave oven safety features involves the integration of artificial intelligence (AI) and advanced sensor technologies. AI can monitor the oven's operation and identify potential hazards, such as the presence of non-microwaveable objects, overheating of food or materials, or incomplete oven door closure. When hazards are detected, the AI system can automatically deactivate the oven or initiate corrective actions. Additionally, advanced sensors can measure internal temperatures, detect gases like steam indicating overheating, and ensure proper door closure, providing critical data for hazard assessment and control. This research area holds significant promise for reducing accidents and injuries associated with microwave oven use, fostering user safety, mitigating product liability concerns, and bolstering consumer trust in microwave oven products [24]. Metamaterials are engineered to focus MW with high precision, reducing uneven heating and the risk of superheating. Nano-optics further refine heating precision by focusing radiation at the nanoscale. Smart sensors monitor material temperature and moisture content, enabling real time feedback control of microwave power output to ensure even heating without damage [24]. Microwaves, falling within the gigahertz (GHz) frequency range and characterized by their relatively short wavelengths, are a subset of electromagnetic waves with extensive scientific and engineering significance. Their generation relies on specialized vacuum tube technologies, including klystrons, magnetrons, and Gunn diodes. Klystrons utilize the principle of electron bunching and velocity modulation within a resonant cavity to produce microwaves, while magnetrons generate microwaves through the interaction of electrons with a magnetic field within a vacuum. Gunn diodes exploit the Gunn effect in semiconductor materials for microwave generation. Understanding the frequency and wavelength properties of microwaves is fundamental for their diverse applications, encompassing communication, scientific research (spectroscopy, material characterization, cosmic microwave background study), and efficient microwave heating, such as in microwave ovens. Ongoing cutting-edge research is dedicated to enhancing microwave technology, including the development of metamaterials for innovative electromagnetic control and the exploration of quantum phenomena in the microwave realm, with the potential to revolutionize quantum computing and communication [115]. Designing a microwave oven that can uniformly and efficiently heat food while minimizing radiation exposure entails several scientific and engineering principles. One fundamental concept involves using advanced materials, such as metamaterials, with unique electromagnetic properties that enable the controlled manipulation and focusing of MW. This allows for more even penetration and heating of food, addressing the challenge of uneven cooking. Additionally, optimizing microwave oven designs, such as incorporating multiple emitters or reflectors to distribute microwaves evenly, contributes to efficient cooking. Moreover, enhancing shielding technologies with innovative materials and cavity designs can significantly reduce MW leakage, thus improving safety. This interdisciplinary field involves physics, materials science, and engineering, with applications in household appliances and commercial [10]. To design a microwave oven that heats food evenly and efficiently while minimizing radiation exposure using the latest advances in materials science, nanotechnology, and quantum physics, several fundamental concepts and principles come into play. Materials science enables the development of advanced materials like metamaterials that possess unique electromagnetic properties, allowing precise manipulation and focusing of MW to ensure even heating. Nanotechnology contributes by offering nanomaterials that can enhance microwave absorption and create highly reflective materials for improved shielding. Additionally, quantum physics plays a role in developing novel microwave generation and detection technologies, as well as materials with high dielectric constants for stronger interactions with MW. These interdisciplinary approaches hold significant potential for creating microwave ovens with reduced cooking time, improved food quality, enhanced safety, and decreased energy consumption, making them valuable innovations for households and commercial kitchens [116]. Designing a new type of microwave device that harnesses the wave behaviors of microwaves, including reflection, transmission, refraction, diffraction, and interference, offers significant technological possibilities. Understanding these behaviors is crucial in creating innovative applications. For instance, microwave imaging utilizes microwaves to produce images for medical diagnostics, security screening, and nondestructive testing. Microwave communications encompass satellite communication, cellular networks, and radar systems. Microwave heating, as exemplified by microwave ovens, is integral to both household and industrial processes like material drying and adhesive curing. Additionally, microwave weapons employ microwaves to disrupt or destroy electronic equipment. Recent research in this field has explored advanced materials like metamaterials to enhance microwave focusing and interference for applications such as cloaking devices, with implications for military and stealth technology [117]. Designing a novel microwave device to harness the unique properties of microwaves, such as their ability to penetrate certain materials and reflect off metal surfaces, presents a formidable challenge. One concept involves a microwave device capable of inspecting the interiors of objects by exploiting microwaves' penetrating properties. Such a device could find applications in security screening, nondestructive testing, and medical imaging. Alternatively, it could be used to create an advanced communication system by capitalizing on microwaves' ability to reflect off metal surfaces, benefiting satellite communication, radar technology, and cellular networks. Another avenue explores using microwaves for industrial processes, harnessing their heating capabilities for tasks like material drying, adhesive curing, and material synthesis [78]. Designing a novel microwave heating device that achieves precise and efficient heating of materials with minimal energy consumption and environmental impact is a multifaceted challenge. One critical issue is addressing uneven heating, which can result in hot spots and cold spots, affecting various applications. To mitigate this, metamaterials, engineered with specific electromagnetic properties, offer a promising solution by enabling more precise and efficient MW control. Additionally, optimizing microwave heating devices to operate at lower frequencies can reduce interference with electronic equipment and potentially improve heating efficiency. Moreover, the integration of AI algorithms can further enhance heating precision and efficiency by dynamically controlling power and frequency. These innovations in microwave heating technologies have the potential to revolutionize various fields, including manufacturing, food processing, and medicine, by minimizing energy consumption and environmental impact while ensuring precise and uniform heating. Examples of ongoing research in this domain include studies on metamaterial based microwave heating devices, as demonstrated, and the application of AI for optimizing microwave food processing [118]. These advancements underscore the significance of microwave heating technology in addressing efficiency and precision challenges across diverse applications [119]. Designing innovative materials and microwave heating devices to harness the fundamental mechanism of dielectric heating is pivotal for achieving precise and energy-efficient material heating with reduced environmental impact. Dielectric heating relies on the interaction between microwave energy and polar molecules within materials, inducing molecular oscillations and generating frictional heat. Additionally, advanced microwave heating devices, such as those employing metamaterials, can focus MW, ensuring more uniform and efficient heating. The integration of AI algorithms further optimizes the process, allowing precise control of power and frequency to minimize energy consumption [120]. These advancements have far-reaching implications across industries, from enhancing food processing to medical applications and manufacturing [99]. In microwave heating applications, a profound comprehension of penetration depth is indispensable for precise control over heating depth. This attribute significantly influences the effectiveness and controllability of microwave heating processes. The essential components of a microwave heating system encompass a microwave generator, such as a magnetron, a waveguide for directing microwaves, and an applicator designed for focusing microwaves on the target material. Diverse applicator configurations, such as cavity, multimode, and single-mode designs, are tailored to meet specific heating requirements [121]. These components collaboratively orchestrate precise and controlled microwave heating across diverse applications. The selection of an applicator design is meticulously tailored to the specific heating process and the unique characteristics of the material in question. For instance, in household microwave ovens, cavity applicators are commonly utilized, reflecting the continued relevance and development of microwave heating methods in modern times [16]. Designing an innovative microwave heating device that efficiently utilizes the fundamental principle of dielectric heating while ensuring safety and environmental responsibility is a complex task. Dielectric heating relies on the interaction between microwaves and polar molecules within materials, causing molecular oscillations and generating heat through friction. To address the challenges and enhance precision, efficiency, and safety, various approaches can be pursued. Developing materials with high dielectric constants and low thermal conductivity is essential. Redesigning microwave heating devices using multiple emitters, reflectors, or metamaterials to focus MW can result in more even and efficient heating. A 2023 study on ResearchGate demonstrated the superior efficiency of a metamaterial-based microwave heating device. Ensuring user and environmental safety is paramount. Incorporating safety features, such as improved shielding materials, robust door seals, and fail-safe mechanisms, can prevent microwave leakage and mitigate potential hazards [122]. Exploiting microwave interaction with substances allows for the development of innovative technologies across diverse fields. Microwave-assisted drying (MAD) enhances food processing, preserving nutrients and flavor. In medicine, microwave ablation efficiently treats tumors with minimal side effects. Manufacturing benefits from microwave sintering for improved material properties, exemplified by titanium nitride coatings. Microwave-assisted synthesis in materials science yields high-purity nanoparticles like graphene. Understanding these principles drives advancements, involving physics, chemistry, engineering, and materials science, with applications in food, medicine, manufacturing, and materials synthesis. These technologies have profound implications, improving processes and products in various industries, and impacting human lives positively. Fundamentally, microwave interactions induce

2.6 Physical Principles of Microwave Heating and Equipment | 51

molecular oscillations, generating heat, and enabling precise and efficient heating in diverse applications. These innovations address industry needs, motivate interdisciplinary research, and continually advance scientific understanding and technological capabilities [123]. Addressing the challenges of microwave heating involves innovating materials, devices, and AI algorithms. New materials with elevated dielectric constants and reduced thermal conductivity enhance heating efficiency, even for highly conductive substances. Innovative microwave heating devices employing metamaterials or multiple emitters/reflections distribute radiation evenly, mitigating issues related to penetration depth and nonpolar materials. Furthermore, controlling microwave polarization through metamaterials improves heating precision. Harnessing AI optimizes heating processes, adjusting power, frequency, and polarization for various material properties. These advancements transcend fields, enhancing heating efficiency and expanding the applicability of microwaves. Fundamentally, microwave heating relies on dielectric loss mechanisms and electromagnetic wave interactions. Innovations motivate multidisciplinary research, encompassing physics, materials science, engineering, and AI. These advancements have far-reaching implications, improving processes and products in various industries while minimizing energy consumption and environmental impact [124]. The fundamental concept underlying the design of new microwave heating technologies is the interaction of microwaves with substances. Microwaves are electromagnetic waves that induce molecular vibrations and generate heat within materials containing polar molecules. This principle is significant because it enables rapid and efficient heating with minimal energy consumption, making microwaves valuable in various applications. Current research focuses on using metamaterials to distribute microwaves evenly, nanotechnology to enhance material properties, and AI to optimize heating processes. These innovations transcend disciplines, involving physics, materials science, engineering, and AI. Fundamentally, microwave heating relies on dielectric loss mechanisms and electromagnetic wave interactions. Innovations motivate multidisciplinary research, encompassing physics, materials science, engineering, and AI. These advancements have far-reaching implications, improving processes and products in various industries while minimizing energy consumption and environmental impact [125]. The fundamental concept underlying the design of a new microwave heating device based on standing waves is the phenomenon of constructive and destructive interference between incident and reflected microwaves. Standing waves are formed when these interference patterns result in regions of high and low microwave intensity within the heating chamber. This concept is significant as it can be harnessed to achieve precise and efficient heating of materials by controlling the distribution of standing waves. Applications of this phenomenon are observed in various fields, including food processing, materials science, and medical treatments, where controlled heating is essential. Multidisciplinary research encompassing physics, engineering, materials science, and safety engineering is engaged to develop innovative devices. Safety precautions, such as effective shielding, fail-safe mechanisms, and user education, ensure the safe operation of these devices. Research in this area is at the forefront of optimizing standing wave-based microwave heating for diverse applications, with implications for enhanced efficiency and safety across industries [126]. The fundamental concept underlying the design of a new microwave heating device based on dielectric heating principles is the interaction between MW and polar molecules within materials. Dielectric heating relies on the phenomenon of polar molecules aligning with the alternating electric field of microwaves, leading to molecular motion and frictional heat generation. This concept is significant because it enables precise and efficient heating of materials, finding applications in various fields like food processing, medicine, manufacturing, and materials science. To improve dielectric heating, approaches involve metamaterials for better microwave distribution, advanced control of microwave beams, development of materials with tailored dielectric properties, and the integration of AI for optimization. Safety measures include shielding, fail-safe mechanisms, and user education. Ongoing research aims to enhance dielectric heating efficiency and overcome its challenges for broader applications [127]. The fundamental concept in designing a microwave heating device lies in the interaction of microwaves with materials. Microwaves, with their specific frequency range, induce molecular motion by interacting with polar molecules within materials, generating frictional heat through dielectric loss. This concept is significant for achieving precise and efficient material heating, with applications spanning various fields. Research focuses on innovative approaches such as metamaterials to control microwave distribution, AI optimization, and tailored materials with enhanced dielectric properties. Ensuring user and environmental safety involves shielding, fail-safe mechanisms, and user education. By addressing microwave heating challenges and safety, novel technologies can transform industries such as food processing, medicine, manufacturing, and materials science, improving global living standards [128]. The fundamental concept involves optimizing MW for precise and efficient heating in microwave ovens. This is achieved through innovations like metamaterials to control radiation distribution, and improved dielectric materials. It is significant for revolutionizing cooking and food processing, finding applications in diverse settings. Research motives include enhancing cooking efficiency and safety. Materials utilized include metamaterials and high-dielectric materials [129]. Microwave heating finds extensive applications in industrial settings for various processes such as drying, sintering, sterilization, and thawing. Industrial microwave systems often employ specialized applicators to ensure uniform heating by efficiently delivering microwaves into the material undergoing processing [129]. The fundamental concept involves optimizing microwave plasma heating for PECVD to achieve precise and efficient thin film deposition while reducing environmental impact. This is achieved by utilizing metamaterials to control radiation distribution and pulsed MW for improved efficiency. It is significant for enhancing thin film manufacturing for various applications. Research motives include improving efficiency and sustainability. The fundamental idea is to improve microwave plasma heating for accurate deposition of thin films in PECVD processes, aiming to boost manufacturing efficiency while minimizing environmental effects. This concept involves the use of metamaterials and draws from engineering, materials science, and environmental science [130]. Safety is a paramount consideration in microwave heating due to the potential for burns and injuries. Proper equipment design and maintenance are crucial to prevent microwave leakage and ensure safe operation [131]. Adhering to the manufacturer's instructions and using only microwave-safe materials is essential for safe microwave heating. Microwave technology offers rapid and effective heating for various applications, including food preparation, material synthesis, and industrial processes. Proper usage and safety precautions are vital to ensure efficient heating and prevent accidents [132]. Microwave heating relies on the rapid rotation and oscillation of polar molecules in response to electromagnetic fields, generating heat within the material. Careful design of microwave heating equipment, including applicators and safety measures, is crucial to ensure effective and controlled heating in various applications [14, 132]. In summary, microwave heating harnesses the dielectric properties of materials to generate heat efficiently. It is employed across various fields and serves as a valuable tool in both household and industrial applications. However, it is essential to prioritize safety precautions when utilizing microwave technology.

2.7 Green Chemistry Through Microwave Heating: Applications and Benefits

In recent years, the field of green chemistry has gained significant attention due to its focus on developing sustainable and environmentally friendly chemical processes. One innovative approach that has emerged as a powerful tool in advancing green chemistry is the utilization of microwave heating. Microwave-assisted reactions offer a range of advantages, such as reduced environmental impact, cost-effectiveness, shorter reaction times, and enhanced yields [133]. This two-page write-up explores the applications of microwave heating in the context of green chemistry and its profound benefits. Microwave heating facilitates energy-efficient chemical reactions by directly delivering energy to reactants. This focused energy transfer reduces heat loss to the surroundings, making reactions more energy-efficient and environmentally friendly. Microwave-assisted reactions often require less solvent or even allow for solvent-free processes [12]. This reduction in solvent usage aligns with green chemistry principles, minimizing the generation of hazardous waste. Microwave heating can enhance the catalytic activity of substances, leading to more efficient reactions. This can reduce the need for large quantities of catalysts, which is in line with the principle of minimizing resource consumption. Precise control of temperature during microwave-assisted reactions can improve the selectivity of chemical transformations [2]. This can lead to fewer byproducts and a higher yield of the desired product. Microwave heating significantly accelerates reaction kinetics, reducing reaction times from hours to minutes or even seconds. This not only conserves energy but also increases laboratory productivity. Microwave-assisted reactions often require lower quantities of reagents and solvents. Additionally, the reduced reaction times translate into cost savings, making it an economically viable choice. The decreased solvent usage and energy consumption in microwave-assisted reactions contribute to a smaller environmental footprint. This aligns with the green chemistry goal of minimizing environmental impact [10]. The unique energy absorption and propagation characteristics of microwave irradiation have opened up new avenues for research in synthetic chemistry. Researchers can explore novel reaction pathways and discover more sustainable synthetic routes. The improved selectivity and reduced byproduct formation in microwave-assisted reactions lead to less waste generation. This is crucial in adhering to the green chemistry principle of waste minimization. Microwave heating can enhance safety in chemical laboratories [134]. Shorter reaction times reduce the exposure of researchers to potentially hazardous materials, contributing to a safer working environment. Microwave heating can be applied to reactions involving renewable feedstocks, supporting the use of sustainable resources in chemical synthesis. Microwave-assisted synthesis has found extensive applications in pharmaceutical and fine chemical industries, where precise control of reaction conditions is crucial for product quality [12], as shown in Figure 2.5. Figure 2.5 highlights the integration of microwave heating with green chemistry principles, emphasizing its applications and the advantages it offers in promoting environmentally friendly and sustainable chemical processes.

Thiazolidin-4-one, a potent moiety found in various approved medications, holds a significant place in contemporary antimicrobial and antiviral chemotherapy. Its broad spectrum and potent activity have made it a focal point of research in the field. Medicinal chemists have been drawn to the diverse medicinal properties exhibited by thiazolidin-4-one-related drugs, motivating them to synthesize an array of novel medicinal substances. MW, an alternative to traditional heating, directly energizes reactions, sparking innovative concepts in synthetic chemistry, particularly in the synthesis of thiazolidin-4-one analogs. It also explores the antiviral and antimicrobial properties of these compounds, offering insights into potential pharmaceutical applications, as shown in Figure 2.6. This evolution

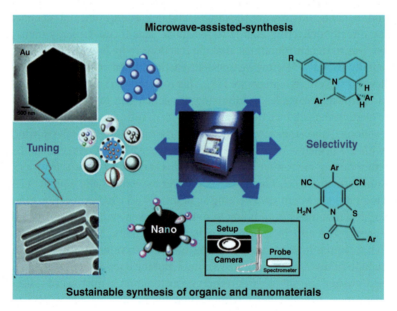

Figure 2.5 Microwave heating with green chemistry in environmentally friendly and sustainable chemical processes. Source: Verma et al. [18]/American Chemical Society.

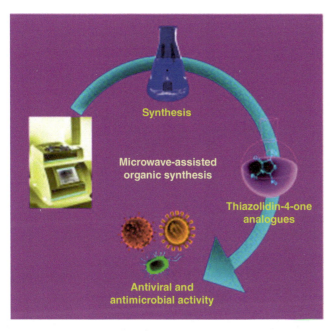

Figure 2.6 Microwave irradiation -based synthetic routes of various thiazolidin-4-one derivatives.

has brought the principles of green chemistry to the forefront in recent years. Microwave heating has emerged as a green protocol in organic synthesis, aligning with these principles. Its advantages encompass a straightforward experimental setup, production of high-purity compounds with excellent yields, solvent-free reactions, reduced reaction times, minimized side product formation, and decreased by-product rates. Heterocyclic compounds, particularly nitrogen-containing ones, hold immense significance not only due to their presence in natural products but also their extensive applications in the pharmaceutical industry.

The multifaceted role of microwaves in heterogeneous catalytic systems is elegantly dissected into four distinct categories, as thoughtfully summarized in Figure 2.7. These categories provide a holistic perspective on the applications of microwave chemistry – First, the operational mode of microwave chemistry is highlighted, where microwaves serve as the direct heat source for substances. This approach offers the potential for automated, robot-assisted chemical synthesis, reminiscent of microwave heating in household ovens. Second, microwaves find a significant niche in green chemistry applications. They remarkably reduce reaction times, even in processes devoid of solvents and catalysts. This environmentally friendly aspect underscores the greener side of microwave-assisted chemistry. The third facet delves into chemical reaction applications. Microwaves are shown to be catalysts themselves, enhancing the kinetics of various chemical reactions. The ability to manipulate microwave frequency and other related phenomena opens new doors for optimizing chemical processes. Lastly, specific heating emerges as a unique application. Microwaves are not just heat sources; they actively contribute to the creation of novel materials [92].

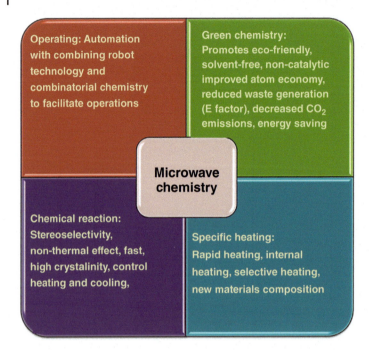

Figure 2.7 The role of microwave radiation in microwave chemistry. Source: Palma et al. [92]/MDPI/Public Domain/CC by 4.0.

The N-heterocycles are integral components of numerous biologically active molecules such as microwave-mediated synthesis of biologically crucial N-heterocycles, highlighting their relevance in contemporary organic synthesis, as shown in Figure 2.8. This application emphasizes the precision and targeted nature of microwave-induced heating. It is essential to acknowledge that while

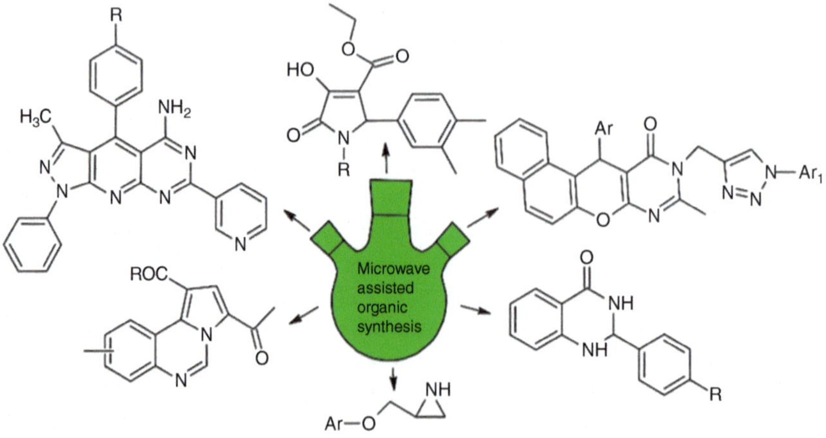

Figure 2.8 Microwave-promoted synthesis of N-based heterocycles. Source: Dinodia [135]/ Bentham Science Publishers.

this classification offers a comprehensive framework, the versatility of microwaves in chemistry and catalysis may extend beyond these categories. Understanding the nuanced differences between microwave and classical heating methods is paramount, and the study of electric and magnetic radiation fields in microwave heating holds significant promise. While many studies employ microwaves as a heat source in chemical reactions, a deeper exploration of MW characteristics remains an intriguing avenue for further research.

2.8 Conclusion

In conclusion, this review chapter has provided a comprehensive examination of the fundamental principles underlying microwave irradiation and its far-reaching implications across various domains. The incorporation of microwave and ultrasonication irradiation within the framework of green chemistry highlights its potential as a revolutionary approach for achieving sustainable chemical processes. By adhering to green chemistry principles and minimizing the use of hazardous substances, this technology has the potential to reshape the chemical industry, promoting greater environmental responsibility. Dielectric and magnetic properties, molecular rotation, and the physical principles of microwave heating have shed light on the intricate mechanisms driving the effectiveness of microwave irradiation. This knowledge serves as the foundation for optimizing interactions in fields ranging from materials science to medical chemistry. The synergy between microwave and ultrasonication irradiation offers a versatile toolkit, enabling a wide range of applications from organic synthesis to environmental remediation. As the field of microwave irradiation continues to advance, its capacity to revolutionize chemical processes while upholding sustainability principles becomes increasingly evident. The fusion of science, innovation, and ethical responsibility underscores its significance in shaping a greener future for the chemical industry. By continually enhancing our comprehension of the foundational principles of microwave irradiation and its alignment with the tenets of green chemistry, we are poised to harness its transformative potential and contribute to a more sustainable and responsible scientific landscape.

References

1 Kumar Mishra, R., Thomas, S., and Karikal, N. (2017). *Micro and Nano Fibrillar Composites (MFCs and NFCs) from Polymer Blends*. Elsevier. https://doi.org/10.1016/B978-0-08-101991-7.09989-1.

2 DeVierno Kreuder, A., House-Knight, T., Whitford, J. et al. (2017). A method for assessing greener alternatives between chemical products following the 12 principles of green chemistry. *ACS Sustainable Chem. Eng.* 5: 2927–2935. https://doi.org/10.1021/acssuschemeng.6b02399.

3 de Marco, B.A., Rechelo, B.S., Tótoli, E.G. et al. (2019). Evolution of green chemistry and its multidimensional impacts: a review. *Saudi Pharm. J.* 27: 1–8. https://doi.org/10.1016/j.jsps.2018.07.011.

4 Anastas, J.C. and Warner, P.T. (1998, Encyclopedia of Toxicology, 1998). *Green Chemistry: Theory and Practice.* New York: Oxford University Press.

5 Trombino, S., Sole, R., Di Gioia, M.L. et al. (2023). Green chemistry principles for nano- and micro-sized hydrogel synthesis. *Molecules* 28: 2107. https://doi.org/10.3390/molecules28052107.

6 Cheng, L. and Ye, X.P. (2010). Recent progress in converting biomass to biofuels and renewable chemicals in sub- or supercritical water. *Biofuels* 1 (1): 109–128. https://doi.org/10.4155/bfs.09.3.

7 Hernandez, A.L.M. and Santos, C.V. (2012). Keratin fibers from chicken-feathers: structure and advances in polymer composites. In: *Keratin Structure, Properties and Applications* (ed. R. Dullaart et al.), 149–211. Nova Science Publishers.

8 Chirayil, C.J., Mishra, R.K., and Thomas, S. (2019). Materials recovery, direct reuse and incineration of PET bottles. In: *Recycling of Polyethylene Terephthalate Bottles*, 37–60. Elsevier. https://doi.org/10.1016/B978-0-12-811361-5.00003-1.

9 Nayak, J., Devi, C., and Vidyapeeth, L. (2016). Microwave assisted synthesis : a green chemistry approach. *Int. Res. J. Pharm. Appl. Sci.* 3: 278–285.

10 Krishnan, R., Shibu, S.N., Poelman, D. et al. (2022). Recent advances in microwave synthesis for photoluminescence and photocatalysis. *Mater. Today Commun.* 32: 103890. https://doi.org/10.1016/j.mtcomm.2022.103890.

11 Dąbrowska, S., Chudoba, T., Wojnarowicz, J., and Łojkowski, W. (2018). Current trends in the development of microwave reactors for the synthesis of nanomaterials in laboratories and industries: a review. *Crystals (Basel).* 8: 379. https://doi.org/10.3390/cryst8100379.

12 Tiwari, S. and Talreja, S. (2022). Green chemistry and microwave irradiation technique: a review. *J. Pharm. Res. Int.* 74–79. https://doi.org/10.9734/jpri/2022/v34i39A36240.

13 Escobedo, R., Miranda, R., and Martínez, J. (2016). Infrared irradiation: toward green chemistry, a review. *Int. J. Mol. Sci.* 17: 453. https://doi.org/10.3390/ijms17040453.

14 Pawełczyk, A., Sowa-Kasprzak, K., Olender, D., and Zaprutko, L. (2018). Microwave (MW), ultrasound (US) and combined synergic MW-US strategies for rapid functionalization of pharmaceutical use phenols. *Molecules* 23: 2360. https://doi.org/10.3390/molecules23092360.

15 Bradu, P., Biswas, A., Nair, C. et al. (2022). Recent advances in green technology and Industrial Revolution 4.0 for a sustainable future. *Environ. Sci. Pollut. Res.* https://doi.org/10.1007/s11356-022-20024-4.

16 Chatel, G. and Varma, R.S. (2019). Ultrasound and microwave irradiation: contributions of alternative physicochemical activation methods to Green Chemistry. *Green Chem.* 21: 6043–6050. https://doi.org/10.1039/C9GC02534K.

17 Raghvendrakumar, M. (2017). Chitosan as promising materials for biomedical application: review. *Res. Dev. Mater. Sci.* 2. https://doi.org/10.31031/RDMS.2017.02.000543.

18 Verma, C., Quraishi, M.A., and Ebenso, E.E. (2018). Microwave and ultrasound irradiations for the synthesis of environmentally sustainable corrosion inhibitors: an overview. *Sustainable Chem. Pharm.* 10: 134–147. https://doi.org/10.1016/j.scp.2018.11.001.

19 Landers, R., Pfister, A., Hübner, U. et al. (2002). Fabrication of soft tissue engineering scaffolds by means of rapid prototyping techniques. *J. Mater. Sci.* 37: 3107–3116. https://doi.org/10.1023/A:1016189724389.

20 Segura-Salazar, J. and Tavares, L. (2018). Sustainability in the minerals industry: seeking a consensus on its meaning. *Sustainability* 10: 1429. https://doi.org/10.3390/su10051429.

21 Taylor, C.J., Pomberger, A., Felton, K.C. et al. (2023). A brief introduction to chemical reaction optimization. *Chem. Rev.* 123: 3089–3126. https://doi.org/10.1021/acs.chemrev.2c00798.

22 Baig, R.B.N. and Varma, R.S. (2012). Alternative energy input: mechanochemical, microwave and ultrasound-assisted organic synthesis. *Chem. Soc. Rev.* 41: 1559–1584. https://doi.org/10.1039/C1CS15204A.

23 Hernández Fernández, F.J. and de los Ríos, A.P. (2021). Special Issue: Green sustainable chemical processes. *Processes* 9: 1097. https://doi.org/10.3390/pr9071097.

24 Jha, A. (2021). Microwave assisted synthesis of organic compounds and nanomaterials. In: *Nanofibers – Synthesis, Properties and Applications*. IntechOpen. https://doi.org/10.5772/intechopen.98224.

25 Bhargava, N., Mor, R.S., Kumar, K., and Sharanagat, V.S. (2021). Advances in application of ultrasound in food processing: a review. *Ultrason. Sonochem.* 70: 105293. https://doi.org/10.1016/j.ultsonch.2020.105293.

26 Wysokowski, M., Luu, R.K., Arevalo, S. et al. (2023). Untapped potential of deep eutectic solvents for the synthesis of bioinspired inorganic–organic materials. *Chem. Mater.* https://doi.org/10.1021/acs.chemmater.3c00847.

27 Martínez-Huitle, C.A., Rodrigo, M.A., Sirés, I., and Scialdone, O. (2023). A critical review on latest innovations and future challenges of electrochemical technology for the abatement of organics in water. *Appl. Catal. B.* 328: 122430. https://doi.org/10.1016/j.apcatb.2023.122430.

28 Anastas, P.T., Kirchhoff, M.M., and Williamson, T.C. (2001). Catalysis as a foundational pillar of green chemistry. *Appl. Catal., A* 221: 3–13. https://doi.org/10.1016/S0926-860X(01)00793-1.

29 Pedro, S.N., Freire, C.S.R., Silvestre, A.J.D., and Freire, M.G. (2020). The role of ionic liquids in the pharmaceutical field: an overview of relevant applications. *Int. J. Mol. Sci.* 21: 8298. https://doi.org/10.3390/ijms21218298.

30 Verma, D.K., Dewangan, Y., Singh, A.K. et al. (2022). Ionic liquids as green and smart lubricant application: an overview. *Ionics (Kiel).* 28. https://doi.org/10.1007/s11581-022-04699-w.

31 Patra, J.K., Das, G., Fraceto, L.F. et al. (2018). Nano based drug delivery systems: recent developments and future prospects. *J. Nanobiotechnol.* 16: 71. https://doi.org/10.1186/s12951-018-0392-8.

32 Mikušová, V. and Mikuš, P. (2021). Advances in chitosan-based nanoparticles for drug delivery. *Int. J. Mol. Sci.* 22: 9652. https://doi.org/10.3390/ijms22179652.

33 Coelho, J.F., Ferreira, P.C., Alves, P. et al. (2010). Drug delivery systems: advanced technologies potentially applicable in personalized treatments. *EPMA J.* 1: 164–209. https://doi.org/10.1007/s13167-010-0001-x.

34 Jiménez-Gómez, C.P. and Cecilia, J.A. (2020). Chitosan: a natural biopolymer with a wide and varied range of applications. *Molecules.* 25: 3981. https://doi.org/10.3390/molecules25173981.

35 Hein, C.D., Liu, X.-M., and Wang, D. (2008). Click chemistry, a powerful tool for pharmaceutical sciences. *Pharm. Res.* 25: 2216–2230. https://doi.org/10.1007/s11095-008-9616-1.

36 Purohit, P., Bhatt, A., Mittal, R.K. et al. (2023). Polymer Grafting and its chemical reactions. *Front. Bioeng. Biotechnol.* 10. https://doi.org/10.3389/fbioe.2022.1044927.

37 Rajendran, R.A., Gudimalla, A.P., Mishra, R.A. et al. (2019). Electrospinning as a novel delivery vehicle for bioactive compounds in food. In: *Innovative Food Science and Emerging Technologies*, 1e (ed. S. Thomas, R. Rajakumari, A. George, and N. Kalarikkal), 426–453. New York: Apple Academic Press.

38 Mishra, R.K. and Kumar, A. (2021). Advanced biopolymer-based composites: an introduction and fracture modeling. In: *Sustainable Biopolymer Composites: Biocompatibility, Self-Healing, Modeling, Repair and Recyclability*, A Volume in Woodhead Publishing Series in Composites Science and Engineering. https://doi.org/10.1016/B978-0-12-822291-1.00001-4.

39 Mohapatra, S.S., Ranjan, S., Dasgupta, N. et al. (2018). *Applications of Targeted Nano Drugs and Delivery Systems: Nanoscience and Nanotechnology in Drug Delivery*. Elsevier. https://doi.org/10.1016/B978-0-12-814029-1.01001-2.

40 Parveen, F. and Kumar Mishra, R. (2019). Different fabrication techniques of aerogels and its applications. In: *Advanced Polymeric Materials for Sustainability and Innovations*. https://doi.org/10.1201/b22326-14.

41 Kumar, A., Mishra, R.K., Verma, K. et al. (2023). A comprehensive review of various biopolymer composites and their applications: from biocompatibility to self-healing. *Mater. Today Sustainability.* https://doi.org/10.1016/j.mtsust.2023.100431.

42 Raghvendra, K.M. (2018). Nanostructured biomimetic, bioresponsive, and bioactive biomaterials. In: *Fundamental Biomaterials: Metals*, 35–65. Elsevier. https://doi.org/10.1016/B978-0-08-102205-4.00002-7.

43 Mishra, R.K. and Rajakumari, R. (2019). Nanobiosensors for biomedical application. In: *Characterization and Biology of Nanomaterials for Drug Delivery*, 1–23. Elsevier. https://doi.org/10.1016/B978-0-12-814031-4.00001-5.

44 Buléon, A., Colonna, P., Planchot, V., and Ball, S. (1998). Starch granules: structure and biosynthesis. *Int. J. Biol. Macromol.* 23: 85–112. https://doi.org/10.1016/S0141-8130(98)00040-3.

45 Domene-López, D., García-Quesada, J.C., Martin-Gullon, I., and Montalbán, M.G. (2019). Influence of starch composition and molecular weight on physicochemical properties of biodegradable films. *Polymers (Basel).* 11: 1084. https://doi.org/10.3390/polym11071084.

46 Sajilata, M.G., Singhal, R.S., and Kulkarni, P.R. (2006). Resistant starch – a review. *Compr. Rev. Food Sci. Food Saf.* 5: 1–17. https://doi.org/10.1111/j.1541-4337.2006.tb00076.x.

47 Peidayesh, H., Ahmadi, Z., Khonakdar, H.A. et al. (2020). Baked hydrogel from corn starch and chitosan blends cross-linked by citric acid: preparation and properties. *Polym. Adv. Technol.* 31: 1256–1269. https://doi.org/10.1002/pat.4855.

48 Fan, Y. and Picchioni, F. (2020). Modification of starch: a review on the application of "green" solvents and controlled functionalization. *Carbohydr. Polym.* 241: 116350. https://doi.org/10.1016/j.carbpol.2020.116350.

49 Chin, S.F., Romainor, A.N.B., Pang, S.C., and Lihan, S. (2019). Antimicrobial starch-citrate hydrogel for potential applications as drug delivery carriers. *J. Drug Delivery Sci. Technol.* 54: 101239. https://doi.org/10.1016/j.jddst.2019.101239.

50 Ma, C., Gerhard, E., Lu, D., and Yang, J. (2018). Citrate chemistry and biology for biomaterials design. *Biomaterials* 178: 383–400. https://doi.org/10.1016/j.biomaterials.2018.05.003.

51 Larrañeta, E., Imízcoz, M., Toh, J.X. et al. (2018). Synthesis and characterization of lignin hydrogels for potential applications as drug eluting antimicrobial coatings for medical materials. *ACS Sustainable Chem. Eng.* 6: 9037–9046. https://doi.org/10.1021/acssuschemeng.8b01371.

52 Larrañeta, E., Lutton, R.E.M., Brady, A.J. et al. (2015). Microwave-assisted preparation of hydrogel-forming microneedle arrays for transdermal drug delivery applications. *Macromol. Mater. Eng.* 300: 586–595. https://doi.org/10.1002/mame.201500016.

53 Afinjuomo, F., Fouladian, P., Parikh, A. et al. (2019). Preparation and characterization of oxidized inulin hydrogel for controlled drug delivery. *Pharmaceutics* 11: 356. https://doi.org/10.3390/pharmaceutics11070356.

54 Imran, M., Hussain, S., Mehmood, K. et al. (2021). Optimization of ecofriendly synthesis of Ag nanoparticles by Linum usitatissimum hydrogel using response surface methodology and its biological applications. *Mater. Today Commun.* 29: 102789. https://doi.org/10.1016/j.mtcomm.2021.102789.

55 Hsueh, Y.-H., Lin, K.-S., Ke, W.-J. et al. (2015). The antimicrobial properties of silver nanoparticles in *Bacillus subtilis* are mediated by released Ag+ ions. *PLoS One* 10: e0144306. https://doi.org/10.1371/journal.pone.0144306.

56 Muthukumar, T., Song, J.E., and Khang, G. (2019). Biological role of gellan gum in improving scaffold drug delivery, cell adhesion properties for tissue engineering applications. *Molecules* 24: 4514. https://doi.org/10.3390/molecules24244514.

57 Yi, Y.T., Sun, J.Y., Lu, Y.W., and Liao, Y.C. (2015). Programmable and on-demand drug release using electrical stimulation. *Biomicrofluidics* 9. https://doi.org/10.1063/1.4915607.

58 Yadav, A., Yadav, K., Ahmad, R., and Abd-Elsalam, K.A. (2023). Emerging frontiers in nanotechnology for precision agriculture: advancements, hurdles and prospects. *Agrochemicals* 2: 220–256. https://doi.org/10.3390/agrochemicals2020016.

59 Bercea, M. (2022). Bioinspired hydrogels as platforms for life-science applications: challenges and opportunities. *Polymers (Basel)*. 14: 2365. https://doi.org/10.3390/polym14122365.

60 Fuku, X., Dyosiba, X., and Iftikhar, F.J. (2023). Green prepared nanomaterials from various biodegradable wastes and their application in energy. *Nano-Struct. Nano-Objects* 35: 100997. https://doi.org/10.1016/j.nanoso.2023.100997.

61 Sarfraz, M.S., Hong, H., and Kim, S.S. (2021). Recent developments in the manufacturing technologies of composite components and their cost-effectiveness in the automotive industry: A review study. *Compos. Struct.* 266: 113864. https://doi.org/10.1016/j.compstruct.2021.113864.

62 Marsh, K. and Bugusu, B. (2007). Food packaging? roles, materials, and environmental issues. *J. Food Sci.* 72: R39–R55. https://doi.org/10.1111/j.1750-3841.2007.00301.x.

63 Luo, X. (2015). Principles of electromagnetic waves in metasurfaces. *Sci. China Phys., Mech. Astron.* 58: 594201. https://doi.org/10.1007/s11433-015-5688-1.

64 Pfeifer, R.N.C., Nieminen, T.A., Heckenberg, N.R., and Rubinsztein-Dunlop, H. (2007). Colloquium: momentum of an electromagnetic wave in dielectric media. *Rev. Mod. Phys.* 79. https://doi.org/10.1103/RevModPhys.79.1197.

65 Tofighi, M.R. and Chiao, J.C. (2012). IEEE transactions on microwave theory and techniques. *IEEE Trans. Microwave Theory Tech.* 60. https://doi.org/10.1109/TMTT.2012.2205624.

66 Zhou, X., Jia, Z., Feng, A. et al. (2020). Dependency of tunable electromagnetic wave absorption performance on morphology-controlled 3D porous carbon fabricated by biomass. *Compos. Commun.* 21: 100404. https://doi.org/10.1016/j.coco.2020.100404.

67 Erkaev, N.V., Weber, C., Grießmeier, J.-M. et al. (2022). Can radio emission escape from the magnetosphere of υ Andromedae b – a new method to constrain the minimum mass of Hot Jupiters. *Mon. Not. R. Astron. Soc.* 512: 4869–4876. https://doi.org/10.1093/mnras/stac767.

68 Liu, Y., Yuan, Z., Wang, J. et al. (2019). Simulation of EM wave propagation along a femtosecond laser plasma filament. *Results Phys.* 14. https://doi.org/10.1016/j.rinp.2019.102359.

69 Sen Zhong, H., Wang, H., Deng, Y.H. et al. (2020). Quantum computational advantage using photons. *Science* (1979): 370. https://doi.org/10.1126/science.abe8770.

70 Sheng, W., Yucel, A.C., Liu, Y. et al. (2023). A Domain-decomposition-based surface integral equation simulator for characterizing em wave propagation in mine environments. *IEEE Trans. Antennas Propag.* 71. https://doi.org/10.1109/TAP.2023.3256579.

71 Yang, X., Wei, B., and Yin, W. (2017). Analysis on the characteristics of em waves propagation in the plasma sheath surrounding a hypersonic vehicle. *IEEE Trans. Plasma Sci.* 45. https://doi.org/10.1109/TPS.2017.2755723.

72 Yang, X., Wei, B., and Yin, W. (2017). A new method to analyze the EM wave propagation characteristics in the hypersonic sheath. *Optik (Stuttg)*. 148. https://doi.org/10.1016/j.ijleo.2017.08.117.

73 Nguyen, B.T., Samimi, A., Vergara, S.E.W. et al. (2019). Analysis of electromagnetic wave propagation in variable magnetized plasma via polynomial chaos expansion. *IEEE Trans. Antennas Propag.* 67. https://doi.org/10.1109/TAP.2018.2879676.

74 Vendik, I., Vendik, O., Pleskachev, V. et al. (2021). Wireless monitoring of biological objects at microwaves. *Electronics (Switzerland)*. 10. https://doi.org/10.3390/electronics10111288.

75 Javandel, V., Akbari, A., Ardebili, M., and Werle, P. (2022). Simulation of negative and positive corona discharges in air for investigation of electromagnetic waves propagation. *IEEE Trans. Plasma Sci.* 50. https://doi.org/10.1109/TPS.2022.3194836.

76 Cheng, J., Zhang, H., Xiong, Y. et al. (2021). Construction of multiple interfaces and dielectric/magnetic heterostructures in electromagnetic wave absorbers with enhanced absorption performance: a review. *J. Materiomics* 7. https://doi.org/10.1016/j.jmat.2021.02.017.

77 Zamorano Ulloa, R., Santiago, M.G.H., and Rueda, V.L.V. (2019). The interaction of microwaves with materials of different properties. In: *Electromagnetic Fields and Waves*. IntechOpen. https://doi.org/10.5772/intechopen.83675.

78 Sun, J., Wang, W., and Yue, Q. (2016). Review on microwave-matter interaction fundamentals and efficient microwave-associated heating strategies. *Materials* 9: 231. https://doi.org/10.3390/ma9040231.

79 Parodi, F. (1989). Physics and chemistry of microwave processing. In: *Comprehensive Polymer Science and Supplements*, 669–728. Elsevier. https://doi.org/10.1016/B978-0-08-096701-1.00258-5.

80 Psarras, G.C. (2018). Fundamentals of dielectric theories. In: *Dielectric Polymer Materials for High-Density Energy Storage*, 11–57. Elsevier. https://doi.org/10.1016/B978-0-12-813215-9.00002-6.

81 Mishra, R.R. and Sharma, A.K. (2016). Microwave–material interaction phenomena: heating mechanisms, challenges and opportunities in material processing. *Compos. Part A Appl. Sci. Manuf.* 81: 78–97. https://doi.org/10.1016/j.compositesa.2015.10.035.

82 Angela, A. and dAmore, M. (2012). Relevance of dielectric properties in microwave assisted processes. In: *Microwave Materials Characterization*. InTech. https://doi.org/10.5772/51098.

83 Mishra, R.K., Dutta, A., Mishra, P., and Thomas, S. (2018). Recent progress in electromagnetic absorbing materials. *Adv. Mater. Electromag. Shiel.* https://doi.org/10.1002/9781119128625.ch7.

84 Feller, J.F., Bruzaud, S., and Grohens, Y. (2004). Influence of clay nanofiller on electrical and rheological properties of conductive polymer composite. *Mater. Lett.* 58: 739–745. https://doi.org/10.1016/j.matlet.2003.07.010.

85 Zhai, W., Zhao, S., Wang, Y. et al. (2018). Segregated conductive polymer composite with synergistically electrical and mechanical properties. *Compos. Part A Appl. Sci. Manuf.* 105: 68–77. https://doi.org/10.1016/j.compositesa.2017.11.008.

86 Yaragalla, S., Thomas, S., Maria, H.J. et al. (2019). *Carbon-based Nanofiller and Their Rubber Nanocomposites*. Elsevier. https://doi.org/10.1016/C2018-0-02522-0.

87 Jayalakshmy, M.S. and Mishra, R.K. (2019). Applications of carbon-based nanofiller-incorporated rubber composites in the fields of tire engineering, flexible electronics and EMI shielding. In: *Carbon-Based Nanofillers and Their Rubber Nanocomposites*. https://doi.org/10.1016/B978-0-12-817342-8.00014-7.

88 S. Yaragalla, R. Mishra, S. Thomas, N. Kalarikkal, H.J. Maria, *Carbon-based nanofillers and their rubber nanocomposites*, 2018. https://doi.org/10.1016/C2016-0-03648-3.

89 Loharkar, P.K., Ingle, A., and Jhavar, S. (2019). Parametric review of microwave-based materials processing and its applications. *J. Mater. Res. Technol.* 8: 3306–3326. https://doi.org/10.1016/j.jmrt.2019.04.004.

90 Mishra, R.K., Thomas, M.G., Abraham, J. et al. (2018). Electromagnetic interference shielding materials for aerospace application. In: *Advanced Materials for Electromagnetic Shielding*, 327–365. Wiley. https://doi.org/10.1002/9781119128625.ch15.

91 Li, K., Xu, J., Li, P., and Fan, Y. (2022). A review of magnetic ordered materials in biomedical field: constructions, applications and prospects. *Compos. B Eng.* 228: 109401. https://doi.org/10.1016/j.compositesb.2021.109401.

92 Palma, V., Barba, D., Cortese, M. et al. (2020). Microwaves and heterogeneous catalysis: a review on selected catalytic processes. *Catalysts* 10. https://doi.org/10.3390/catal10020246.

93 Li, S., Li, C., and Shao, Z. (2022). Microwave pyrolysis of sludge: a review. *Sustainable Environ. Res.* 32. https://doi.org/10.1186/s42834-022-00132-z.

94 Hu, Q., He, Y., Wang, F. et al. (2021). Microwave technology: a novel approach to the transformation of natural metabolites. *Chin. Med.* 16: 87. https://doi.org/10.1186/s13020-021-00500-8.

95 Michalak, J., Czarnowska-Kujawska, M., Klepacka, J., and Gujska, E. (2020). Effect of microwave heating on the acrylamide formation in foods. *Molecules.* 25: 4140. https://doi.org/10.3390/molecules25184140.

96 Albuquerque, H.M.T., Pinto, D.C.G.A., and Silva, A.M.S. (2021). Microwave irradiation: alternative heating process for the synthesis of biologically applicable chromones, Quinolones, and Their Precursors. *Molecules* 26: 6293. https://doi.org/10.3390/molecules26206293.

97 Gawande, M.B., Shelke, S.N., Zboril, R., and Varma, R.S. (2014). Microwave-assisted chemistry: synthetic applications for rapid assembly of nanomaterials and organics. *Acc. Chem. Res.* 47: 1338–1348. https://doi.org/10.1021/ar400309b.

98 Mitev, D.P., Townsend, A.T., Paull, B., and Nesterenko, P.N. (2014). Microwave-assisted purification of detonation nanodiamond. *Diamond Relat. Mater.* 48: 37–46. https://doi.org/10.1016/j.diamond.2014.06.007.

99 de Medeiros, T.V., Manioudakis, J., Noun, F. et al. (2019). Microwave-assisted synthesis of carbon dots and their applications. *J. Mater. Chem. C Mater.* 7: 7175–7195. https://doi.org/10.1039/C9TC01640F.

100 Giri, S.K. and Prasad, S. (2007). Drying kinetics and rehydration characteristics of microwave-vacuum and convective hot-air dried mushrooms. *J. Food Eng.* 78: 512–521. https://doi.org/10.1016/j.jfoodeng.2005.10.021.

101 Hwang, J., Bae, C., Park, J. et al. (2016). Microwave-assisted plasma ignition in a constant volume combustion chamber. *Combust. Flame* 167: 86–96. https://doi.org/10.1016/j.combustflame.2016.02.023.

102 Mehedi, H.A., Achard, J., Rats, D. et al. (2014). Low temperature and large area deposition of nanocrystalline diamond films with distributed antenna array microwave-plasma reactor. *Diamond Relat. Mater.* 47: 58–65. https://doi.org/10.1016/j.diamond.2014.05.004.

103 Singh, R.K., Kumar, R., Singh, D.P. et al. (2019). Progress in microwave-assisted synthesis of quantum dots (graphene/carbon/semiconducting) for bioapplications: a review. *Mater. Today Chem.* 12. https://doi.org/10.1016/j.mtchem.2019.03.001.

104 Jiang, Y., Wang, B., Meng, F. et al. (2015). Microwave-assisted preparation of N-doped carbon dots as a biosensor for electrochemical dopamine detection. *J. Colloid Interface Sci.* 452: 199–202. https://doi.org/10.1016/j.jcis.2015.04.016.

105 Xu, Z., Wang, C., Jiang, K. et al. (2015). Microwave-assisted rapid synthesis of amphibious yellow fluorescent carbon dots as a colorimetric nanosensor for Cr(VI). *Part. Part. Syst. Char.* 32: 1058–1062. https://doi.org/10.1002/ppsc.201500172.

106 Menéndez-Flores, V.M. and Ohno, T. (2014). High visible-light active Ir-doped-TiO_2 brookite photocatalyst synthesized by hydrothermal microwave-assisted process. *Catal. Today* 230: 214–220. https://doi.org/10.1016/j.cattod.2014.01.032.

107 Xu, Y., Li, H., Wang, B. et al. (2018). Microwave-assisted synthesis of carbon dots for "turn-on" fluorometric determination of Hg(II) via aggregation-induced emission. *Microchim. Acta* 185: 252. https://doi.org/10.1007/s00604-018-2781-y.

108 Sharma, B., Thakur, S., Trache, D. et al. (2020). Microwave-assisted rapid synthesis of reduced graphene oxide-based gum tragacanth hydrogel nanocomposite for heavy metal ions adsorption. *Nanomaterials* 10: 1616. https://doi.org/10.3390/nano10081616.

109 Qin, X., Lu, W., Asiri, A.M. et al. (2013). Microwave-assisted rapid green synthesis of photoluminescent carbon nanodots from flour and their applications for sensitive and selective detection of mercury(II) ions. *Sens. Actuators, B* 184: 156–162. https://doi.org/10.1016/j.snb.2013.04.079.

110 Show, Y., Swope, V.M., and Swain, G.M. (2009). The effect of the CH_4 level on the morphology, microstructure, phase purity and electrochemical properties of carbon films deposited by microwave-assisted CVD from Ar-rich source gas mixtures. *Diamond Relat. Mater.* 18: 1426–1434. https://doi.org/10.1016/j.diamond.2009.09.011.

111 Vicente, I., Salagre, P., Cesteros, Y. et al. (2009). Fast microwave synthesis of hectorite. *Appl. Clay Sci.* 43: 103–107. https://doi.org/10.1016/j.clay.2008.07.012.

112 Das, S., Mukhopadhyay, A.K., Datta, S., and Basu, D. (2009). Prospects of microwave processing: an overview. *Bull. Mater. Sci.* 32: 1–13. https://doi.org/10.1007/s12034-009-0001-4.

113 Sharan, N. (2009). Microwave sintering of zirconia and alumina. *Int. J.* 1: 1–4.

114 Li, Z., Yao, Y., Lin, Z. et al. (2010). Ultrafast, dry microwave synthesis of graphene sheets. *J. Mater. Chem.* 20: 4781. https://doi.org/10.1039/c0jm00168f.

115 Mitra, S., Chandra, S., Kundu, T. et al. (2012). Rapid microwave synthesis of fluorescent hydrophobic carbon dots. *RSC Adv.* 2: 12129. https://doi.org/10.1039/c2ra21048g.

116 Zhu, H., Wang, X., Li, Y. et al. (2009). Microwave synthesis of fluorescent carbon nanoparticles with electrochemiluminescence properties. *Chem. Commun.* 5118. https://doi.org/10.1039/b907612c.

117 Mishra, R.K., Loganathan, S., Jacob, J. et al. (2018). Progress in polymer nanocomposites for electromagnetic shielding application. In: *Modern Physical Chemistry: Engineering Models, Materials, and Methods with Applications*, 1e (ed. R.K. Haghi, E. Besalu, M. Jaroszewski, et al.), 198–237. New York: Apple Academic Press.

118 Yang, R. and Chen, J. (2021). Mechanistic and machine learning modeling of microwave heating process in domestic ovens: a review. *Foods* 10: 2029. https://doi.org/10.3390/foods10092029.

119 Zhao, H., Cai, K., Ma, Z. et al. (2018). Synthesis of molybdenum carbide superconducting compounds by microwave-plasma chemical vapor deposition. *J. Appl. Phys.* 123: 053301. https://doi.org/10.1063/1.5010101.

120 Khan, T. (2020). An intelligent microwave oven with thermal imaging and temperature recommendation using deep learning. *Appl. Sys. Innov.* 3: 13. https://doi.org/10.3390/asi3010013.

121 Zhang, Y.M., Wang, P., Han, N., and Lei, H.F. (2007). Microwave irradiation: a novel method for rapid synthesis of D,L-lactide. *Macromol. Rapid Commun.* 28: 417–421. https://doi.org/10.1002/marc.200600668.

122 Voiry, D., Yang, J., Kupferberg, J. et al. (1979). High-quality graphene via microwave reduction of solution-exfoliated graphene oxide. *Science* 353 (2016): 1413–1416. https://doi.org/10.1126/science.aah3398.

123 Chen, W., Lu, X., Yang, Q. et al. (2006). Effects of gas flow rate on diamond deposition in a microwave plasma reactor. *Thin Solid Films* 515: 1970–1975. https://doi.org/10.1016/j.tsf.2006.08.007.

124 Narayanan, T.N., Sunny, V., Shaijumon, M.M. et al. (2009). Enhanced microwave absorption in nickel-filled multiwall carbon nanotubes in the S band. *Electrochem. Solid-State Lett.* 12: K21. https://doi.org/10.1149/1.3065992.

125 Dora, T.L., Owhal, A., Roy, T. et al. (2023). Thermo-physical characteristics of 3C-SiC structure subjected to microwave exposure: a molecular dynamics

126 Liu, X., Li, T., Hou, Y. et al. (2016). Microwave synthesis of carbon dots with multi-response using denatured proteins as carbon source. *RSC Adv.* 6: 11711–11718. https://doi.org/10.1039/C5RA23081K.

127 Shakir, M.F., Abdul Rashid, I., Tariq, A. et al. (2020). EMI shielding characteristics of electrically conductive polymer blends of PS/PANI in microwave and IR region. *J. Electron. Mater.* 49. https://doi.org/10.1007/s11664-019-07631-7.

128 Tsugawa, K., Ishihara, M., Kim, J. et al. (2006). Large-area and low-temperature nanodiamond coating by microwave plasma chemical vapor deposition. *New Diamond Front. Carbon Technol.* 16: 337–346.

129 Wang, X.J. and Yan, C.L. (2010). Synthesis of nano-sized NaY zeolite composite from metakaolin by ionothermal method with microwave assisted. *Inorg. Mater.* https://doi.org/10.1134/S0020168510050146.

130 Marciano, F.R., Bonetti, L.F., Santos, L.V. et al. (2009). Antibacterial activity of DLC and Ag-DLC films produced by PECVD technique. *Diamond Relat. Mater.* 18: 1010–1014. https://doi.org/10.1016/j.diamond.2009.02.014.

131 Zhou, K., Ke, P., Li, X. et al. (2015). Microstructure and electrochemical properties of nitrogen-doped DLC films deposited by PECVD technique. *Appl. Surf. Sci.* 329: 281–286. https://doi.org/10.1016/j.apsusc.2014.12.162.

132 Aldosari, M.A., Othman, A.A., and Alsharaeh, E.H. (2013). Synthesis and characterization of the in situ bulk polymerization of PMMA containing graphene sheets using microwave irradiation. *Molecules* 18: 3152–3167. https://doi.org/10.3390/molecules18033152.

133 Kumar, A., Kuang, Y., Liang, Z., and Sun, X. (2020). Microwave chemistry, recent advancements, and eco-friendly microwave-assisted synthesis of nanoarchitectures and their applications: a review. *Mater. Today Nano* 11: 100076. https://doi.org/10.1016/j.mtnano.2020.100076.

134 Priecel, P. and Lopez-Sanchez, J.A. (2019). Advantages and Limitations of Microwave Reactors: From Chemical Synthesis to the Catalytic Valorization of Biobased Chemicals. *ACS Sustainable Chem. Eng.* 7: 3–21. https://doi.org/10.1021/acssuschemeng.8b03286.

135 Dinodia, M. (2023). Microwave-promoted synthesis of novel bioactive N-based heterocycles. *Mini-Rev. Org. Chem.* 20: 136–155. https://doi.org/10.2174/1570193X19666220420133723.

3

Conventional Versus Green Chemical Transformation: MCRs, Solid Phase Reaction, Green Solvents, Microwave, and Ultrasound Irradiation

Shailendra Yadav[1], Dheeraj S. Chauhan[2], and Mumtaz A. Quraishi[3]

[1] AKS University, Green Chemistry Lab., Department of Chemistry, Basic Science, Satna, Madhya Pradesh 485001, India
[2] Modern National Chemicals, Second Industrial City, Dammam 31421, Saudi Arabia
[3] Banaras Hindu University, Indian Institute of Technology, Department of Chemistry, Varanasi, Uttar Pradesh 221005, India

3.1 Introduction

Green chemistry (GrC) is a comparatively new emerging branch of science striving at the molecular level toward attaining sustainability in the chemical processes [1]. The field has grown from research laboratories and has touched industries, educational environments, and the public. GrC finds a myriad of applications in almost all industries viz. aerospace, cosmetics, automobiles, electronics, energy, household products, pharmaceuticals, and agriculture. Numerous examples of successful applications of GrC in the development of award-winning and economically viable technology have appeared in the past few years. GrC focuses on the design and development of products and processes that could be beneficial and profitable while being safer for human beings as well as the environment.

Over the years, several books and research articles have appeared in this area and, continuously, this field is being enriched. The major goal of this chapter is to focus on the applicative aspects of GrC in chemical synthesis. Emphasis is given to some of the modern methods of chemical synthesis such as multicomponent reactions (MCR), synthesis in the solid phase, and use of green solvents. Green chemical techniques such as microwave (MW) irradiation and ultrasound irradiation (US) are also outlined in this chapter. This chapter is intended to help researchers as well as students who are learning and working in this area.

3.2 A Brief Overview of Green Chemistry

3.2.1 Definition and Historical Background

The term GrC could be described as the "design of chemical products and processes for reduction/elimination of usage and production of hazardous substances."

Green Chemical Synthesis with Microwaves and Ultrasound, First Edition.
Edited by Dakeshwar Kumar Verma, Chandrabhan Verma, and Paz Otero Fuertes.
© 2024 WILEY-VCH GmbH. Published 2024 by WILEY-VCH GmbH.

This definition was coined at the beginning of the 1990s. Since then, hundreds of programs and governmental initiatives on GrC around the world, with leading research programs centered around almost all of the leading institutions, for e.g. in the United States, United Kingdom, and other European countries. These have played a considerable role in obtaining sustainability in chemical synthesis and processes. Notable early programs include the U.S. Presidential Green Chemistry Challenge Awards which was established in 1995, the Green Chemistry Institute founded in 1997, and the publication of the first volume of the now well-established Green Chemistry Journal from the Royal Society of Chemistry in 1999.

An important aspect of GrC is the conceptualization of design. The design would be a statement of human intent that cannot be achieved by accident. The design includes novelty, planning, and systematic conception. The twelve principles of GrC include a set of rules that can assist the chemists in achieving intentional goals of sustainability. GrC could be characterized by careful planning of chemical synthesis and molecular design to achieve the possible reduction of adverse consequences. Through proper design, we could attain sustainable development following the GrC principles.

3.2.2 Significance

Organic synthesis is conventionally accomplished via tedious protocols that involve different synthesis steps, and purification procedures. This renders the whole procedure considerably costly, and time-consuming and makes use of toxic organic solvents that could be detrimental to human beings and to the environment. Herein, the GrC practices have purported a set of principles that could develop organic molecules in agreement with strict environmental regulations [2, 3]. This also restricts the practical applicability of the synthesized products. GrC principles intend to minimize the release of harmful and toxic effluents with the synthesis and application of environment-friendly chemical products [1, 4]. In addition, several metrics to GrC that could elucidate the environmentally benign nature of a synthesis process have been proposed [5, 6]. In recent years, for the synthesis of effective and novel organic molecules, newer synthesis methodologies have been proposed. This includes the use of MCR [7] protocol as a new and effective approach for the synthesis of organic molecules. In addition, organic synthesis is considerably enhanced with the involvement of modern techniques allowing precise control of energy, time, temperature, and selectivity of the reaction [8]. Further, the usage of water, supercritical CO_2 [9, 10], and ionic liquids [10, 12] as green solvents, and the use of phase transfer catalysts and enzymes as bio-catalysts is preferred. In the following sections, we have discussed an overview of some modern green techniques with brief significance and examples of green transformations. Also discussed vide infra are some of the green solvents and chemicals for diverse applications.

List of Good Scientific Journals in Green Chemistry

Green Chemistry (Royal Society of Chemistry) IF: 10.18
ACS Sustainable Chemistry and Engineering (American Chemical Society) IF: 9.224
ChemSusChem (Wiley) IF: 9.18
Green Chemistry Letters and Reviews (Taylor & Francis) IF: 4.99
Sustainable Chemistry and Pharmacy (Elsevier) IF: 4.508
Environmental Progress and Sustainable Energy (Wiley) IF: 2.431

3.3 Multicomponent Reactions

The conventional synthesis procedure requires long synthesis steps, with isolation and purification at each step leading to a considerable loss of reagents, and huge expenditure. One-step MCRs (Figure 3.1), or in other words, multicomponent assembly processes (MCAPs) are a term given to the chemical synthesis process wherein three or more reactive molecules convert to a product following a single-step reaction [7, 13, 15]. MCRs represent an attractive and green strategy as a fruitful alternative to the conventionally utilized multiple step synthesis procedure. This is especially useful for the development of green chemicals for industrial and biological applications. The salient features of MCR include short reaction times, high yields, high chemical selectivity, low operating cost, and a low number of purification steps. Other benefits are less consumption of toxic solvents, and purification reagents that can have a detrimental influence on the environment [16, 17]. Some of the examples of MCRs are given in Figure 3.2.

In a green MCR, the majority of the atoms existing in the reagents should be converted to the product. The major by-product of such green MCR reactions is usually water [13]. An astonishing feature of the MCR is that independent of the preferred

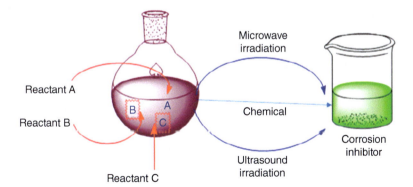

Figure 3.1 Schematic of MCRs facilitated using microwave and ultrasound irradiations. Source: Verma et al. [8]/with permission from Elsevier.

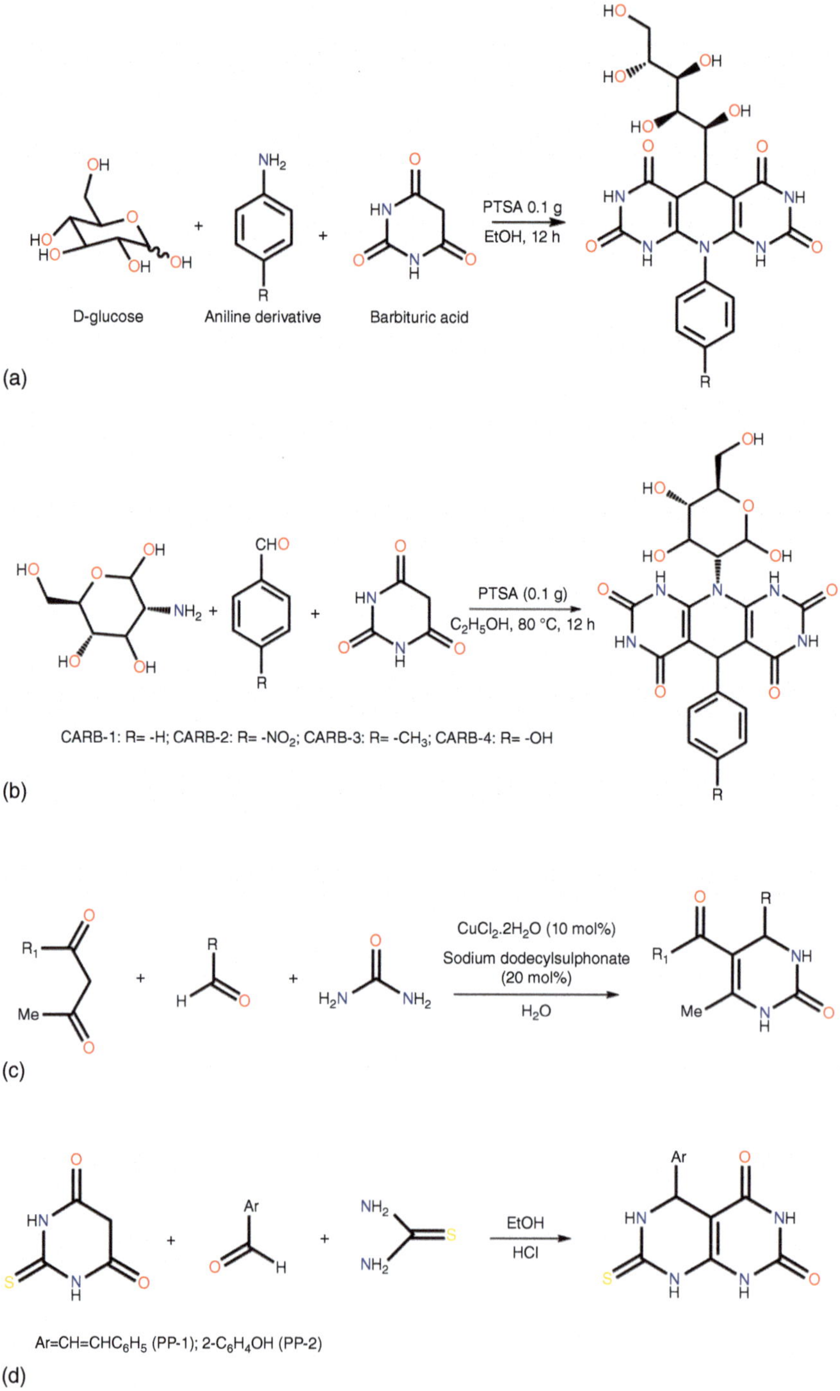

Figure 3.2 Examples of some MCR reactions. (a) Source: Verma et al. [18]/Springer Nature/Licensed under CC BY 4.0., (b) Source: Verma et al. [17]/American Chemical Society, (c) Source: Yadav et al. [19]/with permission from Elsevier, (d) Source: Haque et al. [20]/with permission from Elsevier.

reaction pathway, all mechanisms lead to the same final product. MCRs allow direct and elegant access to bioactive compound libraries and fulfill the requirement for biologically active compound syntheses and discovery.

3.4 Solid Phase Reactions

Aristotle, the great Greek philosopher, concluded that "No Coopora nisi Fluida," which means, "No reaction occurs in the absence of solvent." Such philosophies had a big influence on the evolution of modern sciences in Europe, and this provides one historical reason why most of the organic synthesis reactions are studied in solution. This technique has several advantages – reduced pollution, cost-effectiveness, and simplicity in overall process and handling [21]. These factors are especially important in industry. Figure 3.3 shows the schemes for the solvent-free synthesis of some organic molecules.

Mechanistically, it has been established that molecules can move quite freely in the solid state and even enantioselectively [24]. It has also been proven that organic reactions could proceed via mixing of powdered reactant and reagent in the absence of solvent and with efficient obtainment of reaction products. In some cases, organic reactions also proceed in solvent-free conditions throughout the process of reaction as well as product isolation. Using a chiral host compound for the solid-state reaction of prochiral guests could allow enantioselective thermal and photochemical reactions. However, it should be noted that not all organic syntheses can be performed without the presence of solvent. Some organic reactions proceed explosively in the solid state. In such cases, solvent is required to modulate the rate of reaction.

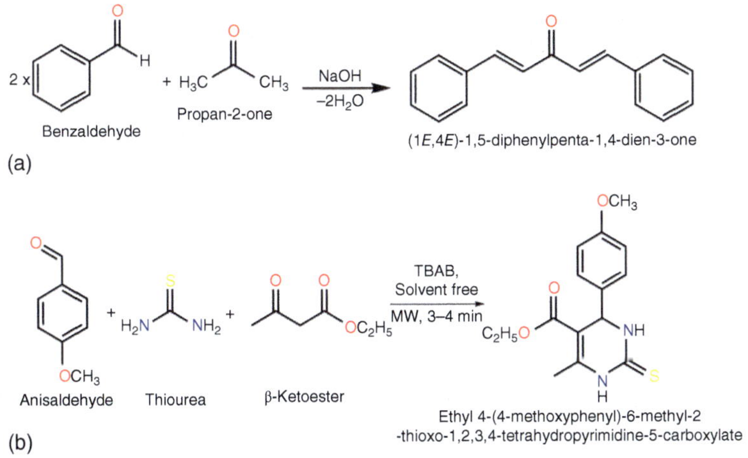

Figure 3.3 Schemes for the solvent-free synthesis of (a) dibenzalacetone. Source: Vanchinathan et al. [22]/with permission from Elsevier, and (b) pyrimidine derivative. Source: Ahmed et al. [23]/with permission from Elsevier.

3.5 Microwave Induced Synthesis

Generally, the organic synthesis is carried out in a reflux-based synthesis set up. This is a simple method; however, the process takes a long time, and the heating process can lead to the development of a localized temperature gradient within the sample. Besides, local overheating could result in the decomposition of the desired product, substrate, and the used reagent. Domestic MW reactor is a common household electrical appliance generally adopted for cooking/re-heating of food items. This is also reported for several organic syntheses, the production of nanomaterials, and even in solid-state science [25].

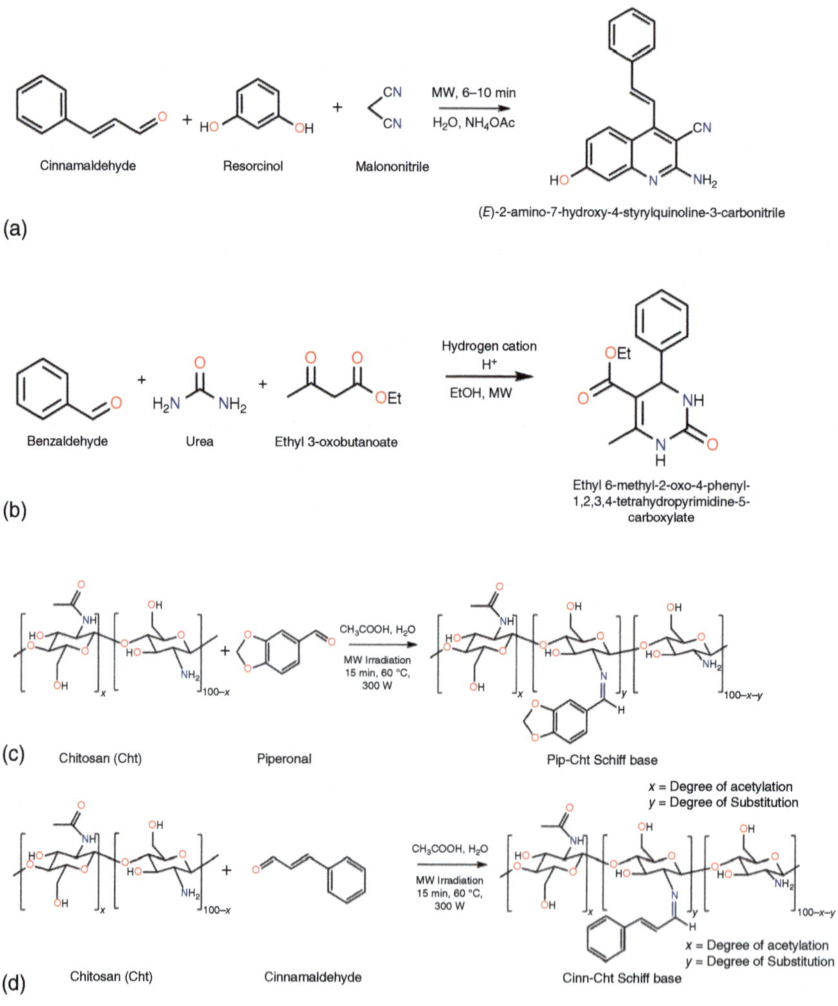

Figure 3.4 Examples of some MW-facilitated synthesis. (a) Source: Onyeachu et al. [35]/with permission from Taylor & Francis, (b) Source: Singh et al. [26]/with permission from Elsevier, (c) Source: Chauhan et al. [31]/with permission from Elsevier, (d) Source: Mouaden et al. [36]/with permission from Elsevier.

In the electromagnetic spectrum, the MW region lies between the infrared (IR) and the radio waves. MW operates between the wavelengths of 1 mm and 1 m, which corresponds to the frequency range of 0.3 and 300 GHz. The radar and telecommunication systems take several bands of frequencies in this region. Therefore, to avoid disturbance with these wavelengths, the MW appliances for domestic and industrial purposes function at 2.45 GHz. Several reports have come up in the literature on the application of the MW technique for the production of nanomaterials [26, 28]. MW works on the principle of dipolar polarization and ionic conduction [29, 30]. MW synthesis affords the acceleration of organic synthesis processes with selectivity and uniform heating upon a careful selection of MW parameters [26, 31, 33]. This can allow notable benefits over conventional syntheses such as faster rates of heating and greater thermal homogeneity. The uniform MW heating of the reaction vessel avoids the development of heating gradients, which are quite commonly observed in the conventional techniques of hot plate and reflux methods [30, 34]. Figure 3.4 shows the schemes for the synthesis of some organic molecules using the MW method.

3.6 Ultrasound Induced Synthesis

Ultrasonic waves afford an alternative energy source and their usage has gained considerable interest in GrC. High-frequency ultrasonic (US) waves (20 kHz–10 MHz) result in a series of compression/rarefaction cycles through a solvent medium through which they are passed. This forms cavitation through the solution wherein the bubbles repeatedly form and collapse in a rapid manner and provide energy that can carry out the US-based synthesis/reactions [25]. The formation of local extremes of high temperature and pressure (hot spots) around/inside the cavitation bubbles, in addition to destroying them, activates the reacting molecules of organic reagents [37]. Moreover, microjet streams and shock waves resulting from cavitation facilitate mass transfer and molecular dispersion in the reaction medium [38]. Figure 3.5 shows the schemes for the synthesis of some organic molecules using the US method.

The most commonly used US instrument is the simple US bath container that is generally used for the cleaning of artificial dentures, jewelry, and electrodes. Although some organic syntheses can also be carried out using this instrument. The US probe processor is a dedicated US instrument, employed in organic synthesis, preparation of nanomaterials, and other scientific applications [8, 38, 42, 44]. The probe is available in different sizes having radii with varying ranges of 3–10 mm. The major difference between the US cleaner and the US probe processor is that in the former, a reaction vessel is inserted in the water. However, in the probe processor, the probe is directly immersed in the reaction vessel. The latter instrument is more sophisticated and possesses considerably greater power compared to the cleaner bath. A careful choice of the US power, pulse period, temperature, energy delivered, frequency, amplitude, and time of sonication parameters and solvent, reaction precursors, and catalysts could provide considerable control over the reaction.

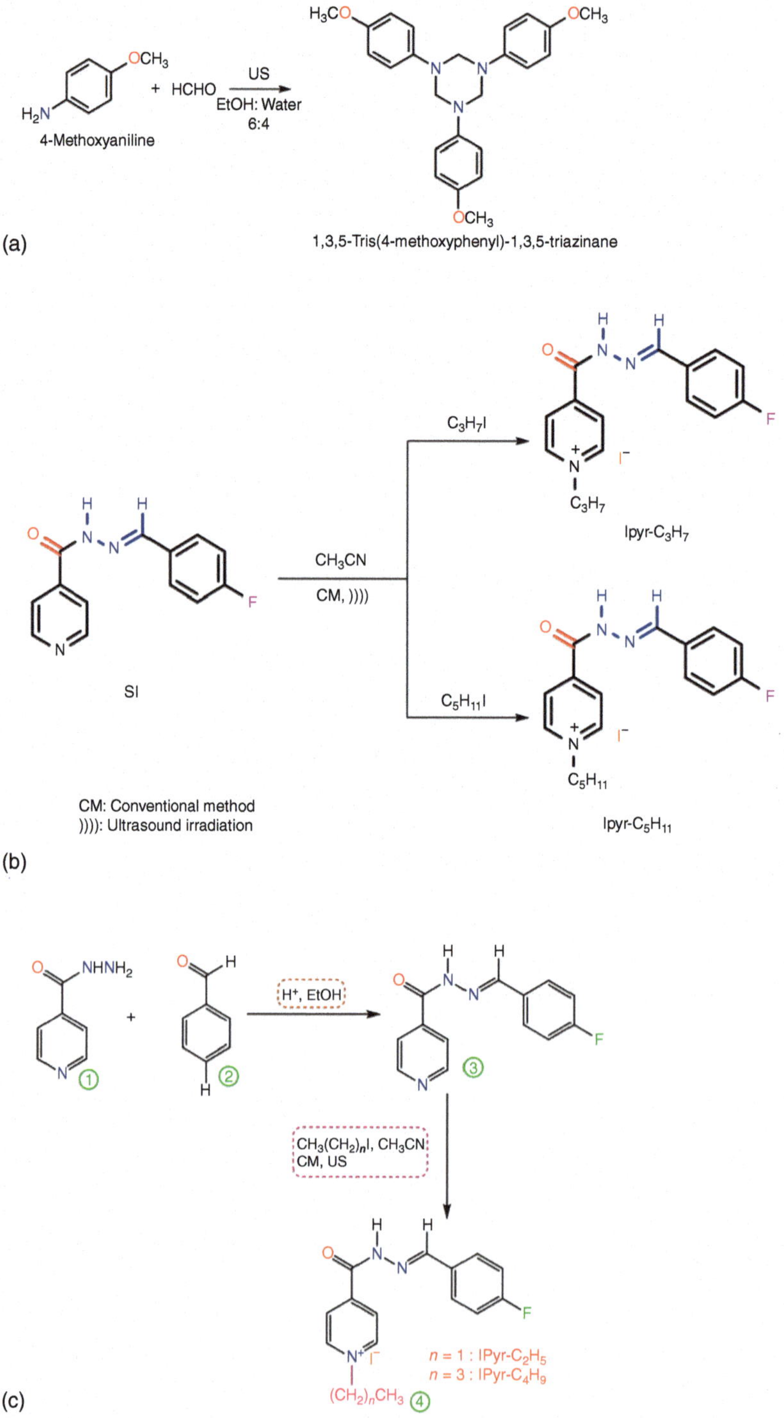

Figure 3.5 Examples of some US-facilitated synthesis. (a) Triazines. Source: Onyeachu et al. [39]/with permission from Taylor & Francis. (b) Pyridinium ionic liquids. Source: El-Hajjaji et al. [40]/with permission from Elsevier. (c) Pyridinium ionic liquids. Source: El-Hajjaji et al. [41]/with permission from Elsevier.

3.7 Green Chemicals and Solvents

In addition to the abovementioned modern techniques for green synthesis, several precursor molecules could be classified as green and environmentally safe. Their application in the synthesis agrees with several provisos of GrC principles of environmentally benign and low toxicity reagents, renewable raw materials, and minimizing the production of waste [1, 45]. The major categories include (i) the extracts from plants and tree parts [46] such as roots, stems, leaves, bark, shell, seed, and flower. These parts contain several phytochemicals that are complex compositions of naturally occurring chemicals and are environmentally safe; (ii) carbohydrates [36, 47] such as glucose, chitin, chitosan, cellulose, starch, and pectin. These are large molecular weight chemicals and contain an abundant number of surface functional groups that allow considerable prospects of reactivity; (iii) amino acids [48] such as glycine, proline, histidine, and others among the important amino acid member category. These provide reactivity, aid in solubility, and also provide prospects for chemical functionalization. In addition, polyamino acids such as polyaspartic acid, and proteins such as soy, gluten, casein, and gelatin have also found diverse applications. It is understood that the abovementioned categories are of natural origin and have considerable biocompatibility; Other classes include (iv) heterocyclic biomolecules [49] such as vitamins, and hormones; (v) pharmaceutical products, viz. drugs, and (vi) ionic liquids (Figure 3.6) [49]. All of these types of chemicals are available in abundance, are environmentally safe, and can be afforded cheap costs.

Figure 3.6 Some of the greener chemicals used for diverse applications. Source: Chauhan et al. [50]/with permission from Elsevier.

Figure 3.7 Greenness of some commonly used analytical solvents.

Supercritical CO_2 [9, 10] and ionic liquids [10, 12] are preferred as green solvents. Additionally, the application of solvent-less or solid phase synthesis, or synthesis in aqueous medium is generally preferred considering the nonflammable nature of water, high dielectric constant, abundant availability, ease of handling, and cost-effectiveness. Additionally, glycerol, ionic liquids, and bio-derived solvents are classified as green solvents. The relative greenness of some commonly used solvents is shown in Figure 3.7.

3.8 Conclusions and Outlook

This chapter presents a brief outline of the significance of GrC with attention to some of the sustainable and modern techniques and sustainable chemicals. A historical background and importance of GrC is provided, with scope of research in this area. Special emphasis is given to MCR, solid-state synthesis, MW, and ultrasonic (US) irradiation as modern synthetic techniques in compliance with the GrC principles. The basic principles governing organic transformations based on MW and US techniques are described in this chapter. Compared to the conventionally used heating of reaction vessel employing hot plate and reflux method, the synthesis could be undertaken much more easily and in a controlled fashion using MW and US techniques. The products could be obtained with high yield and more selectivity, in a quicker time, with precise control over physicochemical properties. The significance and examples of some of the green chemicals for organic synthesis are also discussed. Some examples of green chemicals are highlighted, such as carbohydrates, amino acids, proteins, ionic liquids, and pharmaceutical products. Some of the green solvents used in synthesis and in analytical chemistry are also discussed. Using green techniques in chemical science is likely to address the issues of toxicity and harmful discharges posed by conventional methods and chemicals. In addition, this approach is also likely to address the high-cost aspect of the conventional methodologies.

References

1 Anastas, P. and Eghbali, N. (2010). Green chemistry: principles and practice. *Chem. Soc. Rev.* 39: 301–312.
2 O. Commission (2005). OSPAR protocols on methods for the testing of chemicals used in the offshore oil industry.
3 U. Occupational Safety and Health Administration (OSHA) (2013). Globally harmonized system of classification and labelling of chemicals (GHS). In: *Economic Commission for Europe*, 1–568.
4 Sheldon, R.A. (2012). Fundamentals of green chemistry: efficiency in reaction design. *Chem. Soc. Rev.* 41: 1437–1451.
5 Constable, D.J., Curzons, A.D., and Cunningham, V.L. (2002). Metrics to 'green' chemistry—which are the best? *Green Chem.* 4: 521–527.
6 Verma, C., Quraishi, M.A., and Chauhan, D.S. (2022). *Green Corrosion Inhibition: Fundamentals, Design, Synthesis and Applications*. Royal Society of Chemistry.
7 Verma, C., Haque, J., Quraishi, M.A., and Ebenso, E.E. (2018). Aqueous phase environmental friendly organic corrosion inhibitors derived from one step multicomponent reactions: a review. *J. Mol. Liq.* 275: 18–40.
8 Verma, C., Quraishi, M.A., and Ebenso, E.E. (2018). Microwave and ultrasound irradiations for the synthesis of environmentally sustainable corrosion inhibitors: an overview. *Sustainable Chem. Pharm.* 10: 134–147.
9 Lozowski, D. (2010). Supercritical CO_2: a green solvent. *Chem. Eng.* 117: 15.
10 Dzyuba, S.V. and Bartsch, R.A. (2003). Recent advances in applications of room-temperature ionic liquid/supercritical CO_2 systems. *Angew. Chem. Int. Ed.* 42: 148–150.
11 Zhao, H., Xia, S., and Ma, P. (2005). Use of ionic liquids as 'green' solvents for extractions. *J. Chem. Technol. Biotechnol.* 80: 1089–1096.
12 Yoo, C.G., Pu, Y., and Ragauskas, A.J. (2017). Ionic liquids: promising green solvents for lignocellulosic biomass utilization. *Curr. Opin. Green Sustainable Chem.* 5: 5–11.
13 Alvim, H.G., da Silva Junior, E.N., and Neto, B.A. (2014). What do we know about multicomponent reactions? Mechanisms and trends for the Biginelli, Hantzsch, Mannich, Passerini and Ugi MCRs. *RSC Adv.* 4: 54282–54299.
14 Puripat, M., Ramozzi, R., Hatanaka, M. et al. (2015). The Biginelli reaction is a urea-catalyzed organocatalytic multicomponent reaction. *J. Org. Chem.* 80: 6959–6967.
15 Stadler, A. and Kappe, C.O. (2001). Automated library generation using sequential microwave-assisted chemistry. Application toward the Biginelli multicomponent condensation. *J. Comb. Chem.* 3: 624–630.
16 Singh, P., Ebenso, E.E., Olasunkanmi, L.O. et al. (2016). Electrochemical, theoretical, and surface morphological studies of corrosion inhibition effect of green naphthyridine derivatives on mild steel in hydrochloric acid. *J. Phys. Chem. C* 120: 3408–3419.
17 Verma, C., Olasunkanmi, L.O., Ebenso, E.E. et al. (2016). Adsorption behavior of glucosamine-based, pyrimidine-fused heterocycles as green corrosion

inhibitors for mild steel: experimental and theoretical studies. *J. Phys. Chem. C* 120: 11598–11611.

18 Verma, C., Quraishi, M., Kluza, K. et al. (2017). Corrosion inhibition of mild steel in 1M HCl by D-glucose derivatives of dihydropyrido [2,3-d: 6,5-d′] dipyrimidine-2, 4, 6, 8 (1H, 3H, 5H, 7H)-tetraone. *Sci. Rep.* 7: 44432.

19 Yadav, D.K., Maiti, B., and Quraishi, M.A. (2010). Electrochemical and quantum chemical studies of 3, 4-dihydropyrimidin-2 (1H)-ones as corrosion inhibitors for mild steel in hydrochloric acid solution. *Corros. Sci.* 52: 3586–3598.

20 Haque, J., Ansari, K.R., Srivastava, V. et al. (2017). Pyrimidine derivatives as novel acidizing corrosion inhibitors for N80 steel useful for petroleum industry: a combined experimental and theoretical approach. *J. Ind. Eng. Chem.* 49: 176–188.

21 Tanaka, K. and Toda, F. (2000). Solvent-free organic synthesis. *Chem. Rev.* 100: 1025–1074.

22 Vanchinathan, K., Bhagavannarayana, G., Muthu, K., and Meenakshisundaram, S. (2011). Synthesis, crystal growth and characterization of 1,5-diphenylpenta-1,4-dien-3-one: an organic crystal. *Physica B* 406: 4195–4199.

23 Ahmed, B., Khan, R.A., and Keshari, M. (2009). An improved synthesis of Biginelli-type compounds via phase-transfer catalysis. *Tetrahedron Lett.* 50: 2889–2892.

24 Tanaka, K. (2009). *Solvent-free Organic Synthesis*. Wiley.

25 Quraishi, M.A., Chauhan, D.S., and Saji, V.S. (2020). *Heterocyclic Organic Corrosion Inhibitors: Principles and Applications*. Amsterdam: Elsevier Inc.

26 Singh, P., Chauhan, D.S., Chauhan, S.S. et al. (2019). Bioinspired synergistic formulation from dihydropyrimdinones and iodide ions for corrosion inhibition of carbon steel in sulphuric acid. *J. Mol. Liq.* 298: 112051.

27 Moseley, J.D. and Kappe, C.O. (2011). A critical assessment of the greenness and energy efficiency of microwave-assisted organic synthesis. *Green Chem.* 13: 794–806.

28 Chauhan, D.S., Gopal, C.S.A., Kumar, D. et al. (2018). Microwave induced facile synthesis and characterization of ZnO nanoparticles as efficient antibacterial agents. *Mater. Discover* 11: 19–25.

29 Kappe, C.O. (2004). Controlled microwave heating in modern organic synthesis. *Angew. Chem. Int. Ed.* 43: 6250–6284.

30 Polshettiwar, V. and Varma, R.S. (2008). Microwave-assisted organic synthesis and transformations using benign reaction media. *Acc. Chem. Res.* 41: 629–639.

31 Chauhan, D.S., Mazumder, M.J., Quraishi, M.A. et al. (2020). Microwave-assisted synthesis of a new Piperonal-Chitosan Schiff base as a bio-inspired corrosion inhibitor for oil-well acidizing. *Int. J. Biol. Macromol.* 158: 231–243.

32 Tierney, J. and Lidström, P. (2009). *Microwave Assisted Organic Synthesis*. Wiley.

33 Chauhan, D.S., Mazumder, M.J., Quraishi, M.A., and Ansari, K. (2020). Chitosan-cinnamaldehyde Schiff base: a bioinspired macromolecule as corrosion inhibitor for oil and gas industry. *Int. J. Biol. Macromol.* 158: 127–138.

34 Kappe, C.O. (2008). Microwave dielectric heating in synthetic organic chemistry. *Chem. Soc. Rev.* 37: 1127–1139.

35 Onyeachu, I.B., Chauhan, D.S., Quraishi, M.A. et al. (2021). (E)-2-amino-7-hydroxy-4-styrylquinoline-3-carbonitrile as a novel inhibitor for oil and gas industry: influence of temperature and synergistic agent. *J. Adhes. Sci. Technol.* 1–25.

36 Mouaden, K.E., Chauhan, D.S., Quraishi, M. et al. (2020). Cinnamaldehyde-modified chitosan as a bio-derived corrosion inhibitor for acid pickling of copper: microwave synthesis, experimental and computational study. *Int. J. Biol. Macromol.*

37 Penteado, F., Monti, B., Sancineto, L. et al. (2018). Ultrasound-assisted multicomponent reactions, organometallic and organochalcogen chemistry. *Asian J. Org. Chem.* 7: 2368–2385.

38 Kimura, T. (2015). Application of ultrasound to organic synthesis. In: *Sonochemistry and the Acoustic Bubble*, 171–186. Elsevier.

39 Onyeachu, I.B., Chauhan, D.S., Quraishi, M., and Obot, I. (2020). Influence of hydrodynamic condition on 1,3,5-tris (4-methoxyphenyl)-1,3,5-triazinane as a novel corrosion inhibitor formulation for oil and gas industry. *Corros. Eng. Sci. Technol.* 56: 154–161.

40 El-Hajjaji, F., Salim, R., Taleb, M. et al. (2021). Pyridinium-based ionic liquids as novel eco-friendly corrosion inhibitors for mild steel in molar hydrochloric acid: experimental & computational approach. *Surf. Interfaces* 22: 100881.

41 El-Hajjaji, F., Ech-chihbi, E., Rezki, N. et al. (2020). Electrochemical and theoretical insights on the adsorption and corrosion inhibition of novel pyridinium-derived ionic liquids for mild steel in 1 M HCl. *J. Mol. Liq.* 314: 113737.

42 Yang, H., Li, H., Zhai, J. et al. (2014). Simple synthesis of graphene oxide using ultrasonic cleaner from expanded graphite. *Ind. Eng. Chem. Res.* 53: 17878–17883.

43 Kumar, R., Kumar, V.B., and Gedanken, A. (2020). Sonochemical synthesis of carbon dots, mechanism, effect of parameters, and catalytic, energy, biomedical and tissue engineering applications. *Ultrason. Sonochem.* 64: 105009.

44 Saleem, M., Naz, M., Shukrullah, S. et al. (2021). One-pot sonochemical preparation of carbon dots, influence of process parameters and potential applications: a review. *Carbon Lett.* 1–17.

45 Kurian, M. and Paul, A. (2021). Recent trends in the use of green sources for carbon dot synthesis – a short review. *Carbon Trends* 3: 100032.

46 Chaubey, N., Qurashi, A., Chauhan, D.S., and Quraishi, M. (2020). Frontiers and advances in green and sustainable inhibitors for corrosion applications: a critical review. *J. Mol. Liq.* 114385.

47 El Mouaden, K., El Ibrahimi, B., Oukhrib, R. et al. (2018). Chitosan polymer as a green corrosion inhibitor for copper in sulfide-containing synthetic seawater. *Int. J. Biol. Macromol.* 119: 1311–1323.

48 Chauhan, D.S., Quraishi, M., Srivastava, V. et al. (2021). Virgin and chemically functionalized amino acids as green corrosion inhibitors: influence of molecular structure through experimental and in silico studies. *J. Mol. Struct.* 1226: 129259.

49 Quraishi, M.A., Chauhan, D.S., and Saji, V.S. (2021). Heterocyclic biomolecules as green corrosion inhibitors. *J. Mol. Liq.* 117265.

50 Chauhan, D.S., Quraishi, M., and Qurashi, A. (2021). Recent trends in environmentally sustainable sweet corrosion inhibitors. *J. Mol. Liq.* 326: 115117.

4

Metal-Catalyzed Reactions Under Microwave and Ultrasound Irradiation

Suresh Maddila[1], Immandhi S.S. Anantha[1], Pamerla Mulralidhar[1], Nagaraju Kerru[2], and Sudhakar Chintakula[1]

[1]GITAM University, GITAM School of Sciences, Department of Chemistry, Rushikonda, Visakhapatnam, Andhra Pradesh 530045, India
[2]GITAM University, GITAM School of Science, Department of Chemistry, Bengaluru Campus, Karnataka 561203, India

4.1 Ultrasonic Irradiation

The dissolution and formation of chemical bonds are both key aspects of sonochemistry [1]. A phenomenon known as sonolysis can cause chemical bonds to break, whenever a liquid is subjected to ultrasonic vibrations that operate at frequencies higher than 20 kHz. This causes the production of free radicals, which are crucial in starting a series of chemical alterations [2]. H· and ·OH radicals are produced when water is used as a solvent in reduction–oxidation processes. Throughout the 1980s, the Luche group began researching the impact of ultrasound on chemical reactions and fundamental mechanistic pathways in the field of sonochemistry. Significantly, three different groups of sonochemical reactions are classified in aseminal article from 2021 by Vinatoru and Mason [3].

- Homogeneous sonochemistry is a subcategory in which reactions follow a radical route, and intermediate products, such as radicals or radical ions, arise by cavitation in uniform mediums. As a result, whereas ionic processes are unaffected by such exposure, sonication can accelerate reactions by harnessing radicals. This area can be considered "authentic" sonochemistry.
- **Heterogeneous sonochemistry**: This category includes processes that are impacted mechanically by waves moving across various settings. In other words, the impacts of cavitational disruption accelerate processes that involve ionic intermediates. The term "misleading" sonochemistry may be used to describe this kind.
- **Sonocatalysis**: This classification includes heterogeneous reactions with diverse mechanisms, including ionic and radical ones. Through the incorporation of the total mechanical effect typical of heterogeneous sonochemistry, sonication considerably improves the radical pathway.

Green Chemical Synthesis with Microwaves and Ultrasound, First Edition.
Edited by Dakeshwar Kumar Verma, Chandrabhan Verma, and Paz Otero Fuertes.
© 2024 WILEY-VCH GmbH. Published 2024 by WILEY-VCH GmbH.

The hot spots, or extreme conditions, that exist at the center of cavitation bubbles, become extremely severe upon their collapse. Significant gradients in temperature, pressure, and shear are all part of these circumstances. A variety of radical species, including hydrogen radicals (H·), hydroxyl radicals (·OH), and hydroperoxyl radicals (·OOH), are produced by the presence of gases in the bubbles, or at the interfacial area between the bubble's core and the surrounding liquid. Through the action of the radicals ·OH and H·, the sonolysis of water results in additional products including hydrogen peroxide (H_2O_2) and hydrogen gas (H_2). Although oxygen improves sonochemical reactions, it is not a prerequisite for water sonolysis. Any gas can cause sonochemical oxidation and reduction to take place. However, oxygen may scavenge H· radicals, resulting in the formation of ·OOH, an oxidizing agent.

Researchers have looked at the generation of radicals (·OH) during different frequency water sonolysis; by injecting luminol close to cavitation bubbles, sonochemi-luminescence was produced, allowing for the detection of these radicals [4]. Similar to this, the Kavanal group investigated the mechanistic facets of the 2-aminopurine sonochemical oxidation in an aqueous setting [5]. In order to clarify the function of hydroxyl radicals (·OH) as process mediators, this work combined DFT calculations with the analysis of end products, using the LC-QTOF-MS/MS technique. These incidents highlight how extremely reactive radicals can start a variety of sonochemical processes.

The relationship between sonochemistry and catalysis demonstrates that the cavitation collapse of the bubbles process enhances the catalyst's dispersion throughout the medium. This process results in nonpassivated and noticeably reactive regions on the catalyst particle surface. Furthermore, the reagent dispersion throughout the catalyst's surface is enhanced. Many chemical reactions are successfully sped up by the combination of these factors, and certain reactions that would ordinarily be slow owing to low kinetic energy may even be made possible. In some cases, a synergistic interaction between ultrasound and the catalyst is seen, especially when solid catalysts are present. Heterogeneous nucleation is a process that primarily causes cavitation bubbles to develop on the surface of the particle. The subsequent asymmetric collapsing of these bubbles results in chemical and physical effects that, in turn, trigger the development or stimulation of catalytic sites [6].

The confinement of nanoparticles (NPs) onto solid foundations is a strategy that successfully adheres to the clean chemistry norms and has grown in popularity within the heterogeneous catalysis area in recent years. Notably, these solid supports include carbon-based substances including graphene, GO, GNSs, and CNTs as well as porous silica, zeolites, polymers, and carbon-based materials. Naeimi and Shaabani proposed a facile and ecologically friendly prototype for the preparation of triazole scaffolds. The research shows how effective a substituted graphene oxide Cu(I) complex is as heterogenous catalytic particles, and how ultrasonic irradiation helps to speed up the process [7]. Nguyen et al. offer a fascinating method for using brickyard ash as a precursor for the synthesis of mesoporous ZnO/SBA-15 and silica SBA-15 substances. Due to their capacity to effectively use solar energy for different environmental remediation processes, these materials might potentially be used as photo-catalysts, which is an important field of research [8]. With an emphasis on their uses in heterogeneous green catalysis, Lima et al. investigated the use of zeolites as catalysts for multi-component processes [9]. Kosinov et al. presented a thorough examination of the engineering issues surrounding transition metal catalysts immobilized inside zeolite frameworks. Zeolites provide ideal hosts for confining transition metal species, improving their catalytic efficacy and selectivity due to their well-defined microporous structures and adjustable characteristics [10]. The study by Romanazzi et al. offers an in-depth analysis of the synthesis and characterization of metal-containing copolymers. The method of copolymerization is described, and the authors go through how different metal species affect the characteristics and catalytic efficiency of the resultant heterogeneous catalysts [11]. Dzhardimalieva et al. thoroughly examined the catalytic activity of the metal nanoparticles and polymer-immobilized clusters in a variety of processes, such as oxidation, hydrogenation, and other organic transformations [12]. The study by Yue represents an important development in the design of metal-free chiral catalysts by demonstrating the effectiveness and the selectivity of amino acid-functionalized CNTs nanotubes in the electro-reduction of aromatic ketones, providing potential applications in the synthesis of chiral compounds [13]. Rasal et al. comprehensively analyzed GNSAgNPs' ability to support the production of -aminophosphonates using ultrasound. The research illustrates the mechanism of the reaction and the function of the GNSAgNPs catalyst in accelerating the synthesis of -aminophosphonates [14]. Achary illustrates the effectiveness and long-term viability of the graphene oxide catalyst coated with copper oxide nanoparticles for the production of amino carbonyl compounds.

The work clarifies the reaction's process as well as how the phosphate-functionalized graphene oxide catalyst enhanced by copper oxide nanoparticles aids in the synthesis of -amino carbonyl compounds. [15]. The intrinsic surface modifiability of several of these supports, including graphene, is a remarkable trait that may be linked to the existence of significant nucleation or stabilizing spots. In addition to these qualities, these materials also have great mechanical strength, little contact with the supporting materials, and porosity, all of which significantly increase their appeal in the catalysis industry. Mo examines the numerous elements that have an impact on the ability of graphene oxide membranes to accomplish separation. They concentrate on the impact of mechanical support, the characteristics of graphene oxide, and the presence of unusual species on the membrane's capacity for separation [16]. On the other hand, these composite catalytic environments continue to have the advantage of easy extraction from the reaction media, appreciating their significance in the quest for environment-friendly synthetic methods. Furthermore, the use of ultrasonic waves in conjunction with these heterogeneous materials has proven to be an effective method for raising reaction yields. This effect can be primarily attributable to better magnetic nanoparticle (MNP) dispersion throughout the reaction brought on by ultra-sonication, which results in an increase in the active surface area for catalyzing the reaction. R. Taheri-Ledari investigates the use of the synergistic catalytic effect of ultrasonic waves and MNP, that have been functionalized with pyrimidine-2,4-diamine for the manufacture of 1,4-dihydropyridine medicinal derivatives. This section provides a thorough discussion of the use of heterogeneous catalysts in the ultrasonic-aided preparation of numerous kinds of heterocycles.

4.1.1 Iron-Based Catalysts

Taheri-Ledari et al. developed a unique hybrid catalytic system to create extremely efficient magnetic nanocatalysts, made of Fe_3O_4, which could be incorporated onto the silica surface. As a substrate, they used [Fe_3O_4/SiO_2-PDA], which stands for MNPs mixed with silica, functionalized by pyrimidine-2,4-diamine (PDA). For the synthesis of 1,4-dihydropyridine (1,4-DHP) derivatives, this compound was used at a proportion of 2 w/w% with EtOH as the solvent. Ultrasonic irradiation, at a frequency of 50 kHz and a power of 250 WL-1, in a bath running at room temperature, promoted the reaction. Only trace quantities of the required product were produced without the nanocatalyst. The process was meticulously optimized before the researchers looked at the combined impact of using [Fe_3O_4/SiO_2-PDA] and ultrasonic irradiation in the same reaction. The results demonstrated a cooperative relationship between the ultrasonic waves and the NH_2 groups, which served as active catalytic centers. Both devices ran at 50 kHz and 250 W of electricity, respectively. The authors explained that the increased nitrogen atom binding capacity is what gives the nano-composite its improved catalytic efficiency. This improvement is ascribed to a beneficial synergistic interaction between the ultrasonic waves and the PDA-NH_2 groups, which serve as active catalytic sites. This interaction quickens the electronic resonance of the charged heterocycle's conjugated system. Magnetic separation made it simple to recover the nanocatalyst, which showed reusability for up to seven cycles without noticeably losing any of its catalytic efficiency (Figure 4.1) [17].

A modern and reusable heterogeneous catalyst made of Fe_3O_4 nanoparticles inserted into a sulfonated graphitic carbon nitride surface, known as Fe_3O_4@g-C_3N_4-AS (0.96 mol%), was used by the Edrisi research team to create a series of 2-amino-3-cyanopyridine compounds. The robustness against both

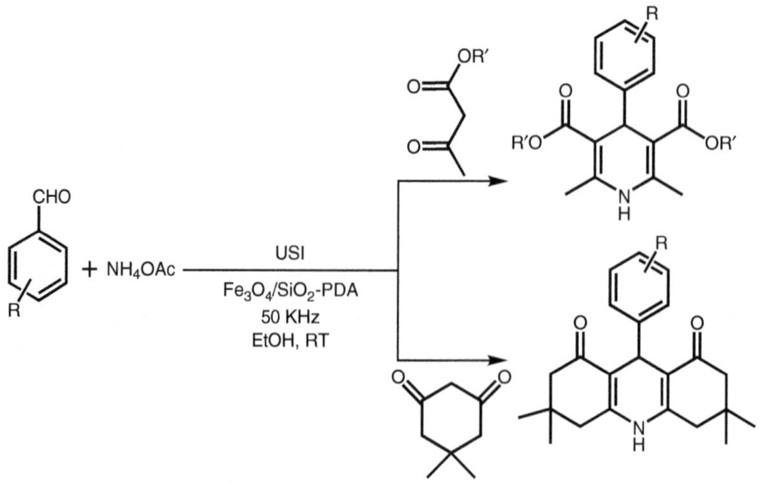

Figure 4.1 Synthesis of 1,4-DHP derivatives, co-catalyzed by Fe_3O_4/SiO_2-PDA NPs and USW irradiation. Source: Taheri-Ledari et al. [17].

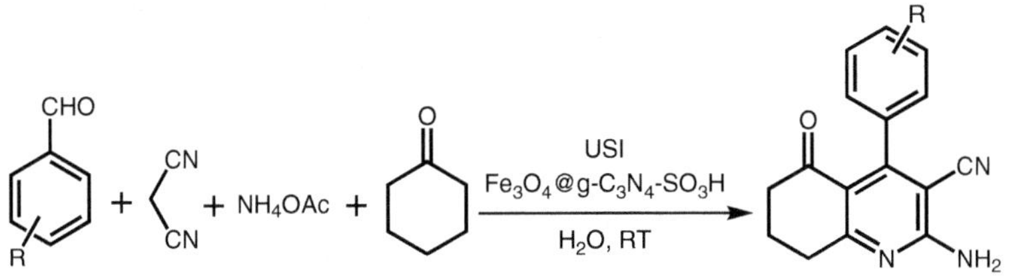

Figure 4.2 Synthesis of 1,4-DHPs analogs by Fe_3O_4@g-C_3N_4-SA nanocatalyst. Source: Edrisi and Azizi [18].

acidic and basic conditions as well as the low toxicity of carbon nitride surfaces, are advantages that this catalytic framework uses. Additionally, it makes use of the extraordinary properties of Fe_3O_4 nanoparticles as powerful catalysts. When comparing the effect of ultrasound on the reaction to that of conventional heating at 60 °C, the former showed a response time that was 50 times shorter. An ultrasonic homogenizer device with a 24 kHz frequency and 250 W of nominal power was used to carry out the reaction. The reaction was repeated under the exact same settings, but with the irradiation power changed to 50%, 80%, and 100% to further investigate the impact of ultrasound. With a 100% irradiation power, the best yield-related findings were obtained. In addition to recyclability, the created catalytic system maintained its catalytic effectiveness even after five cycles (Figure 4.2) [18].

As demonstrated by the work of Verma et al. recent discoveries have demonstrated the potential of polysaccharides as agents for the fictionalization of magnetic nanoparticles (MNPs). In their study, a brand new hybrid bionanocatalyst, made of F_3O_4/MNPs and immobilized within starch (referred to as s-Fe_3O_4/MNPs), was created. This bionanocatalyst was used to create benzoimidazo-pyrimidine derivatives in an environmentally friendly one-pot method. The synthesis method uses ultrasound to aid in the condensation of malononitrile, 2-aminobenzimidazole, and aromatic aldehydes in water at room temperature. The reaction was repeated over both reflux and ultrasonic irradiation conditions to evaluate its effectiveness. Surprisingly, under ultrasound, the reaction moved forward at an amazing 40 times faster speed and produced around 1.22 times more than it did using the reflux approach. The use of a small amount of the nanocatalyst (0.075 w/w%), simple product separation, and chemoselectivity are important aspects of this technique. The nanocatalyst also showed exceptional recyclability, enabling it to be utilized again for up to six cycles without noticeably losing its catalytic effectiveness (Figure 4.3) [19].

In a similar line, the Esmaeil research team began the multi component reaction (MCR) synthesis of tetrahydrobenzo-[b]-pyran derivatives using the Fe_3O_4@SiO_2-Imid-PMAn magnetic nanocatalyst. The relative effectiveness of ultrasonic irradiation compared to traditional heating was investigated in the quest for optimization. Notably, this study demonstrated how effective ultrasonic irradiation is in producing the desired chemicals with faster reaction times and higher yields. An ultrasonic cleaning bath, with a frequency of 40 Hz and a voltage of 220 V, was used in the experimental setup. This method led to the successful synthesis of 24 different

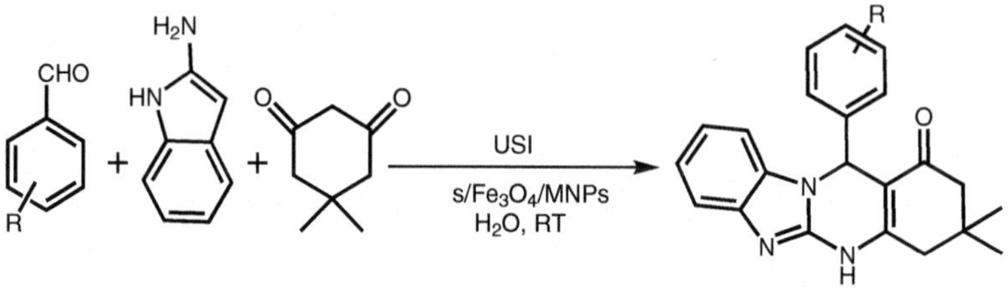

Figure 4.3 Sonochemical preparation of benzimidazoloquinazolines mediated by heterogeneous s/Fe$_3$O$_4$/MNPs catalyst. Source: Verma et al. [19].

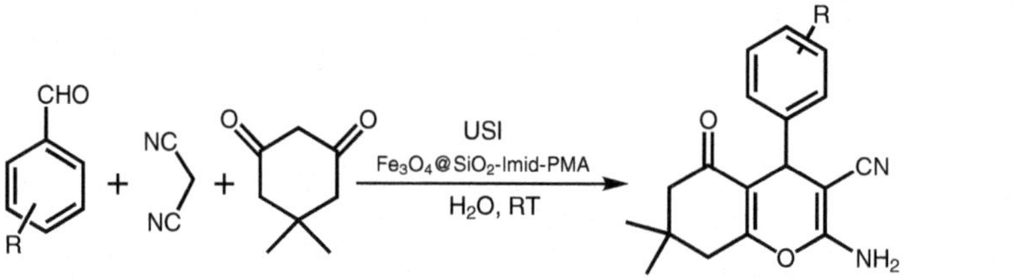

Figure 4.4 Synthesis of tetrahydrobenzo[b]pyrans under US irradiation and Fe$_3$O$_4$@SiO$_2$-Imid-PMA catalyst. Source: Esmaeilpour et al. [20].

tetrahydrobenzo[b]pyran examples, with yields above 89%. With excellent yields, quick reaction times, and the use of water as the solvent, the examined nanocatalyst impressively proved reusability across eight cycles (Figure 4.4) [20].

Recently, it has been shown that certain proteinaceous substances may efficiently support nanoparticles, which has sparked the creation of adaptable heterogeneous catalysts. By immobilizing Fe$_3$O$_4$ nanoparticles (NPs) in Isinglas (IG), a collagen produced from the swimming bladders of tropical fish, the Pourian group achieved a hybrid catalytic system. Because of its helical shape, which contains several acidic and basic groups that may act as active sites when properly functionalized, IG is of great importance in catalysis. The nanocatalyst was created utilizing an ultrasonic-assisted technique, which involved 30 minutes of ultrasonication at a power of 150 W and a frequency of 35 kHz. Several methods were used to describe the core-shell structure of Fe$_3$O$_4$@GA@Isinglass that resulted. Fe$_3$O$_4$ nanoparticle production was verified by X-ray diffraction (XRD) analysis, with recognizable peaks seen at 2θ values of 30.2°, 35.6°, 43.1°, 57.1°, and 62.5°. Specific functional groups, including OH stretching at 3442 cm^{-1}, C=O stretching at 1713 cm^{-1}, and C—O stretching at 1226 cm^{-1}, were detected by Fourier-transform infrared spectroscopy (FTIR) analysis. The morphology of the Fe$_3$O$_4$@GA@IG in glass nanocatalyst was revealed by scanning electron microscopy (SEM) images, which revealed a core-shell structure with well-dispersed nanoparticles. The Fe$_3$O$_4$@GA@IG singlass nanocatalyst's catalytic efficiency was assessed during the production of 1,4-dihydropyridine and 4H-pyran derivatives. An ultrasonic apparatus with a probe sonicator was used to perform the reactions. Unfortunately, the cited study did not offer any more information on the particular ultrasonic device. At a power of 400 W and a frequency of 20 kHz, the gadget was in operation. Using an external

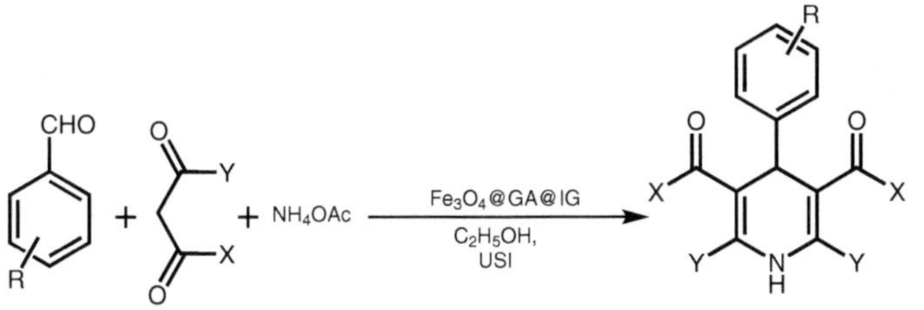

Figure 4.5 A simple protocol for the preparation of 1,4-DHPs mediated by innovative Fe_3O_4@GA@Isinglass catalytic systems and ultrasonic irradiation. Source: Pourian et al. [21].

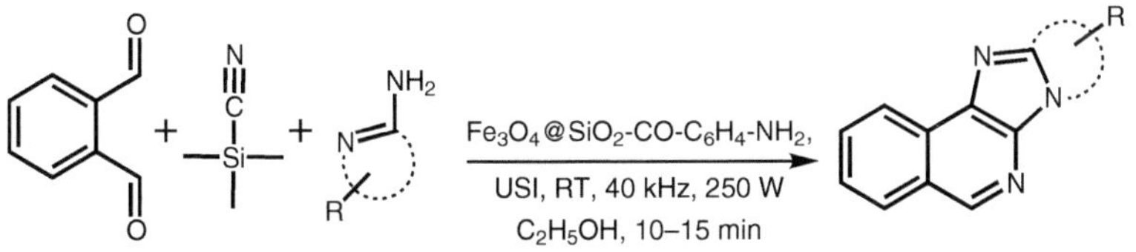

Figure 4.6 Synthesis of pyridoimidazoisoquinolines-imidazoles under US irradiation and Fe_3O_4@SiO_2-CO-C_6H_4-NH_2 nanocatalyst. Source: Maleki and Aghaei [22].

magnetic field, the Fe_3O_4@GA@IG catalyst was recovered, then washed with ethyl acetate, and dried. Surprisingly, the catalyst may be re-used up to six times before suffering a noticeable decline in catalytic performance (Figure 4.5) [21].

Fe_3O_4/NPs bolstered on SiO_2 altered with aminobenzoylchloride (Fe_3O_4@SiO_2-CO-C_6H_4-NH_2), a new nanocatalyst created by Maleki et al. was effectively used in the synthesis of pyridoimidazo-isoquinolines, aiding ultrasonication. Phthalaldehyde, trimethylsilylcyanide, and aminopyridines were condensed in ethanol. Initially, a frequency of 40 kHz along with 250 W power was used to irradiate the reaction with an initial catalytic quantity (5 mol%) of Fe_3O_4@SiO_2-CO-C_6H_4-NH_2. After only 10 minutes of reaction time, an astonishing 90% yield of the target product was attained. The reaction was also carried out with several additional catalysts, including protic solid acids, Lewis acids, liquid acids, Fe_3O_4, and Fe_3O_4@SiO_2, under the same circumstances to evaluate the activity of the newly created nano-material as a catalyst. Fe_3O_4@SiO_2-CO-C_6H_4-NH_2 demonstrated the strongest catalytic performance, yielding the highest reaction yield when all examples were compared. The Fe_3O_4@SiO_2-CO-C_6H_4-NH_2 nano-composite's simplicity in magnetic separation was another benefit. The nanocatalyst also showed amazing recyclability, being able to be utilized up to five times without significantly losing its catalytic activity (Figure 4.6) [22].

4.1.2 Copper-Based Catalysts

CuI@amine-functionalized halloysite (CuI@HNTs-2N), a novel method for the one-pot synthesis of propargylamines, was developed by the Sadjadi research team. The reaction was first carried out using reflux in EtOH with the nanocomposite

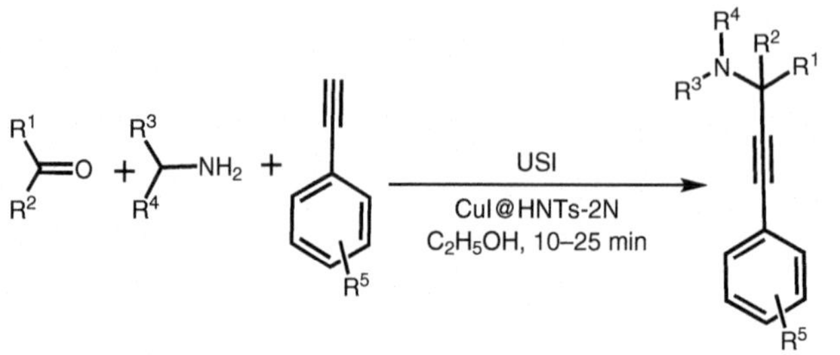

Figure 4.7 Sonicated protocol for the synthesis of propargylamines mediated by heterogeneous CuI@HNTs-2N catalyst. Source: Sadjadi et al. [23].

(present at 9.2 w/w%) acting as a catalyst, which produced a remarkable 93% of the product. Surprisingly, the intended product was produced with an even greater yield and a reaction time that was lowered by 1.5 times by integrating ultrasound while leaving all other variables the same. With a 150 W output power and a TT13 tip, a device was used to accomplish the ultra-sonication procedure. CuI@HNTs-2N also outperformed other catalysts for the same transformation that had been previously observed in the literature. This nano-composite showed remarkable catalytic effectiveness. Notably, there was no appreciable degradation in the catalyst's catalytic performance after recycling the catalyst and using it in up to five consecutive reaction cycles (Figure 4.7) [23].

Utilizing metal-supported nanocatalysts, Safa and colleagues carried out an inventive, one-pot Hantzsch DHP synthesis in an aqueous medium. They investigated a variety of transition metal catalysts (Mn, Fe, Co, and Cu) in their research, together with alumina (Al_2O_3), the aluminosilicate zeolite Zeolite Socony Mobil-5 (ZSM-5), and the silicoaluminophosphate molecular sieve SAPO-34. The experiment included a variety of solvents including ethanol, ethanol/water, acetonitrile, water, dichloromethane (DCM), toluene, and solvent-free conditions (120 °C) at ambient temperature. With a Cu–Fe/ZSM-5 bimetallic catalyst (at 3 w/w%) in a water medium under ultrasonic irradiation at room temperature, working at 20% power of the sonication equipment, the best results were obtained. Notably, the nanocatalyst performed quite well under these circumstances. Regarding sustainability, the catalyst's reusability was investigated, and encouragingly, even after experiencing three reaction cycles, no decline in its catalytic activity was found (Figure 4.8) [24].

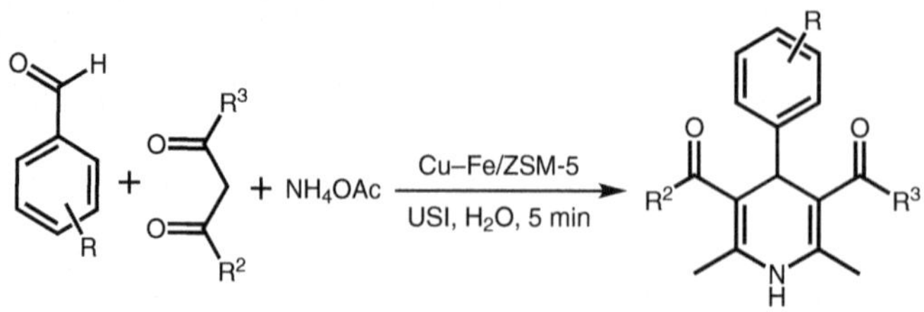

Figure 4.8 Protocol for the preparation of 1,4-DHPs mediated by green solvents, innovative Cu-Fe/ZSM-5 catalytic systems, and ultrasonic irradiation. Source: Safa et al. [24].

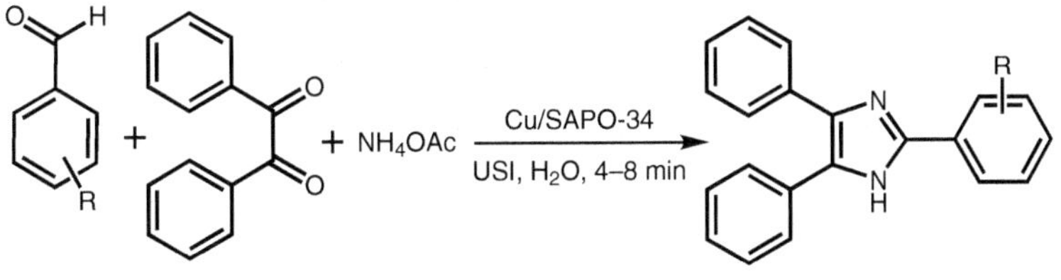

Figure 4.9 One-pot synthesis of substituted imidazoles mediated by ultrasound and Cu/SAPO-34 nanocatalysts. Source: Safa et al. [25].

Safa and colleagues developed a method based on silicoaluminophosphate (SAPO-n) zeolites, employing MCRs, with the aim of developing a streamlined and ecologically friendly method for the synthesis of imidazoles. They were able to effectively coordinate the synthesis of 16 different 1,2,4,5-tetraaryl-imidazoles. A Cu/SAPO-34 nanocatalyst (present at 5 wt%) in a water medium, served as the catalyst for this novel process, which also benefited from ultrasonic irradiation at room temperature. Using an ultrasonic homogenizer probe instrument, 20% power level ultrasonic irradiation was applied. This strategy demonstrates the potential of nanocatalysts in ecologically responsible synthesis methods and represents an effective and sustainable technique for synthesizing a variety of imidazole derivatives (Figure 4.9) [25].

4.1.2.1 Dihydropyrimidinones by Cu-Based Catalysts

A work of Ghasemzadeh and Ghaffarian demonstrates the creation and characterization of a $CoFe_2O_4$/OCMC/Cu(BDC) nanostructure (core/shell/shell) as a heterogeneous magnetic catalyst for the preparation of substituted quinazolines, acridines, and xanthenes under ultrasonic irradiation (Figure 4.10). The (220), (311), (400), (422), and (511) planes, respectively, correspond to the diffraction peaks detected at 2θ values of roughly 30.1°, 35.6°, 43.1°, 53.6°, and 57.1°, which corroborate the development of the $CoFe_2O_4$ core. The shape of the core/shell/shell nanostructure was visible in scanning electron microscopy (SEM) pictures, which also showed well-defined particles with a size range of around 60–80 nm. The existence of Co, Fe, C, O, and Cu elements in the nanostructure was validated by EDX analysis, confirming the system's composition of $CoFe_2O_4$/OCMC/Cu(BDC) (core/shell/shell). The nanostructure's magnetic behavior was shown by vibrating sample magnetometry (VSM) studies, with a saturation magnetization value of

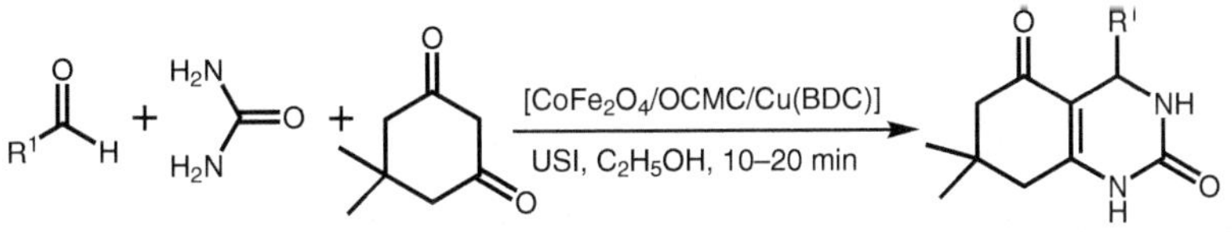

Figure 4.10 Sonicated synthesis of dihydropyrimidinone derivatives mediated by heterogeneous [$CoFe_2O_4$/OCMC/Cu(BDC)] catalyst. Source: Ghasemzadeh and Ghaffarian [26].

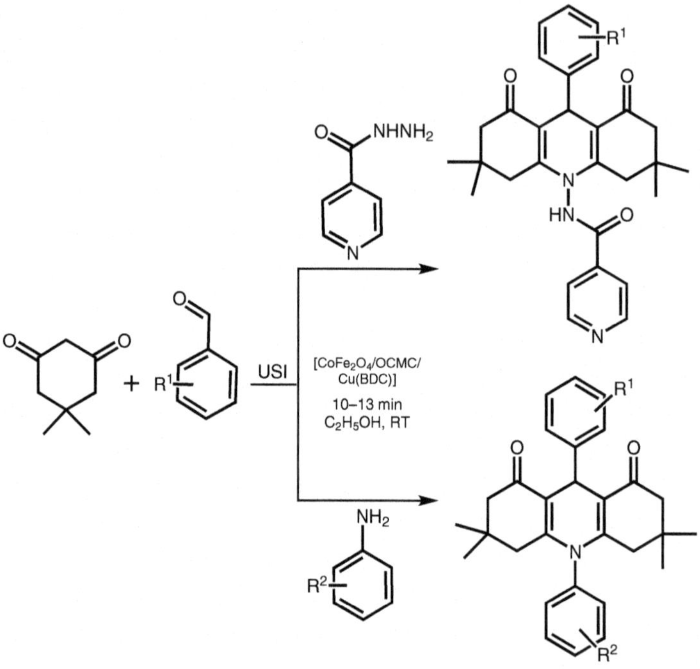

Figure 4.11 Sonicated synthesis of acridine derivatives mediated by heterogeneous [CoFe$_2$O$_4$/OCMC/Cu(BDC)] catalyst. Source: Ghasemzadeh and Ghaffarian [26].

about 30 emu/g, indicating the possibility of magnetic separation. Under ultrasonic irradiation, the CoFe$_2$O$_4$/OCMC/Cu(BDC) (core/shell/shell) nanostructure demonstrated effective catalytic activity for the production of substituted xanthenes, quinazolines, and acridines. A frequency of 35 kHz of ultrasonic irradiation was used to conduct the reactions. 80–95% of modified quinazolines, acridines, and xanthenes were produced in high yields, demonstrating the nanostructure's potency as a catalyst for these reactions. The effectiveness of the catalytic method was demonstrated by the quick (20–30 minutes) reaction time required to complete the reactions (Figure 4.11) [26].

4.1.2.2 Dihydroquinazolinones by Cu-Based Catalysts

Dihydroquinazolinones have been synthesized using ultrasound-mediated techniques that use adaptable and highly effective heterogeneous catalysts. Due to the vast range of biological activities that this bicyclic structure is connected with, such as its antibacterial and anticonvulsant capabilities, it has attracted a lot of attention. The study of Vasudhevan and Joel Karunakaran adds to the evaluation of the synthetic 2,3-dihydroquinazolinone derivatives' antibacterial properties against a panel of microorganisms, including bacteria and fungus [27]. The objective of the work by White et al. was to create novel pyrido[2,3-*d*]pyrimidinone compounds and evaluate their potential as anticonvulsant drugs. Measuring variables, including seizure threshold, length, and intensity, allowed researchers to evaluate the anticonvulsant

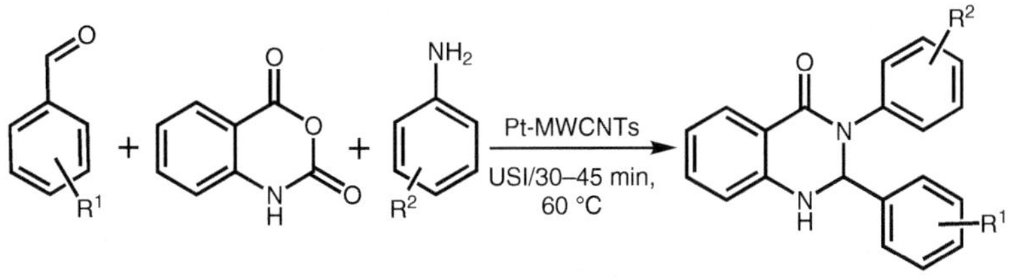

Figure 4.12 Platinum carbon nanotube/NPs hybrid catalyzed one-pot US-assisted synthesis of dihydroquinazolinones.

activity [28]. The multi-component reactions (MCRs) that involve aldehydes, isatoic anhydride, substituted anilines, and a nitrogen source are the topic of discussion in this section. The 2,3-dihydroquinazolin-4(1H)-one synthesis was promoted in the first method by a new hybrid nanocomposite of Pt-MWCNTs (0.93% – w/w) that was exposed to ultrasonic irradiation. The reaction was conducted independently with ultrasonic irradiation and reflux under the same circumstances through optimization studies. The outcomes showed that under ultrasonic irradiation, the nano-composite displayed quicker reaction times and greater yields. Further, the nanocomposite had shown cyclability, permitting up to four reaction cycles without noticeably losing catalytic activity. These results highlight the applicability of hybrid catalytic systems based on Pt-MWCNTs for the efficient synthesis of dihydroquinazolinones as well as the possibility of ultrasound-assisted reactions (Figure 4.12). CNTs and MWCNTs provide excellent support materials for heterogeneous catalysis due to their chemical stability, high surface area, excellent electron conductivity, and recyclability.

In their study, Maleki et al. describe the synthesis of dihydroquinazolinones using an ultrasonic-treated hybrid magnetic composite nanostructure made of $CoFe_2O_4@B_2O_3-SiO_2$. The characteristic diffraction peaks observed at 2θ values of approximately 30.1°, 35.6°, 43.1°, 53.6°, and 57.1°, corresponding to the (220), (311), (400), (422), and (511) planes, respectively, provided conclusive evidence that the $CoFe_2O_4@B_2O_3-SiO_2$ nanostructure had formed. The morphology of $CoFe_2O_4@B_2O_3-SiO_2$ was shown in scanning electron microscopy (SEM) pictures, which showed well-defined nanoparticles with diameters ranging from 40 to 80 nm. The morphology of the $CoFe_2O_4@B_2O_3-SiO_2$ nanostructure was disclosed by transmission electron microscopy (TEM) pictures, which showed well-defined particles with a size range of around 10–20 nm. Functional groups including hydroxyl (–OH) and boron oxide (B O) groups that are present in $CoFe_2O_4@B_2O_3-SiO_2$ were identified by FTIR research. Under ultrasonic irradiation, the $CoFe_2O_4@B_2O_3-SiO_2$ nanostructure demonstrated effective catalytic activity for the production of dihydroquinazolinones. The ultrasonic treatment was applied to the reactions at a 35 kHz frequency. The reactions were carried out under ideal circumstances, which included a reaction duration of 60 minutes and a temperature of 90 °C. With yields ranging from 78% to 93%, the catalyst produced high yields of dihydroquinazolinones as a result of its good performance. The synergistic impacts of $CoFe_2O_4$'s

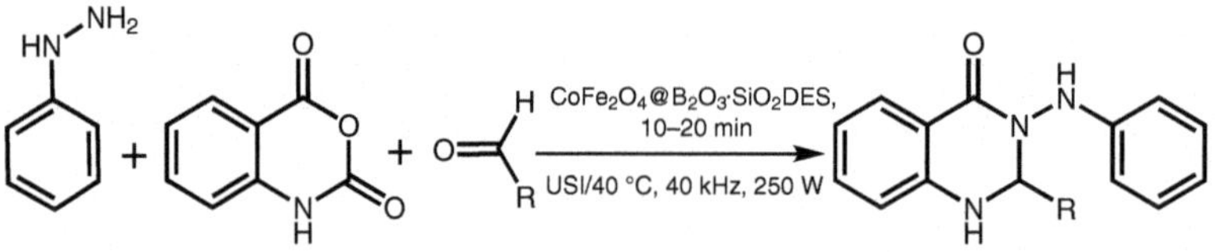

Figure 4.13 Ultrasound-promoted preparation of dihydroquinazolinones catalyzed by $CoFe_2O_4@B_2O_3.SiO_2$. Source: Maleki et al. [29].

magnetic qualities and the catalytic abilities of B_2O_3–SiO_2 allowed the reactions to continue successfully (Figure 4.13) [29].

4.1.3 Misalliances Metal-Based Catalysts

The research by Jadhav et al. reveals the use of $ZnFe_2O_4$ nanoparticles as a catalyst for the multi-component synthesis of tetrahydropyranoquinoline in a single pot under clean ultrasonication. The $ZnFe_2O_4$'s morphology was disclosed by SEM pictures, which showed well-defined particles with a size range of around 50–70 nm. Zn and Fe elements were identified in the nanoparticles by EDX research, verifying the $ZnFe_2O_4$ composition. The crystalline nature of the $ZnFe_2O_4$ nanoparticles was verified by X-ray diffraction (XRD) analysis. Diffraction peaks were found at 2θ values of about 30.1°, 35.6°, 43.1°, 53.6°, and 57.1°, which correspond to the (220), (311), (400), (422), and (511) planes, respectively. Under neat ultrasonic irradiation, the $ZnFe_2O_4$ nanoparticles showed effective catalytic activity for the single-pot reaction as a multi-component synthesis of substituted tetrahydropyranoquinoline. The reactions were conducted in a clean environment, without the use of extra solvents, at room temperature. An ultrasonic bath of 5.5 l liquid holding capacity, running at a frequency of 50 Hz, and a temperature of 70 °C was used to carry the reaction. The Jadhav group used a heterogeneous and recyclable $ZnFe_2O_4$ (10 mol%) nanocomposite as the catalyst to enable the efficient production of THQ compounds. It is noteworthy that without the catalyst, the model process would stall. The nano-$ZnFe_2O_4$ catalyst was retrieved after the process was finished and was discovered to still have catalytic activity. It might be used once again up to five times without significantly losing catalytic performance. With yields of substituted tetrahydropyranoquinoline compounds ranging from 83–95%, illustrate the efficiency of $ZnFe_2O_4$ nanoparticles as a catalyst. The reactions moved along quickly, taking only 15–25 minutes to complete (Figure 4.14) [30].

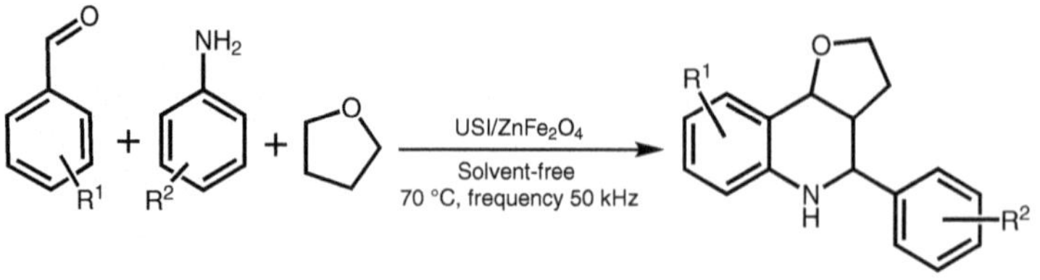

Figure 4.14 Ultrasound-mediated MCR for the synthesis of tetrahydropyranquinolines using a $ZnFe_2O_4$ nanocatalyst. Source: Jadhav et al. [30].

In the study of Sadeghzadeh, the use of a dipyridine complex immobilizing gold(III) on SBA-15 as a nanocatalyst for the environmentally friendly production of thiazoloquinolines at room temperature is highlighted. The SEM investigation revealed that the well-defined particles in the SBA-15 support material had a size range of around 150–300 nm. Gold (Au) was identified by EDX analysis in the immobilized gold(III) dipyridine complex on SBA-15, demonstrating the effectiveness of the immobilization procedure. The SBA-15 support material was also connected with additional elements, such as silicon (Si) and oxygen (O), which are essential components of silica-based materials, according to the EDX study. A dipyridine complex immobalizing-Au(III) on silica SBA-15 SBA-15/[AubpyCl$_2$]Cl is used as a nanocatalyst in production, which takes place in an aqueous medium at room temperature. Aldehydes, dimedone, -enolicdithioesters, and cysteamine, among others, all participate in the process. It was discovered throughout the optimization process that the sonicated reaction produced no product when the nanocatalyst was absent. Additionally, due to the creation of by-products, using a free gold-pyridine combination under the same circumstances only produced modest quantities of the desired product. The current methodology and various thiazoloquinoline synthesis options were compared in comparative research. It was determined that the current approach, especially at lower temperatures, shows superiority in catalytic quantity and the duration of the reaction. Additionally, the recovered SBA-15/[AubpyCl$_2$]Cl nanocatalyst demonstrated sustained catalytic activity across seven successive cycles, showing its potential for long-term reuse with little to no efficiency loss (Figure 4.15) [31].

Safari et al. investigated MgAl$_2$O$_4$/NPs as an efficient heterogeneous catalyst in ethanol for the simple and safe preparation of imidazoles. The reaction proceeded under identical circumstances utilizing both reflux and ultrasonic irradiation, employing a temperature of 60 °C at 50 kHz and 200 W. This allowed us to compare this method to more traditional methods. In all instances, compared to conventional procedures, the application of ultrasonic waves made it possible to generate the necessary material at a faster rate with high yields, and with faster reaction times. Notably, the method was cost-effective because of the extremely low catalyst loading need (0.035 mol%). The moderate reaction conditions, straightforward operation, and quick work-up procedure, all contributed to the protocol's extra benefits. It is crucial to note, however, that the authors made no mention of the recycling and

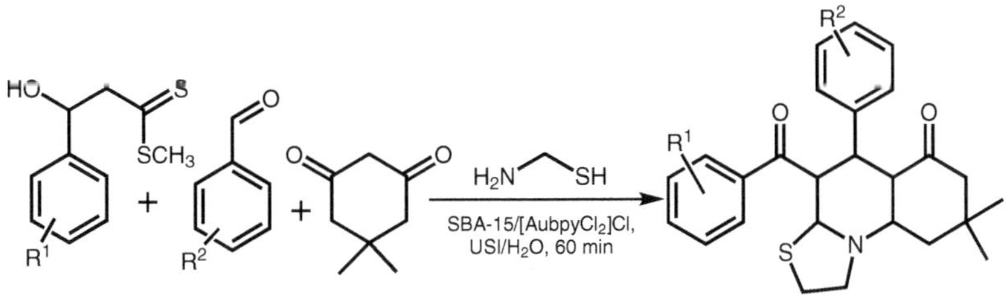

Figure 4.15 On water preparation of thiazoloquinolines using SBA-15/[AubpyCl$_2$]Cl nanocatalyst under ultrasonic irradiation. Source: Sadeghzadeh [31].

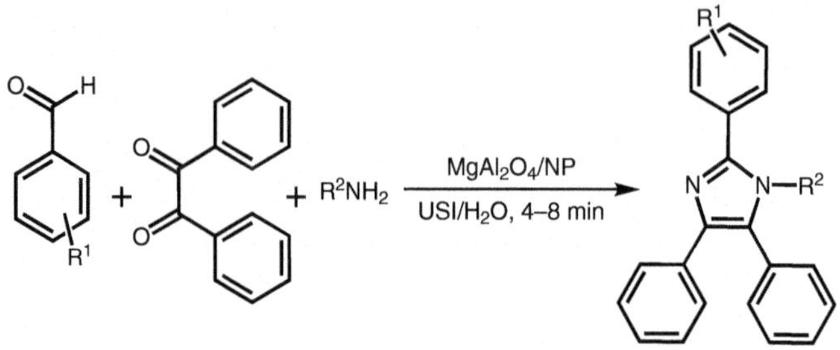

Figure 4.16 Synthesis of substituted imidazoles mediated by $MgAl_2O_4$/NP and USI. Source: Safari et al. [32].

reuse of the $MgAl_2O_4$/NPs nanocomposite catalyst, which is an important issue to look into for the method's overall sustainability and viability. The catalyst's capacity to be recycled would increase its cost-effectiveness, lower waste production, and increase its potential as an ecologically benign method for synthesizing imidazole derivatives (Figure 4.16) [32].

$NiFe_2O_4$/FMNPs were used by the Hajizadeh group to create a cutting-edge green nanocatalyst, by integrating them into a bentonite geopolymer substrate and studying the catalytic characteristics of these ferromagnetic nanoparticles. They used this nanocatalyst to create imidazoles when ethanol (EtOH) was present. The researchers looked at each element independently in order to comprehend the distinct catalytic activities of bentonite, geopolymer, and $NiFe_2O_4$/FMNPs. According to the findings, the geopolymer had modest catalytic activity whereas bentonite showed no discernible activity. However, under the same reaction environment, it was discovered that the catalytic ability of the hybrid nanocomposite $NiFe_2O_4$/FMNPs geopolymer was 3.25 times higher(at a concentration of 6.8 w/w%) than that of the isolated geopolymer. The researchers hypothesized that a harmonious interaction among the OH groups found in the geopolymer assembly and the $NiFe_2O_4$/FMNPs was responsible for the improved efficiency of this catalytic material. When compared to employing the geopolymer or $NiFe_2O_4$/FMNPs independently, the catalytic activity was likely enhanced by the synergistic relationship, resulting in improved performance. Following the reaction, a straightforward magnetic separation procedure could be used to separate the nanocomposite, which could then be washed with EtOH and dried. The nanocomposite's extraordinary recyclability and stability were demonstrated by the fact that it maintained its catalytic activity even after eight further reaction cycles (Figure 4.17) [33].

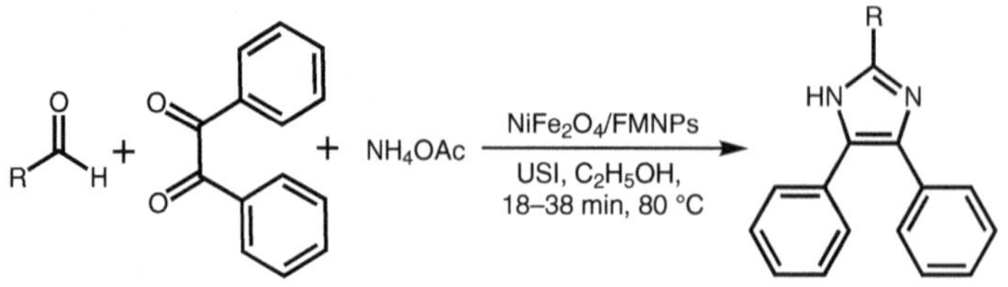

Figure 4.17 Sonochemical preparation of imidazoles catalyzed by $NiFe_2O_4$/FMNPs. Source: Hajizadeh et al. [33].

4.2 Microwave-Assisted Reactions

Due to the distinctive interactions involving specific substances and high-frequency electromagnetic waves, the use of MWs has an array of advantages. Water, alcohols, ionic liquids, and carbon are examples of materials that have a strong ability to absorb MW radiation [34]. On the other hand, MWs can hardly penetrate materials like plastics, oils, and nonpolar solvents. The materials' dielectric characteristics are the only factor influencing this behavior. As a result, MWs may penetrate a mixture deeply and interact in various ways with its many phases. When compared to non-sensitive materials, components susceptible to MW radiation suffer rapid and selective heating, resulting in considerable temperature differences [35]. Therefore, the energy is concentrated inside a mixture's microwave-responsive components, limiting direct heating of the nearby elements. The volumetric character of MW heating, which addresses heat transfer restrictions frequently experienced in systems, depending on heat conduction and convection to transfer heat across solid barriers, is a key feature of this technology. Therefore, if the energy input is kept constant, MW heating may result in orders of magnitude shorter heating durations than traditional heating. Chemists have made considerable use of this quality, and various literature articles demonstrate how MW radiation may speed up chemical processes [36].

The Arrhenius law states that rate constants vary exponentially with the reciprocal of temperature, causing significant changes in reaction speed in response to small temperature changes. However, it can be difficult to precisely measure temperature fluctuations at the microscopic level within chemical phases during reactions. Discussions on the precise interactions between waves and materials have been generated by this inability to quantify temperature accurately [37]. However, it is well-accepted that localized temperatures within reaction mixtures can noticeably exceed the bulk temperature. The acceleration of microwave-assisted reactions, seen in many circumstances, is probably explained by these phenomena [38, 39].

The electric field interacts with charged particles like free charges and induced dipoles that are present in the material to produce MW heating in nonmagnetic materials [40]. These species' polarization is crucial to this interaction [41]. Material heating at radiofrequency and MW frequencies is primarily caused by two methods.

- In the GHz frequency range, dipolar reorientation or polarization is the main process. Induced dipoles are affected by the electric field, forcing them to rotate until electrostatic interactions bring them into equilibrium. Although molecules' orientation can vary, dipole–dipole interactions limit how quickly they can move, limiting full relaxation or reorientation. Because of the friction caused by the movement of the species, heat is produced, creating a thermal effect. Because it takes more time to completely rotate the whole molecule and the polar group than it does to transmit energy, a disturbed system does not quickly return to equilibrium before the next energy input [42].
- The second process, which is more noticeable at lower frequencies, results from the existence of ionic species in solid materials or ionic liquids. Anions

and cations move in opposite directions as a result of the applied electric field, producing a net dipole moment. An extra heating mechanism linked to interfacial or Maxwell–Wagner polarization may appear in heterogeneous systems. This polarization results from the build-up of charges at interfaces or contact points between several components. It happens because different substances' conductivity and dielectric characteristics differ at these contacts. Charge build-up therefore causes field distortions and consequences of dielectric heating [42].

4.2.1 Solid Acid and Base Catalysts

4.2.1.1 Condensation Reactions

The use of heterogeneous catalysts that are transition metal-free, in combination with MW assistance, has been a continuing and active field in organic synthesis due to the tremendous benefits of these systems. The extraordinary MW absorption characteristics of solid acid catalysts, such as metal oxides, zeolites, and clays, make them suitable for microwave-assisted reactions. Due to the fact that the catalytic material itself acts as a quick internal heating source, this compatibility produces a synergistic effect. These catalysts also have the added advantage of being safe for the environment and, in most cases, easily recyclable [43].

A common family of transformations in organic chemistry, known as condensation reactions, is defined by the successive occurrence of addition and elimination events. Depending on the nature of the reaction, either acids or bases can catalyze it. A technique for the synthesis of -amino-, -unsaturated esters and ketones was developed by Marvi and colleagues using MW irradiation and montmorillonite K-10 as the catalyst. This method demonstrated outstanding selectivity, efficiency, and speed under moderate and solvent-free reaction conditions. When different dicarbonyl compounds were put through reactions with a variety of aromatic amines, the ensuing -amino-, -unsaturated esters or ketones were produced with yields of 90% or higher in a matter of minutes (Figure 4.18) [44].

Montmorillonite KSF exhibited effective catalysis of cross-aldol condensation processes using MW activation. Following tests on substituted acetophenones of various kinds, chalcones were produced with yields ranging from fair to outstanding. The yields that were attained were equivalent to those that were attained utilizing traditional methods that used strong acids or bases. Surprisingly, the technique showed tolerance for both electron-donating and electron-withdrawing substitutes. Reaction

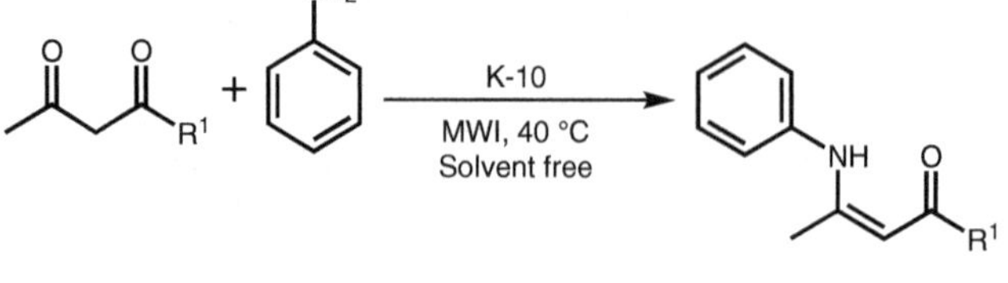

Figure 4.18 Montmorillonite K-10 assisted condensation reaction assisted by MWI. Source: Marvi et al. [44].

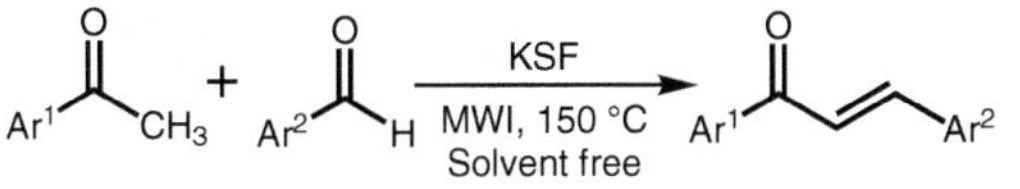

Figure 4.19 Montmorillonite KSF catalyzed cross-aldol condensation reaction by MWI. Source: Adapted from [45, 46].

times were sped up, and conversion was improved, thanks to the use of MW irradiation. Additionally, a greener version of this reaction using homogeneous catalysis was also described, with the same tidy conditions and temperature ranges of 100–150 °C used as in the KSF-catalyzed aldol condensation (Figure 4.19) [45, 46].

The condensation process between trifluoromethyl acetophenones and methylbenzylamines was made easier by the use of montmorillonite K-10 as a catalyst. After the solvent employed for mixing under vacuum was eliminated, this microwave-aided reaction took place in the solid phase. The dry mix was subsequently cooked for 20–50 minutes at 175 °C, producing products with high yields. In comparison, over a lengthy duration of 168 hours, the typical process using oil bath heating only produced 75–80%. It is noteworthy that this approach worked well whether there were electron-donating or electron-withdrawing substituents present (Figure 4.20) [47].

Utilizing both their unique qualities and their MW absorption capabilities, certain infrequent bio-traced compounds were used as solid heterogeneous catalysts. Notably, the condensation reaction between furfural and acetone has been carried out using chitosan beads. Even after numerous reuse cycles, the catalyst continued to function, although it is important to note that the yields steadily decreased (Figure 4.21) [48].

Synthetic polymers have found widespread use in the field of heterogeneous catalysis because of their beneficial traits of being simple to handle and recyclable, especially in reactions carried out in a single-phase media. Polymers are valuable provision resources that may be made to have basic or acidic properties to aid catalytic reactions. A technique using a catalyst based on polystyrene, functionalized with

Figure 4.20 Montmorillonite K-10 assisted condensation reaction assisted by MW irradiation. Source: Singh and Sharma [47].

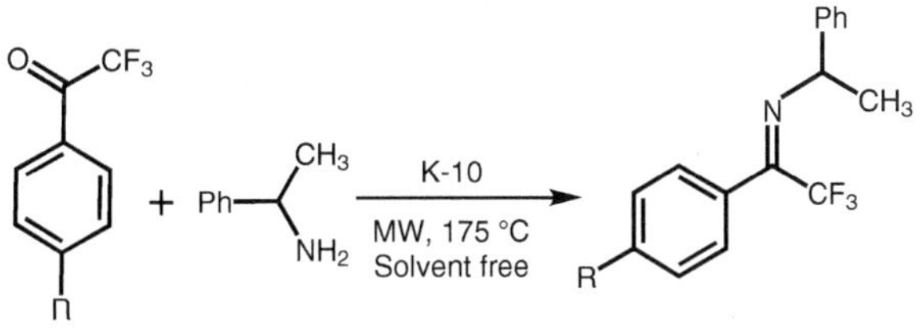

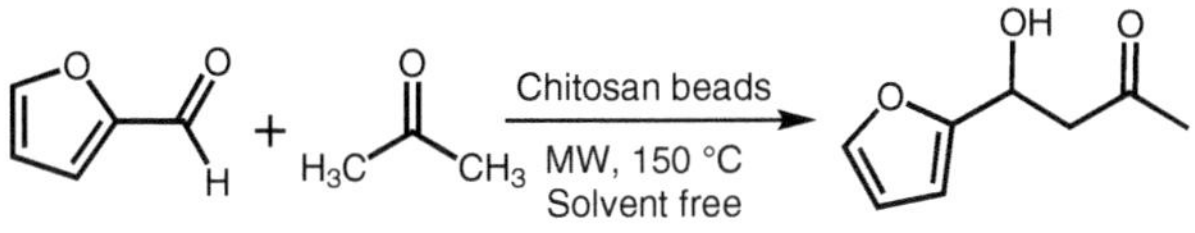

Figure 4.21 Chitosan beads assisted condensation reaction under microwave irradiation. Source: Dasgupta et al. [48].

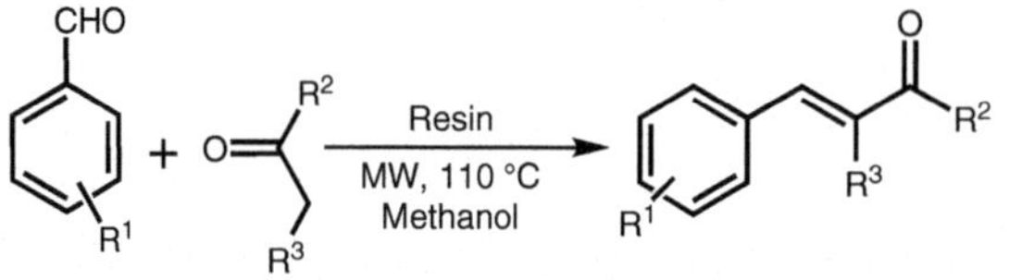

Figure 4.22 Resin-assisted condensation reaction by microwave irradiation. Source: Liao et al. [49].

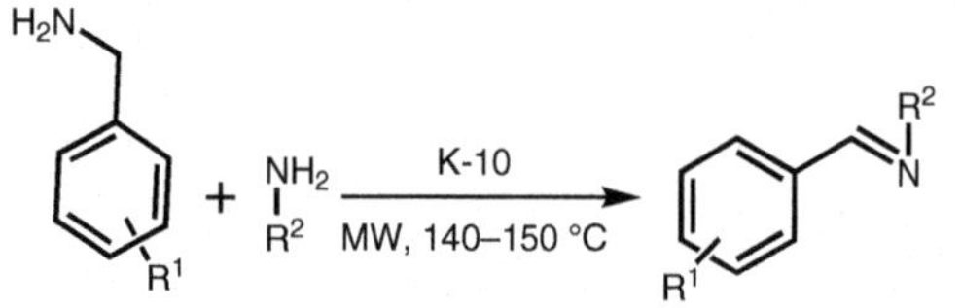

Figure 4.23 Montmorillonite K-10 assisted condensation reaction under microwave irradiation. Source: Atanassova et al. [50].

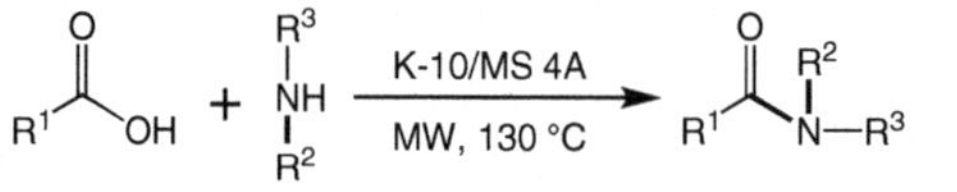

Figure 4.24 Montmorillonite K-10 catalyzed amidation under microwave irradiation. Source: Kumar et al. [51].

an imino group for the aldol reaction, was presented by Zhu et al. Diverse carboxyls were transformed into the respective products via aldol condensation under the reaction conditions at 110 °C, with yields ranging from moderate to good. The imino group most probably contributes to the beginning of the nucleophilic ketonic carbon attack on the aldehyde's carbonyl group. The yield considerably improved by aiding MW, with a faster reaction rate (Figure 4.22) [49].

Montmorillonite K-10 was appreciated as a catalyst for transforming amines to imines via oxidative coupling under the influence of MWs. At a temperature of 150 °C, benzylamines easily performed homocoupling reactions, and also extended to various anilines or alkylamines, with good to exceptional yields. Notably, the self-coupling product of 1° benzylamine was regularly found and played a key role as an intermediate in the transamination pathway that led to the creation of mixed products. Ammonia was found to be released during the reaction, according to thermo-gravimetry analysis of the reaction process. Additionally, DRIFT spectroscopy demonstrated that benzylamine underwent an initial oxidation/hydrolysis that produced benzaldehyde, which then underwent additional reactions with the remaining benzylamine to produce the crucial intermediate (Figure 4.23) [50].

With the aid of a groundbreaking microwave-assisted technique, the wide range of uses for montmorillonite K-10 may now be expanded to include the straight amidation of carboxylic acids among amines. With regard to a wide range of carboxylic acids, including substituted, aromatic, aliphatic, aliphatic conjugated, and unsaturated acids, this approach performed exceptionally well. The adaptability and potential utility of this method are further increased by the fact that the chirality of asymmetric substrates was successfully maintained throughout the course of the reactions (Figure 4.24) [51].

4.2.1.2 Cyclization Reactions

A wide variety of chemical transformations known as cyclization processes are used to produce compounds with just carbon or heteroatom-containing rings. They are

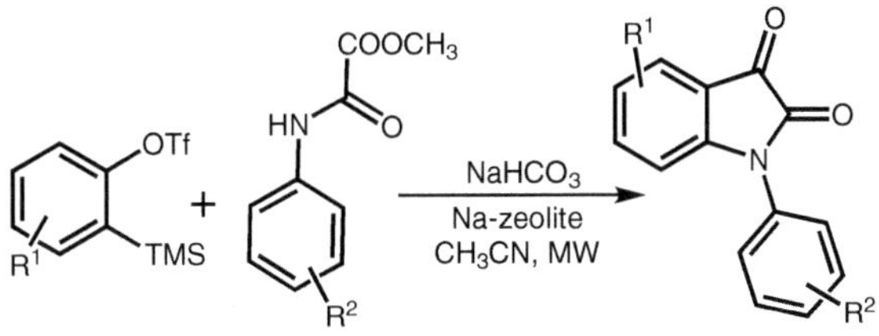

Figure 4.25 Na-zeolite assisted cyclization reaction under microwave irradiation. Source: Singh and Kumar [53].

distinguished by a variety of mechanisms [52]. Although many of these reactions, particularly condensation reactions, include numerous phases, this section emphasizes the final result of these instances to underline their relevance.

A zeolite was used as the catalyst in Singh et al.'s easy microwave-assisted synthesis of N-arylisatins. Comparing this strategy to more traditional, often drawn out, time-consuming procedures, it showed a dramatically decreased reaction time. With good yields and in a matter of minutes, arynes and methyl-2-oxo-2-arylaminoacetate were transformed into N-arylisatins. It is hypothesized that the heterogeneous catalyst makes it easier for the nitrogen to attack the aryne nucleophilically, boosting the cyclization of the molecule by pushing out the methyl group and producing the desired isatin product. The yield of the matching isatin was slightly impacted by electron-deficient groups, but the reaction's overall effectiveness was still rather promising (Figure 4.25) [53].

In a different study, it was discovered that the use of zeolites and MW irradiation to produce 1,5-benzodiazepines was quite efficient. Particularly, under solvent-free and moderate temperature circumstances, the use of HY zeolite as a catalyst made it easier to produce benzodiazepines from substituted ketones and o-phenylenediamines [54]. With an 86% yield still present, even after the sixth cycle, this ecologically friendly technology produced outstanding product yields and allowed the catalyst to be recycled across numerous subsequent reactions. A new catalyst-free method has also been described for the synthesis of comparable benzodiazepines utilizing o-phenylenediamines and alkynones, which is noteworthy given that this technique generates the required benzodiazepines with favorable efficiency (Figure 4.26) [55].

Several cyclization reactions were examined in thorough research on the energy competence of heterogeneous catalytic reactions under the influence of MWs in organic processes. For instance, using MW irradiation, montmorillonite K-10 was utilized for anilines to be cyclized with 2,5-hexanedione yielding virtually measurable yields in under one minute while using just 0.01 kWh. While using four times with much energy, the identical process carried out under standard heating

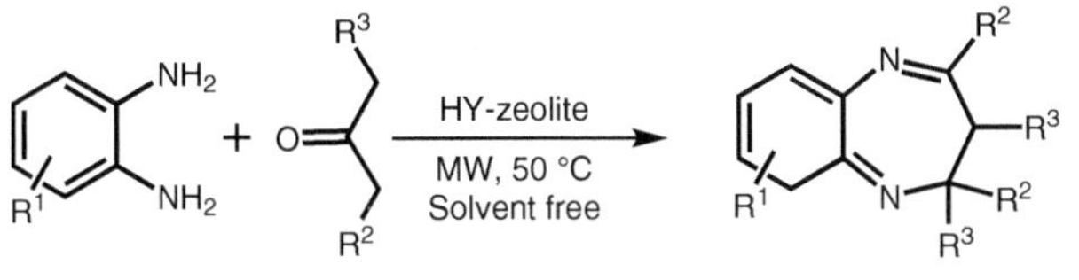

Figure 4.26 HY-zeolite assisted cyclization reaction under microwave irradiation. Source: Jeganathan and Pitchumani [55].

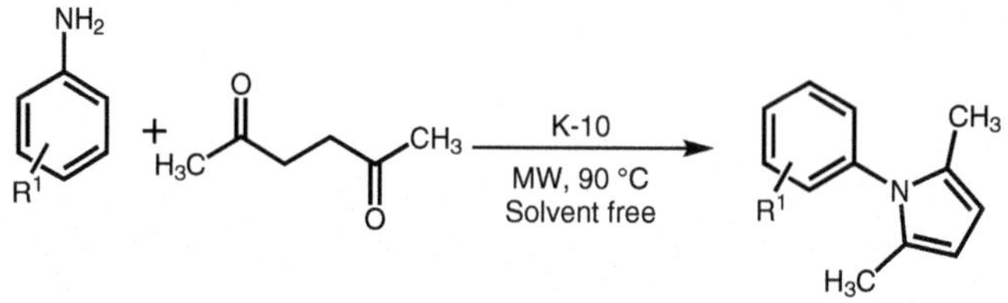

Figure 4.27 Montmorillonite K-10 assisted cyclization reaction under microwave irradiation. Source: Solan et al. [56].

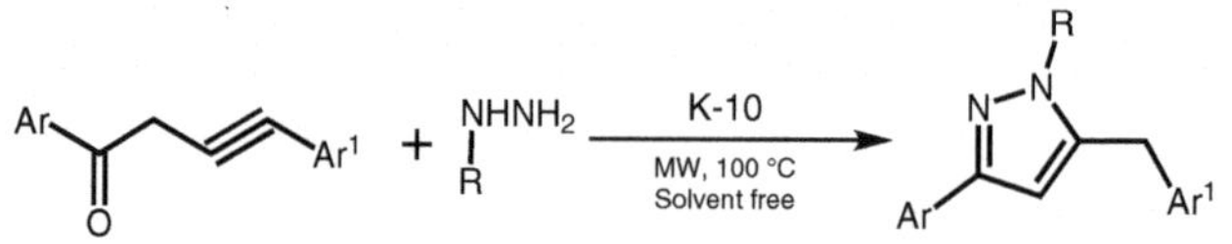

Figure 4.28 Montmorillonite K-10 assisted cyclization reaction under microwave irradiation. Source: Borkin et al. [57].

settings produced a 79% yield after 30 minutes. In terms of energy efficiency, the majority of the study's cases preferred MW irradiation over traditional heating (Figure 4.27) [56].

The use of Montmorillonite K-10 in the production of trisubstituted pyrazoles was also proven. This procedure involved reacting propargyl ketones and hydrazines, leading to the production of pyrazoles by condensation followed by a cyclization mechanism. By making the carbonyl group of the ketone more electrophilic, the catalyst is thought to enhance the mechanism's first step. MW irradiation was used to help produce remarkably high yields in solvent-free settings. Further evidence of the reaction's effectiveness came from the observation of substantial region selectivity (Figure 4.28) [57].

The use of silica organic transformations, including metal catalysts as an espousal material has become very common. Additionally, it has been effectively coupled with organo catalysts to offer synthesis methods without the need for metals. When salicylaldehydes and dialkylacetylenedicarboxylates were combined, Valizadeh et al. created hydroxychromenes in a way that is ecologically friendly. A dehydration–rehydration sequence was suggested to be responsible for the production of two separate regioisomers, with a common intermediate produced during dehydration, acting as the precursor. Imidazole-functionalized silica nanoparticles were found to be more efficient than metal-based catalysts like ZnO or $TiCl_4$, after testing a variety of catalysts. The contact between the salicylaldehyde's hydroxyl group and the nitrogen of the restrained imidazole, which increased the reactant's nucleophilicity, was thought to be the catalyst's mechanism of action. Furthermore, because the catalyst was solid, it was simple to recover and reuse it over numerous subsequent cycles without seeing a material loss in activity up to the fifth cycle. The use of MW irradiation significantly sped up the process and produced high starting material conversion (Figure 4.29) [58].

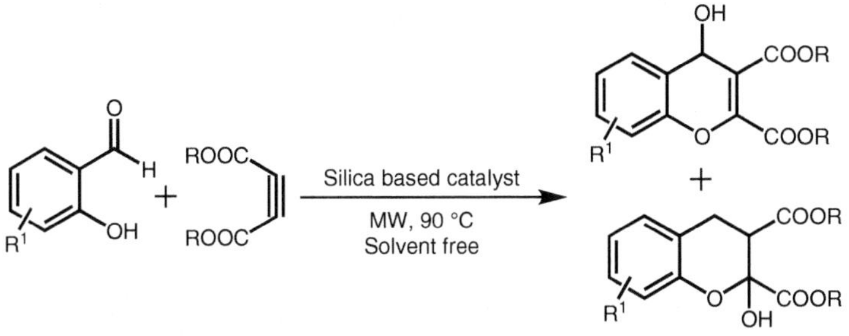

Figure 4.29 Montmorillonite K-10 assisted cyclization reaction under microwave irradiation. Source: Valizadeh et al. [58].

For organic transformations, silica nanoparticles have proven to be efficient catalysts, demonstrating improved activity at the nanoscale without the requirement for additional catalytic materials to be immobilized. Quinolines were synthesized utilizing silica nanoparticles by the reaction among aminoaryl ketones and carbonyl compounds using a process that was both remarkably effective and solvent-free. With this approach, large yields of quinolines were produced in 15 minutes or less reaction time. Due to their natural acidity and capability to establish H^+ bonds, which are essential for promoting the transformation, silica nanoparticles significantly contributed to the process (Figure 4.30) [59].

In order to successfully create pyrrole derivatives, K-10 catalysis was activated under MW. These created compounds were looked at as possible fructose 1,6-bisphosphatase inhibitors. In particular, carbomethoxy-phenyl-sulfonamide and dimethoxytetrahydrofuran were combined to create (carbomethoxy-phenylsulfonyl)-pyrrole. This process produced high-quality product yields in a matter of minutes, proving its effectiveness (Figure 4.31) [60].

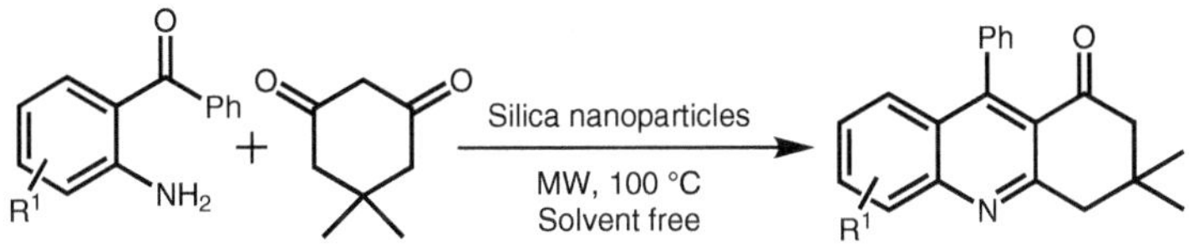

Figure 4.30 Montmorillonite K-10 assisted cyclization reaction under microwave irradiation. Source: Hasaninejad et al. [59].

Figure 4.31 Montmorillonite K-10 assisted cyclization reaction under microwave irradiation. Source: Rudnitskay et al. [60].

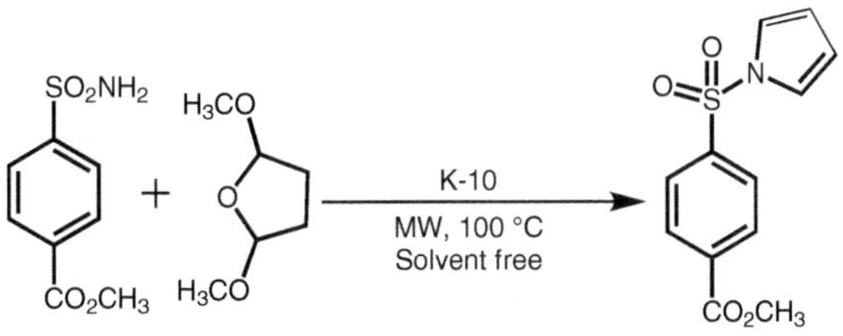

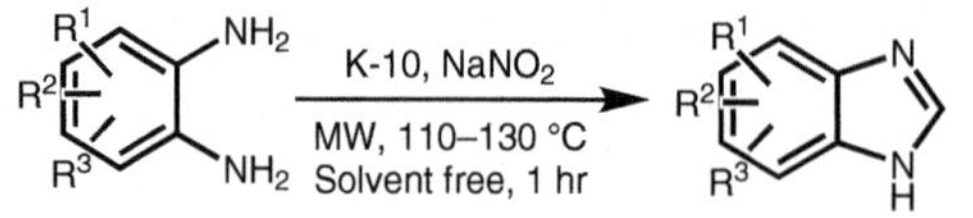

Figure 4.32 Montmorillonite K-10, NaNO$_2$ assisted cyclization under microwave irradiation. Source: Kokel and Török [61].

A solid-phase diazotation method, catalyzed by montmorillonite K-10, has been established for the fabrication of benzotriazoles by means of substituted phenylenediamines. The reaction consistently delivered brilliant yields at a faster reaction rate of just one hour, while exhibiting minimal substituent effects. Remarkably, the catalyst displayed recyclability, maintaining its activity, and providing yields of over 95% across several repeated reactions. Additionally, the authors conducted an assessment of the energy competence of the procedure, revealing that the microwave-aided protocol consumed around half the energy associated with the conservative oil-bath activation (Figure 4.32) [61].

4.2.1.3 Multi-component Reactions

MCRs have a lengthy historical past, similar to many reactions included in this article. The Hantzsch and Biginelli reactions, for example, were originally documented in the late 1800s, but the Ugi reaction was more recently discovered and has been well-known since the late 1950s. Although simple, the idea of multicomponent reactions is quite appealing to green synthetic chemists. Integrating many starting materials in a single reaction vessel enables the synthesis of reasonably complicated structures without the need for time-consuming, stepwise syntheses that can include several isolation and purification stages [62].

Utilizing montmorillonite K-10 under a microwave-assisted process, substituted quinolines were effectively produced. This synthesis produced quinolines by generating an intermediate imine through an MCR. The imine then interacts with phenylacetylene, goes through intra-molecular cyclization, and eventually goes through oxidation to become an aromatized compound. By considerably reducing the reaction time through the use of MW irradiation, a more effective synthesis procedure was made possible (Figure 4.33) [63].

Using chitosan as the catalyst, Khan and Siddiqui created an eco-friendly, single-pot synthesis in a multi-component system for functionalized imidazoles. The stability of chitosan in both acidic and basic environments allows for pre-processing to improve its catalytic activities. In this procedure, chitosan was pre-treated with sulfuric acid. A broad variety of aldehydes and amines could be

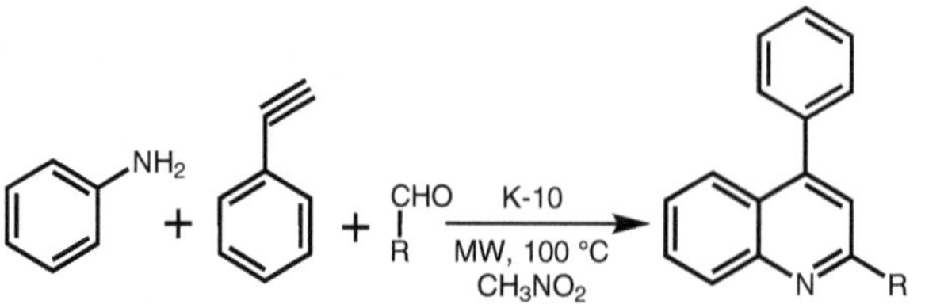

Figure 4.33 Montmorillonite K-10 assisted MCR reaction under microwave irradiation. Source: Biggs-Houck et al. [63].

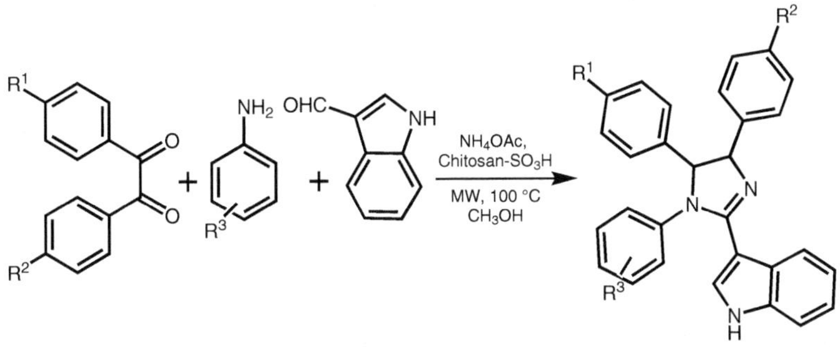

Figure 4.34 Chitosan-SO$_3$H assisted MCR reaction under microwave irradiation. Source: Kulkarni and Török [64].

converted with this method, and the resulting substituted imidazoles were produced with magnificent yields. The catalyst also proved to be reusable for up to five cycles. Although MWs increased the product yields in this specific synthesis, they did not considerably shorten the reaction durations (Figure 4.34) [64].

A desirable material for catalyst support, graphene oxide demonstrates fascinating qualities, including strong mechanical strength and great physicochemical stability. Its abundance of functional groups that include oxygen also makes functionalization simple. In a specific application, pyridin-pyrazolo-pyridine-5-carbonitriles were synthesized in a eutectic medium, influenced by MWs using GO-SO$_3$H as a catalyst. The addition of a magnetic component improved catalyst recovery. When different benzaldehydes with nucleophilic and electrophilic substituents were combined with oxo-pyridinyl-propanenitrile and phenyl-pyridinylpyrazol-5-amine, the desired compounds were produced in large quantities. With yields surpassing 90%, magnetic removal of the catalyst made it easy for reusability till eight cycles (Figure 4.35) [65, 66].

The use of MgAl$_2$O$_4$ nano-crystals as a catalyst for the production of 2,4,6-triarylpyridines proved to be quite successful. When compared to other conventional catalysts used in this MCR, including NaOEt or K$_2$CO$_3$, this unique catalyst demonstrated more activity. Aromatic aldehydes, ammonium acetate, and acetophenone derivatives were used as the reactants in the single-pot system with

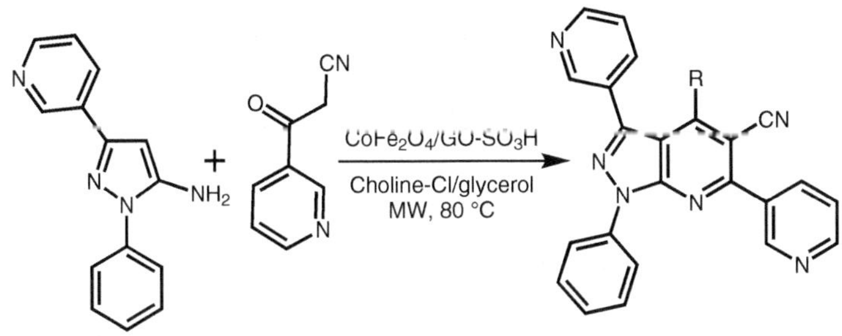

Figure 4.35 CoFe$_2$O$_4$/GO-SO$_3$H assisted MCR reaction under microwave irradiation. Source: Adapted from [65, 66].

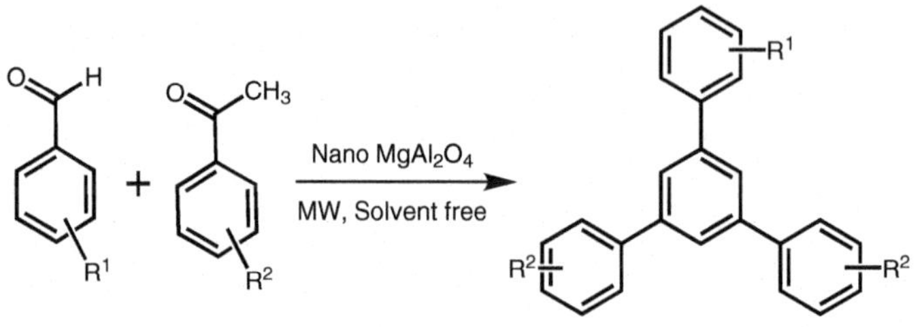

Figure 4.36 Nano MgAl$_2$O$_4$ assisted MCR reaction under microwave irradiation. Source: Safari et al. [67].

a no-solvent environment under microwave-assisted conditions. All the stages in the reaction pathway, such as aldol condensation followed by Michael addition and cyclization, ultimately leading to oxidation, were shown to be made easier by MgAl$_2$O$_4$ nanocrystals. While an unsubstituted precursor was used as a model substrate, the catalyst astonishingly showed recyclability by sustaining yields of over 90% even after the fifth consecutive cycle. The benefit of quick reaction times was further enhanced by the use of MW activation (Figure 4.36) [67].

4.2.1.4 Friedel–Crafts Reactions

With roots in the late 1800s, the Friedel–Crafts chemistry is an established and fundamental field of study in organic synthesis. Over the last century, there has been a thorough examination of electrophilic halogenation, electrophilic nitration, and aromatic acylation processes. Notably, Olah's book series is a thorough and wide resource in this area, providing insightful information and understanding on these subjects [68].

In a research on the acylated of *p*-cresol, supported by montmorillonite clay serving as the catalyst, Wu et al. High levels of Lewis/Bronsted acid sites as well as its sizable surface area make it a suitable catalyst. The use of MW was essential in avoiding catalyst deactivation, which frequently happened as a result of the acylium ions' reactivity when they were supposed to mix with cresol to produce the desired product. The acylium ions were oriented in such a way by the electromagnetic field produced by MW irradiation that it inhibited undesirable side reactions of the catalyst to favor the cresol reaction (Figure 4.37) [69].

Zeolites have proven to be excellent catalysts to favor Friedel–Crafts reactions. In particular, the MW irradiation of zeolites has effectively produced aromatic ketones

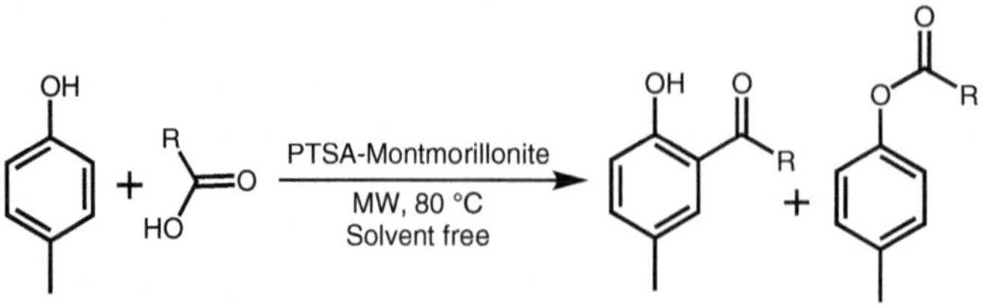

Figure 4.37 PTSA-Montmorillonite assisted Friedel–Crafts reaction under microwave irradiation. Source: Wu et al. [69].

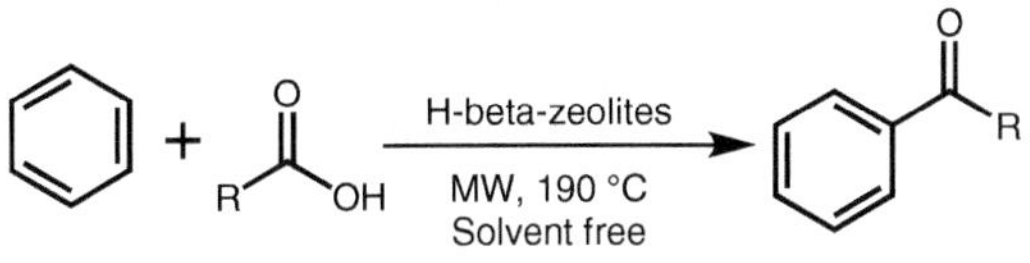

Figure 4.38 H-beta-zeolites assisted Friedel–Crafts reaction under microwave irradiation. Source: Venkatesha et al. [70].

via the acylation of aromatic compounds. H-beta-type zeolites accommodate aromatic compounds and carboxylic acid at the same time near the active sites. Notably, a significant decline in yield was seen when the reaction was heated conventionally. To clarify the effect of MW irradiation in facilitating the reaction, more research was carried out. The chemical route and activation energies under MW irradiation were comparable to those under conventional heating settings, according to Arrhenius plots. It was determined that MWs most likely made it easier for water to be ejected from the catalyst surface, hastening the acetylation process. Furthermore, MWs could specifically raise the active polar temperatures on the catalyst surface, speeding up the reaction (Figure 4.38) [70].

4.2.1.5 Reaction Involving Catalysts of Biological Origin

A mesoporous bi-functionalized organosilica (MPBOS) catalyst was effectively used to catalyze the preparation of 5-hydroxymethylfurfural (HMF) from carbohydrates under the MW environment. SBA-15 was used in a two-step procedure to create the catalyst. The first stage included functionalization using (3-chloropropyl)-triethoxy-silane, which was followed by a substitution reaction involving the chloro-group of the newly grafted molecule and the amino group of 5-aminoisophthalic acid. The resultant bi-functional catalyst demonstrated notable effectiveness, yielding 74% of HMF from glucose. The catalyst's potential for reuse was highlighted by an investigation into its reusability, which revealed a progressive, if modest, decline in yield after five consecutive cycles (Figure 4.39) [71].

Using a particular kind of zeolite beta catalyst, Chen and the team produced biodiesel via triolein-transesterification. According to their research, high Si/Al ratio zeolites that were subjected to a weak solution of NaOH were shown to be effective in catalyzing trans-esterification reactions. Glycerol and fatty acid methyl esters were produced from a triglyceride, triolein by MW irradiation at 110 °C for an hour in the presence of methanol. In addition to the zeolite's basic sites, which aided in the creation of the active sodium methoxide intermediate, the presence of sodium cations on the surface of the zeolite was credited with catalytic activity. Notably, the catalyst showed the capability for reusing without noticeably losing activity, emphasizing its usefulness and sustainability (Figure 4.40) [72].

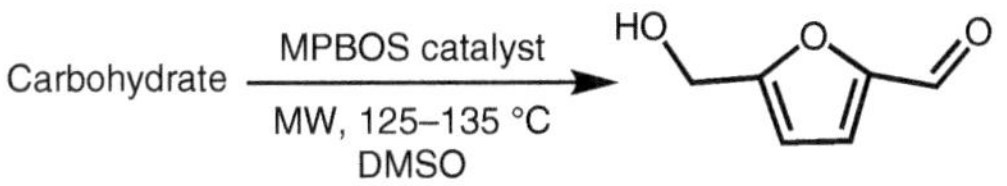

Figure 4.39 MPBOS-assisted transformation reaction under microwave irradiation. Source: Bhanja et al. [71].

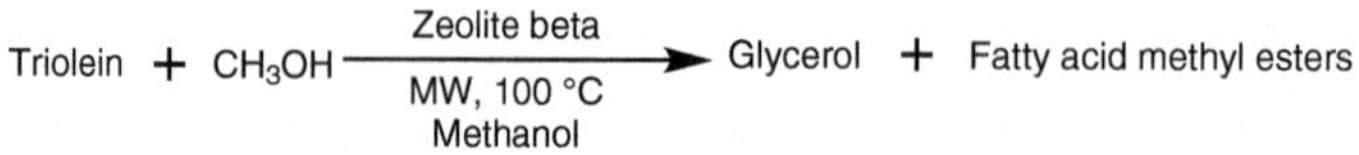

Triolein + CH₃OH $\xrightarrow[\text{Methanol}]{\text{Zeolite beta} \atop \text{MW, 100 °C}}$ Glycerol + Fatty acid methyl esters

Figure 4.40 Zeolite Beta assisted transformation reaction under microwave irradiation. Source: Marwan [72].

An affordable and simple procedure was provided by Jhung and colleagues for making an acidic sulfonated silica catalyst. Tetraethyl orthosilicate and chlorosulfuric acid were used to prepare the catalyst in place, obviating the requirement for supplement materials such as surfactants. Oleic acid esterification by the catalytic presence resulted in promising results and reusability. The number of anchoring acid sites had a significant impact on the production of methyl oleate, demonstrating the crucial importance of the medium's addition of chlorosulfuric acid. Additionally, the reaction time was significantly shortened by using MW irradiation while still producing yields that were comparable to those attained with traditional heating techniques [73].

A new trans-esterification catalyst made from seashells of Cyrtopleuracostata proved to be remarkably efficient. Under MW irradiation, and in the presence of methanol, biodiesel was effectively produced from palm oil using this catalyst. The surface hydroxyl groups of the catalyst interact with the solvent to generate a proactive intermediate called a methoxide, which is thought to catalyze the trans-esterification process. Microwave-assisted reactions showed definite benefits over traditional reactors as per conversion and rate. Several researchers also investigated the purpose of dolomitic rocks and shells of mussels as catalysts for related processes. These substances were ground and calcined to produce CaO or CaO–MgO powder, which was then reacted with CH_3OH. The active species produced by the mussel shell-based catalyst in this instance was calcium methoxide. Within minutes, the microwave-assisted process utilizing this catalyst produced methyl esters of fatty acid made from Jatropha curcas oil, with yields exceeding 90%. The catalysts also showed great reusability, which further increased their usefulness and sustainability (Figure 4.41) [74, 75].

Pei et al. studied microalgae trans-esterification employing a heterogeneous catalyst made of KF/CaO. Notably, the study showed that excellent yields for the trans-esterification products required MW along with ultrasonic irradiation (US) activations. As a result of the combined use of MW and US activation, the mass

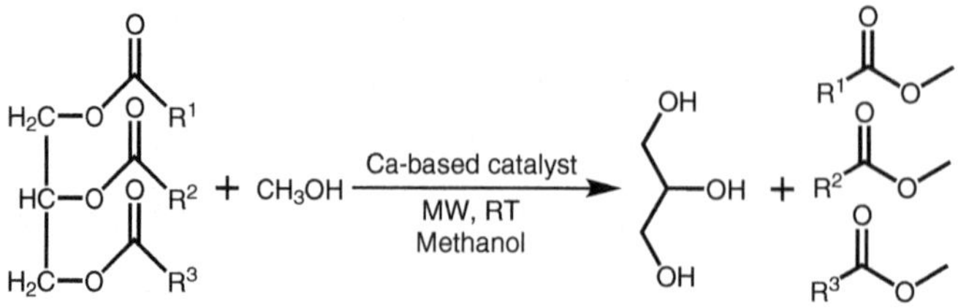

Figure 4.41 Ca-based catalyst-assisted transformation reaction under microwave irradiation. Source: Adapted from [74, 75].

transfer of the reactants was improved, allowing for effective mixing, and catalyzing effective interactions between the reactants and catalyst. Additionally, the completion of the trans-esterification process was made possible by the combined MW-US activation, which supplied the requisite energy [76].

Tangy et al. created a new heterogeneous catalyst by functionalizing strontium oxide with carbon quantum dots that were subsequently used for a trans-esterification process. The catalyst was created utilizing polyethylene glycol and the SrO precursor $Sr(NO_3)_2$ through a sonochemical alteration followed by a calcination procedure. Lipid transesterification from Chlorella vulgaris was assisted by a strontium-based catalyst under a MW environment to assess its catalytic potential. Surprisingly, lipid conversion reached 97%, yielding a fatty acid methyl ester yield of about 50%. The strontium oxide modified with carbon quantum dots showed increased catalytic efficiency when compared to SrO, which is readily accessible commercially, according to a comparison between the new catalyst and SrO. The synergistic interaction between the basic SrO sites and the acidic carbon quantum dot sites was essential in accelerating the trans-esterification process [77].

Metal-containing catalysts of transitional series origin have extremely high levels of toxicity. Therefore, even if these catalysts frequently demonstrate amazing effectiveness, it is necessary to approach their employment from a green chemistry perspective with the utmost caution. These metals are used in a variety of ways, such as in their elemental state (e.g. as hydrogenation catalysts), or as compounds like oxides, halides, or complexes with organic ligands. These catalysts must be given specific attention in our discussion because of the inherent dangers connected with their toxicity and their participation in complex catalytic processes. We shall discuss their uses based on the main categories of reactions they assist; similar to how solid acid/base catalysts are classified [78].

4.2.1.6 Reduction

In organic synthesis, the reduction of different functional groups is a frequent and essential reaction [79]. Due to its enormous practical value, this topic has been around for more than a century, yet research in it is still quite active. Exploration and development of new strategies continue to fuel developments in the field of organic synthesis as conventional techniques are being replaced by contemporary ones [80].

Using MW heating under comparatively moderate circumstances (125 °C and 10 bar H_2), has effectively reduced furfural to furfuryl alcohol with a remarkable selectivity of 99%. Cu nanoparticles supported on TiO_2 are the catalyst used in this procedure. Although the selectivity of the reaction is not dramatically affected by the use of microwave heating, the conversion efficiency is greatly improved. It should be noted that this heightened impact does not appear at higher temperatures (175 °C), which suggests that MW heating may produce isolated flashpoints with greater temperatures than the temperature that was recorded for the majority of the area (Figure 4.42) [79].

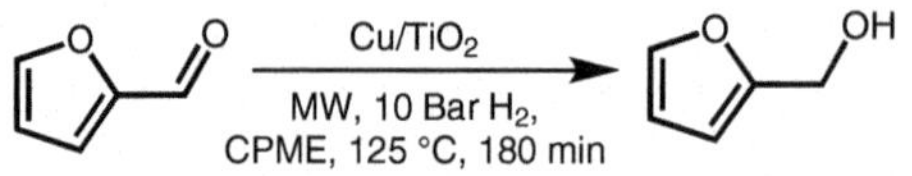

Figure 4.42 Cu/TiO$_2$ catalyst-assisted reduction reaction under microwave irradiation. Source: Romano et al. [79].

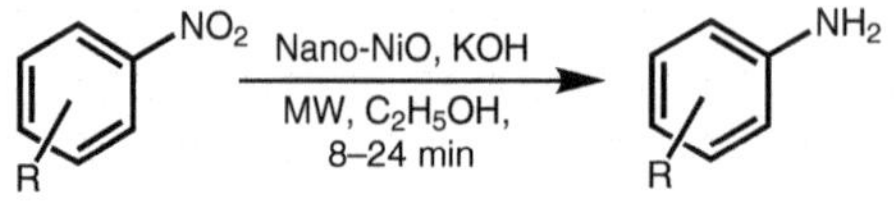

Figure 4.43 Nano-NiO catalyst assisted reduction reaction under microwave irradiation. Source: Farhadi et al. [81].

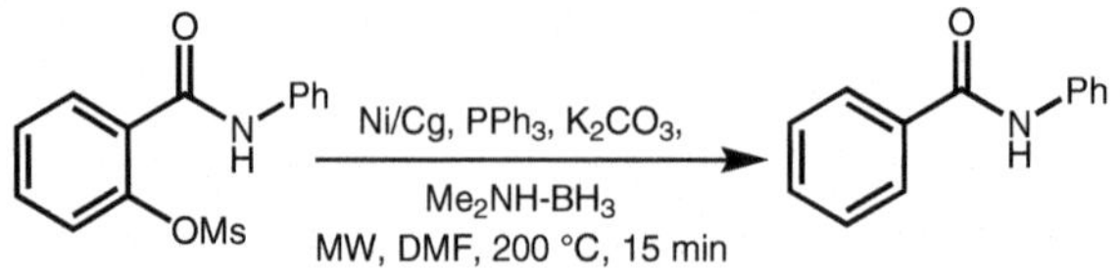

Figure 4.44 Ni/Cg catalyst assisted reduction reaction under microwave irradiation. Source: Lipshutz et al. [83].

As shown (Figure 4.43), the reduction of nitro-compounds to amines by hydrogen transfer was accomplished effectively using a NiO nanocatalyst. The catalyst was created by thermally breaking down a Ni-(II)-complex, which produced nanoparticles with more activity than the NiO that is typically found on the market. Ethanol, which served as both the solvent and the source of hydrogen, was used to perform the reactions. Although microwave irradiation significantly sped up reactions, it was found that the technique of heating had no impact on the reaction's yield. In other words, the ultimate yield of the target products was unaffected by the use of MW irradiation (Figure 4.43) [82]. The catalytic ability of SmFeO$_3$ was investigated using the same reaction. The most efficient hydrogen donor in the reduction of aromatic nitro compounds to their respective amines was found to be isopropanol under these particular circumstances [81].

Mesylates and aryl tosylates were reduced to their corresponding aryl-H compounds using a catalyst made of graphene with nickel (Ni/Cg) in another study by Lipshutz et al. The scientists showed that, for some chosen cases, the reaction time may be greatly reduced when using microwave irradiation (15 minutes versus 14 hours), even if conventional heating was mostly employed. Notably, after switching from conventional heating to microwave irradiation, the product yield remained unaffected, showing that the chosen heating technique had no effect on the ultimate yield of the intended goods (Figure 4.44) [83].

4.2.1.7 Oxidation

Similar to reduction, oxidation is one of the basic chemical processes. In spite of the fact that traditional oxidation techniques have a wide range of applications, many of them employ hazardous chemicals and reagent-based oxidants. There has been a movement in recent years toward catalytic methods that use air or oxygen as the oxidizing agents. This transition to catalytic oxidation techniques seeks to lessen dependency on potentially dangerous chemicals and promote more sustainable and eco-friendly procedures [84].

A highly sought-after subject is the ecologically friendly or "green" oxidation of primary alcohols to aldehydes. In order to do this, Stiegman et al. investigated the use of Fe, Co, and Cu spinel catalysts in the oxidation of alcohol to aldehyde with water as

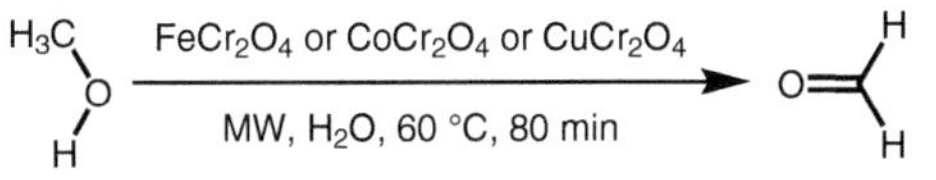

4.2 Microwave-Assisted Reactions | **111**

Figure 4.45 FeCr$_2$O$_4$ or CoCr$_2$O$_4$ or CuCr$_2$O$_4$ catalyst-assisted oxidation reaction under MWI. Source: Crosswhite et al. [85].

the solvent. Even though the technique only achieved very modest conversion rates (varying from 13% to 25%) when MW heating at 60 °C was used, it showed absolute selectivity and had no other feasible options, making it a promising method. Notably, when using conventional heating under the same circumstances, the reaction failed to yield any product, indicating that the use of MW heating was crucial. This demonstrates the important role that MW heating plays in accelerating the generation of formaldehyde, making the proposed process a possible development in the area of green oxidation of primary alcohols (Figure 4.45) [85].

In a study, Baruah et al. showed how effective cellulose-supported copper nanoparticles (Cu/cellulose) are as oxidation catalysts. By using an aqueous solution and MW heating, they were able to demonstrate how effectively cinnamyl and benzyl alcohols may be converted to the corresponding aldehydes with high yields. Their findings made it clear that the use of microwaves was necessary in order to obtain these advantageous yields within realistic reaction times. In order to properly facilitate the oxidation process, microwave heating was required. The catalyst's potential to be recycled was also examined, and the findings showed that it could do so up to four cycles with just a minimal reduction of activity, ranging from 2% to 6%. The Cu/cellulose catalyst's potential application and sustainability in oxidation processes are further increased by its recyclability (Figure 4.46) [86].

An intriguing method for the quick conversion preparation of esters from aldehydes and alcohols is oxidative esterification. In this procedure, the aerobic oxidation of alcohols and aldehydes might be assisted by Pd/C, and microwaves were used to speed up the reaction. The outcomes demonstrated the effortless translation of several substituted benzaldehydes and benzyl alcohols into their corresponding esters. Even with lengthy response durations, the reaction did not take place when using conventional heating; hence the usage of microwave heating was crucial. Because it is ecologically safe, oxygen was chosen as the oxidant in this procedure. However, more pressure was required for the efficient use of oxygen. The addition of halogen-substituted molecules resulted in decreased yields because both the starting

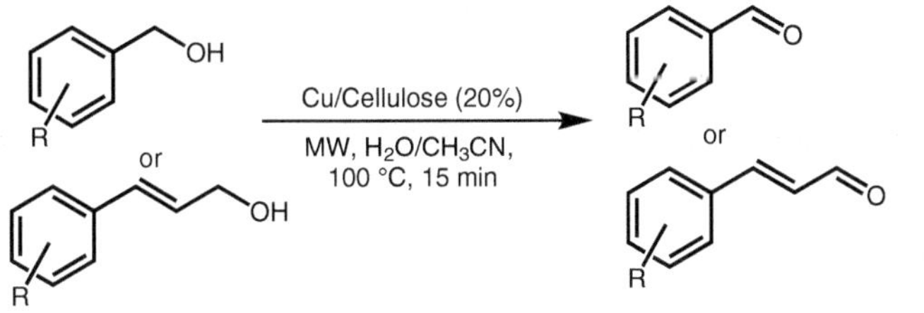

Figure 4.46 Cu/Cellulose (20%) catalyst-assisted oxidation reaction under MWI. Source: Baruah et al. [86].

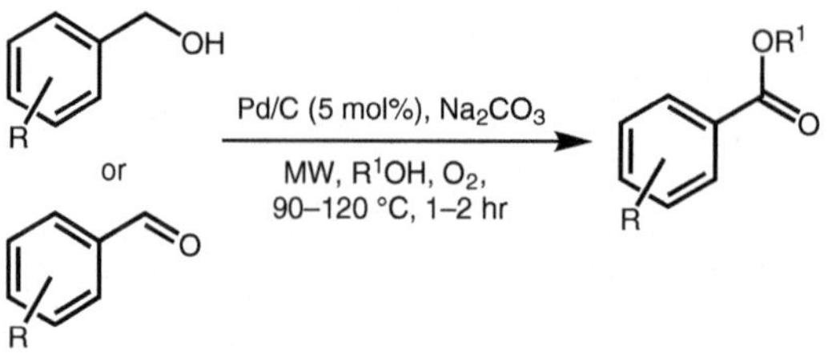

Figure 4.47 Pd/C (5 mol%) catalyst-assisted oxidation reaction under microwave irradiation. Source: Caporaso et al. [87].

material and the finished product were dehalogenated, even though the reaction showed tolerance towards oxygen, nitrogen, and sulfur-containing substituents. In conclusion, the oxidative esterification of alcohols and aldehydes into esters using oxygen as the green oxidant is made possible by using Pd/C as the catalyst in combination with microwave heating. To prevent unfavorable side effects, caution must be used while working with halogen-substituted substances (Figure 4.47) [87].

According to research, heterogeneous tin coordination polymers can be used as catalysts in Baeyer–Villiger oxidations without the need for solvents. Tin is coupled with a three-dentate ligand to create the catalyst, which produces the catalytically active species. Without the need for an additional solvent, the reaction is carried out utilizing H_2O_2 as the oxidant. When using MW heating, a variety of ketones, including linear aliphatic, aromatic, and cyclic ketones, may be effectively converted into the appropriate esters or lactones with high yields. Notably, only modest transformations of the starting material were seen when ordinary heating was used. However, by using ultrasonic activation rather than microwave heating, respectable conversion rates may still be attained. Although ultrasonic heating had a little lower conversion than microwave heating, it was accompanied by a rise in selectivity that may be useful in some circumstances. In conclusion, converting different ketones into esters or lactones under solvent-free conditions is made possible by the heterogeneous tin coordination polymer used as a catalyst. Ultrasound activation provides a different option with improved selectivity, but at somewhat lower conversion rates, whereas microwave heating is seen to be very successful in attaining high conversions. This catalyst's noteworthy quality is its capacity to be recycled; it may be used up to five times without suffering any yield or selectivity losses (Figure 4.48) [88].

The production of sulfate radicals ($SO_4^{\cdot-}$) by MW heating has proven useful in accelerating the breakdown of p-nitrophenol [89]. In research by Pang and Lei, peroxymonosulfate (PMS) was used as the oxidant while $MnFe_2O_4$ was used as a catalyst for radical oxidation. In comparison to other known systems, the performance of the PMS and $MnFe_2O_4$ combination with microwave heating was superior. Surprisingly, this method allowed for virtually full p-nitrophenol degradation within a remarkably short time frame of about two minutes. The quick and extensive degradation

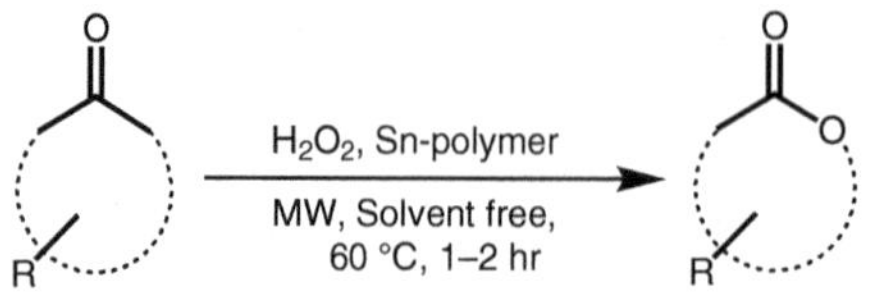

Figure 4.48 Sn-polymer catalyst assisted oxidation reaction under microwave irradiation. Source: Martins et al. [88].

of p-nitrophenol by the PMS/MnFe$_2$O$_4$ combination demonstrated by this study makes it a potential technique for environmental remediation and water treatment applications [90].

The decomposition of p-nitrophenol using a flow system and MWs has shown to be quite successful. A CuO/Al$_2$O$_3$ catalyst proved effective in this process under the circumstances by assisting the production of hydroxyl radicals (OH·) from hydrogen peroxide (H$_2$O$_2$). It normally takes less than six minutes for these OH· radicals to completely convert p-nitrophenol into poly-phenols, which is a remarkably rapid response time. This technology provides a quick and effective way to degrade p-nitrophenol, which is crucial for applications such as water purification and environmental remediation. The use of microwaves in tandem with the flow system improves the efficiency of the procedure by establishing precise and regulated reaction conditions and enabling speedy and complete p-nitrophenol degradation [91].

4.2.1.8 Coupling Reactions

The initial papers on coupling reactions catalyzed by Pd may date back to 1968, when Heck's groundbreaking study resulted in their discovery [92]. Later, this topic saw a major scope growth and was crowned to be the most intriguing subfield of organic synthesis. The culmination of these developments was symbolized by Heck, Suzuki, and Negishi receiving the renowned Nobel Prizes in 2010 [93]. These honors acknowledged their groundbreaking work in the progress of Pd-catalyzed coupling processes, which have had a significant influence on the field of chemistry.

Sonogashira Coupling Palladium nanoparticles distributed within an organosilica matrix have been shown to be efficient catalysts for the coupling of various bromo- and iodoarenes with phenylacetylene. Notably, this catalytic system may be used with MW heating as well as traditional heating, with the latter showing the benefit of having significantly quicker reaction times to produce equivalent yields. A modest catalyst loading of 0.5% was used throughout the research, and it turned out to be adequate for the intended reactions. Surprisingly, the coupling reactions with iodoarenes produced results that were superior to those with the comparable bromoarenes in terms of both reaction rate and total yield. These results emphasized the possible importance of this catalytic system in synthetic organic chemistry, especially in the context of coupling reactions involving aryl halides and alkynes (Figure 4.49) [94].

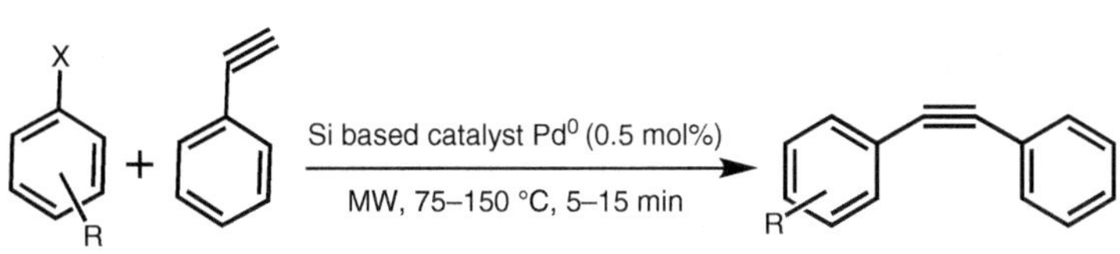

Figure 4.49 Si-based Pd^0 (0.5 mol%) catalyst-assisted coupling reaction under MWI. Source: Ciriminna et al. [94].

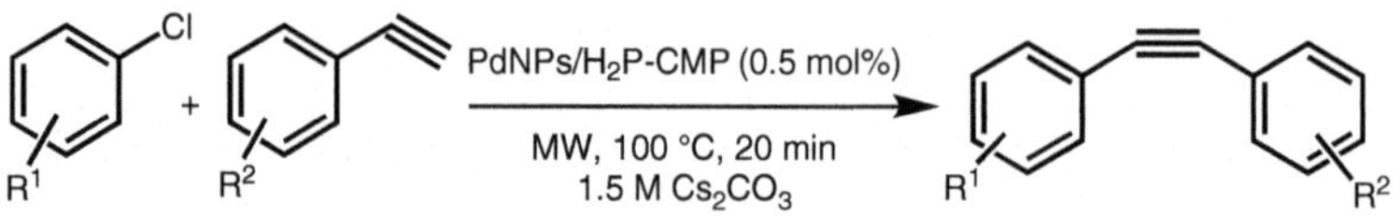

Figure 4.50 PdNPs/H$_2$P-CMP (0.5 mol%) catalyst-assisted coupling reaction under MWI. Source: Huang et al. [95].

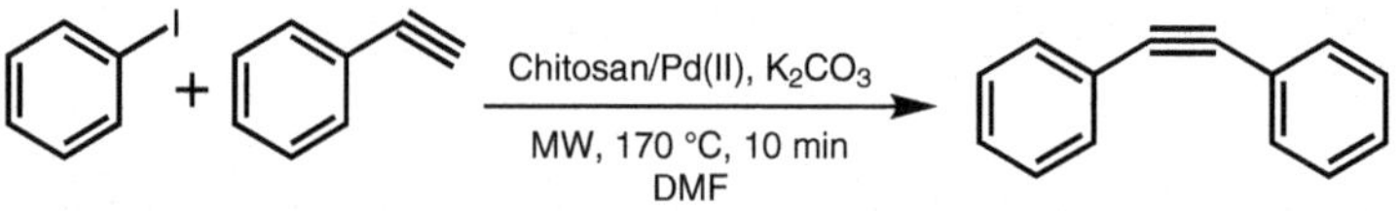

Figure 4.51 Chitosan/Pd(II) catalyst-assisted coupling reaction under microwave irradiation. Source: Leonhardt et al. [96].

Palladium nanoparticles have successfully been used in Sonogashira couplings thanks to their integration into the PdNPs/H$_2$P-CMP conjugated microporous/mesoporous polymer matrix. High yields of the required products may be produced by this catalytic system's coupling reaction of various substituted aryl chlorides and phenylacetylenes in an aqueous medium. Notably, the total yield of the reaction was unaffected by the use of MW or traditional heating techniques. Instead of taking eight hours to reach the necessary results under normal heating settings, the use of MW heating significantly shortened the response time. This expedient feature emphasizes the beneficial function of PdNPs/H$_2$P-CMP in boosting the effectiveness of Sonogashira couplings, therefore improving its prospective usage in diverse synthetic organic chemistry applications (Figure 4.50) [95].

Chitosan, a naturally occurring carbohydrate-based polymer, was used in the study by Ondruschka et al. as a supporting component for a palladium catalyst. They effectively created an active catalyst to aid in the Sonogashira coupling of phenylacetylene and iodobenzene by co-precipitating chitosan and PdCl$_2$. Surprisingly, when MW irradiation was used as the heating mechanism, this coupling process was accomplished in a relatively short time of 10 minutes. However, the dehalogenation of iodobenzene to produce benzene in the used circumstances was a notable side reaction that the authors detected during the procedure. To guarantee the necessary selectivity and efficiency in Sonogashira coupling procedures combining aryl halides and alkynes, consideration must be given to this undesirable side reaction in practical applications, despite the catalytic system's remarkable reactivity and speed. To reduce the incidence of this side reaction and improve the overall usability of the chitosan-supported palladium catalyst for such coupling reactions, more optimization and fine-tuning of the reaction conditions may be required (Figure 4.51) [96].

Suzuki–Miyaura Coupling An immobilized (N-heterocyclic carbene) NHC-Pd catalyst was used in a new heterogeneous catalytic approach to produce biphenyls from aryl boronic acids and halides. Triethoxysilylpropyl-tether was aided in grafting NHC on silica gel, and then the active catalyst was produced by treating the material

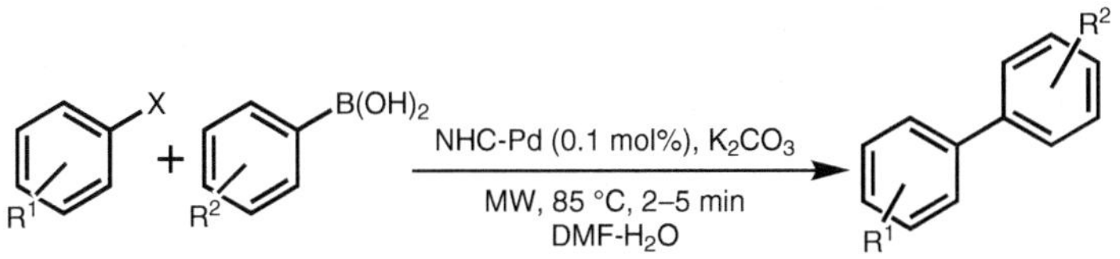

Figure 4.52 NHC-Pd catalyst assisted Suzuki–Mayaura coupling with microwave irradiation. Source: Taher et al. [97].

with Pd(OAc)$_2$. With exceptional effectiveness, this NHC-Pd catalyst produced the necessary coupling products from a variety of substituted arylboronic acids using arylhalides. Notably, using microwave heating was favorable and produced better results than using traditional heating techniques in terms of both rate and quantity. Additionally, the catalytic system successfully worked with substrates including bromo-pyridine and -thiophene, proving its adaptability and use in a variety of reaction situations. The synthesis of substituted biphenyls using this immobilized NHC-Pd catalyst is novel and has the potential to advance synthetic organic chemistry techniques in the future (Figure 4.52) [97].

Shen and the research group utilized Suzuki–Miyaura coupling assisted by dipyridylamine ligand as a very efficient ligand for a Pd(II) catalyst in their investigation. A natural semi-hard resin, modified gum resin was encapsulated by the ligand. The researchers discovered that the use of MW heating produced significant gains in both response time and product yield. Aryl chlorides, -bromides, and -iodides were successfully used as substrates in the Suzuki–Miyaura coupling reactions carried out with this catalytic system. Notable was the better performance shown when bromo- and iodoarenes reacted with these aryl halides, leading to noticeably higher product yields. This catalytic system's recyclability provided an additional benefit. With just a 20% reduction recorded after the fourth cycle; the catalyst could be reused up to three times before the product yield started to noticeably decline. This recyclability property underlines the 2,2′-dipyridylamine tethered to modified gum resin ligand-based Pd(II) catalyst's potential sustainability and cost-effectiveness, making it a good candidate for real-world use in Suzuki–Miyaura coupling processes (Figure 4.53) [98].

Pd(II) was immobilized using a similar technique on a chitosan/cellulose support. By using MW heating to produce rapid reaction rates, this supported Pd(II) catalyst was successful in aiding the synthesis of biphenyl compounds under solvent-free

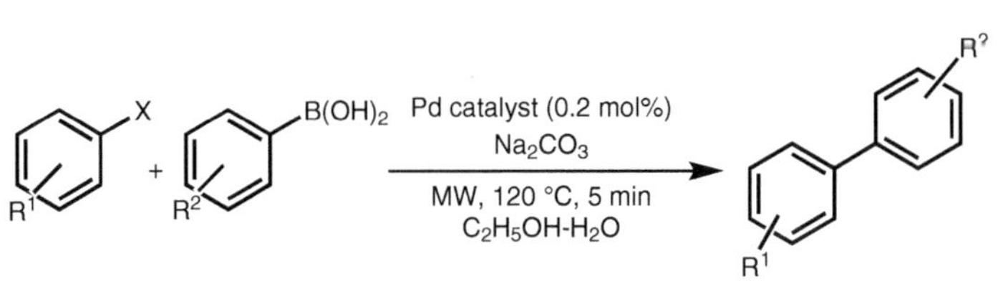

Figure 4.53 Pd catalyzed Suzuki–Miyaura coupling reaction under microwave irradiation. Source: Pan et al. [98].

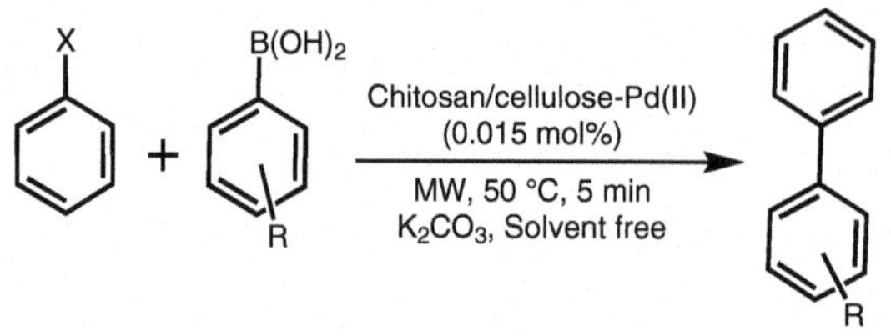

Figure 4.54 Chitosan/cellulose-Pd(II) catalyzed Suzuki–Miyaura coupling with MWI. Source: Baran et al. [99].

conditions. MW heating produced greater yields and noticeably faster response times as compared to conventional heating in toluene. This demonstrates the benefits of using MW irradiation in this catalytic system to promote effective and quick synthesis of biphenyl. Although aryl chlorides were capable of being used as substrates in the reaction, it was found that their usage produced noticeably lower yields than aryl bromides and iodides. This substrate-dependent selectivity highlights the specific catalytic setup's affinity for aryl bromides and iodides even more. The overall findings point to the Pd(II)-anchored chitosan/cellulose-supported catalyst as an effective and ecologically benign method for producing biphenyls, one that has the potential for wider use in quick and efficient synthetic processes (Figure 4.54) [99].

Magnaldehyde B and E were synthesized by the Suzuki–Miyaura coupling, which was successfully proved by Schmidt and Riemer to employ Pd/C as a catalyst. It is interesting to note that the intended product could not be successfully produced when attempting to combine the two aryl systems under standard heating settings. A breakthrough was made, nevertheless, when they combined MW heating with Pd/C as the catalyst and NBu4F as an addition. The intended product was produced with an outstanding 81% yield under these optimal circumstances, demonstrating the efficacy and efficiency of this catalytic system. This effective application emphasizes how crucial it is to use the proper heating technique and catalyst in Suzuki–Miyaura couplings since these factors have a big impact on how the reaction turns out. Magnaldehyde B and E were produced using a potent and effective method using Pd/C as the catalyst, the necessary addition, and MW heating, bringing up new opportunities for the synthesis of important chemical molecules (Figure 4.55) [100].

In Suzuki–Miyaura reactions, nanostructured Pd(0) immobilized on the organosilica xerogel SiliaCat is shown to be an efficient catalyst. Notably, when MW heating was used, this catalyst demonstrated the capacity to effectively link aryl bromides and aryl chlorides with phenylboronic acid, producing good yields. The reactions

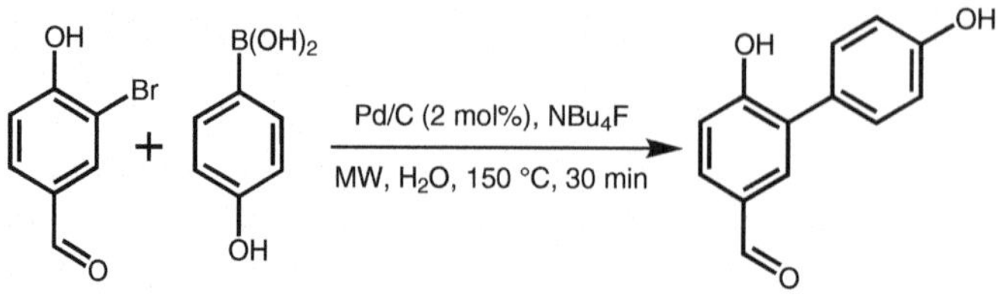

Figure 4.55 Pd/C (2 mol%) catalyzed Suzuki–Miyaura with microwave irradiation. Source: Schmidt and Riemer [100].

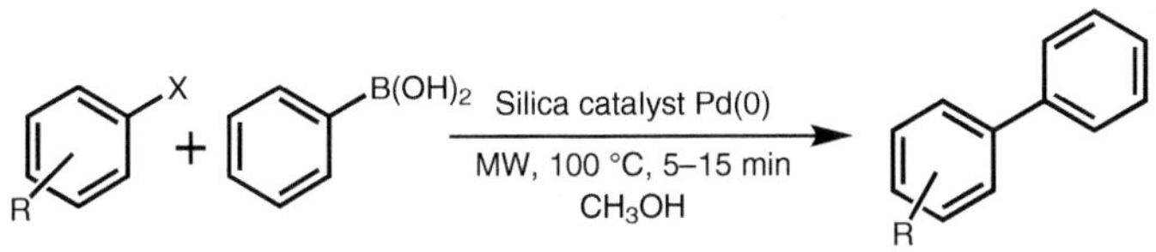

Figure 4.56 SiO$_2$ catalyst Pd(0) assisted Suzuki–Miyaura coupling with microwave irradiation. Source: Ciriminna et al. [101].

moved quite quickly under the ideal MW heating settings, taking only 5–15 minutes to complete. Contrarily, when ordinary heating was used, the reactions took much longer to complete – up to six hours. In addition, conventional heating was less successful in converting aryl chlorides, yielding just a 35% conversion even after 22 hours of continuous heating. This suggests that SiliaCat when employed as the catalyst under MW heating circumstances, not only permits quick reactions but also performs better than traditional heating, especially when it comes to converting aryl chlorides. These results demonstrate the efficacy of nanostructured Pd(0) anchored on SiliaCat as a promising Suzuki–Miyaura reaction catalyst, providing improved efficiency and adaptability for a range of chemical synthesis applications (Figure 4.56) [101].

With the explicit intention of using them as an active catalyst in Suzuki cross-coupling processes, CuPd nanoparticles were effectively produced. When MW heating was used, this inventive catalyst showed astounding effectiveness when combining different substituted aryl iodides with arylboronic acids, producing high yields. The researchers also looked at the CuPd nanoparticle catalyst's potential to be recycled. Although the catalyst was found to be recyclable, it was noted that after the third run, performance started to somewhat deteriorate. In particular, production decreased by 10% in successive recycling cycles. Reusing the CuPd nanoparticle catalyst is a valuable feature that adds to the sustainability and affordability of the Suzuki cross-coupling process, despite the minor efficiency loss with each recycling cycle. Overall, CuPd nanoparticles' effective production and use as a catalyst in Suzuki cross-coupling reactions show great potential for developing synthetic procedures in organic chemistry (Figure 4.57) [102].

Das and his research team discovered the first Suzuki–Miyaura coupling using Amberlite IRA 900 supportedrhodium(0) nano-/microparticles (SS-Rh). Surprisingly, the activity of this heterogeneous catalyst was ligand-free. PEG-400 was used as the solvent for the coupling process, which hastened the production of the

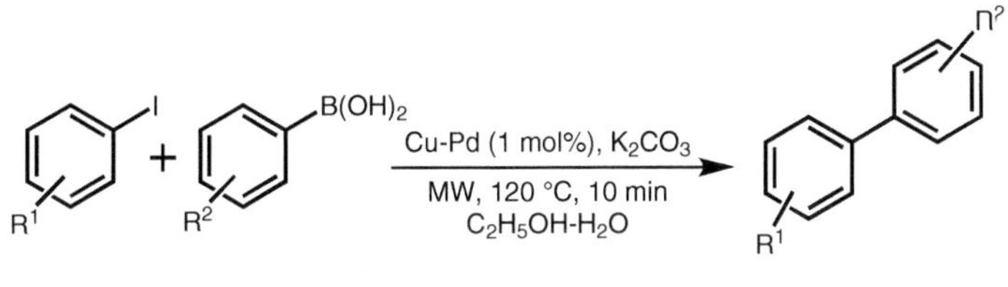

Figure 4.57 Cu-Pd (1 mol%) catalyzed Suzuki–Miyaura coupling with microwave irradiation. Source: Smith et al. [102].

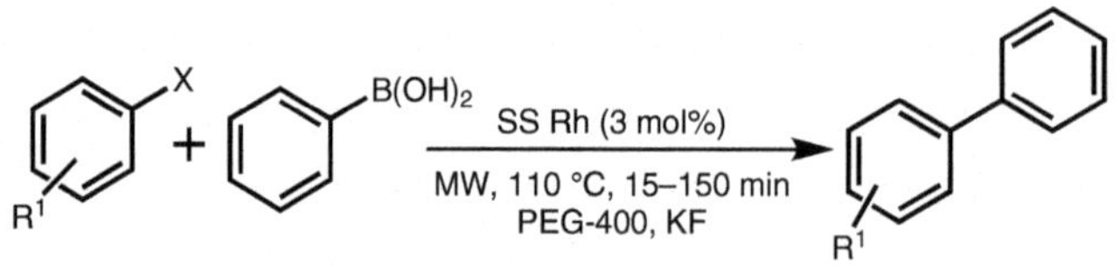

Figure 4.58 SS-Rh catalyzed Suzuki–Miyaura coupling under microwave irradiation. Source: Guha et al. [103].

products. This catalytic system has the unique ability to tolerate highly reactive functional groups, including keto, aldehyde, and nitrile groups. These groups were satisfactorily accommodated under the SS-Rh catalysis even though they are normally incompatible with the conventional conditions used in Suzuki–Miyaura couplings. The catalyst's potential to be recycled was also investigated, and it demonstrated exceptional stability and reusability. The SS-Rh catalyst was easily recovered by straightforward filtering and could be recycled up to 12 times without suffering any appreciable activity loss. After several recycling cycles, just a 3% decrease in catalytic activity was noticed. Suzuki–Miyaura couplings have advanced significantly with the effective use of Amberlite IRA 900 supportedrhodium(0) nano-/microparticles (SS-Rh) as a ligand-free heterogeneous catalyst. The system has the potential to be used in efficient and sustainable organic synthesis procedures due to its compatibility with a variety of reactive functional groups and its cyclability (Figure 4.58) [103].

Heck Coupling Pietro and team have used sphere shaped silica supported Pd-coated as a catalyst to carry out Heck coupling with the aid of microwaves. Due to its suitability for both conventional and microwave-assisted continuous-flow of organic synthesis (MACOS), this catalyst has shown exceptional flexibility. The Pd-coated silica macrospheres catalyst efficiently enabled the synthesis of coupling products with high yields from aryl iodides and esters. The MACOS system's flow capillaries were investigated for their potential for reuse, and it was discovered that they may be reused up to five times without significantly reducing the reaction yield. The Pd-coated silica macrosphere catalyst is a good choice for sustainable and effective synthetic applications because of its reusability feature, which emphasizes the strength and stability of the catalyst. This catalyst's successful use in batch and continuous-flow environments shows that it has the potential to be a useful and effective tool for conducting Heck coupling reactions, enabling quick and high-yield production with a variety of substrates (Figure 4.59) [104].

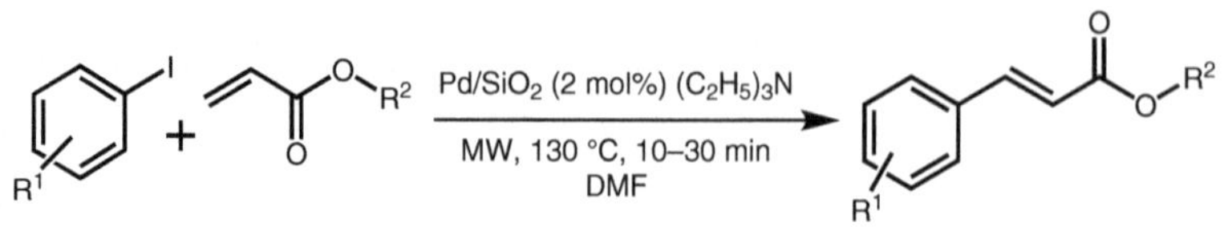

Figure 4.59 Pd/SiO$_2$ (2 mol%) catalyst Heck coupling with microwave irradiation. Source: Schruder et al. [104].

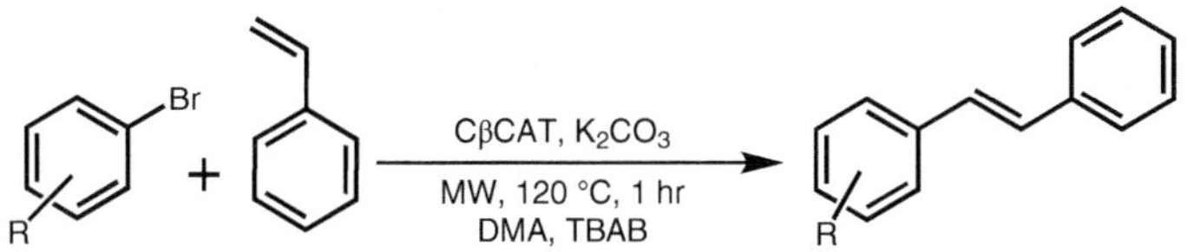

Figure 4.60 CβCAT catalyst assisted heck coupling reaction under microwave irradiation. Source: Cravotto et al. [105].

Cyclodextrin-supported Pd (II) (CβCAT@Pd(II)) was used as a catalyst to successfully carry out Heck coupling reactions. Notably, this catalyst showed promise in the efficient styrene coupling with aryl bromides, resulting in enhanced product yields, and outstanding E/Z selectivities. The scientists discovered that either conventional heating or MW irradiation may be used to carry out the Heck coupling reaction. When MW was used, however, the reaction showed a clear benefit, producing high yields in noticeably less time. As a catalyst for Heck coupling, Pd(II) immobilized -cyclodextrin (CβCAT) offers a potential method for producing efficient and selective synthesis of coupling products. The robustness of using both conventional and MW heating techniques, together with the excellent yields and selectivities seen in the reaction, highlights the potential value of this catalytic system in a range of synthetic organic chemistry applications (Figure 4.60) [105].

A glass/polystyrene-supported Pd(II)-catalyst was used to react to a variety of arylbromides with butylacrylate/styrene. When the reactions were conducted in water while being MW irradiated, the products were produced with enhanced yields and in a short amount of time. The key disadvantage was the much longer time needed to attain the necessary product yields, even though the yields that were reached with conventional heating were equivalent to those obtained with MW irradiation. In contrast to conventional heating, which takes around five hours, MW irradiation significantly sped up the response time to just five minutes. The catalyst reusability of the microwave-assisted process, however, was shown to have a significant constraint. The glass/polystyrene-supported Pd(II)-catalyst appeared to decompose more quickly under the influence of MW irradiation, leaving it unusable. The catalyst's activity could not be sustained over numerous reaction cycles as a result. The problem of catalyst degradation may need to be addressed in further research to improve the overall long-term viability and effectiveness of the catalytic system for such Heck coupling reactions, despite the quicker reaction durations and high product yields made possible by MW irradiation (Figure 4.61) [106].

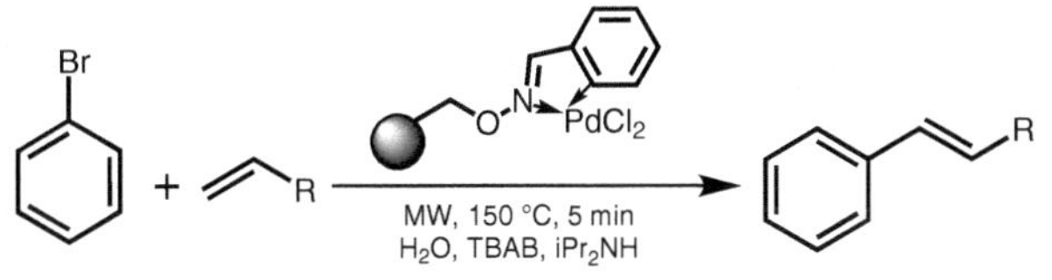

Figure 4.61 Glass/polystyrene-supported Pd(II) catalyst-assisted heck coupling reaction under microwave irradiation. Source: Dawood et al. [106].

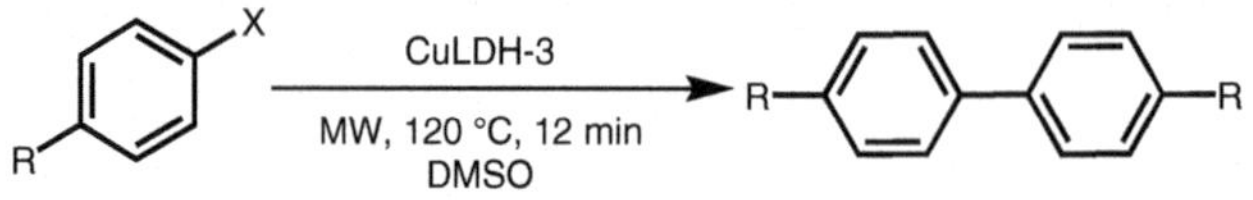

Figure 4.62 CuLDH-3 catalyst assisted Ullmann coupling under microwave irradiation. Source: Narasimharao et al. [107].

Ullmann Coupling In their study, Narasimharao and team showed microwaving may be used in conjunction with a Cu-hydrotalcite catalyst to speed up the homo-coupling of substituted halides in Ullmann style. Reduced reaction temperature, enhanced rate of reaction, and less base and metal were used thanks to this novel technique, among other important benefits. The Cu–Mg–Al hydrotalcite catalyst (CuLDH-3) used by the authors has a particular composition (Cu: Mg: Al = 2: 0: 1). With the aid of this catalyst, biphenyls may be effectively formed from a variety of aryl halides in noticeably less time. Surprisingly, the homocoupling reaction's yield and selectivity were unaffected by the use of MW heating. However, compared to traditional heating, the response time was significantly shortened; just 12 minutes as opposed to a laborious 10 hours were needed under MW heating. A potential method to increase the effectiveness and durability of Ullmann-type homocoupling reactions uses MW heating in conjunction with the Cu-hydrotalcite catalyst. The advantages of quicker reaction times and less resource usage help organic synthesis techniques become more environmentally friendly and time-effective (Figure 4.62) [107].

Buchwald–Hartwig Coupling Under solvent-free circumstances, N-substituted anilines were synthesized by Buchwald–Hartwig coupling with $NiCl_2$ acting as a heterogeneous catalyst. This catalytic system enabled the synthesis of the appropriate secondary amines by coupling various substituted aryl iodides among substituted amines. After just 20 minutes of microwaving the reaction, medium to high yields of the products were achieved. The nickel chloride catalyst, in particular, showed exceptional reusability. It is a practical and sustainable choice for repeated reaction runs since it can be recycled for as many as three times with almost any activity loss. The research also showed that aryl bromides and chlorides might be included in the Buchwald–Hartwig coupling process. However, compared to aryl iodides, stronger conditions were required to accomplish effective reactions with these substrates. Importantly, even after extended response periods (one hour as opposed to 20 minutes), conversions were much lower when conventional heating was used as opposed to MW irradiation. This demonstrates the potential of MW heating to speed up and enhance the synthesis of N-substituted anilines and highlights the significant benefits of using it for quick and effective Buchwald–Hartwig coupling processes (Figure 4.63) [108].

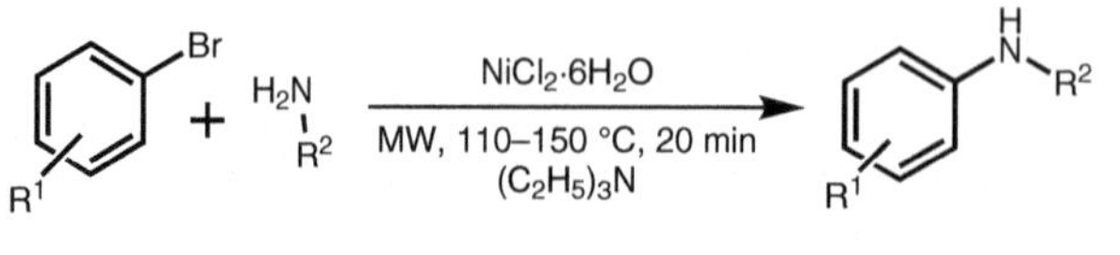

Figure 4.63 $NiCl_2 \cdot 6H_2O$ catalyzed Buchwald–Hartwig coupling under microwave irradiation. Source: Gupta et al. [108].

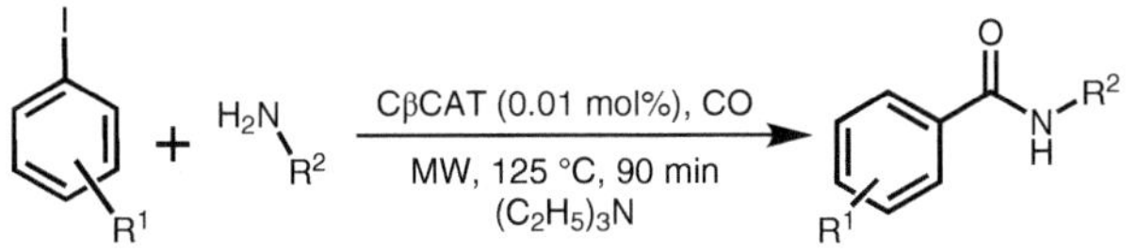

Figure 4.64 CβCAT (0.01 mol%) catalyzed carbonylation/carboxylation reaction under microwave irradiation. Source: Gaudino et al. [109].

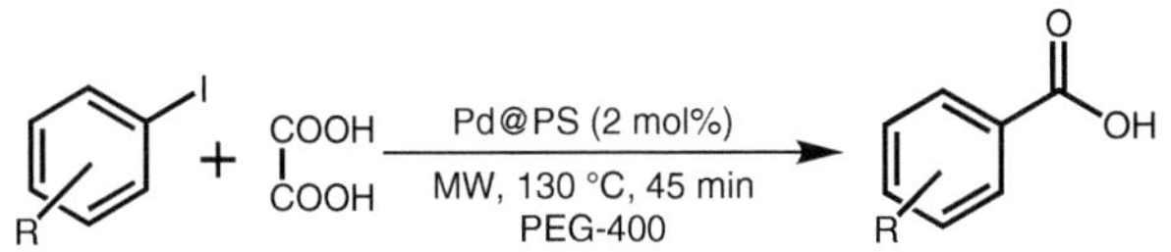

Figure 4.65 Pd@PS (2 mol%) catalyst-assisted carbonylation/carboxylation reaction under microwave irradiation. Source: Shil et al. [110].

Carbonylation/Carboxylation It has been suggested that using palladium catalysis, the reaction between amines and CO can transform aryl iodides into aryl amides. Cross-linked -cyclodextrin (CCAT) supported Pd(II) serves as the catalyst for this reaction. Notably, good yields were observed when starting with aryl iodides. Aryl chlorides, on the other hand, were discovered to be inert under the same circumstances, whereas the process with aryl bromides produced lower yields. When compared to traditional heating, the use of MW heating considerably increased reaction rates and decreased reaction times. Interestingly, while contrasting reactions carried out with MW heating to those done with traditional heating, no difference was seen in the total conversion and selectivity. These results show that cross-linked -cyclodextrin-supported Pd(II) has the potential to be a useful catalyst for the production of aryl amides from aryl iodides and amines. In addition, MW heating speeds up reactions, which might be useful in real-world settings even if it does not seem to have an impact on the reaction's overall conversion and selectivity (Figure 4.64) [109].

Shil et al. have carboxylated aryl iodides with the help of polystyrene assisted palladium nanoparticles (Pd@PS) as the catalyst and $(COOH)_2$ as the carbon source. The distinctive feature of this carboxylation process is oxalic acid's contribution. Oxalic acid breaks down under the effect of MW heating, releasing CO_2, which is used as the carbon source for the carboxylation process. The efficient production of carboxylated products is made possible by this process. However, even with the addition of more external CO_2, no product is produced when the same reaction is carried out under standard heating conditions. This striking disparity in reactivity emphasizes how important microwave heating and the in situ production of CO_2 from oxalic acid are to successfully drive the carboxylation process. Overall, the direct carboxylation of aryl iodides using this catalytic system based on Pd@PS nanoparticles and oxalic acid shows the importance of MW heating in enabling this transformation and highlights its potential application in organic synthesis (Figure 4.65) [110].

4.2.1.9 Micelliances Reactions

The majority of MCRs are classified as solid acid-catalyzed reactions because they frequently employ transition metal compounds as catalysts (such as oxides and salts). It is crucial to recognize that these transition metals' toxicity calls for extra care and awareness. While transition metals are useful catalysts in many processes, their poisonous qualities make them potentially dangerous to the environment and human health. When dangerous transition metal compounds are exposed to and released during chemical reactions, the environment may become contaminated, and serious health consequences may result. Researchers are currently looking into alternative catalytic systems that can produce equivalent or even greater catalytic activity while avoiding or eliminating the usage of hazardous transition metals, given the significance of sustainable and green chemistry concepts. The objective is to create greener, more environmentally friendly catalytic methods that can perform effective MCR without compromising environmental impact or safety. As a result, the search for non-toxic, eco-friendly catalysts has received a lot of attention lately, and a number of alternative catalytic systems, including organic catalysts and bio-inspired catalysts, are being investigated to allay worries about the toxicity of conventional transition metal catalysts.

It has been successfully performed to couple diamines, aldehydes, and diethyl phosphites in a single pot, producing -diaminophosphonates. Microwave-assisted heterogeneous catalysis was used to speed up this transition, and the catalyst used was silica-supported cerium(III) chloride. The reaction performed best when heated by MWs without the use of solvents, which facilitated the production of products quickly and effectively. For the synthesis of -diaminophosphonates, MW irradiation was a useful and efficient method since it increased the reaction rate and shortened the overall reaction time. Notably, the catalyst made of cerium-(III) chloride showed good reusability. With little activity loss, it may be recycled up to five times. This reusability feature emphasizes the heterogeneous catalyst's durability and robustness, making it a suitable choice for environmentally friendly and commercially feasible synthesis of -diaminophosphonates in industrial applications. As a whole, the one-pot three-component coupling, microwave-assisted heterogeneous catalysis, and the capacity for recycling of the cerium(III) chloride supported on silica catalyst make this methodology an appealing and effective route for the synthesis of -diaminophosphonates with potential applications in the pharmacological industries (Figure 4.66) [111].

Using a heterogeneous catalytic copper, Gupta et al. described a three-component Biginelli reaction for the production of dihydropyrimidinones. In this reaction, silica-supported copper-(II) chloride ($SiO_2/CuCl_2$) served as the catalyst. It was discovered that the reaction could be carried out satisfactorily utilizing either conventional heating with an injection of diphenic acid as an additive or MW heating in acetonitrile. While the reaction moved along more quickly when diphenic acid was used instead of standard heating methods, the yields were marginally higher. The reaction pace and product yields may be affected by choosing between MW heating and traditional heating. Shorter reaction times are a benefit of MW heating; however diphenic acid heating traditionally results in somewhat higher yields. Depending on the particular needs of the synthesis, such as the desired reaction rate, total yield, and convenience of handling, the heating technique may be chosen. Overall, the synthesis of dihydropyrimidinones, utilizing Biginelli

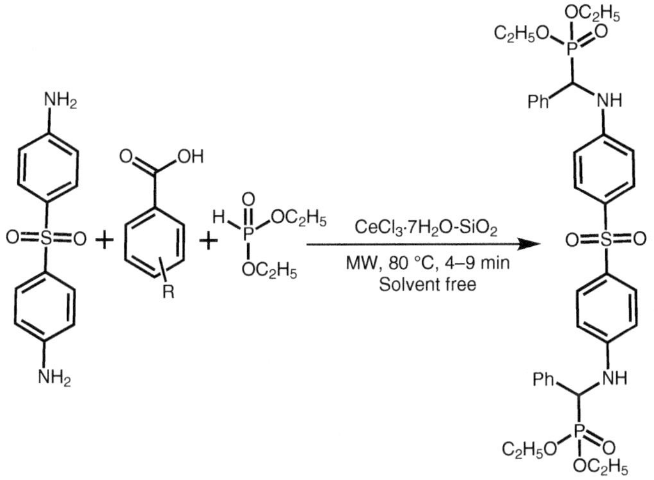

Figure 4.66 CeCl$_3$ · 7H$_2$O-SiO$_2$ catalyzed multicomponent reaction under microwave irradiation. Source: Devineni et al. [111].

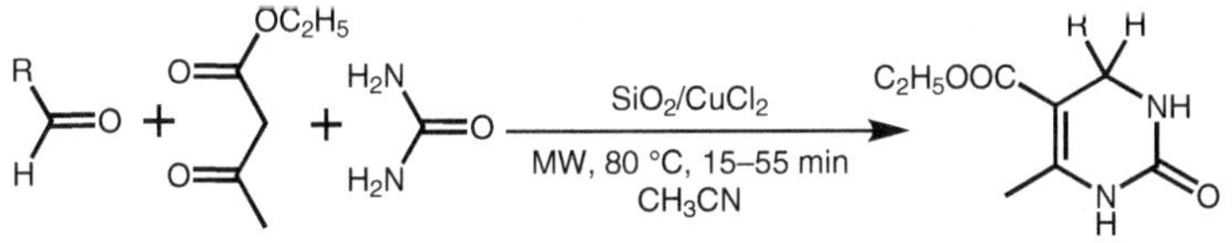

Figure 4.67 SiO$_2$/CuCl$_2$ catalyst-assisted multicomponent reaction under microwave irradiation. Source: Kour et al. [112].

reaction with a heterogeneous catalytic copper, offers a flexible and effective method, with the ability to use various heating techniques to produce desired results (Figure 4.67) [112].

Similar dihydropyrimidinones were made using effective catalytic nickel-(II) oxide. When MW heating was used, the use of NiO as a catalyst was shown to be more efficient than other known catalysts, allowing the reaction to continue in high yields and at an astonishingly rapid rate. The process was completed in MW heating in a very short period, requiring only four to eight minutes to generate the required products in high yields. In comparison, the process needed longer reaction durations of 60–90 minutes and had somewhat lower product yields when conventional heating was applied. The dihydropyrimidinone synthesis using nickel-(II) oxide under MW heating demonstrated its exceptional catalytic activity, highlighting its promise as a strong and effective catalyst for this transformation. The ability to accomplish quick reactions with excellent yields emphasizes the benefit of using MW heating to speed up chemical transformations and highlights the significance of choosing the right catalysts to improve synthetic processes [113].

TiO$_2$ was used by Safari and Gandomi–Ravandi as the catalyst for the previously mentioned process of the synthesis of dihydropyrimidinones. Particularly under no-solvent circumstances employing MW heating, this catalytic system

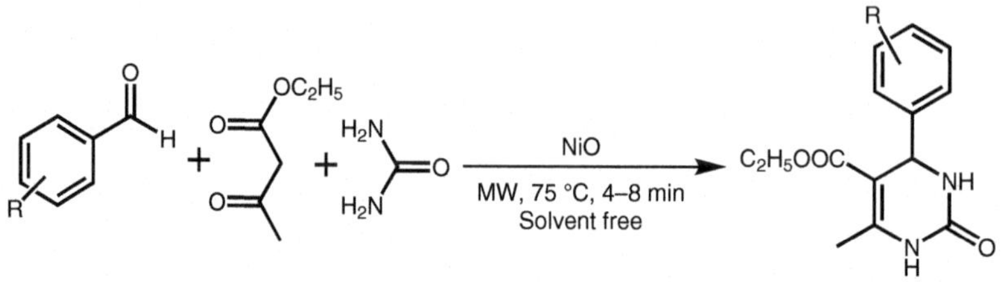

Figure 4.68 NiO catalyst-assisted multicomponent reaction under microwave irradiation. Source: Safari and Gandomi-Ravandi [114].

demonstrated remarkable efficacy, resulting in enhanced product yields within quick periods. The product yields that were attained varied from 75% to 96%. Outstanding reusability was shown by the TiO_2/multi-walled carbon nanotubes catalyst, which could be used up to five counts in a row without any significant loss in its catalytic ability. Due to its increased cost-effectiveness and sustainability, this reusability property is essential for the practical use of a catalyst in large-scale synthesis. TiO_2 supported on multi-walled carbon nanotubes has been successfully used as a catalyst, demonstrating the promise of such hybrid materials in heterogeneous catalysis, particularly for microwave-assisted processes. These nanomaterials may be combined to provide effective catalytic performance and recyclable materials, which makes them interesting candidates for a variety of organic transformations such as the production of dihydropyrimidinones (Figure 4.68) [114].

The efficacy of silica-supported cobalt-(II)-diimine(Co/SBA-15) was evaluated in a study examining Strecker-type reactions. Trimethylsilyl cyanide, aromatic aldehydes, and secondary amines were coupled together in the process. The experimenters discovered that both conservative and microwave heating techniques could be used to perform the reaction. However, compared to traditional heating, the use of MWs significantly enhanced the reaction rates and sped up the production of the final product. Contrary to popular belief, the technique of heating had no effect on the product yield despite variations in reaction speeds. This suggests that choosing between standard and microwave heating may have an impact on the reaction's speed, but not on the Strecker-type reaction's overall efficiency. The three-component coupling process was successfully enabled by the silica-supported cobalt-(II)-diimine (Co/SBA-15) catalyst, offering an effective and adaptable catalytic system for Strecker-type reactions. The versatility of being able to use both conventional and MW heating techniques allows for the optimization of the reaction conditions to meet particular synthetic needs without sacrificing the product yield (Figure 4.69) [115].

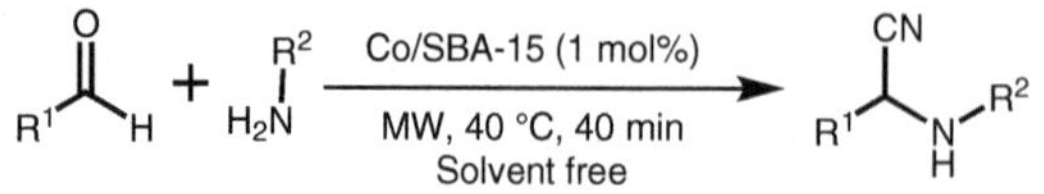

Figure 4.69 Co/SBA-15 (1 mol%) catalyst assisted MCR under microwave irradiation. Source: Rajabi et al. [115].

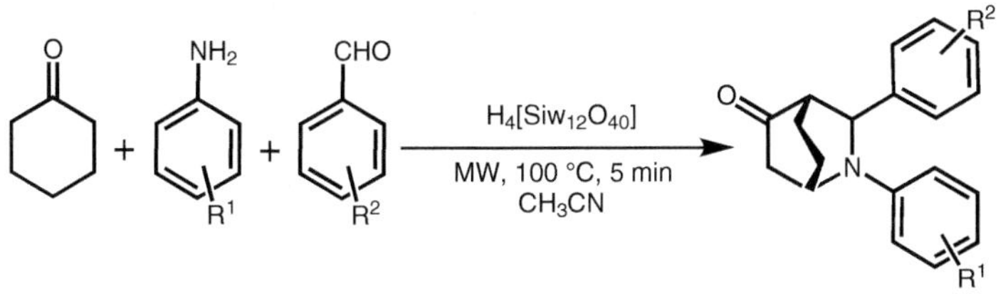

Figure 4.70 $H_4[SiW_{12}O_{40}]$ catalyst assisted multi-component reaction under microwave irradiation. Source: Borkin et al. [116].

In particular, azabicyclo-[2.2.2]octan-5-ones, which are prospective multi-component anti-Alzheimer's disease drugs, were effectively synthesized employing a microwave-assisted three-component process. When other acid catalysts were tested for the reaction's efficacy, it was discovered that silico-tungstic acid performed best. The heteropolyacid catalyst was successful in enabling the multi-component coupling as evidenced by the reaction's moderate to excellent yields of the desired products. Furthermore, the endo products formed with exceptional diastereo-selectivity, suggesting that the reaction was under strong stereochemical control. The intended azabicyclo-[2.2.2]-octan-5-ones may be produced in this MCR more quickly and effectively, thanks to the use of microwave-assisted catalysis, making it a viable strategy for the creation of possible anti-Alzheimer's disease medicines. In conclusion, this study emphasizes the importance of acid catalysts, particularly silico-tungstic acid, in facilitating effective MCRs for the production of biologically relevant compounds and highlights the possible application of microwave-assisted catalysis for speeding up drug discovery efforts. The method's utility for creating complex compounds with particular stereochemical configurations is further enhanced by the remarkable diastereo-selectivity attained (Figure 4.70) [116].

4.2.1.10 Click Chemistry

Nobel laureate Barry Sharpless first used the phrase "Click Chemistry" to refer to a class of chemical reactions that are distinguished by their effectiveness, high yields, wide range of compound applications, and excellent selectivity. Chemistry reactions are appealing for green and sustainable chemistry operations because they frequently feature straightforward reaction setups and ecologically safe solvents. Numerous scientific disciplines, including materials sciences, polymer chemistry, and organic/medical chemistry, have found widespread use for click chemistry. Click chemistry reactions are important instruments for the quick and effective synthesis of complex compounds and functional materials because of their adaptability and dependability. In the field of materials science, click chemistry is used to precisely regulate the characteristics and functions of complex molecular structures, such as dendrimers and nanomaterials. Click chemistry makes it possible to synthesize well-defined, functional polymers with specialized structures for a variety of applications in polymer chemistry. Click chemistry is essential to the effective synthesis of complex potential drugs and bioactive compounds in organic and medicinal chemistry. Drug discovery processes are streamlined by the high yields and selectivity of click chemistry reactions, which also enable the production of varied chemical libraries for screening and optimization [117].

There have been reports of 1,2,3-triazole synthesis in heterogeneous circumstances aided by MWs. An inert backbone was aided in bolstering Cu-(I) species to assist in the reaction. In an MCR, Jayaram et al. studied copper hydroxyapatite (Cu-HAP) as the catalyst to effectively produce triazoles from alkyl halides, alkynes, and sodium azide. The reaction may be carried out with a variety of substituted benzyl and alkyl halides with diverse functional groups because of the catalytic system's impressive adaptability. The study's usage of terminal alkynes, however, was restricted to phenyl and non-functionalized alkyl groups. In terms of the reduced reaction time needed, the use of MW irradiation offered significant benefits over traditional heating. The yields for 1,2,3-triazole production were the same for both heating techniques, illustrating the effectiveness of microwave-assisted reactions in producing equal product yields more quickly. A practical and eco-friendly three-component reaction using Cu-HAP as a heterogeneous catalyst and MW heating offers an effective and sustainable technique for the production of triazoles. The potential value of this approach in the synthesis of many 1,2,3-triazole derivatives with significant applications in medicinal chemistry and materials science is enhanced by the wide substrate range and simple reaction conditions [118].

Mnasri et al. used silica rods doped with copper particles as the heterogenous catalyst in a process involving alkyl bromides to produce 1,2,3-triazoles. Surprisingly, despite only utilizing a minimal catalytic usage of 1.25 mol%, the triazoles have been produced in remarkable yields, ranging from 69% to 96%. Under MW irradiation, the reaction progressed well, and the necessary products were produced in a brief period of time, usually between one and two hours. The catalytic material may be reprocessed up to 10 times without suffering a substantial reduction in its ability, thanks to the "silica rods doped with copper particles" exceptional recyclability. The recycling method resulted in just a 10% reduction in product yield, proving the catalyst's durability and stability over several reaction cycles. This catalytic system is appealing for the synthesis of 1,2,3-triazoles because it has a number of benefits, such as good yields, quick reaction times, and minimal catalyst loading. The feasibility and sustainability of this technology for prospective large-scale applications in the synthesis of various triazole derivatives are further improved by the recyclability of the copper-containing rod-shaped silica particles (Figure 4.71) [119].

Megia–Fernandez et al. provided an unusual modification of the aforementioned method in which cyclic sulfates and sulfamidates were used in a multi-component synthesis. The cyclic sulfate/sulfamidate is nucleophilically ring-opened by the azide in this reaction, which is followed by a cycloaddition among alkynes. With faster yields within 15 minutes, the reaction demonstrated outstanding efficiency,

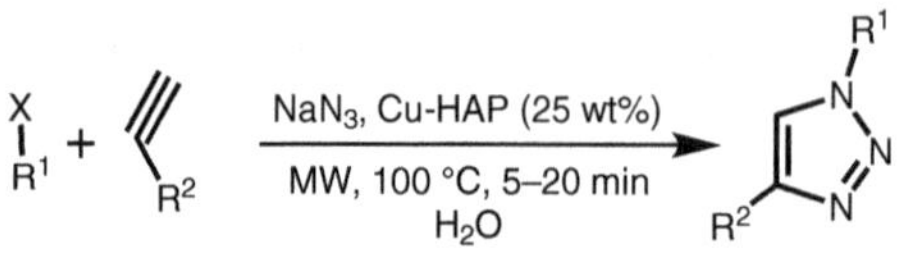

Figure 4.71 Cu-HAP (25 wt%) catalyst assisted click reaction under microwave irradiation. Source: Mnasri et al. [119].

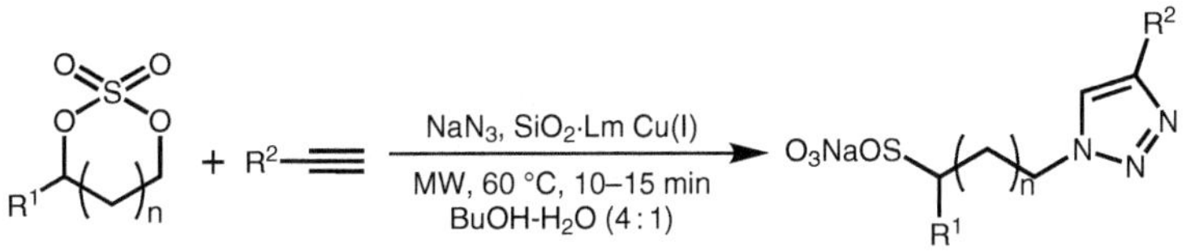

Figure 4.72 SiO$_2$ · Lm Cu(I) catalyst assisted click reaction under microwave irradiation. Source: Megia-Fernandez et al. [120].

Figure 4.73 Cu(OCH$_3$)$_2$/porous glass catalyzed click reaction under microwave irradiation.

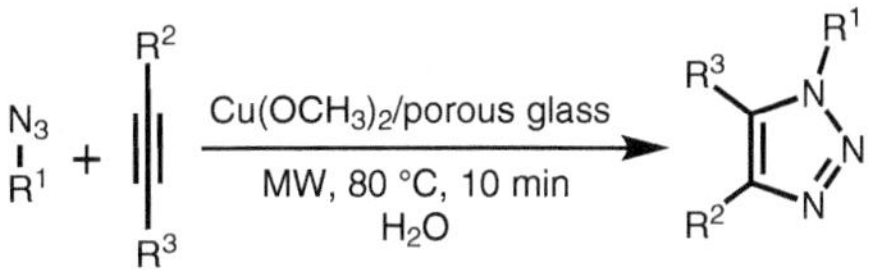

producing high yields. Additionally, isolating the products was simple and practical; it just required the catalyst to be filtered before the solvent evaporated. Utilizing cyclic sulfates and sulfamidates, the tandem nucleophilic ring-opening and cycloaddition procedure provides a quick and efficient way to access various 1,2,3-triazole derivatives. This technology is especially promising for the quick synthesis of triazole-containing compounds with potential applications in drug discovery and materials research due to the effective reaction conditions and simple product isolation (Figure 4.72) [120].

Copper acetate immobilized on porous glass was investigated by Jacob and Stolle as a potential catalyst for the preparation of triazoles. In their research, organic azides and alkynes were used as starting materials to start the reaction. Under microwave heating, the reaction moved forward well, producing medium to high product yields in comparatively fast reaction times. The reaction was also tried to be carried out in situ using sodium azide and alkyl chlorides to create azide. Even though the intended outcome might have been produced, the yields were substantially inferior in the reaction that started with an organic azide. This finding implies that employing organic azides directly as starting materials is preferable for producing greater yields of triazoles when using catalytic copper acetate immobilized on porous glass. Overall, this work offers insightful information on the potential of catalytic copper acetate immobilized on porous glass for the effective synthesis of 1,2,3-triazoles. When employing organic azides as the starting materials, the process offers a flexible way for the quick production of a variety of triazole derivatives (Figure 4.73) [121].

4.3 Conclusion

There is no question that the combination of MW activation/ultrasonic activation and metal-based heterogeneous catalysis opens up a wide variety of opportunities for developing green synthesis methods. This formerly unorthodox heating process has developed into the preferred approach for the majority of synthetic chemists as MW and ultrasonic instruments have become common equipment in research labs. It is not unexpected that solid catalysts are widely used in microwave-assisted or ultra-sound irradiated organic synthesis. The past several years have seen a lot of research toward developing more environmentally friendly synthetic techniques with the goal of producing unique molecules with distinctive structural features and noteworthy biological activities. The multi-component processes used in the synthesis of N- and O-heterocyclic, as well as acyclic compounds, are examined in further detail in this thorough overview. It is important to apply eco-friendly practices, which include heterogeneous catalysis, solvent recycling, MW irradiation, and ultrasonic irradiation. Shorter reaction times, less energy use, and the ability to efficiently recycle heterogeneous catalysts are just a few benefits of using such techniques to promote ecologically friendly protocols, as shown by the variety of applications shown in this chapter.

Acknowledgments

The authors are grateful to GITAM, deemed to be university, Visakhapatnam and Bangalore, India, for financial and research support.

References

1 Li, Z., Dong, J., Zhang, H. et al. (2021). *Nanoscale Adv.* 3 (1): 41–72.
2 Ahmadian-Fard-Fini, S., Ghanbari, D., Amiri, O., and Salavati-Niasari, M. (2021). *RSC Adv.* 11: 22805–22811.
3 Vinatoru, M. and Mason, T.J. (2021). *Molecules* 26: 755–765.
4 Ji, R., Pflieger, R., Virot, M., and Nikitenko, S.I. (2018). *J. Phys. Chem. B.* 122 (27): 6989–6994.
5 Prasanthkumar, K.P., Rayaroth, M.P., and Alvarez-Idaboy, J.R. (2020). *J. Phys. Chem. B* 124 (29): 6245–6256.
6 Amaniampong, P.N. and Jerome, F. (2020). *Curr. Opin. Green Sustainable Chem.* 22: 7–12.
7 Naeimi, H. and Shaabani, R. (2017). *Ultrason. Sonochem.* 34: 246–254.
8 Nguyen, Q.N.K., Yen, N.T., Hau, N.D., and Tran, H.L. (2020). *J. Chem.* 1–8.
9 Lima, C.G.S., Moreira, N.M., Paixao, M.W., and Corrêa, A.G. (2018). *Curr. Opin. Green Sustainable Chem.* 15: 7–12.
10 Kosinov, N., Liu, C., Hensen, E.J.M., and Pidko, E.A. (2018). *Chem. Mater.* 30 (10): 3177–3198.
11 Romanazzi, G., Mastrorilli, P., Latronico, M. et al. (2018). *Open Chem.* 16: 520–534.
12 Dzhardimalieva, G.I., Zharmagambetova, A.K., Kudaibergenov, S.E., and Uflyand, I.E. (2020). *Kinet. Catal.* 61 (2): 198–223.
13 Yue, Y.N., Meng, W.J., Liu, L. et al. (2018). *Electrochim. Acta* 260: 606–613.
14 Rasal, S.A., Tamore, M.S., and Shimpi, N.G. (2019). *ChemistrySelect* 4 (8): 2293–2300.

15 Achary, L.S.K., Nayak, P.S., Barik, B. et al. (2020). *Catal. Today* 348: 137–147.
16 Mo, Y., Zhao, X., Yuan, T., and He, B. (2018). *J. Chem. Technol. Biotechnol.* 93 (5): 1388–1393.
17 Taheri-Ledari, R., Rahimi, J., and Maleki, A. (2019). *Ultrason. Sonochem.* 59: 104737–104746.
18 Edrisi, M. and Azizi, N. (2020). *J. Iran. Chem. Soc.* 17 (4): 901–910.
19 Verma, P., Pal, S., Chauhan, S. et al. (2020). *J. Mol. Struct.* 1203: 127410–127420.
20 Esmaeilpour, M., Javidi, J., Dehghani, F., and Nowroozi Dodeji, F. (2015). *RSC Adv.* 5 (34): 26625–26633.
21 Pourian, E., Javanshir, S., Dolatkhah, Z. et al. (2018). *ACS Omega* 3 (5): 5012–5020.
22 Maleki, A. and Aghaei, M. (2017). *Ultrason. Sonochem.* 38: 115–119.
23 Sadjadi, S. and Bahri-Laleh, N. (2018). *J. Porous Mater.* 25 (3): 821–833.
24 Safa, K.D., Esmaili, M., and Allahvirdinesbat, M. (2016). *J. Iran. Chem. Soc.* 13 (2): 267–277.
25 Safa, K.D., Allahvirdinesbat, M., Namazi, H., and Panahi, P.N. (2015). *C.R. Chim.* 18: 883–890.
26 Ghasemzadeh, M.A. and Ghaffarian, F. (2020). *Appl. Organomet. Chem.* 34: 1–10.
27 Vasudhevan, S. and Joel Karunakaran, R. (2013). *Int. J. Chem. Tech Res.* 5: 2844–2853.
28 White, D.C., Greenwood, T.D., Downey, A.L. et al. (2004). *Bioorg. Med. Chem.* 12 (21): 5711–5717.
29 Maleki, A., Aghaei, M., Hafizi-Atabak, H.R., and Ferdowsi, M. (2017). *Ultrason. Sonochem.* 37: 260–266.
30 Jadhav, S., Farooqui, M., Chavan, P. et al. (2020). *Polycyclic Aromat. Compd.* 1–9.
31 Sadeghzadeh, S.M. (2015). *RSC Adv.* 5: 68947–68952.
32 Safari, J., Gandomi-Ravandi, S., and Akbari, Z. (2013). *J. Adv. Res.* 4 (6): 509–514.
33 Hajizadeh, Z., Radinekiyan, F., Eivazzadeh-keihan, R., and Maleki, A. (2020). *Sci. Rep.* 10: 1–11.
34 Metaxas, A.C. and Meredith, R.J. (1983). *IEEE Power Eng. Series* 4.
35 Adlington, K., Joe Jones, G., El Harfi, J. et al. (2013). *Macromolecules* 46 (10): 3922–3930.
36 Kappe, C.O. and Dallinger, D. (2009). *Mol. Diversity* 13 (2): 71–193.
37 Kappe, C.O., Pieber, B., and Dallinger, D. (2013). *Angew. Chem. Int. Ed.* 52 (4): 1088–1094.
38 Hayes, B.L. (2002). *Microwave Synthesis – Chemistry at the Speed of Light*. Metthews: CEM Publishing.
39 Stuerga, D. and Gaillard, P. (1996). *Tetrahedron* 52 (15): 5505–5510.
40 Meredith, R. (1998). *Engineers' Handbook of Industrial Microwave Heating*. Stevenage: Lightning Source UK Ltd.
41 Baig, R.B.N. and Varma, R.S. (2014). *Chem. Soc. Rev.* 41 (4): 1559–1584.

42 Whittles, G. (1997). *A Basic Introduction to Microwave Chemistry*. Oxford: Oxford University Press.
43 Mackie, R.K., Smith, D.M., and Aitken, R.A. (1999). *Guidebook to Organic Synthesis*, vol. 5, 69. Harlow: Pearson.
44 Marvi, O., Giahi, M., Ayub, P.P., and Nikpasand, M. (2014). *J. Serb. Chem. Soc.* 79: 921.
45 Rocchi, D., Gonzalez, J.F., and Menendez, J.C. (2014). *Molecules* 19: 7317.
46 Limnios, D. and Kokotos, C.G. (2013). *RSC Adv.* 3: 4496.
47 Singh, V. and Sharma, Y.C. (2017). *Energy Convers. Manage.* 138: 627.
48 Dasgupta, S., Morzhina, E., Schäfer, C. et al. (2016). *Top. Catal.* 59: 1207.
49 Liao, F.X., Wang, Y.G., and Zhu, Q. (2014). *Synth. Commun.* 44: 161.
50 Atanassova, V., Ganno, K., Kulkarni, A. et al. (2011). *Appl. Clay Sci.* 53: 220.
51 Kumar, M., Sharma, S., Thakur, K. et al. (2017). *Asian J. Org. Chem.* 6: 342.
52 Mackie, R.K., Smith, D.M., and Aitken, R.A. (1999). *Guidebook to Organic Synthesis*, vol. 7, 120. Harlow: Pearson.
53 Singh, R. and Kumar, R. (2013). *Asian J. Chem.* 25: 4935.
54 Cho, H., Török, F., and Török, B. (2014). *Green Chem.* 16: 3623.
55 Jeganathan, M. and Pitchumani, K. (2014). *ACS Sustainable Chem. Eng.* 2: 1169.
56 Solan, A., Nişanci, B., Belcher, M. et al. (2014). *Green Chem.* 16: 1120. J. Young, C. Schäfer, A. Solan, A. Baldrica, M. Belcher, B. Nişanci, K. A. Wheeler, E. Trivedi, B. Törökand R. Dembinski, *RSC Adv.*, 6 (2016) 107081.
57 Borkin, D.A., Puscau, M., Carlson, A. et al. (2012). *Org. Biomol. Chem.* 10: 4505.
58 Valizadeh, H., Dinparast, L., Noorshargh, S., and Heravi, M.M. (2016). *C.R. Chim.* 19: 395.
59 Hasaninejad, A., Shekouhy, M., and Zare, A. (2012). *Catal. Sci. Technol.* 2: 201.
60 Rudnitskaya, A., Borkin, D.A., Huynh, K. et al. (2010). *ChemMedChem* 5: 384.
61 Kokel, A. and Török, B. (2017). *Green Chem.* 19: 2515.
62 Khan, K. and Siddiqui, Z.N. (2015). *Ind. Eng. Chem. Res.* 54: 6611.
63 Biggs-Houck, J.E., Younai, A., and Shaw, J.T. (2010). *Curr. Opin. Chem. Biol.* 14: 371.
64 Kulkarni, A. and Török, B. (2010). *Green Chem.* 12: 875.
65 Nişancı, B., Ganjehyan, K., Metin, Ö. et al. (2015). *J. Mol. Catal. A: Chem.* 409: 191.
66 Zhang, M., Liu, P., Liu, Y. et al. (2016). *RSC Adv.* 6: 106160.
67 Safari, J., Gandomi-Ravandi, S., and Borujeni, M.B. (2013). *J. Chem. Sci.* 125: 1063.
68 Olah, G.A. (ed.) (1964). *Friedel-Crafts and Related Reactions*. New York: Interscience; Olah, G.A., Prakash, G.K.S., Molnár, Á., and Sommer, J. (2009). *Superacid Chemistry*, vol. 2. Hoboken: Wiley.
69 Wu, Y., Zhang, C., Liu, Y. et al. (2012). *BioResources* 7: 5950.
70 Venkatesha, N.J., Chandrashekara, B.M., Prakash, B.S.J., and Bhat, Y.S. (2014). *J. Mol. Catal. A: Chem.* 392: 181.
71 Bhanja, P., Modak, A., Chatterjee, S., and Bhaumik, A. (2017). *ACS Sustainable Chem. Eng.* 5: 2763.
72 Marwan, I.E. (2016). *Energy Convers. Manage.* 117: 319.

73 Buasri, A., Lukkanasiri, M., Nernrimnong, R. et al. (2016). *Korean J. Chem. Eng.* 33: 3388.
74 Wang, Y. and Chen, B. (2016). *Catal. Today* 278: 335.
75 Hasan, Z., Yoon, J.W., and Jhung, S.H. (2015). *Chem. Eng. J.* 278: 105.
76 Ma, G., Hu, W., Pei, H. et al. (2015). *Energy Convers. Manage.* 90: 41.
77 Tangy, A., Kumar, V., Pulidindi, I. et al. (2016). *Energy Fuels* 30: 10602.
78 Tchounwou, P.B., Yedjou, C.G., Patlolla, A.K., and Sutton, D.J. (2012). Heavy metal toxicity and the environment. In: *Molecular, Clinical and Environmental Toxicology* (ed. A. Luch) ExperientiaSupplementum 101, 133. Basel: Springer.
79 Romano, P.N., de Almeida, J.M.A.R., Carvalho, Y. et al. (2016). *ChemSusChem* 9: 3387.
80 Schäfer, C., Ellstrom, C.J., Cho, H., and Török, B. (2017). *Green Chem.* 19: 1230.
81 Farhadi, S., Siadatnasab, F., and Kazem, M. (2011). *J. Chem. Res.* 35: 104.
82 Farhadi, S., Kazem, M., and Siadatnasab, F. (2011). *Polyhedron* 30: 606.
83 Lipshutz, B.H., Frieman, B.A., Butler, T., and Kogan, V. (2006). *Angew. Chem., Int. Ed.* 45: 800.
84 Mackie, R.K., Smith, D.M., and Aitken, R.A. (1999). *Guidebook to Organic Synthesis*, vol. 9, 183. Harlow: Pearson.
85 Crosswhite, M., Hunt, J., Southworth, T. et al. (2013). *ACS Catal.* 3: 1318.
86 Baruah, D., Saikia, U.P., Pahari, P., and Konwar, D. (2015). *Tetrahedron Lett.* 56: 2543.
87 Caporaso, M., Cravotto, G., Georgakopoulos, S. et al. (2014). *Beilstein J. Org. Chem.* 10: 1454.
88 Martins, L.M.D.R.S., Hazra, S., Guedes da Silva, M.F.C., and Pombeiro, A.J.L. (2016). *RSC Adv.* 6: 78225.
89 Pang, Y. and Lei, H. (2016). *Chem. Eng. J.* 287: 585.
90 Guan, Y.H., Ma, J., Ren, Y.M. et al. (2013). *Water Res.* 47: 5431.
91 Pan, W., Zhang, G., Zheng, T., and Wang, P. (2015). *RSC Adv.* 5: 27043.
92 Heck, R.F. (1968). *JACS* 90: 5518.
93 Molnár, Á. (2013). *Palladium-Catalyzed Coupling Reactions: Practical Aspects and Future Developments*. Weinheim: Wiley-VCH.
94 Ciriminna, R., Pandarus, V., Gingras, G. et al. (2013). *ACS Sustainable Chem. Eng.* 1: 57.
95 Huang, N., Xu, Y., and Jiang, D. (2014). *Sci. Rep.* 4: 7228.
96 Leonhardt, S.E.S., Stolle, A., Ondruschka, B. et al. (2010). *Appl. Catal., A* 379: 30.
97 Taher, A., Yoo, H.-Y., Oh, P. I. et al. (2016). *Bull. Korean Chem. Soc.* 37: 1908.
98 Pan, D., Wu, A., Li, P. et al. (2014). *J. Chem. Res.* 38: 715.
99 Baran, T., Sargin, I., Kaya, M., and Mentes, A. (2016). *Carbohydr. Polym.* 152: 181.
100 Schmidt, B. and Riemer, M. (2015). *Eur. J. Org. Chem.* 3760.
101 Ciriminna, R., Pandarus, V., Gingras, G. et al. (2012). *RSC Adv.* 2: 10798.
102 Smith, S.E., Siamaki, A.R., Gupton, B.F., and Carpenter, E.E. (2016). *RSC Adv.* 6: 91541.

103 Guha, N.R., Reddy, C.B., Aggarwal, N. et al. (2012). *Adv. Synth. Catal.* 354: 2911.
104 Schruder, C.W., Organ, M.G., and Pietro, W.J. (2016). *Curr. Nanosci.* 12: 448.
105 Cravotto, G., Gaudino, E.C., Tagliapietra, S. et al. (2012). *Green Process. Synth.* 1: 269.
106 Dawood, K.M., Solodenke, W., and Kirschning, A. (2007). *ARKIVOC* 5: 104.
107 Narasimharao, K., Al-Sabban, E., Saleh, T.S. et al. (2013). *J. Mol. Catal. A: Chem.* 379: 152.
108 Gupta, A.K., Rao, G.T., and Singh, K.N. (2012). *Tetrahedron Lett.* 53: 2218.
109 Gaudino, E.C., Carnaroglio, D., Martina, K. et al. (2015). *Org. Process Res. Dev.* 19: 499.
110 Shil, A.K., Kumar, S., Reddy, C.B. et al. (2015). *Org. Lett.* 17: 5352.
111 Devineni, S.R., Doddaga, S., Donka, R., and Chamarthi, N.R. (2013). *Chin. Chem. Lett.* 24: 759.
112 Kour, M., Gupta, S., Paul, R., and Gupta, V.K. (2014). *J. Mol. Catal. A: Chem.* 392: 260.
113 Lal, J., Sharma, M., Sahu, P.K., and Agarwal, D.D. (2013). *Proc. Natl. Acad. Sci., India* 83: 187.
114 Safari, J. and Gandomi-Ravandi, S. (2014). *New J. Chem.* 38: 3514.
115 Rajabi, F., Nourian, S., Ghiassian, S. et al. (2011). *Green Chem.* 13: 3282.
116 Borkin, D., Morzhina, E., Datta, S. et al. (2011). *Org. Biomol. Chem.* 9: 1394.
117 Kolb, H.C., Finn, M.G., and Sharpless, K.B. (2001). *Angew. Chem., Int. Ed.* 40: 2004.
118 Kale, S., Kahandal, S., Disale, S., and Jayaram, R. (2012). *Curr. Chem. Lett.* 1: 69.
119 Mnasri, N., Nyalosaso, J.L., Colacino, E. et al. (2015). *ACS Sustainable Chem. Eng.* 3: 2516.
120 Megia-Fernandez, A., Ortega-Muçoz, M., Hernandez-Mateo, F., and Santoyo-Gonzalez, F. (2012). *Adv. Synth. Catal.* 354: 1797.
121 Jacob, K. and Stolle, A. (2014). *Synth. Commun.* 44: 1251.

5
Microwave- and Ultrasonic-Assisted Coupling Reactions

Sandeep Yadav[1,2], Anirudh P.S. Raman[1,2], Kashmiri Lal[3], Pallavi Jain[2], and Prashant Singh[1]

[1] University of Delhi, Atma Ram Sanatan Dharma College, Department of Chemistry, Dhaula Kuan, Delhi 110021, India
[2] SRM Institute of Science and Technology, Department of Chemistry, Faculty of Engineering and Technology, NCR Campus, Uttar Pradesh 201204, India
[3] Guru Jambheshwar University of Science and Technology, Department of Chemistry, Hisar, Haryana 125001, India

5.1 Introduction

Over the years, there have been major breakthroughs in the field of organic synthesis, leading to the development of new and effective methods for the synthesis of compounds. Microwave- and ultrasound-assisted coupling reactions have emerged as powerful tools in modern synthetic chemistry, offering accelerated reaction rates, enhanced yields, and improved selectivity compared to traditional heating methods [1]. These innovative techniques harness the unique properties of microwave and ultrasound energy to facilitate the formation of chemical bonds, revolutionizing the way chemists approach coupling reactions. Microwave irradiation and ultrasound waves have distinct mechanisms of action but share common features that make them valuable in organic synthesis. Microwave irradiation utilizes electromagnetic waves ranging from 300 MHz to 300 GHz, causing the rotation and vibration of polar molecules within the reaction mixture [2]. This excitation of molecular motion generates localized heating and significantly increases the collision frequency and energy of the reactant, arising in accelerated reaction kinetics [3]. Ultrasound waves, on the other hand, are mechanical vibrations with frequencies above the audible range (typically 20 kHz to several MHz) [4]. When ultrasound waves propagate through a liquid medium, they produce compression and rarefaction cycles, resulting in the formation of small gas bubbles called cavitation bubbles. The violent collapse of these bubbles creates localized hotspots, high pressures, and intense shear forces, which enhance mass transfer, promote mixing, and induce chemical reactions [5]. Both microwave- and ultrasound-assisted coupling reactions offer several advantages over conventional heating methods [6]. Firstly, they provide rapid and uniform heating throughout the reaction mixture, allowing for faster

Green Chemical Synthesis with Microwaves and Ultrasound, First Edition.
Edited by Dakeshwar Kumar Verma, Chandrabhan Verma, and Paz Otero Fuertes.
© 2024 WILEY-VCH GmbH. Published 2024 by WILEY-VCH GmbH.

reaction rates and shorter reaction times. Additionally, the localized energy delivery reduces undesired side reactions and enhances selectivity, enabling the synthesis of complex molecules with higher yields and fewer byproducts [7]. Moreover, the noncontact nature of these techniques prevents contamination and facilitates the scale-up of reactions. In recent years, microwave- and ultrasound-assisted coupling reactions have found applications in various types of synthetic transformations, including carbon–carbon, and carbon–heteroatom bond formation. These innovative methods have demonstrated their efficiency and versatility in various chemical disciplines, from synthesizing pharmaceuticals and natural products to preparing advanced materials and polymers [8, 9].

The article provides an overview of the principles, advantages, and applications of microwave- and ultrasound-assisted coupling reactions in organic synthesis. By exploring the underlying mechanisms and showcasing representative examples, we will highlight the transformative impact of these techniques on modern synthetic chemistry and shed light on their potential for future developments in the field.

5.2 Microwave

Microwaves are a type of nonionizing electromagnetic radiation that falls between radio waves and infrared (IR) wavelengths. Microwaves with wavelengths between 1 mm and 1 m interact with materials in a number of ways [10]. They can be reflected by metals, absorbed by materials (resulting in heating), or transmitted through good insulators without causing significant heating. The unique characteristic of microwaves to act as an energy source has been utilized for heating food for several decades [11]. In recent years, its application has expanded to chemical processes, particularly organic synthesis. Microwave-assisted organic synthesis takes advantage of the selective absorption of energy from microwaves by polar molecules as shown in Figure 5.1 [12]. When exposed to microwaves, polar molecules rapidly absorb the energy and convert it into heat, promoting accelerated chemical reactions. The first microwave heating studies used household microwave ovens and strong dielectric solvents like dimethylformamide (DMF) and dimethylsulfoxide (DMSO). These early studies demonstrated enhanced reaction rates, attributed to the heating effect of polar solvents and pressure effects [13]. Whereas liquid-phase reactions using high dielectric solvents faced limitations such as increased pressure and the need for specialized sealed vessels, nonpolar solvents, such as hydrocarbons (e.g. toluene, hexane) or chlorinated solvents (e.g. dichloromethane), have low dielectric constants and generally exhibit lower microwave absorption compared to polar solvents. Nonpolar solvents with lower dielectric constants may selectively heat specific reactants or intermediates with high dipole moments or polar functionalities [14]. This can result in localized heating and promote specific reaction pathways or transformations. It can control reaction selectivity by providing different reaction environments compared to polar solvents. They can influence reaction equilibrium, alter the solvation of reactive species, or favor specific conformations, affecting reaction outcomes [15].

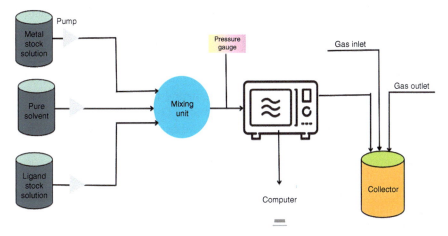

Figure 5.1 Diagram of continuous flow of microwave reaction. Source: Adapted from Gupta et al. [12].

Some reactions may involve hazardous or highly reactive species. Nonpolar solvents can provide a safer environment by reducing the chances of side reactions or undesired thermal effects, as they typically have lower boiling points and lower reactivity under microwave irradiation.

5.2.1 Microwave-Assisted Coupling Reactions

In recent years, there has been a growing trend in adopting economical chemical processes that utilize MW irradiation and water as a preferred solvent. In the realm of green synthesis, there is considerable interest in solvent-free protocols, which offer significant advantages [7]. Das et al. developed a one-step microwave-assisted approach for synthesizing pyrrole derivatives four linked with coumarins. The strategy involved utilizing a catalyst, $InCl_3$, at a concentration of 30 mol%. The reaction involved coumarin, dialkyl acetylene dicarboxylate, and nitrostyrene producing the required product. The methodology was discovered to only work with 6-amino coumarin ($R_1 = R_3 = H$; $R_2 = NH_2$), underlining the procedure's limits (Scheme 5.1) [16]. 2,4-Dihalogenated pyrimidine with hetero boronic acid

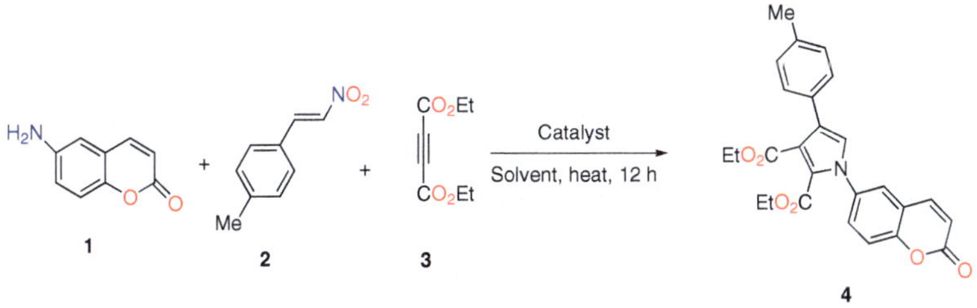

Scheme 5.1 One-pot synthesis of coumarins-containing pyrrole.

undergoes Suzuki coupling reaction with the application of microwave irradiation giving an excellent yield of C4-substituted pyrimidines [17].

A novel diaryletherification protocol using MW irradiation and copper catalysts was developed. With the help of this technique, aryl halides can be cross-coupled with ortho-prenylated phenols that are sterically inhibited. The ideal conditions for this procedure allow for microwave irradiating a Cu and bidentate ligand combination composed of 10 mol% of CuI and N-(2-fluorophenyl)picolinamide, 2 equiv. of K_3PO_4, and CH_3CN for 0.5 hours [9]. Mishra et al. presented a metal-free method for producing a wide range of coumarin-linked pyrrole derivatives. This approach involved a multicomponent reaction between compounds (**5**), (**6**), and different amines (**7**, **9**, **11**, or **13**). The reaction, carried out under MW heating (130 °C), demonstrated excellent efficiency, resulting in the products (**8**, **10**, **12**, and **14**) with high yields in 30 minutes as in Scheme 5.2 [18].

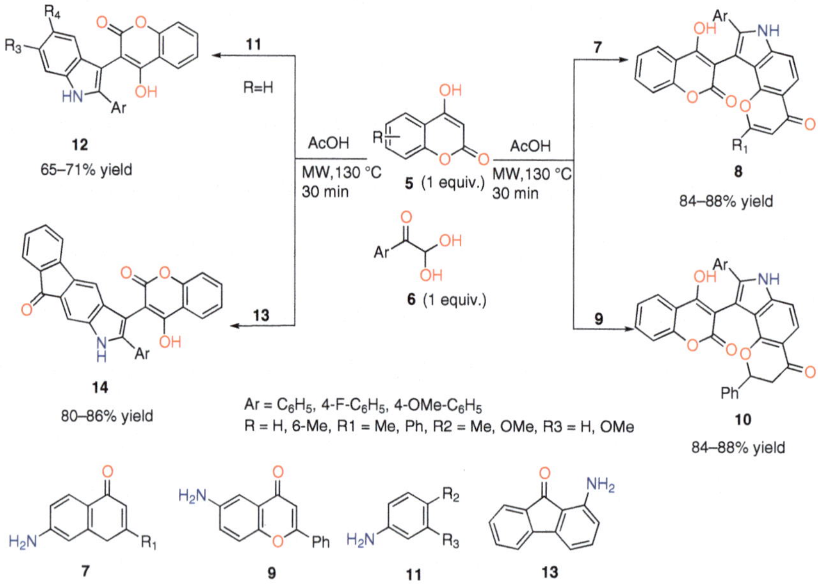

Scheme 5.2 One-pot multicomponent method.

The azide-alkyne Huisgen (3 + 2) cycloaddition is a highly versatile synthetic approach widely utilized in heterocyclic chemistry. It involves a 1,3-dipolar cycloaddition reaction between an organoazide (R–N3) and a terminal or internal alkyne (C≡C) to yield a mixture of two regioisomers, particularly when using asymmetric alkynes. Consequently, the classical Huisgen (3 + 2) cycloaddition does not fulfill the criteria of a true click reaction. However, an alternative copper-catalyzed variation enables the reaction to be conducted in aqueous conditions, even at room temperature. Moreover, unlike the classical Huisgen 1,3-dipolar cycloaddition, the copper-catalyzed variant selectively affords the 1,4-disubstituted regioisomers [19, 20]. Saptal and his colleagues utilized amine-functionalized graphene oxide as a host material to stabilize Pd nanoparticles (Pd@APGO). The Suzuki–Miyaura

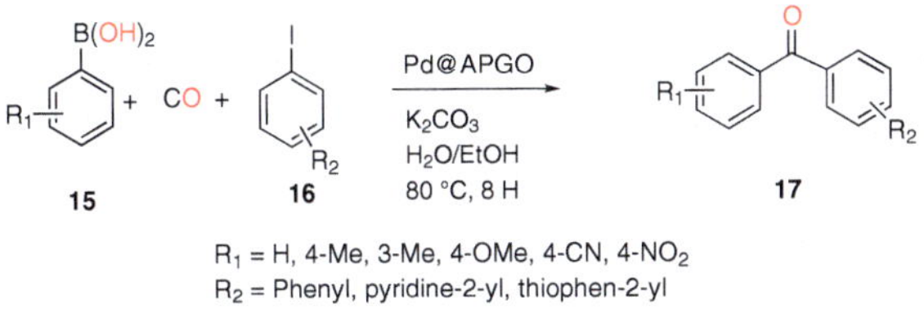

Scheme 5.3 SMC reaction for the formation of biaryls.

cross-coupling reaction (SMCR), involving aryl halides and arylboronic acids, was then carried out using this catalyst, as depicted in Scheme 5.3. K_2CO_3 was used as the base, and water and ethanol (H_2O/EtOH) as solvents. The reaction was carried out at 80 °C for six hours. The desired results of this reaction were biaryls. Pd nanoparticles were added to the amine-functionalized graphene oxide (Pd@APGO), which resulted in improved stability and increased catalytic activity. The researchers successfully completed cross-coupling reactions to create biaryls by tuning the reaction parameters, such as the base, solvent, and temperature [21].

Akiyama and colleagues synthesized sulfur-modified ruthenium nanoparticles anchored on gold (SARu) and evaluated them as a catalyst for the SMCR. Arylboronic acids and aryl halides are combined to produce biphenyls under moderate conditions. In particular, the investigators used DME as the solvent and K_3PO_4 as the base, at 80–100 °C for 6–10 hours [22]. Sonogashira reactions were employed to synthesize derivatives of 4-formyl-2-arylbenzofurans, with $PdCl_2(PPh_3)_3$ serving as the catalyst [23]. The prevalence can be attributed to the favorable electron characteristics of the pyrrole ring, which contribute to enhanced binding with enzymes and receptors. Additionally, the presence of the pyrrole ring allows for further modifications to the molecular scaffold, enabling the attainment of the desired activity profile [24, 25]. The compound **20** and its diazo ligand were synthesized by the Jha group and are shown in Scheme 5.4. In the first technique, ethylene glycol was present while **18** and **19** were exposed to 300 W of radiation

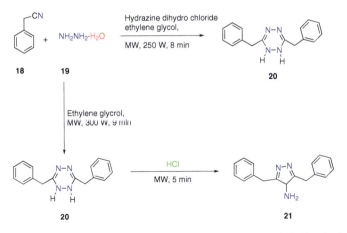

Scheme 5.4 Pathway to design 3,5-dibenzyl-4-amino1,2,4-triazole derivative.

for nine minutes. The result of the reaction was the dark pink compound **20**. The obtained compound **20** was microwave irradiated for five minutes while being in the presence of HCl in the second step, which caused a rearrangement and the creation of the compound (**21**). The needed ligand was synthesized using yet another microwave technique [26].

Sayed and group conducted the synthesis of bioactive pyran derivatives, as shown in Scheme 5.5. In the initial step, compound **24** was formed by 2.5 minutes of irradiation on a 1 : 1 ratio of compound **23** and urea. Another microwave-asserted reaction involved mixing an equimolar amount of phenacyl bromide and compound **23** in N,N-DMF for two minutes, leading to the synthesis of compound **26**. Compound **25** was prepared by irradiating an equimolar mixture of compound **23** with **22**, may take place through the attack of the carbanion of malononitrile on the diazonium salt [27].

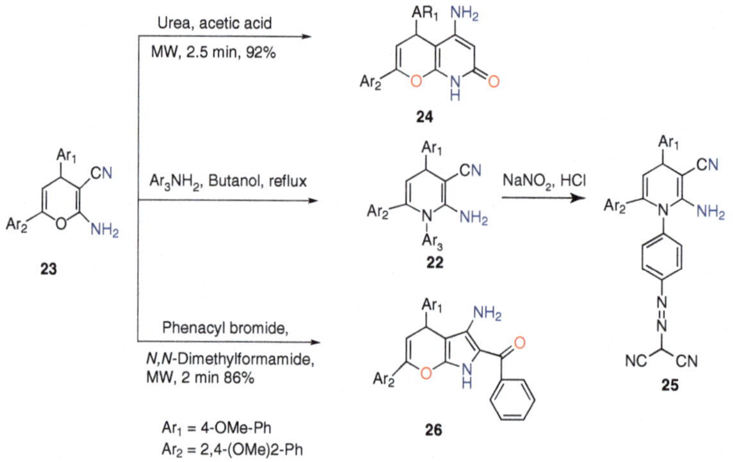

Scheme 5.5 Synthesis of dihydropyridine derivatives.

Filho et al. conducted a reaction for the production of compound **29**, which was successfully synthesized with an impressive 80% yield. The reaction involved a mixture of compounds **28** and **27**, which were combined with 4.5 mmol of NaOH in ethanol (Scheme 5.6). The reaction was subjected to 50 W MW irradiation for a duration ranging from 10 minutes to 3 hours [28].

Shaikh et al. reported the synthesis of triazolothiadizepinylquinolines **32** using MW-assisted methods and highlighted their pharmacological significance, as shown in Scheme 5.7. In this process, a 1 : 1 ratio of compounds **30** and **31** were reacted K_2CO_3 under microwaves at 120 W and 110–120 °C temperature range for 5–10 minutes. This reaction yielded compound **32** of 84–92% [20].

Conventional C–H activation method or transmetallation of organo mercuric compound is quite difficult to attain. Zhou et al. employed a MW method to synthesize Au(III) complexes as TADF emitters using C–H activation as shown in Scheme 5.8 [29].

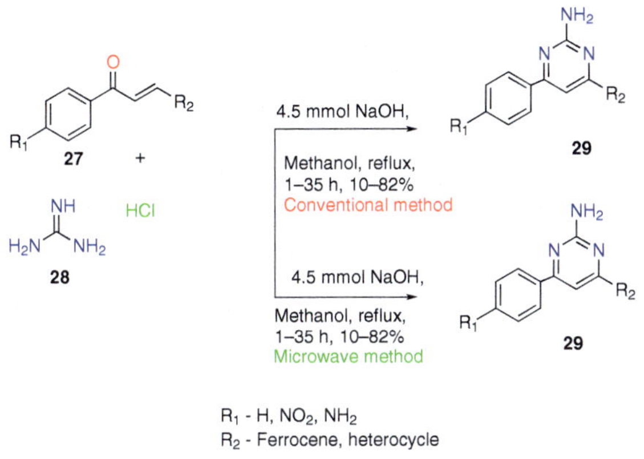

Scheme 5.6 Synthesis of 2-aminopyrimidine.

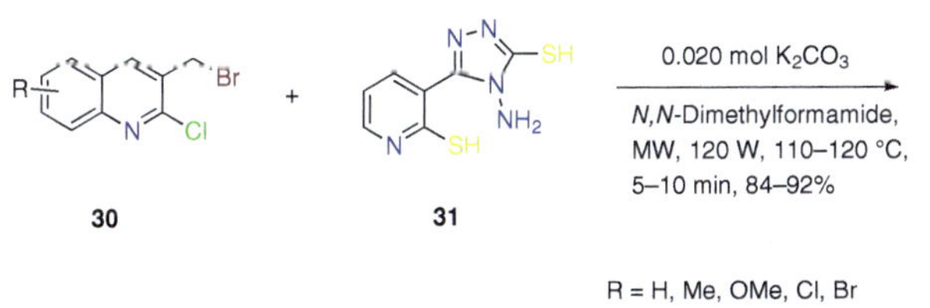

Scheme 5.7 Synthesis of triazolothiadizepinylquinolines.

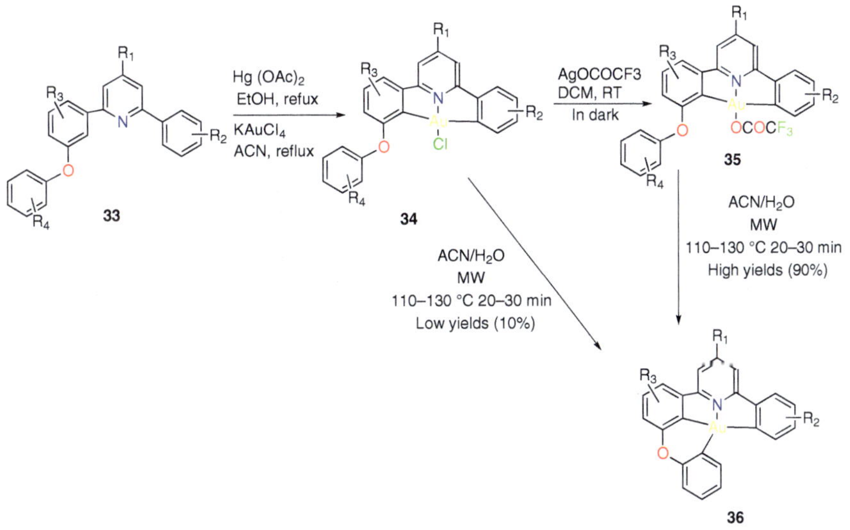

Scheme 5.8 Synthesis of Au(II) complex.

Shyamlal and colleagues introduced a pioneering MW-applied approach for the C–H activation of methyl xanthines. The compound **39** was produced by reacting the cyclic ethers with R–OH (Scheme 5.9). Under microwave conditions, the reaction was performed without the need for a transition metal catalyst. Instead, *tert*-butyl hydroperoxide (TBHP) was employed as the oxidant, while ethanol served as the solvent, operating at a temperature of 120 °C [30].

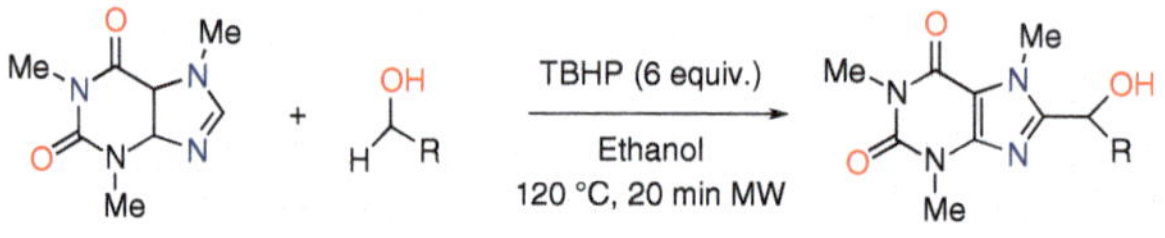

Scheme 5.9 Synthesis of hydroxy methyl xanthine derivatives.

Bora et al. synthesized mono-substituted, regioselective hydroxylated β-carboline derivatives. The catalytic system consisted of Cu(OAc)$_2$ in a solvent mixture of acetonitrile and acetic acid in a 1 : 1 ratio depicted in Scheme 5.10 [31].

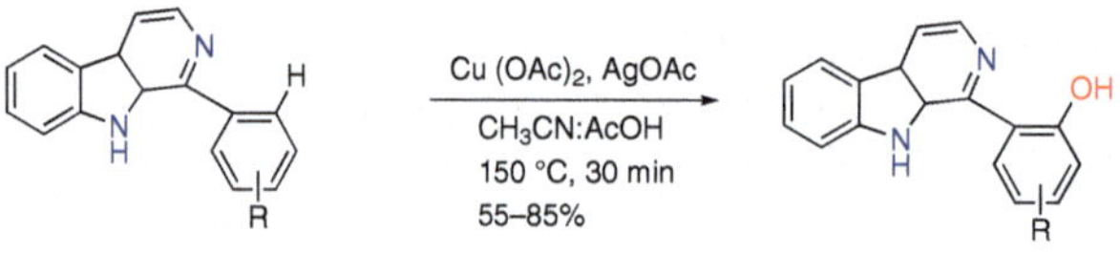

Scheme 5.10 Synthesis of hydroxylated β-carboline derivatives.

The Ugi cascade reaction was used by the Zhang group to generate substituted pyrrole-imidazole (Scheme 5.11). The target product **45** was produced by the MW irradiated processes in a 92% yield at 130 °C in 2 : 1 K$_2$CO$_3$ in CH$_3$CN [32].

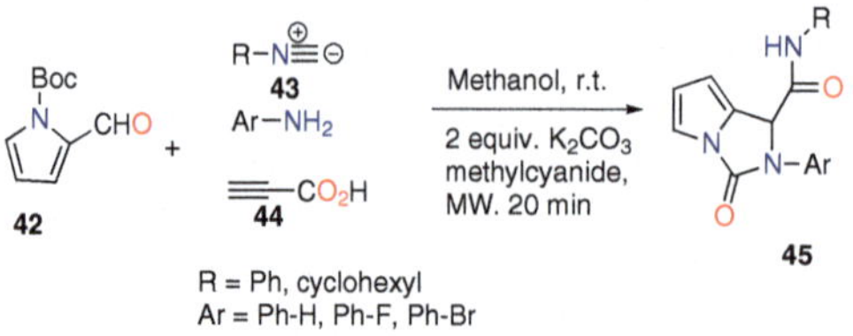

Scheme 5.11 Synthesis of pyrrole-imidazole.

Khormi et al. developed product **49**, and its complexity and potential as a catalyst for (SMC) reactions have been examined. First, the precatalyst – was chosen for its reactivity and because it was an affordable bromide that could be utilized in a reaction in Scheme 5.12 [33].

During the reaction, desired coumarin derivatives were obtained with yields ranging from 75% to 89% when compounds **50**, **51**, and **53** were being used. The

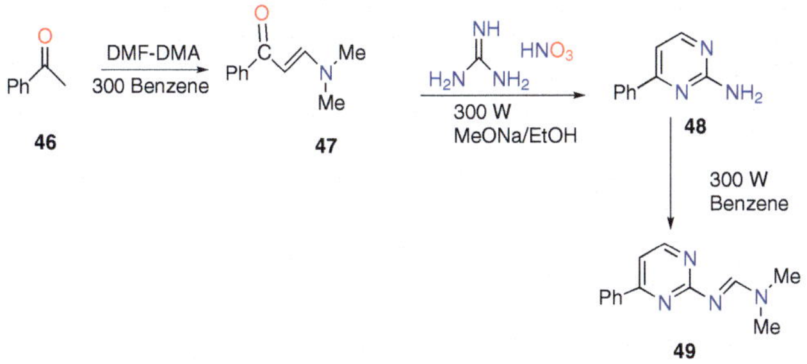

Scheme 5.12 Synthesis of the formamidine-based ligand.

reaction was carried out at 120 °C for 15 minutes (Scheme 5.13). In the presence of compounds **50** and **51**, a nucleophilic substitution takes place, resulting in the formation of an intermediate called substituted thiourea [34].

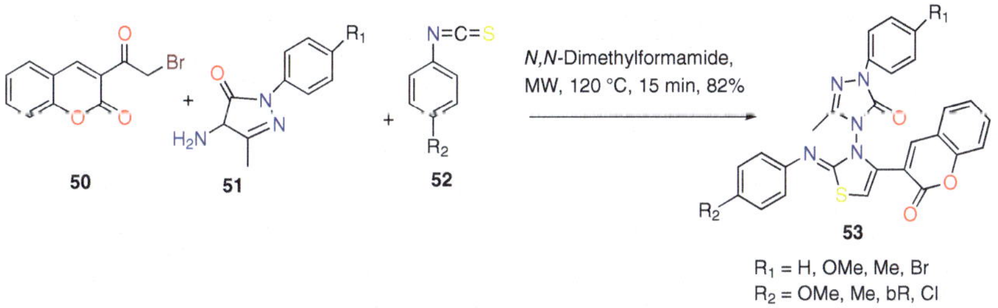

Scheme 5.13 Synthesis of coumarin-3-yl-1,2,4-triazolin-3-ones.

One pot synthesis of compound **57**, under the traditional heating method, revealed that only a maximum yield of 78% was attained in Scheme 5.14, proving that MW leads the process to continue successfully and produce more environmentally friendly goods at excellent yields [35].

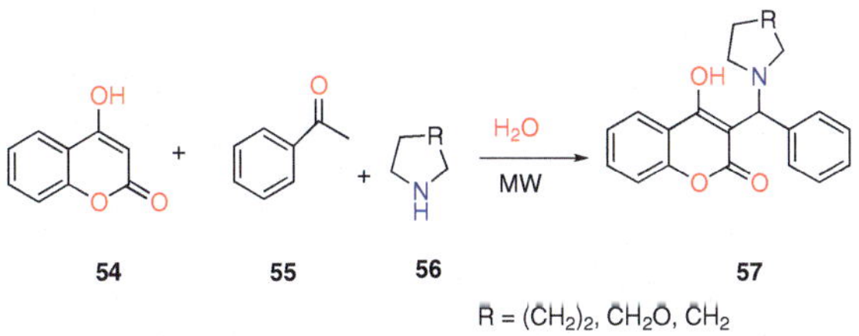

Scheme 5.14 Synthesis of 4-hydroxy-3[aryl(piperidin-1-yl/morpholino/pyrrolidin-1-l)methyl]-2H-2-chromenones.

Katla et al. employed K_2CO_3 as a base, and an MW-aided C–Se cross-coupling reaction was conducted; acetonitrile was used as the solvent at 200 W, for 25 to 35 minutes, yielding the required products in excellent amounts (Scheme 5.15) [36].

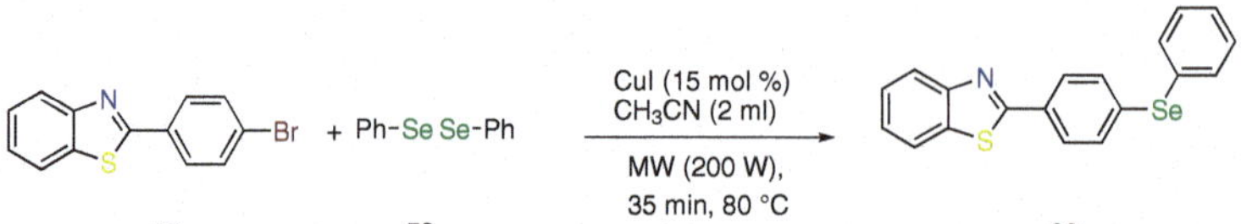

Scheme 5.15 C–S cross-coupling reaction from thiols and 2-(4-bromo phenyl)-benzothiazole.

The future potential of microwave-assisted reactions lies in their ability to replace expensive heavy metal catalysts with cheaper metals or metal-free procedures. This advancement would enable the synthesis of bioactive molecules in a manner that is economical in terms of atoms, time, and cost. A few more studies of microwave-assisted reactions are listed in Table 5.1. Currently, the microwave method is undergoing continuous development and is expected to find broader applications in medicinal chemistry and drug discovery in the coming years.

5.2.2 Ultrasound-Assisted Coupling Reactions

Ultrasound-assisted coupling reactions have emerged as a versatile and potent tool within synthetic chemistry, revolutionizing the field of organic synthesis. These reactions exploit the unique properties of high-frequency sound waves to expedite and enhance the efficiency of various coupling processes. In contrast to conventional coupling methods that heavily rely on heat, catalysts, or extended reaction times, ultrasound-assisted coupling reactions offer distinct advantages by enabling rapid and selective transformations under mild reaction conditions.

The application of ultrasound in coupling reactions has garnered significant attention in recent years due to its remarkable ability to boost reaction rates, increase yields, and facilitate the synthesis of complex molecules. The underlying principle of ultrasound-assisted coupling reactions revolves around the phenomenon of acoustic cavitation. This cavitation process generates highly reactive species, including free radicals, ions, and transient high-energy species, which effectively activate chemical bonds and facilitate coupling reactions [42, 43].

Furthermore, ultrasound irradiation induces sonochemical effects such as acoustic streaming and microstreaming, contributing further to the overall enhanced efficiency of these reactions.

A notable advantage of ultrasound-assisted coupling reactions lies in their broad applicability across a wide range of coupling reactions. Researchers have successfully employed this technique in various coupling processes involving the formation of C—C, C—N, C—O, C—S, and C—heteroatom bonds. These reactions encompass well-known cross-coupling methodologies like the Suzuki–Miyaura, Heck, Sonogashira, and Buchwald–Hartwig reactions, as well as other important coupling transformations such as the Ullmann, Hiyama, and Stille reactions. These diverse coupling methodologies enable the construction of intricate organic molecules, including pharmaceuticals, agrochemicals, and functional materials.

Table 5.1 Microwave-assisted reactions.

Reaction	Catalyst and reaction temperature	Yield using MW	Difference	References
Conversion of fructose	HCl, 80 °C	85%	Reaction time reduced by 30 minutes	[37]
Degradation of dye	$CoFe_2O_4$–SiC foam	95.01%	SiC skeleton acts as a good absorber	[38]
Conversion of ethane to ethylene	Fe/ZSM-5; 375 °C	80%	Higher selectivity	[39]
SMC reaction	MeONa/EtOH, DMF-DMA	85% at 300 MW	Excellent yield	[2]
Dehydrogenation	Pt/AC and Pt-Sn/AC; 250–340 °C	98%	Enhanced transport and diffusion	[40]
Synthesize 5-arylthiobenzo-2,1,3-thiadiazole derivatives	CuI (10 mol%), DMF (2 ml)	65% with 200 W	Excellent yield	[2]
SMC reaction	Pd/AC; 120 °C	40% with 1000 with 500 W	Hot spots during MWH	[41]

Moreover, ultrasound-assisted coupling reactions offer several advantages over traditional coupling methods. The rapid and efficient nature of ultrasound irradiation allows for reduced reaction times, thereby increasing overall productivity. Additionally, the mild reaction conditions and absence of harsh reagents or catalysts often result in improved chemoselectivity, effectively minimizing undesirable side reactions. The simplicity of the technique, combined with its environmentally friendly and cost-effective nature, has contributed to its growing popularity in both academic and industrial settings.

Please note that while the content remains the same, this version provides a different writing style by utilizing alternative sentence structures and word choices. If you need any further assistance, feel free to let me know [44].

Kargar et al. synthesized Fe_3O_4@CNF@Cu nanocomposites as a catalyst for O-arylation and C–C cross-coupling reactions, as shown in Scheme 5.16, are magnetic and reusable. The reactions were done using ultrasonic assistance in water as the green reaction medium with a yield of more than 80% in 10–60 minutes. The use of ultrasounds and the magnetic separation feature of the catalyst offered an alternative and promising technology for aqueous cross-coupling reactions, inspiring further research in pharmaceutical synthesis applications [45].

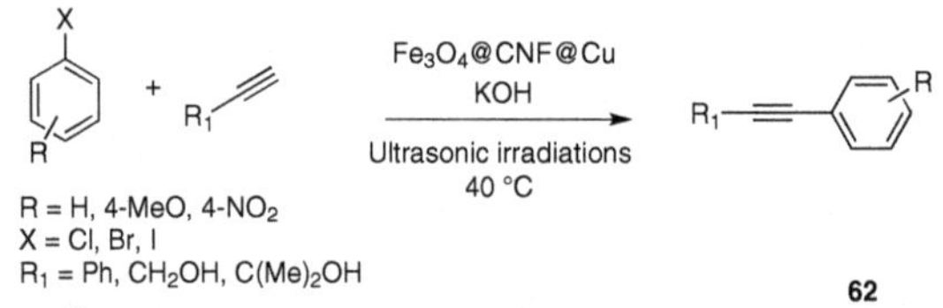

R = H, 4-MeO, 4-NO$_2$
X = Cl, Br, I
R$_1$ = Ph, CH$_2$OH, C(Me)$_2$OH
61

62

Scheme 5.16 Fe_3O_4@CNF@Cu catalyzed the Ullmann reaction of various halides with phenols.

El-bendary et al. synthesized a complex $[Pd(paOH)_2]\cdot 2Cl$, which formed a 3D network structure due to the presence of H bonds, lone pair–π, and lone pair–lone pair interactions 75–80 °C. The sonochemical reaction proceeded till the reactants were no longer traced. Ultrasound increased the rate of reaction and also maintained the production of a zero-valent Pd-colloidal species which acted as a catalyst, leading to enhanced reaction rates [46]. Scheme 5.17 illustrates the coupling reactions performed in the ultrasonication bath.

A novel nanocomposite **66** was successfully developed by Taghizadeh and group for cross-coupling reactions (Scheme 5.18). Spectroscopic analyses confirmed the successful attachment of the Crypt-222 and Pd ions onto the Fe_3O_4@PEG. For cross-coupling reactions, the catalyst exhibited remarkable activity under ultrasonication. It displayed the potential for industrial applications, maintaining its activity after five cycles of reuse. The catalyst demonstrated a synergistic effect in the presence of ultrasound irradiation, enabling a significant reduction in reaction time and temperature. This effect was attributed to the supply of activation energy through cavitation phenomena. The findings suggest that the method holds potential for the emergence of other catalysts, including Rh, Ru, Ni, Pt, and rare

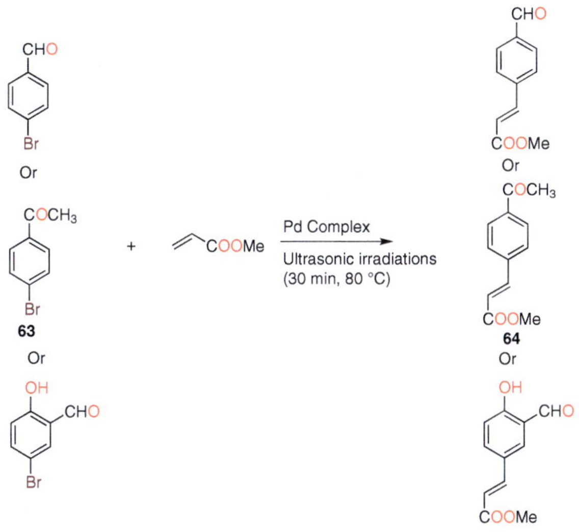

Scheme 5.17 Heck reaction of different substrates with methyl acrylate.

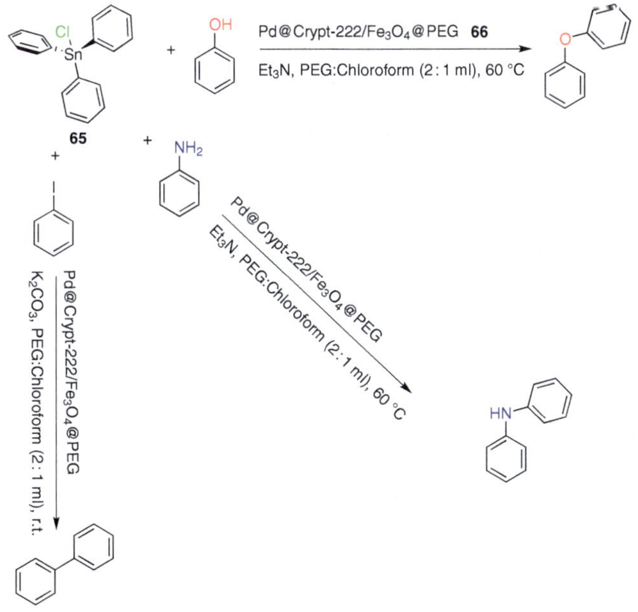

Scheme 5.18 Coupling of iodobenzene, aniline, and phenol with Ph_3SnCl.

earth metals-based catalysts, for environmentally friendly organic synthesis under ultrasonication conditions.

Wang et al. synthesized a catalyst and performed solvent-free oxidative coupling reactions of amines successfully, utilizing $Na_2S_2O_8$ as the oxidant with the aid of ultrasound. $Na_2S_2O_8$ exhibited exceptional oxidative performance, while ultrasound played a vital role in facilitating this transformation. The combination of $Na_2S_2O_8$

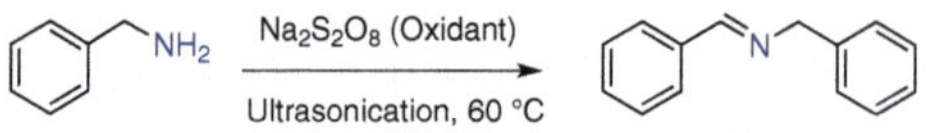

Scheme 5.19 Coupling reaction of Benzylamine.

and ultrasound enabled the reaction to proceed under mild reaction conditions as mentioned in Scheme 5.19 [47].

Veisi et al. successfully synthesized Fe_3O_4@CS-Starch/Pd nanocomposite using sonochemical promotion and in situ Pd-nanoparticle synthesis. The green synthesis method avoided toxic chemicals. The core–shell Fe_3O_4@CS-Starch composite acted as a functionalized mesh, anchoring Pd ions and stabilizing small Pd NPs (5–6 nm). The nanocomposite was tested under sonication in biaryl derivative synthesis via Suzuki coupling and reduction of 4-nitrophenol (Scheme 5.20) for its catalytic activity. Both methods offered advantages such as moderate reaction conditions, easy work-up, and high yield. With the aid of a magnet, the catalyst was recovered and could be recycled 11 times without losing any activity [48].

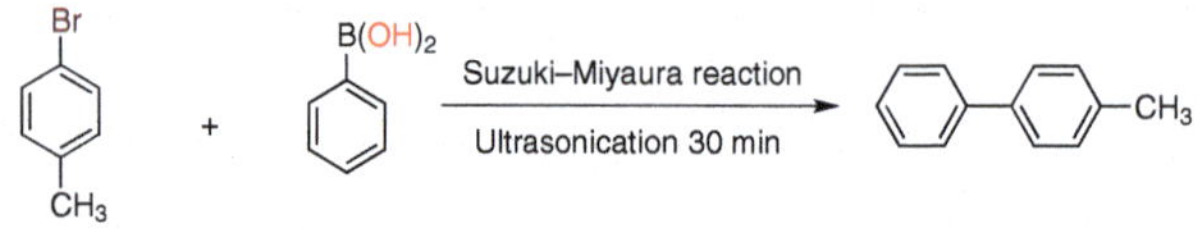

Scheme 5.20 Suzuki reaction of 4-methylbromobenzene with phenyl boronic acid.

A methodology was developed by Ganapathisivaraja et al. to synthesize 72 through a Cu(I)-catalyzed decarboxylative cross-coupling reaction (Scheme 5.21) under ultrasonication. The optimized reaction conditions were established by investigating the reaction of compound **71** acid in DMSO. Different temperatures and ultrasound frequencies were tested, but satisfactory results were only achieved at an elevated temperature (~80 °C) for 60 minutes. Attempts to enhance the yield by increasing the amounts of catalyst, base, and reactants separately did not yield better outcomes. Notably, the reaction without ultrasound resulted in a significantly lower product yield [49].

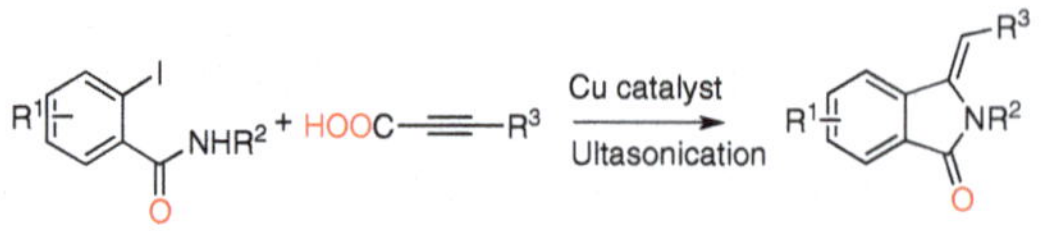

Scheme 5.21 Synthesis of 3-heteroarylmethylene substituted isoindolin-1-one derivatives.

A new class of compounds combining the structural characteristics of **73**, and **74** was formed by Kumar et al. To synthesize these rosuvastatin-based azaindoles, a one-pot sonochemical approach was established as shown in Scheme 5.22. The

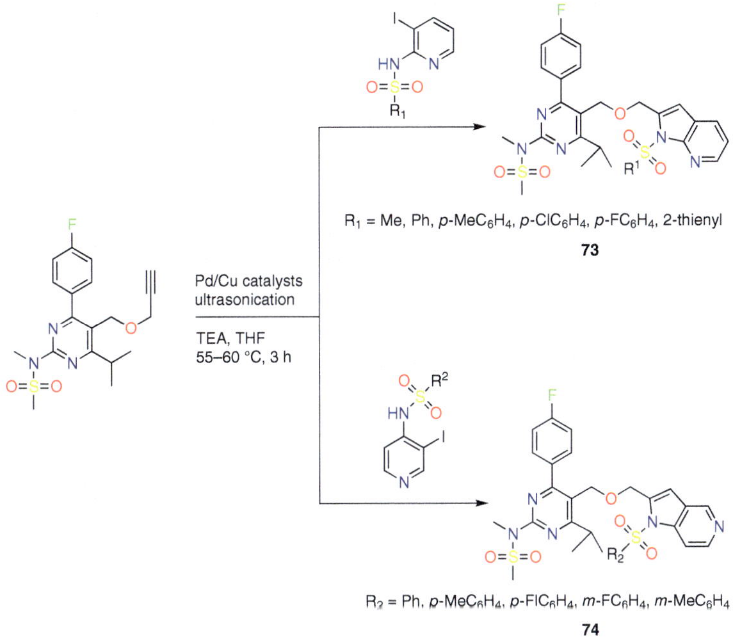
Pd/Cu catalysts
ultrasonication
TEA, THF
55–60 °C, 3 h
R₁ = Me, Ph, p-MeC₆H₄, p-ClC₆H₄, p-FC₆H₄, 2-thienyl
73
R₂ = Ph, p-MeC₆H₄, p-FlC₆H₄, m-FC₆H₄, m-MeC₆H₄
74

Scheme 5.22 Ultrasound-assisted synthesis of rosuvastatin-based azaindole derivatives.

reaction included cyclization and coupling of a rosuvastatin-derived alkyne with suitable derivatives of 3-iodopyridine, using Pd/Cu-catalysts [50].

A few desired compounds were synthesized by Manikanttha and group using a one-pot method, combining Ullmann–Goldberg coupling and cyclization, under ultrasound irradiation. The reaction involved the CuI-catalyzed reaction of 75 with quinolin-2-amine as shown in Scheme 5.23. The laboratory ultrasonic bath was used, and it had a fixed frequency of 35 kHz. Initially, the reaction did not progress, and on increasing the reaction temperature (80 °C), the desired product was obtained with an acceptable yield within 30 minutes. Reducing the time of reaction to 15 minutes did not cause any change in the yield percentage [51].

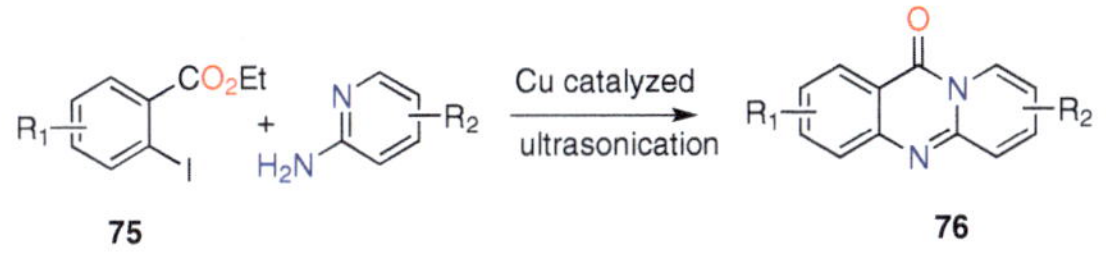
Cu catalyzed
ultrasonication
75
76

Scheme 5.23 Cu-catalyzed one-pot synthesis of 11*H*-pyrido[2,1-*b*]quinazolin-11-one derivatives.

In order to synthesize 3-Br-indazoles, Ying et al. developed an ultrasound-assisted, effective, and quick bromination method using DBDMH as a source of bromine (Scheme 5.24). One of the few C—Br bond construction reactions under ultrasonic waves, it underwent a C—H bond cleavage and C—Br bond formation phase. Additionally, a variety of modified indazoles can be used under moderate reactions [52].

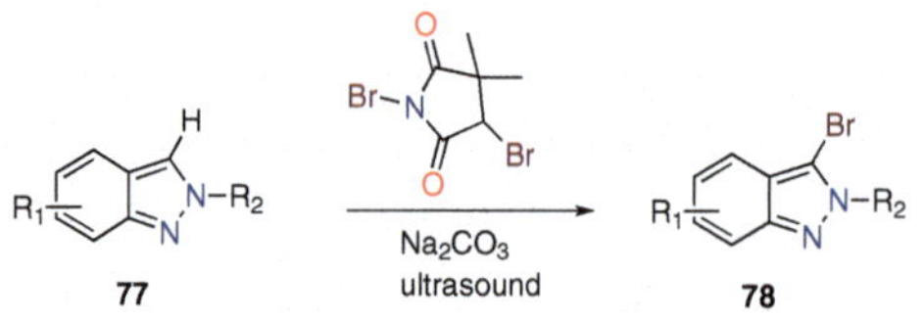

Scheme 5.24 Synthesis of 3-Br-indazoles.

Meeniga and group used environment-friendly ionic liquids (ILs) as the precursors for synthesizing 2-aryl benzimidazoles under ultrasonication (Scheme 5.25). ILs based on imidazoles and benzimidazoles were used as the catalyst in the Phillips–Ladenburg reaction, and the results included a very short reaction time, good yields, and a high level of substrate sensitivity [53].

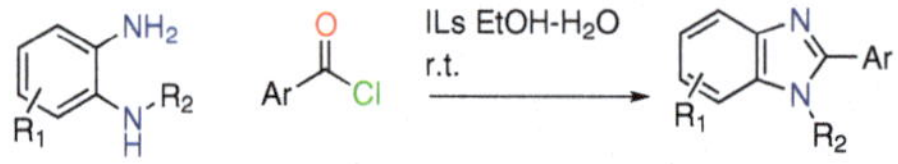

Scheme 5.25 Synthesis of 2-aryl benzimidazoles.

Sreenivasulu et al. prepared compound **81** by interacting reactants **79** and **80** at 50–55 °C (Scheme 5.26). The use of sonication produced a significantly faster reaction time which included ultrasonication (36–53 minutes) and reflux time (300–540 minutes), and higher productivity than the traditional thermal approach [54].

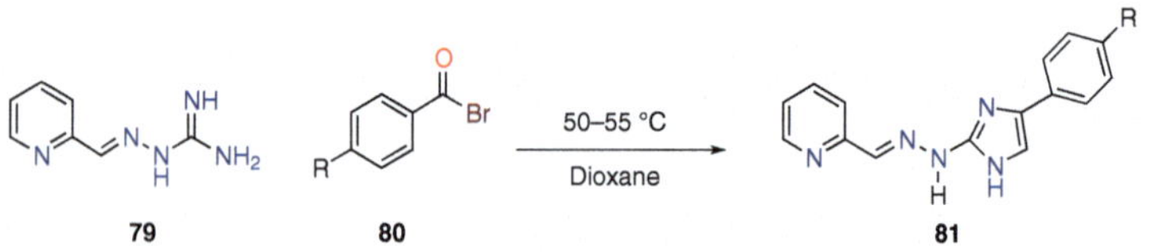

Scheme 5.26 Synthesis of pyridine-linked hydrazinylimidazoles.

A synthetic procedure for the synthesis of product **84** employed by Venkateshwarlu and group included a one-pot reaction utilizing Pd/C–Cu catalysts. In the first step, the reaction involved the interaction between aryl iodide and **82**, resulting in the formation of the C—N bond. A C—C bond was formed between the **83** as intermediate and a terminal alkyne, later followed by the establishment of a C—N bond (Scheme 5.27) [55].

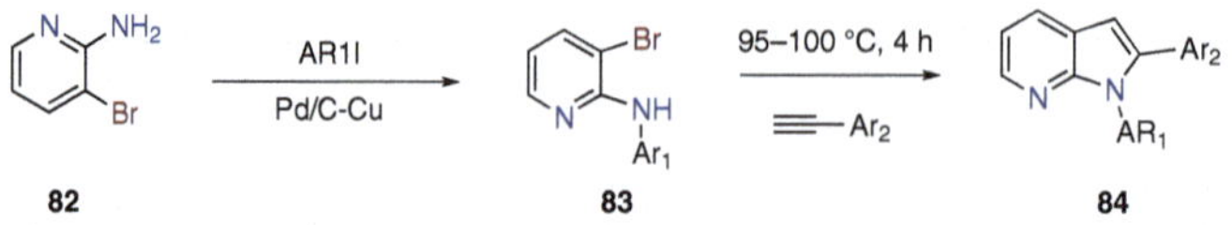

Scheme 5.27 One-pot synthesis of 1,2-diaryl azaindoles under Pd/C–Cu catalysis.

Shahinshavali et al. synthesized **86** derivatives by using a Cu-catalyzed coupling of **85** with cheaper terminal alkynes under the influence of ultrasound irradiation (35 kHz). The method involved using PEG-400 solvent with mild reaction conditions and no co-catalyst was required. The combination of CuI, PPh$_3$, and K$_2$CO$_3$ in PEG-400 at 50 °C (Scheme 5.28) with ultrasonication was found to be optimum and was utilized for the production of other derivatives [56].

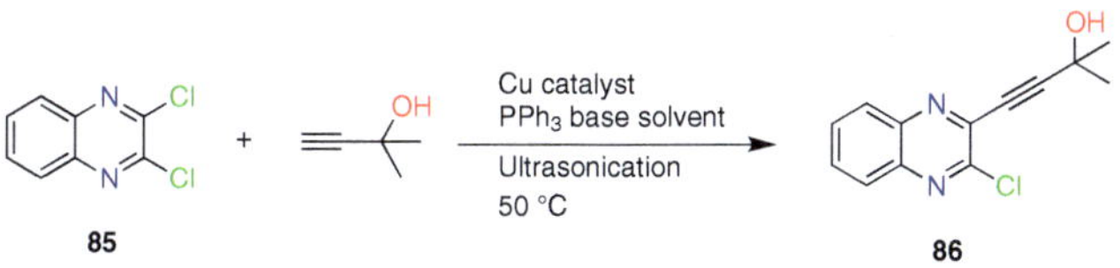

Scheme 5.28 Ultrasonication assisted reaction of 2,3-dichloroquinoxaline with terminal alkyne.

Verma et al. undertook the synthesis of CdS and Ag-doped CdS nanoparticles (Scheme 5.29), followed by an evaluation of their catalytic capabilities. To

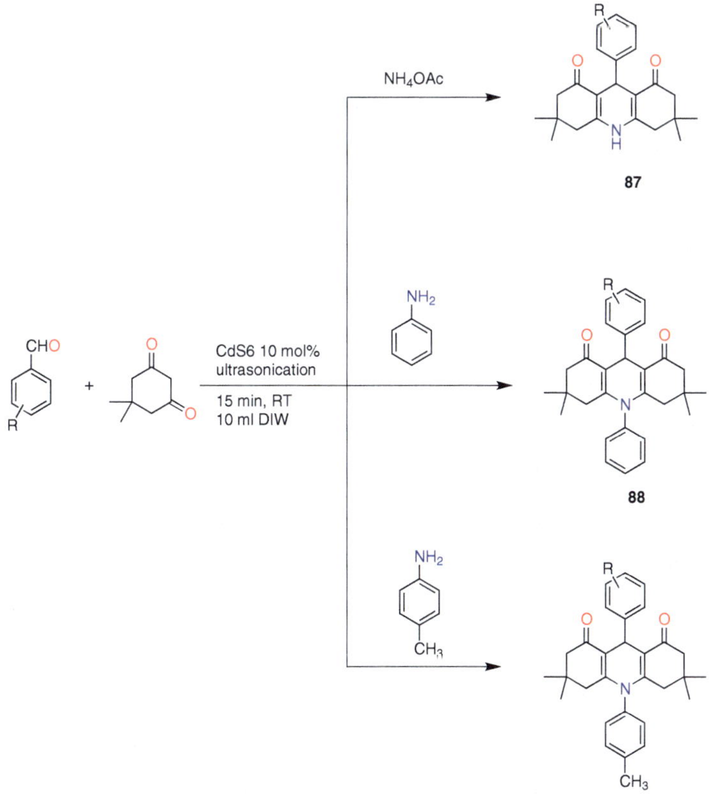

Scheme 5.29 Schematic illustration for the formation of DODHAs using dimedone and aryladehyde with ammonium acetate or aniline or *p*-toluidine.

assess their effectiveness, the NPs were employed in multi-component reactions (MCRs) for the synthesis of 14 distinct types of 1,8-dioxodecahydroacridine (DODHAs) derivatives using ultrasounds. MCRs involved the combination of dimedone, aryl aldehyde, and either ammonium acetate or aniline, or *p*-toluidine, all dissolved in distilled water at room temperature within a remarkably short duration of 15 minutes. The obtained outcomes revealed exceptional yields of the desired products, demonstrating the catalyst's remarkable reusability for up to six consecutive runs, with each run yielding an impressive 89% product yield [57].

Safari et al. proposed a synthesis of compound **93**. A three-component reaction involving reactants **90**, **91**, and **92** in the presence of a $Co_3O_4/CS/PWA$ catalyst with ultrasonication as shown in Scheme 5.30. The reaction resulted in good yields of the products. Furthermore, the catalyst employed in the synthesis was easily taken back from the product mixture by using an external magnet and found to be reusable multiple times with no major loss in its efficiency. This underscores the catalyst's practicality and potential for future applications [58].

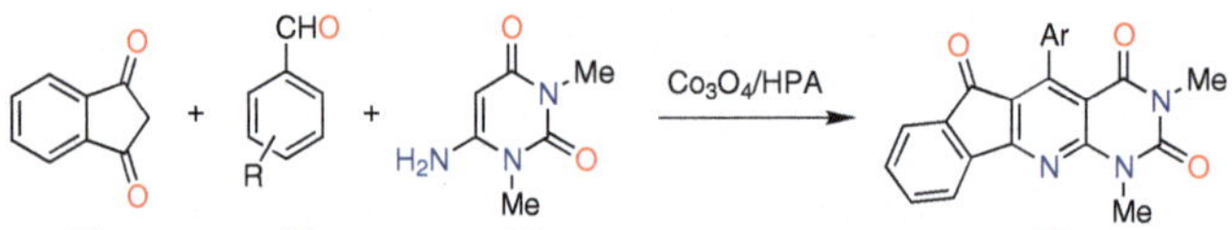

Scheme 5.30 Synthesis of Indeno[2′,1′:5,6]pyrido[2,3-*d*]pyrimidines.

5.3 Conclusion

Microwave- and ultrasound-assisted coupling reactions have significantly transformed the landscape of organic synthesis, revolutionizing the field with their unique capabilities. These techniques have demonstrated exceptional potential in expediting reactions, improving yields, and enabling milder reaction conditions. By harnessing the power of high-frequency waves and acoustic cavitation, these methods enhance the efficiency and selectivity of various coupling processes, offering faster reaction rates and higher yields compared to traditional methods. Moreover, their versatility is evident in their broad applicability to diverse coupling reactions, spanning a wide range of bond formations. The ability to perform reactions in environmentally friendly solvents further contributes to the sustainable nature of these methods. As we look to the future, the continued exploration and refinement of microwave- and ultrasound-assisted coupling reactions hold great promise for advancing synthetic chemistry. The potential for further developments, optimization, and application in complex synthesis scenarios suggests a bright future for these innovative techniques. With their significant contributions to reaction efficiency, yield improvement, and green chemistry practices, microwave, and ultrasound-assisted coupling reactions are set to continue driving progress in the field of organic synthesis.

References

1 De Souza, L.F., Silva, L.C., Oliveira, B.L., and Antunes, O.A.C. (2008). Microwave- and ultrasound-assisted Suzuki – Miyaura cross-coupling reactions catalyzed by Pd/PVP. *Tetrahedron Lett.* 49: 3895–3898. https://doi.org/10.1016/j.tetlet.2008.04.061.

2 Dhanush, P.C., Veetil, P., and Anilkumar, G. (2021). Microwave assisted C-H activation reaction: an overview. *Tetrahedron* 105 (2022): 132614. https://doi.org/10.1016/j.tet.2021.132614.

3 Gawande, M.B., Shelke, S.N., Zboril, R., and Varma, R.S. (2014). Microwave-assisted chemistry: synthetic applications for rapid assembly of nanomaterials and organics. *Acc. Chem. Res.* 47 (4): 1338–1348. https://doi.org/10.1021/ar400309b.

4 Mittersteiner, M., Farias, F.F.S., Bonacorso, H.G. et al. (2021). Ultrasonics sonochemistry ultrasound-assisted synthesis of pyrimidines and their fused derivatives: a review. *Ultrason. Sonochem.* 79 (July): 105683. https://doi.org/10.1016/j.ultsonch.2021.105683.

5 Kaping, S., Daioo, H., Bankynmaw, L., and Kandasamy, A. (2023). Ultrasound assisted regioselective synthesis of novel adamantyl-pyrazolo [1,5-*a*] pyrimidines in aqueous media and molecular docking and drug likeness studies. *J. Mol. Struct.* 1288 (November): 135766. https://doi.org/10.1016/j.molstruc.2023.135766.

6 Al-mutairi, A.A., El-baih, F.E.M., and Al-hazimi, H.M. (2009). Microwave versus ultrasound assisted synthesis of some new heterocycles based on pyrazolone moiety. *JSCS* 14 (3): 287–299. https://doi.org/10.1016/j.jscs.2010.02.010.

7 Borah, B., Dwivedi, K.D., Kumar, B., and Chowhan, L.R. (2021). Recent advances in the microwave- and ultrasound-assisted green synthesis of coumarin-heterocycles. *Arabian J. Chem.* 15 (3): 103654. https://doi.org/10.1016/j.arabjc.2021.103654.

8 Yousaf, M., Zahoor, A.F., Akhtar, R. et al. (2020). Development of green methodologies for Heck, Chan–Lam, Stille and Suzuki cross-coupling reactions. *Mol. Diversity* 24 (3): 821–839. https://doi.org/10.1007/s11030-019-09988-7.

9 Jo, S., Kang, B., and Jung, J.-W. (2023). Microwave-assisted Cu-catalyzed diaryletherification for facile synthesis of bioactive prenylated diresorcinols. *Molecules* 28 (1). https://doi.org/10.3390/molecules28010062.

10 Damera, T., Pagadala, R., Rana, S., and Jonnalagadda, S.B. (2023). A concise review of multicomponent reactions using novel heterogeneous catalysts under microwave irradiation. *Catalysts* 13 (7). https://doi.org/10.3390/catal13071034.

11 Martina, K., Cravotto, G., and Varma, R.S. (2021). Impact of microwaves on organic synthesis and strategies toward flow processes and scaling up. *J. Organomet. Chem.* 86 (20): 13857–13872. https://doi.org/10.1021/acs.joc.1c00865.

12 Gupta, D., Jamwal, D., Rana, D., and Katoch, A. (2018). *Microwave Synthesized Nanocomposites for Enhancing Oral Bioavailability of Drugs*. Elsevier Inc. https://doi.org/10.1016/B978-0-12-813741-3.00027-3.

13 Pal, A. and Gayen, K.S. (2021). The impact of microwave irradiation reaction in medicinal chemistry: a review. *Orient. J. Chem.* 37 (1): 01–24. https://doi.org/10.13005/ojc/370101.

14 Izza, H.F., Susanti, D.Y., Mariyam, S., and Saputro, A.D. (2023). Performance of microwave-assisted extraction of proanthocyanidins from red sorghum grain in various power and citric acid concentration. *J. Saudi Soc. Agric. Sci.* https://doi.org/10.1016/j.jssas.2023.05.002.

15 Kang, J., Ko, Y., Kim, J.P. et al. (2023). Microwave-assisted design of nanoporous graphene membrane for ultrafast and switchable organic solvent nanofiltration. *Nat. Commun.* 14 (1): 901. https://doi.org/10.1038/s41467-023-36524-x.

16 Das, A., Roy, H., and Ansary, I. (2018). Microwave-assisted, one-pot three-component synthesis of 6-(pyrrolyl) coumarin/quinolone derivatives catalyzed by In(III) chloride. *ChemistrySelect* 3 (33): 9592–9595. https://doi.org/10.1002/slct.201801931.

17 Dolšak, A., Mrgole, K., and Sova, M. (2021). Microwave-assisted regioselective suzuki coupling of 2,4-dichloropyrimidines with aryl and heteroaryl boronic acids. *Catalysts* 11 (4). https://doi.org/10.3390/catal11040439.

18 Mishra, R., Jana, A., Panday, A.K., and Choudhury, L.H. (2018). Synthesis of fused pyrroles containing 4-hydroxycoumarins by regioselective metal-free multicomponent reactions. *Org. Biomol. Chem.* 16 (17): 3289–3302. https://doi.org/10.1039/c8ob00161h.

19 Breugst, M. and Reissig, H.U. (2020). The Huisgen reaction: milestones of the 1,3-dipolar cycloaddition. *Angew. Chem. Int. Ed.* 59 (30): 12293–12307. https://doi.org/10.1002/anie.202003115.

20 Shaikh, S.K.J., Kamble, R.R., Bayannavar, P.K. et al. (2020). Triazolothiadizepinylquinolines as potential MetAP-2 and NMT inhibitors: microwave-assisted synthesis, pharmacological evaluation and molecular docking studies. *J. Mol. Struct.* 1203: 127445. https://doi.org/10.1016/j.molstruc.2019.127445.

21 Saptal, V.B., Saptal, M.V., Mane, R.S. et al. (2019). Amine-functionalized graphene oxide-stabilized Pd nanoparticles (Pd@APGO): a novel and efficient catalyst for the Suzuki and carbonylative Suzuki-Miyaura coupling reactions. *ACS Omega* 4 (1): 643–649. https://doi.org/10.1021/acsomega.8b03023.

22 Akiyama, T., Taniguchi, T., Saito, N. et al. (2017). Ligand-free Suzuki–Miyaura coupling using ruthenium(0) nanoparticles and a continuously irradiating microwave system. *Green Chem.* 19 (14): 3357–3369. https://doi.org/10.1039/C7GC01166K.

23 Vo, D.D. and Elofsson, M. (2017). Synthesis of 4-formyl-2-arylbenzofuran derivatives by PdCl(C_3H_5)dppb-catalyzed tandem Sonogashira coupling-cyclization under microwave irradiation - application to the synthesis of viniferifuran analogues. *ChemistrySelect* 2 (22): 6245–6248. https://doi.org/10.1002/slct.201701490.

24 Raimondi, M.V., Presentato, A., Li Petri, G. et al. (2020). New synthetic nitro-pyrrolomycins as promising antibacterial and anticancer agents. *Antibiotics* 9 (6). https://doi.org/10.3390/antibiotics9060292.

25 Li Petri, G., Spanò, V., Spatola, R. et al. (2020). Bioactive pyrrole-based compounds with target selectivity. *Eur. J. Med. Chem.* 208: 112783. https://doi.org/10.1016/j.ejmech.2020.112783.

26 Yadagiri, D., Rivas, M., and Gevorgyan, V. (2020). Denitrogenative transformations of pyridotriazoles and related compounds: synthesis of N-containing heterocyclic compounds and beyond. *J. Organomet. Chem.* 85 (17): 11030–11046. https://doi.org/10.1021/acs.joc.0c01652.

27 Sayed, G.H., Azab, M.E., and Anwer, K.E. (2019). Conventional and microwave-assisted synthesis and biological activity study of novel heterocycles containing pyran moiety. *J. Heterocyclic Chem.* 56 (8): 2121–2133. https://doi.org/10.1002/jhet.3606.

28 Filho, E.V., Pina, J.W., Antoniazi, M.K. et al. (2021). Synthesis, docking, machine learning and antiproliferative activity of the 6-ferrocene/heterocycle-2-aminopyrimidine and 5-ferrocene-1*H*-Pyrazole derivatives obtained by microwave-assisted Atwal reaction as potential anticancer agents. *Bioorg. Med. Chem. Lett.* 48: 128240. https://doi.org/10.1016/j.bmcl.2021.128240.

29 Zhou, D., Pong To, W., Tong, G. et al. (2020). Tetradentate Gold(III) complexes as thermally activated delayed fluorescence (TADF) emitters: microwave-assisted synthesis and high-performance OLEDs with long operational lifetime. *Angew. Chem.* 59: 6375–6382. https://doi.org/10.1002/anie.201914661.

30 Shyamlal, B.R.K., Mathur, M., Yadav, D.K., and Chaudhary, S. (2020). Microwave-assisted modified synthesis of C8-analogues of naturally occurring methylxanthines: synthesis, biological evaluation and their practical applications. *Fitoterapia* 143: 104533. https://doi.org/10.1016/j.fitote.2020.104533.

31 Bora, D., Tokala, R., John, S.E. et al. (2020). β-Carboline directed regioselective hydroxylation by employing $Cu(OAc)_2$ and mechanistic investigation by ESI-MS. *Org. Biomol. Chem.* 18 (12): 2307–2311. https://doi.org/10.1039/d0ob00250j.

32 Zhang, M., Ding, Y., Qin, H.X. et al. (2020). One-pot synthesis of substituted pyrrole–imidazole derivatives with anticancer activity. *Mol. Diversity* 24 (4): 1177–1184. https://doi.org/10.1007/s11030-019-09982-z.

33 Khormi, A.Y., Farghaly, T.A., and Shaaban, M.R. (2020). New palladium(II)-complex based on nitrogen rich ligand efficient precatalyst for C–C cross-coupling in water under microwaves irradiation. *J. Inorg. Organomet. Polym. Mater.* 30 (12): 5133–5147. https://doi.org/10.1007/s10904-020-01620-8.

34 Shaikh, S.K.J., Sannaikar, M.S., and Kumbar, M.N. (2018). Microwave-expedited green synthesis, photophysical, computational studies of anticancer activity. *ChemistrySelect* 4448–4462. https://doi.org/10.1002/slct.201702596.

35 Muthusaravanan, S., Teju, E., and Thangamani, A. (2019). A greener and microwave-mediated synthesis and spectral studies of β-aminomethylhydroxylcoumarins. *Chem. Data Collect.* 20: 100199. https://doi.org/10.1016/j.cdc.2019.100199.

36 Katla, R. and Katla, R. (2022). Microwave assisted C–S cross-coupling reaction from thiols and 2-(4-bromo phenyl)-benzothiazole employed by CuI in acetonitrile. *New J. Chem.* 46 (29): 13918–13923. https://doi.org/10.1039/D2NJ02065C.

37 Breeden, S.W., Clark, J.H., Farmer, T.J. et al. (2013). Microwave heating for rapid conversion of sugars and polysaccharides to 5-chloromethyl furfural. *Green Chem.* 15 (1): 72–75. https://doi.org/10.1039/C2GC36290B.

38 Mao, Y., Yang, S., and Xue, C. (2018). Subject category: subject areas: rapid degradation of malachite green by $CoFe_2O_4$ – SiC foam under microwave radiation. *R. Soc. Open Sci.* 5 (6): 180085.

39 Robinson, B., Caiola, A., Bai, X. et al. (2020). Catalytic direct conversion of ethane to value-added chemicals under microwave irradiation. *Catal. Today* 356: 3–10. https://doi.org/10.1016/j.cattod.2020.03.001.

40 Suttisawat, Y., Sakai, H., Abe, M. et al. (2012). Microwave effect in the dehydrogenation of tetralin and decalin with a fixed-bed reactor. *Int. J. Hydrogen Energy* 37: 3242–3250. https://doi.org/10.1016/j.ijhydene.2011.10.111.

41 Horikoshi, S., Kamata, M., Mitani, T., and Serpone, N. (2014). Control of microwave-generated hot spots. 6. Generation of hot spots in dispersed catalyst particulates and factors that affect catalyzed organic syntheses in heterogeneous media. *Ind. Eng. Chem. Res.* 53 (39): 14941–14947. https://doi.org/10.1021/ie502169z.

42 Penteado, F., Monti, B., Sancineto, L. et al. (2018). Ultrasound-assisted multicomponent reactions, organometallic and organochalcogen chemistry. *Asian J. Org. Chem.* 7 (12): 2368–2385. https://doi.org/10.1002/AJOC.201800477.

43 Baig, R.B.N. and Varma, R.S. (2012). Alternative energy input: mechanochemical, microwave and ultrasound-assisted organic synthesis. *Chem. Soc. Rev.* 41 (4): 1559–1584. https://doi.org/10.1039/C1CS15204A.

44 Banakar, V.V., Sabnis, S.S., Gogate, P.R. et al. (2022). Ultrasound assisted continuous processing in microreactors with focus on crystallization and chemical synthesis: a critical review. *Chem. Eng. Res. Des.* 182: 273–289. https://doi.org/10.1016/J.CHERD.2022.03.049.

45 Ghamari Kargar, P., Len, C., and Luque, R. (2022). Cu/cellulose-modified magnetite nanocomposites as a highly active and selective catalyst for ultrasound-promoted aqueous O-arylation Ullmann and sp-sp2 Sonogashira cross-coupling reactions. *Sustainable Chem. Pharm.* 27: 100672. https://doi.org/10.1016/J.SCP.2022.100672.

46 El-bendary, M.M., Saleh, T.S., and Al-Bogami, A.S. (2021). Synthesis and structural characterization of a palladium complex as an anticancer agent, and a highly efficient and reusable catalyst for the Heck coupling reaction under ultrasound irradiation: a convenient sustainable green protocol. *Polyhedron* 194: 114924. https://doi.org/10.1016/J.POLY.2020.114924.

47 Wang, J., Wang, X., Liu, X. et al. (2022). Efficient and solvent-free oxidation coupling of amines to imines using persulfate as oxidant with ultrasound assistance. *Polycyclic Aromat. Compd.* 42 (9): 6282–6289. https://doi.org/10.1080/10406638.2021.1977350.

48 Veisi, H., Joshani, Z., Karmakar, B. et al. (2021). Ultrasound assisted synthesis of Pd NPs decorated chitosan-starch functionalized Fe_3O_4 nanocomposite catalyst towards Suzuki–Miyaura coupling and reduction of 4-nitrophenol. *Int. J. Biol. Macromol.* 172: 104–113. https://doi.org/10.1016/J.IJBIOMAC.2021.01.040.

49 Ganapathisivaraja, P., Rao, G.V., Ramarao, A. et al. (2022). Ultrasound assisted Cu-catalyzed decarbonylative Sonogashira coupling-cyclization strategy: synthesis and evaluation of 3-heteroarylmethylene isoindolin-1-ones against SIRT1. *J. Mol. Struct.* 1250: 131788. https://doi.org/10.1016/J.MOLSTRUC.2021.131788.

50 Kumar, J.S., Reddy, G.S., Medishetti, R. et al. (2022). Ultrasound assisted one-pot synthesis of rosuvastatin based novel azaindole derivatives via coupling–cyclization strategy under Pd/Cu-catalysis: their evaluation as potential cytotoxic agents. *Bioorg. Chem.* 124: 105857. https://doi.org/10.1016/J.BIOORG.2022.105857.

51 Manikantha, M., Deepti, K., Tej, M.B. et al. (2023). Ultrasound assisted Cu-catalyzed Ullmann-Goldberg type coupling-cyclization in a single pot: synthesis and in silico evaluation of 11*H*-pyrido[2,1-*b*]quinazolin-11-ones against SARS-CoV-2 RdRp. *J. Mol. Struct.* 1280: 135044. https://doi.org/10.1016/J.MOLSTRUC.2023.135044.

52 Ying, S., Liu, X., Guo, T. et al. (2022). Ultrasound-assisted bromination of indazoles at the C3 position with dibromohydantoin. *RSC Adv.* 13 (1): 581–585. https://doi.org/10.1039/d2ra06867b.

53 Meeniga, I., Gokanapalli, A., and Peddiahgari, V.G.R. (2022). Synthesis of environmentally benign new ionic liquids for the preparation of 2-aryl/heteroaryl benzimidazoles/benzoxazoles under ultrasonication. *Sustainable Chem. Pharm.* 30: 100874. https://doi.org/10.1016/j.scp.2022.100874.

54 Sreenivasulu, T., Reddy, G.M., Sravya, G. et al. (2019). Synthesis, characterization and antimicrobial activity of pyridine linked hydrazinyl oxazoles/imidazoles. *AIP Conf. Proc.* 2063 (January): 1–6. https://doi.org/10.1063/1.5087387.

55 Venkateshwarlu, R., Nath Singh, S., Siddaiah, V. et al. (2019). Ultrasound assisted one-pot synthesis of 1,2-diaryl azaindoles via Pd/C-Cu catalysis: identification of potential cytotoxic agents. *Tetrahedron Lett.* 60 (52): 151326. https://doi.org/10.1016/j.tetlet.2019.151326.

56 Shahinshavali, S., Hossain, K.A., Kumar, A.V. et al. (2020). Ultrasound assisted synthesis of 3-alkynyl substituted 2-chloroquinoxaline derivatives: their in silico assessment as potential ligands for N-protein of SARS-CoV-2. *Tetrahedron Lett.* 61 (40): 152336. https://doi.org/10.1016/J.TETLET.2020.152336.

57 Verma, D., Sharma, V., Jain, S., and Singh Okram, G. (2020). Ultrasound-assisted synthesis of 1,8-dioxodecahydroacridine derivatives in presence of Ag doped CdS nanocatalyst. *J. Dispersion Sci. Technol.* 41 (8): 1145–1158. https://doi.org/10.1080/01932691.2019.1614460.

58 Safari, J., Tavakoli, M., and Ghasemzadeh, M.A. (2019). Ultrasound-promoted an efficient method for the one-pot synthesis of indeno fused pyrido[2,3-*d*] pyrimidines catalyzed by $H_3PW_{12}O_{40}$ functionalized chitosan@Co_3O_4 as a novel and green catalyst. *J. Organomet. Chem.* 880: 75–82. https://doi.org/10.1016/J.JORGANCHEM.2018.10.028.

6

Synthesis of Heterocyclic Compounds Under Microwave Irradiation Using Name Reactions

Sheryn Wong[1] and Anton V. Dolzhenko[1,2]

[1] Monash University Malaysia, School of Pharmacy, Jalan Lagoon Selatan, Bandar Sunway, Selangor 46150, Malaysia
[2] Curtin University, Curtin Health Innovation Research Institute, Curtin Medical School, GPO Box U1987, Perth, Western Australia 6845, Australia

6.1 Introduction

With microwave irradiation as an efficient method of delivering activation energy, microwave reactors have become valuable tools in the arsenal of synthetic organic chemists. The dielectric energy transfer under microwave irradiation makes heating more energy-efficient, contributing to the green character of microwave-promoted reactions. The history of synthesis under microwave irradiation started with two reports published in *Tetrahedron Letters* in 1986 [1, 2]. Starting with simple domestic microwave ovens as reactors, microwave irradiation applications for organic synthesis have been growing in geometrical progression. However, the development of commercial dedicated microwave reactors with precise control of reaction parameters made an impact that was recognized as "a quiet revolution in chemical synthesis" [3]. Although reports using domestic microwave ovens for organic synthesis are still published occasionally, the current standards of safe and reproducible synthesis under microwave irradiation dictate the application of specialized microwave reactors, which are mainly represented by equipment produced by Anton Paar (Austria), CEM (USA), Biotage (Sweden), and Milestone (Italy) manufacturers. This chapter discusses reactions performed using the systems dedicated to microwave-assisted synthesis with controlled reaction conditions.

Microwave irradiation has found an extensive application in the synthesis of various heterocyclic systems, and many efficient approaches have been developed and discussed in numerous reviews [4–11]. In this chapter, we illustrate the benefits of microwave-assisted synthesis of five- and six-membered heterocyclic aromatic compounds using classical named reactions.

Green Chemical Synthesis with Microwaves and Ultrasound, First Edition.
Edited by Dakeshwar Kumar Verma, Chandrabhan Verma, and Paz Otero Fuertes.
© 2024 WILEY-VCH GmbH. Published 2024 by WILEY-VCH GmbH.

6.2 Classical Methods for Heterocyclic Synthesis Under Microwave Irradiation

6.2.1 Piloty–Robinson Pyrrole Synthesis

Microwave irradiation was effectively applied to improve traditional methods for the synthesis of pyrroles. The classical Piloty–Robinson pyrrole synthesis is based on the thermal [3,3] sigmatropic rearrangement of tautomerizable ketazines in the presence of zinc chloride [12, 13]. The Piloty–Robinson pyrrole synthesis under microwave irradiation was used for the preparation of N-acyl 3,4-disubstituted pyrroles via the reaction of azines, obtained from hydrazine and aliphatic aldehydes, with acid chlorides [14]. For example, the reaction of propionaldehyde (**1**) with hydrazine resulted in the formation of azine **2**, which reacted with benzoyl chloride (**3**) rearranging under microwave irradiation to afford 1-benzoyl-3,4-dimethylpyrrole (**4**) (Scheme 6.1) [14].

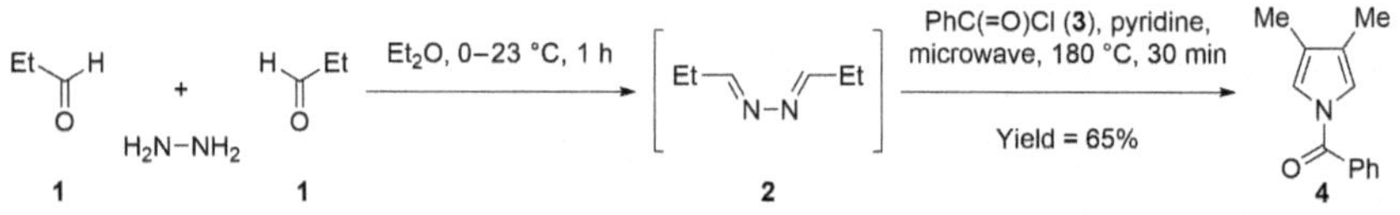

Scheme 6.1 Piloty–Robinson pyrrole synthesis under microwave irradiation.

6.2.2 Clauson–Kaas Pyrrole Synthesis

N-Substituted pyrroles can be prepared using the Clauson–Kaas pyrrole synthesis, i.e. reaction of primary amines with 2,5-dimethoxytetrahydrofuran (**5**) [15, 16]. This reaction can be conveniently carried out under microwave irradiation using a variety of N-nucleophiles, including aliphatic and aromatic amines, amides, and sulfonamides. The microwave-assisted reaction was successful in an aqueous medium [17], but higher yields were obtained in acetic acid [17] or in the presence of a solid acid catalyst [18]. For example, heating **5** with aniline (**6**) in water at 170 °C for 10 minutes under microwave irradiation in an Emrys Creator reactor resulted in the formation of 1-phenylpyrrole (**7**) in 63% yield (Scheme 6.2) [17]. By changing the reaction medium to acetic acid, the yield under the same conditions

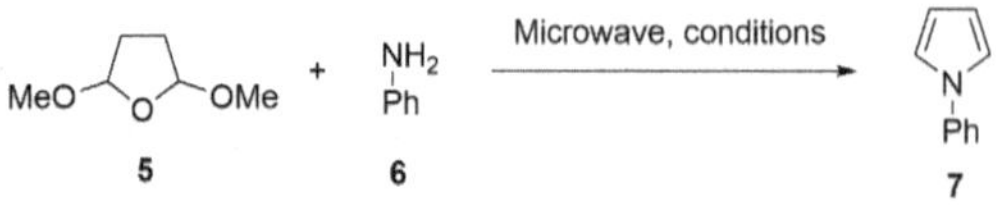

Conditions	Yield, %	Ref.
Water, 170 °C, 10 min	63	[17]
AcOH, 170 °C, 10 min	70	[17]
K-10, 100 °C, 4 min	90	[18]
$Mn(NO_3)_2 \cdot 4H_2O$ (10 mol%), 120 °C, 20 min	89	[19]

Scheme 6.2 Clauson–Kaas synthesis of 1-phenylpyrrole (**7**) under microwave irradiation.

was improved to 70%. When this reaction was performed with reactants absorbed on montmorillonite K-10, a 90% yield was achieved after four minutes of irradiation at 100 °C (Discover Benchmate, CEM) [18]. Magnesium nitrate was also an effective catalyst in the solvent-free Clauson–Kaas reaction under microwave irradiation in Monowave 300 (Anton Paar) [19].

The reaction of 2,5-dimethoxytetrahydrofuran (**5**) with sulfonamides was also efficient under microwave irradiation as exemplified by the synthesis of 1-(phenylsulfonyl)pyrrole (**9**) from phenylsulfonamide (**8**) (Scheme 6.3) [17, 18].

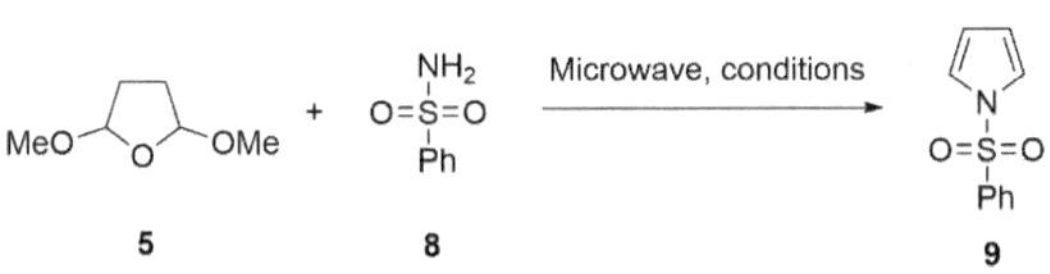

Conditions	Yield, %	Ref.
Water, 170 °C, 10 min	70	[17]
AcOH, 170 °C, 10 min	96	[17]
K-10, 100 °C, 3 min	93	[18]

Scheme 6.3 Clauson–Kaas pyrrole synthesis under microwave irradiation.

6.2.3 Paal–Knorr Pyrrole Synthesis

The Paal–Knorr pyrrole synthesis is based on the reaction of 1,4-diketones with ammonia or primary amines. It is one of the oldest and most explored methods for the preparation of pyrroles. In 1884, Ludwig Knorr briefly mentioned the pyrrole ring formation in the reactions of diacetosuccinic acid ester with ammonia or primary arylamines [20]. Next year, he reported [21] several detailed examples of pyrrole synthesis using the reaction of 1,4-diketones with ammonia or primary amines. Concurrently, Paal expanded his method of the synthesis of 2-methyl-5-phenylfuran from acetophenonacetone (**10**) [22] to the preparation of 2-methyl-5-phenylpyrrole (**11**) using the reaction of acetophenonacetone (**10**) with ammonia [23].

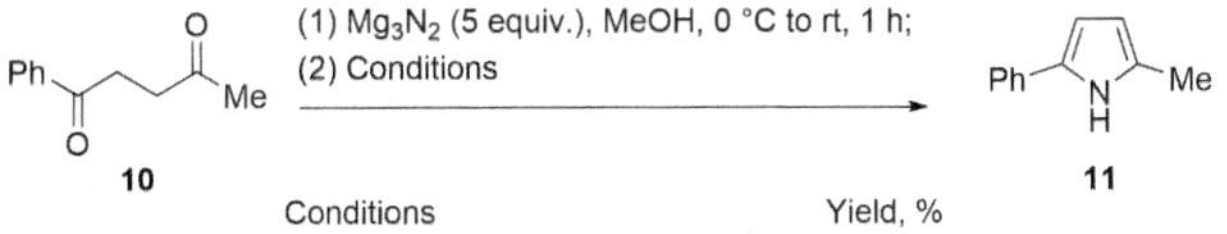

Conditions	Yield, %
Microwave, 120 °C, 1 h	99
Conventional heating, 80 °C, 24 h	95

Scheme 6.4 Paal–Knorr pyrrole synthesis under microwave irradiation and conventional heating.

Magnesium nitride in alcohol was found to be a convenient source of ammonia for the Paal–Knorr pyrrole synthesis [24]. The reaction of acetophenonacetone (**10**) with ammonia formed in situ from magnesium nitride resulted in the formation of 2-methyl-5-phenylpyrrole (**11**). The reaction afforded good yield under microwave irradiation and conventional heating, but reaction time was substantially reduced when microwave irradiation was applied (Scheme 6.4). The relative efficiency of the microwave and conventional heating protocol was found to depend on the diketone structure.

The Paal–Knorr pyrrole synthesis using the reaction of 1,4-diketones with various primary amines was also successfully performed under microwave irradiation [25, 26]. For example, microwave irradiation of diketone **12** with benzylamine (**13**) in acetic acid using an open-vessel mode in a Discover microwave reactor (CEM, USA) resulted in the formation of pyrrole **14** in 70% yield (Scheme 6.5).

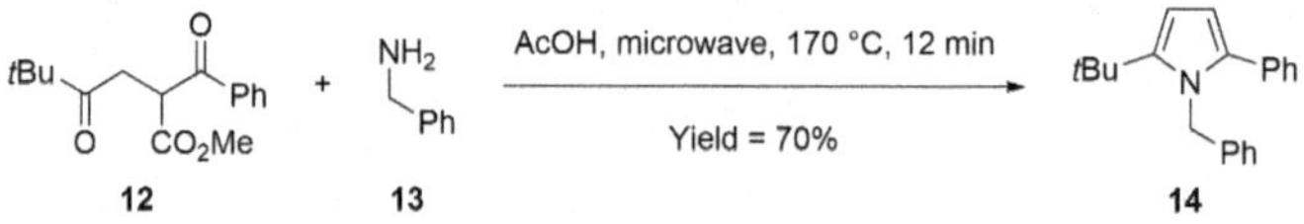

Scheme 6.5 Paal–Knorr pyrrole synthesis under microwave irradiation.

The Paal–Knorr pyrrole synthesis under microwave irradiation was utilized in the total synthesis of natural products, such as Marineosin A [27].

6.2.4 Paal–Knorr Furan Synthesis

The scope of the Paal–Knorr reaction also includes the synthesis of substituted furans via the cyclization of 1,4-dicarbonyl compounds in an acidic medium [22]. The Paal–Knorr furan synthesis under microwave irradiation was found to be highly efficient as illustrated by the microwave-promoted cyclization of **12** in the presence of acid (Scheme 6.6) [25, 26]. An excellent yield of furan **15** was obtained after heating **12** under microwave irradiation in a CEM Discover at 140 °C for just three minutes [26].

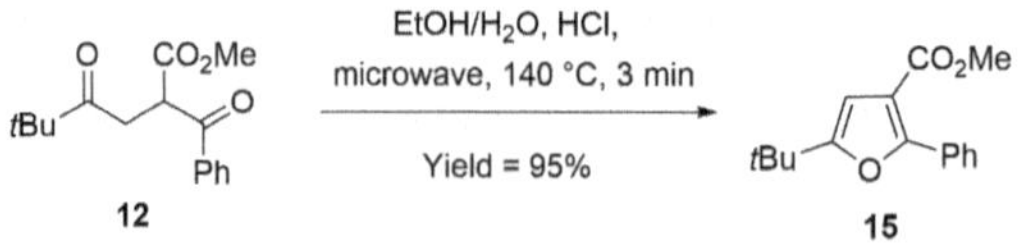

Scheme 6.6 Paal–Knorr furan synthesis under microwave irradiation.

6.2.5 Paal–Knorr Thiophene Synthesis

The Paal–Knorr reaction, extended to the thiophene synthesis, involves condensation of 1,4-dicarbonyl compounds with Lawesson's reagent as a source of sulfur, and can also be efficiently performed under microwave irradiation [26]. For example, heating diketone **12** with an excess of Lawesson's reagent in toluene under microwave irradiation in a CEM Discover (open-vessel mode) for six minutes afforded thiophene **16** (Scheme 6.7).

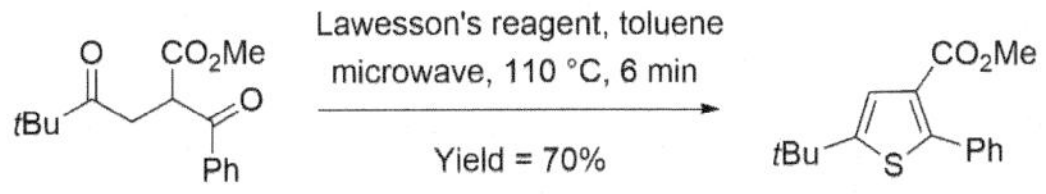

6.2 Classical Methods for Heterocyclic Synthesis Under Microwave Irradiation | 161

Scheme 6.7 Paal–Knorr thiophene synthesis under microwave irradiation.

6.2.6 Gewald Reaction

Another classical method for thiophene synthesis was developed by Karl Gewald [28] and is based on the reaction of enolizable aldehydes or ketones, methylene-active nitriles, and elemental sulfur. Microwave irradiation was demonstrated to substantially improve the Gewald reaction efficiency. The reaction time for the synthesis of ethyl 2-amino-5-phenyl-3-thiophenecarboxylate (**19**) from phenacetaldehyde (**17**), elemental sulfur, and ethyl cyanoacetate (**18**) was reduced from four hours of conventional heating to 20 minutes of controlled microwave irradiation at the same temperature, using an open-vessel mode in a CEM Discover microwave reactor (Scheme 6.8) [29]. Moreover, the yield of **19** was improved from 80% to nearly quantitative.

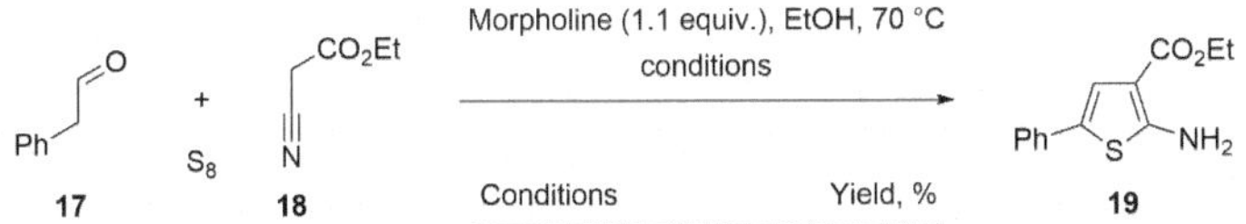

Conditions	Yield, %
Microwave, 20 min	99
Conventional heating, 4 h	80

Scheme 6.8 Gewald reaction under microwave irradiation and conventional heating.

A substantial reaction acceleration under microwave irradiation was also demonstrated when ketones were used as substrates [30]. When cyclohexanone (**20**) was involved in the Gewald reaction, using potassium fluoride immobilized on alumina as a catalyst, microwave irradiation conditions (closed-vessel mode, CEM Discover) reduced the reaction time required for the formation of thiophene **21** to 8 minutes (from 5.5 hours of conventional heating) without any yield loss (Scheme 6.9).

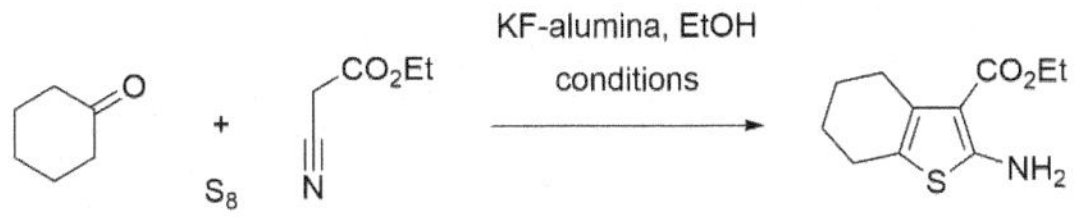

Conditions	Yield, %
Microwave, 100 °C, 8 min	91
Reflux, 5.5 h	89

Scheme 6.9 Cyclohexanone (**20**) in Gewald reaction under microwave irradiation and conventional heating.

6.2.7 Fischer Indole Synthesis

Microwave irradiation effectively promoted syntheses of indoles via various classical methods. The Fischer indole synthesis, which is an equivalent of the Piloty–Robinson pyrrole synthesis, utilizes the cyclization of hydrazones formed from phenylhydrazines and aldehydes or ketones [31, 32]. The microwave irradiation conditions were reported to promote the Fisher indole synthesis using various catalytic systems. For example, 2,3-dimethylindole (**24**) was successfully prepared from a microwave-assisted reaction of phenylhydrazine (**22**) and 2-butanone (**23**) using propylphosphonic acid cyclic anhydride (T3P) [33] or sulfonic acid-functionalized ionic liquid [34] to facilitate the reaction, which was also effective even in the absence of any catalysts and additives [35] (Scheme 6.10).

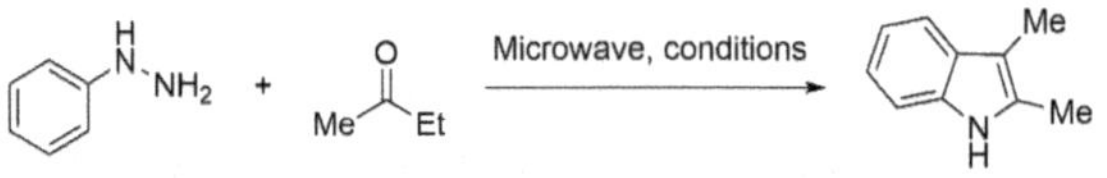

Conditions	Yield, %	Ref.
T3P (1.25 equiv.), EtOAc, 110 °C, 15 min	76	[33]
[(HSO$_3$-p)$_2$im][CF$_3$SO$_3$] (5 mol%), water, 100 °C, 15 min	93	[34]
THF, 150 °C, 10 min	96	[35]

Scheme 6.10 Microwave-assisted Fisher indole synthesis.

6.2.8 Bischler–Möhlau Indole Synthesis

The reaction between α-bromoacetophenones and anilines to produce 2-arylindoles is known as the Bischler–Möhlau indole synthesis [36]. Microwave irradiation was reported [37, 38] to promote this reaction. For example, *p*-anisole (**25**) and α-bromoacetophenone **26** readily reacted under microwave irradiation in a CEM Discover reactor to afford indole **27** (Scheme 6.11). No desired product **27** was formed when conventional heating was applied (the same temperature and reaction time). Substantially longer conventional heating at a higher temperature had to be utilized to initiate the reaction, but the yield remained lower than that obtained under microwave irradiation conditions (Scheme 6.11).

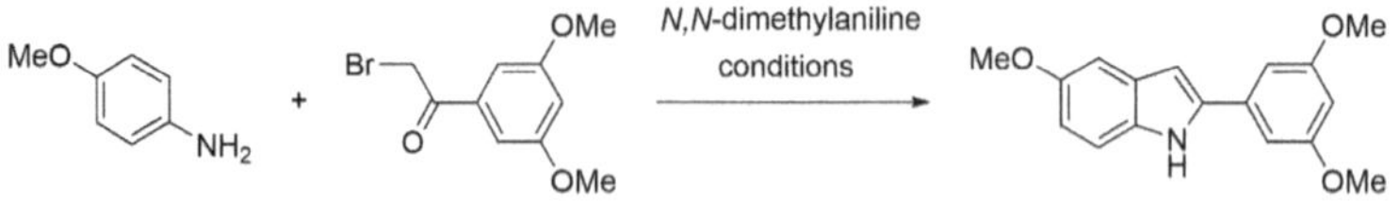

Conditions	Yield, %
Microwave, 100 W, 150 °C, 10 min	48
Xylene, reflux, 170 °C, 3 h	19
Conventional heating, 150 °C, 10 min	0

Scheme 6.11 Bischler–Möhlau indole synthesis under microwave irradiation and conventional heating.

6.2.9 Hemetsberger–Knittel Indole Synthesis

The thermolysis of 3-aryl-2-azidopropenoates with the indole ring closure was developed by Hemetsberger and Knittel [39]. This reaction, known as the Hemetsberger–Knittel indole synthesis, was reported [40] to benefit from microwave irradiation conditions, which shortened the reaction time and increased yields. For example, the thermolysis of azide **28** under microwave irradiation in toluene afforded indole **29** in better yield and much faster than the same reaction under conventional heating (Scheme 6.12).

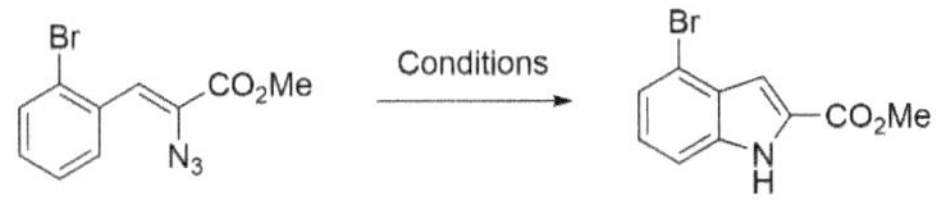

Conditions	Yield, %
Toluene, microwave, 200 °C, 10 min	94
Xylene, conventional heating, 140 °C, 2 h	70

Scheme 6.12 Hemetsberger–Knittel indole synthesis under microwave irradiation and conventional heating.

6.2.10 Leimgruber–Batcho Indole Synthesis

The reaction sequence, starting from the condensation of *o*-nitrotoluenes with formamide acetals and followed by the reductive indole ring closure, was developed by Batcho and Leimgruber [41] and known as the Leimgruber–Batcho indole synthesis. The first step of this synthesis can be efficiently promoted by microwave irradiation, as illustrated in Scheme 6.13 [42]. Microwave irradiation in an Emrys Liberator microwave reactor allowed a complete conversion of *o*-nitrotoluene (**30**) into enamine **32** upon treatment with dimethyl formamide dimethyl acetal (**31**) with convenient isolation of the product in 95% yield (Scheme 6.13). The presence of small quantities of dimethyl formamide (DMF) was found to enhance the reaction rate, presumably by improving the microwave absorption profile. The catalytic hydrogenation of **32** afforded indole (**33**).

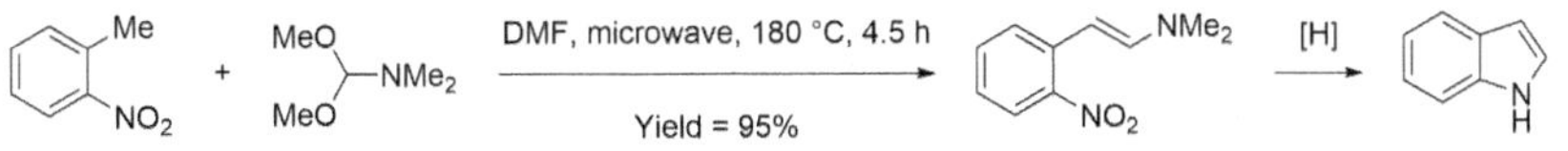

Scheme 6.13 Leimgruber–Batcho indole synthesis under microwave irradiation.

6.2.11 Cadogan–Sundberg Indole Synthesis

Another reductive indole ring closure approach, known as the Cadogan–Sundberg indole synthesis, involves the cyclization of *o*-nitrostyrenes and their analogs via nitrene intermediates generated upon the treatment with trialkyl phosphites [43, 44]. It was reported [45] that the cyclization of **34** in the presence of triethyl phosphite resulted in the formation of **35** and its alkylated derivative **36** with considerable time-saving and yield improvements when the reaction was performed under microwave irradiation in a CEM Discover reactor (Scheme 6.14). By the replacement of triethyl phosphite with triphenylphosphine, selectivity towards **35** was achieved and the microwave-assisted protocol was also more efficient.

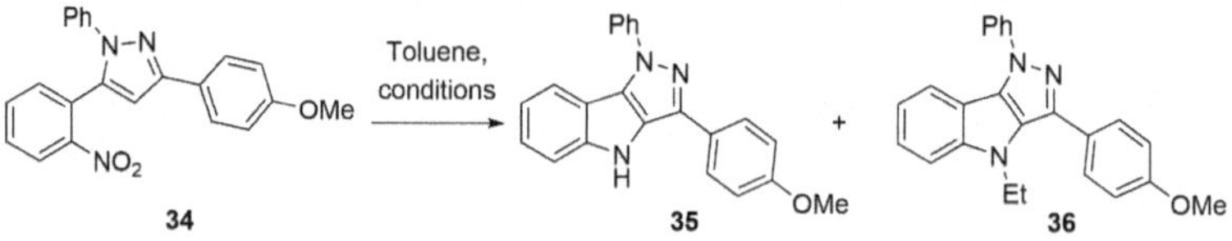

Conditions	Yield (**35/36**), %
P(OEt)$_3$ (3 equiv.), microwave, 210 °C, 30 min	52/18
P(OEt)$_3$ (3 equiv.), reflux, 24 h	16/9
PPh$_3$ (3 equiv.), microwave, 180 °C, 20 min	61/0
PPh$_3$ (3 equiv.), reflux, 24 h	22/0

Scheme 6.14 Cadogan–Sundberg indole synthesis under microwave irradiation and conventional heating.

6.2.12 Pechmann Pyrazole Synthesis

The synthesis of pyrazoles via a 1,3-dipolar cycloaddition of diazomethane to alkynes was developed by Hans von Pechmann in 1895 [46]. This reaction, as well as other important heterocyclizations proceeding via 1,3-dipolar cycloadditions (e.g. Huisgen cycloaddition and Finnegan tetrazole synthesis described below), benefited from microwave irradiation conditions [47]. For example, a higher yield of pyrazole **39** and time-saving were achieved when ethyl diazoacetate (**37**) was irradiated in an excess of phenylacetylene (**38**) (Scheme 6.15). Conventional heating was considerably less efficient.

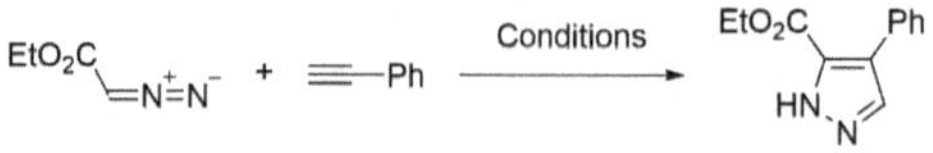

Conditions	Yield, %
Microwave, 120 °C, 10 min	97
Conventional heating, 24 h	81

Scheme 6.15 Pechmann pyrazole synthesis under microwave irradiation and conventional heating.

6.2.13 Debus–Radziszewski Reaction

The imidazole synthesis by condensation of glyoxal and ammonia was first reported by Debus in 1858 [48] and modified for the synthesis of substituted imidazoles by

Radziszewski [49, 50], and then by Drefahl and Herma [51], who expanded the scope of this reaction. Nowadays, the Debus–Radziszewski reaction refers to the multi-component reaction of a 1,2-dicarbonyl compound, an aldehyde, and two molecules of ammonia or one molecule of ammonia and a primary amine. The reaction was successfully performed under microwave irradiation with good yields of products and short reaction times. For example, in the microwave-facilitated synthesis of natural alkaloid Lepidiline B (**43**), an intermediate 2,4,5-trimethylimidazole (**42**) was prepared from diacetyl (**40**), acetaldehyde (**41**), and ammonium acetate using the Debus–Radziszewski reaction under microwave irradiation in a Smithsynthesizer reactor (Scheme 6.16) [52]. Similarly, an antiaggregant agent trifenagrel (**46**) was prepared with excellent yield from benzil (**44**), aldehyde **45**, and ammonium acetate (Scheme 6.17) [52].

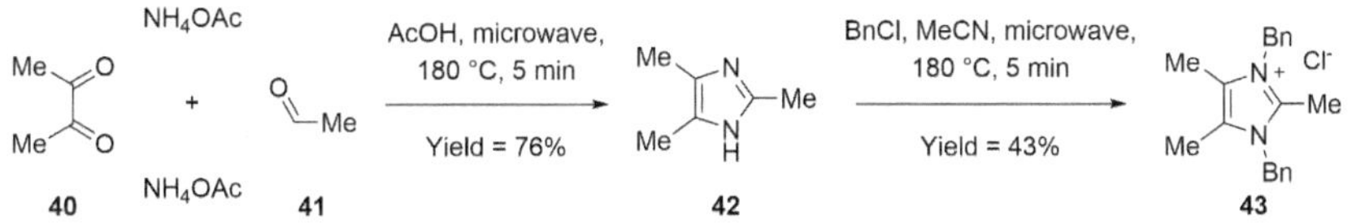

Scheme 6.16 Microwave-assisted synthesis of Lepidiline B (**43**).

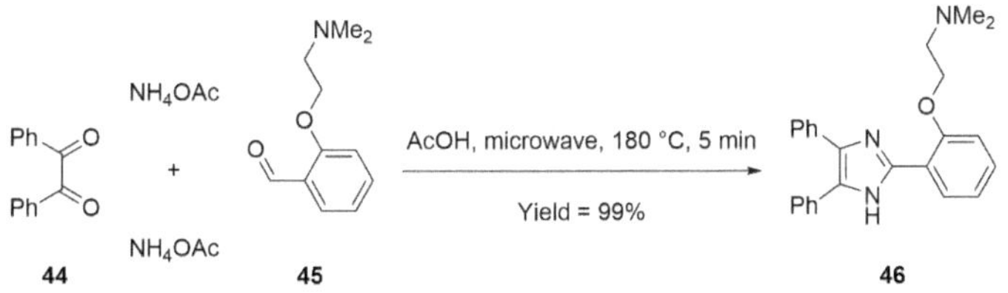

Scheme 6.17 Debus–Radziszewski reaction under microwave irradiation for the synthesis of trifenagrel (**46**).

The microwave reactor design often provides opportunities for parallel synthesis or for automated synthesis of compound libraries using compatible autosampler assemblies. For example, a combinatorial library of imidazoles was prepared using an Explorer auto-sampler for the CEM Discover reactor [53]. The model reaction of benzil (**44**), benzaldehyde (**47**), benzylamine (**13**), and ammonium acetate was used for condition optimization (Scheme 6.18). The parallel microwave-assisted synthesis using the Debus–Radziszewski reaction was conveniently applied for educational

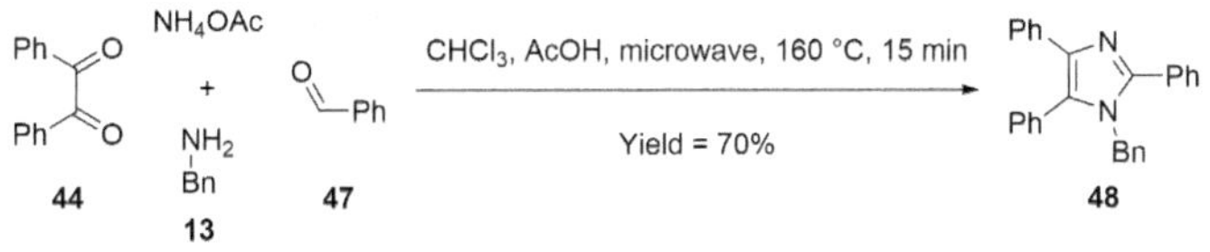

Scheme 6.18 Synthesis of tetrasubstituted imidazole **48** using the microwave-assisted Debus–Radziszewski reaction.

purposes when the short reaction time under microwave irradiation made this reaction suitable for undergraduate student teaching [54]. Sixteen reactions for the synthesis of liphine (2,4,5-triphenylimidazole) were performed simultaneously in a Milestone START Microwave Lab Station using benzil, benzaldehyde, and ammonium acetate.

6.2.14 van Leusen Imidazole Synthesis

The van Leusen imidazole synthesis involves the formation of substituted imidazoles through the base-catalyzed cycloaddition of tosylmethyl isocyanide (**50**) to a Schiff base [55].

It was reported [56, 57] that van Leusen imidazole synthesis benefited from microwave irradiation conditions. For example, the reaction time in the synthesis of bioactive imidazole **51** from imine **49** and tosylmethyl isocyanide (**50**) was reduced from three days of conventional heating to seven hours of microwave irradiation in a CEM Discover reactor (Scheme 6.19) [56]. Moreover, the yield of **51** had also improved.

Conditions	Yield, %
Microwave, 90 °C, 7 h	52
Reflux, 72 h	44

Scheme 6.19 van Leusen imidazole synthesis under microwave irradiation and conventional heating.

The microwave-assisted protocol was also effectively adopted for the solid-phase van Leusen imidazole synthesis using the cycloaddition of tosylmethyl isocyanide (**50**) to polymer-bound imines [57]. The products were formed in good yields after 20 minutes of microwave irradiation in DMF at 130 °C using a Biotage Initiator reactor.

6.2.15 van Leusen Oxazole Synthesis

Tosylmethyl isocyanide (**50**) can also be employed for the synthesis of oxazoles and the reaction of **50** with carbonyl compounds in the process known as the van Leusen oxazole synthesis [58]. This reaction was reported [59] to be particularly efficient under microwave irradiation. The reaction time for the synthesis of 5-phenyloxazole (**52**) from benzaldehyde (**47**) and tosylmethyl isocyanide (**50**), in the presence of potassium phosphate, was reduced from one hour to eight minutes

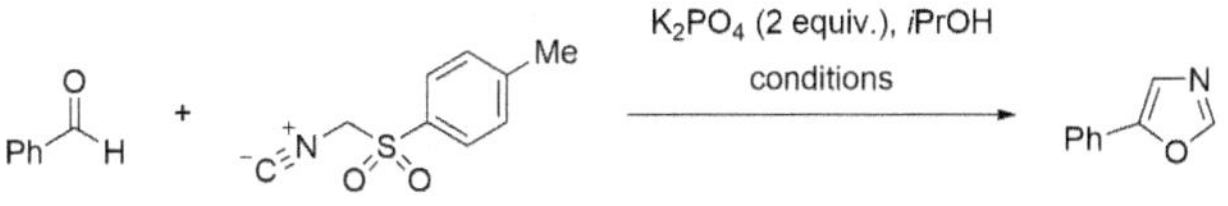

Scheme 6.20

Reaction: Ph-CHO (**47**) + TosMIC reagent (**50**) → 5-phenyloxazole (**52**)

Conditions: K₂PO₄ (2 equiv.), iPrOH, conditions

Conditions	Yield, %
Microwave, 65 °C, 8 min	96
Conventional heating, 60 °C, 1 h	95

Scheme 6.20 van Leusen oxazole synthesis under microwave irradiation and conventional heating.

when the reaction was performed under microwave irradiation in the open vessel using a CATA R (Catalyst System) microwave reactor (Scheme 6.20).

6.2.16 Robinson–Gabriel Reaction

The Robinson–Gabriel reaction was developed in the early eighteenth century and was extensively used for the preparation of oxazoles via cyclocondensation of α-acylaminoketones in the presence of a dehydrating agent [60, 61]. This reaction was found to benefit from microwave irradiation conditions. For example, 2-methyl-5-phenyloxazole (**54**) was prepared from phenacylacetamide (**53**) by using the Burgess reagent as a mild dehydrating agent [62]. The desired product **54** was isolated in an 80% yield after microwave irradiation of **53** in a Labwell MW10 reactor at 100 W for two minutes (Scheme 6.21). The same reaction under conventional heating (reflux in THF for two hours) resulted in 68% conversion (by LC–MS analysis).

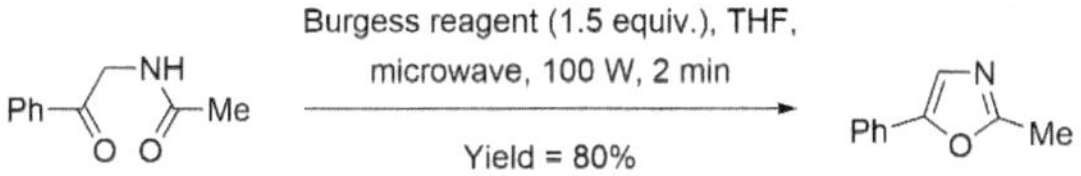

Scheme 6.21 Robinson–Gabriel reaction under microwave irradiation.

The microwave-assisted Robinson–Gabriel oxazole synthesis was also applied for the preparation of some indole alkaloids [63]. The cyclization of **55** was performed in the presence of propylphosphonic anhydride (T3P) under microwave irradiation in a MicroSYNTH T640 reactor and afforded products **56** in excellent yields (Scheme 6.22).

6.2.17 Hantzsch Thiazole Synthesis

The condensation of α-haloketones with thioamides (or their analogs) is known as the Hantzsch thiazole synthesis [64]. The microwave irradiation was reported [65] to accelerate this reaction as exemplified by the synthesis of thiazole **59** from

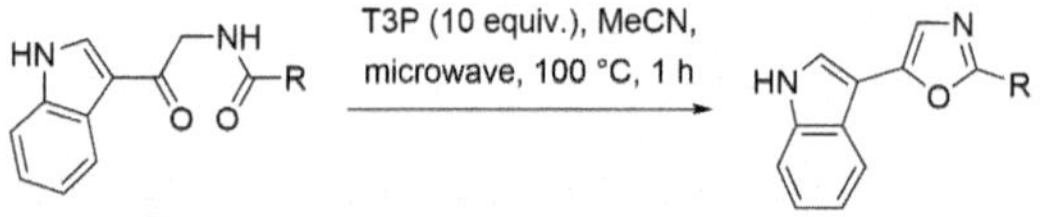

Scheme 6.22 Robinson–Gabriel under microwave irradiation for the alkaloid synthesis.

56	R	Yield, %
Pimprinine	Me	100
Pimpriethine	Et	100
Pimprinaphine	Bn	100[a]
Labradorin 1	iBu	100
Labradorin 2	Pent	100

[a] 2 h reaction time

α-bromoacetophenone **57** and thiourea (**58**), under controlled irradiation in an Emrys Optimizer reactor for 20 minutes (Scheme 6.23). The same reaction using conventional heating required 90 minutes to achieve a comparable yield.

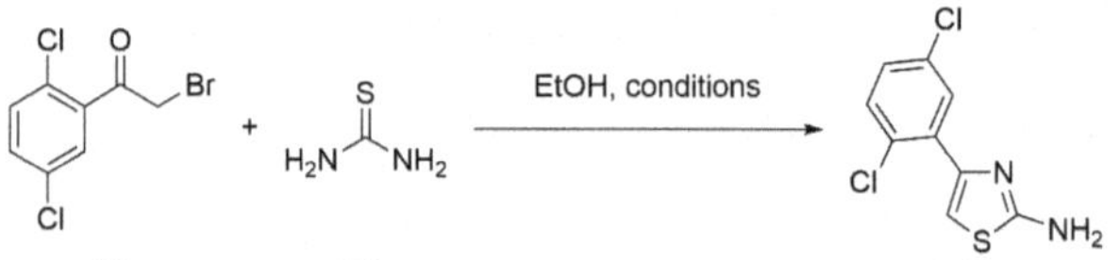

Conditions	Yield, %
Microwave, 150 W, 80 °C, 20 min	96
Reflux, 1.5 h	93

Scheme 6.23 Hantzsch thiazole synthesis under microwave irradiation and conventional heating.

6.2.18 Einhorn–Brunner Reaction

The 1,2,4-triazole formation upon heating imides with hydrazines is known as the Einhorn–Brunner reaction [66, 67]. This reaction, performed under microwave irradiation, resulted in substantially improved yields while reducing the reaction time manifolds. For example, the reaction of imide **60** with phenylhydrazine hydrochloride (**61**) under microwave irradiation (200 °C, 300 W) was reported [68] to produce triazole **62** in an 84% yield within one minute (Scheme 6.24). The conventional heating was less efficient affording **62** in a 63% yield after four hours of heating under reflux in pyridine.

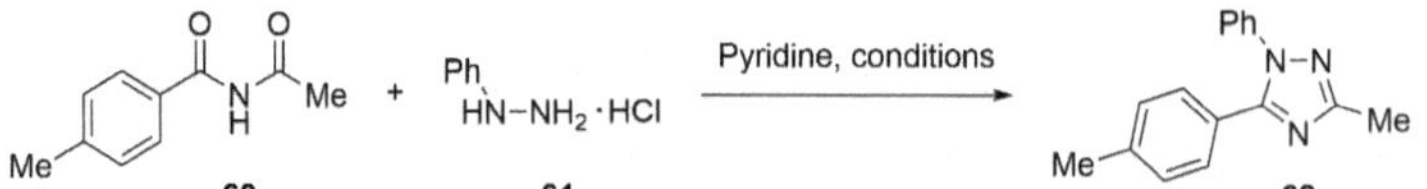

Conditions	Yield, %
Microwave, 200 °C, 1 min	84
Reflux, 4 h	63

Scheme 6.24 Einhorn–Brunner reaction under microwave irradiation and conventional heating.

The microwave-assisted synthesis of triazole **62** was also successfully carried out in a one-pot manner by the initial N-acetylation of *p*-toluamide (**63**) and

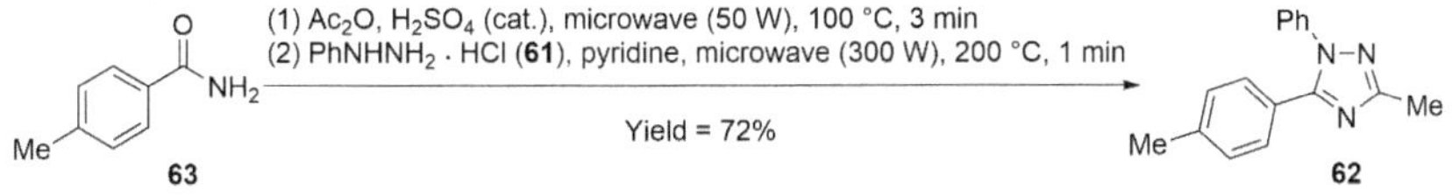

Scheme 6.25 One-pot synthesis of 1,2,4-triazole **62** under microwave irradiation.

the subsequent Einhorn–Brunner reaction of the intermediate imide **60** with phenylhydrazine hydrochloride (**61**) (Scheme 6.25) [68].

The Einhorn–Brunner reaction under microwave irradiation demonstrated a broad scope tolerating a wide range of imides and hydrazines. The one-pot protocol also proved its efficiency in the reactions of several benzamides [68].

6.2.19 Pellizzari Reaction

The Pellizzari reaction utilizes cyclodehydration of amides with hydrazides to construct the 1,2,4-triazole ring [69]. The reaction requires a high temperature and often suffers from low yields. Microwave irradiation improves the reaction efficiency and expands the reaction scope. 3,4,5-Trisubstituted 1,2,4-triazoles were prepared using a microwave-assisted reaction of N-substituted amides with hydrazides [70]. The amides were activated by trifluoromethanesulfonic anhydride (Tf$_2$O) in the presence of 2-fluoropyridine (2-FPyr) as a base. Thus, the Pellizzari reaction between amide **64** and benzhydrazide (**65**) under microwave irradiation in a Biotage Initiator Classic reactor resulted in the formation of triazole **66** (Scheme 6.26) [70]. A wide range of amides with hydrazides was effectively employed under these conditions in producing analogs of **66** in good yields.

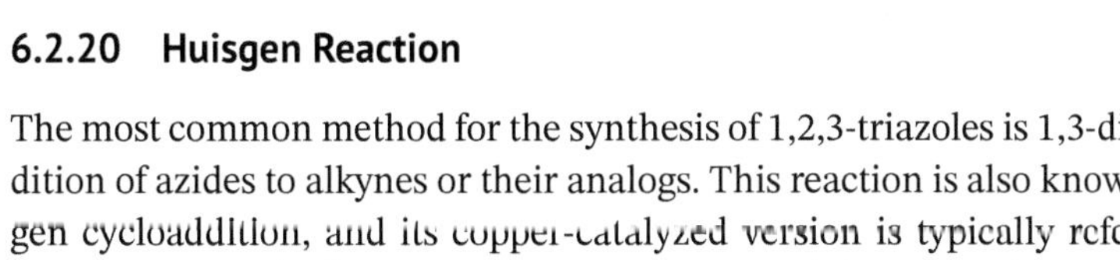

Scheme 6.26 Pellizzari reaction under microwave irradiation.

6.2.20 Huisgen Reaction

The most common method for the synthesis of 1,2,3-triazoles is 1,3-dipolar cycloaddition of azides to alkynes or their analogs. This reaction is also known as the Huisgen cycloaddition, and its copper-catalyzed version is typically referred to as the click chemistry [71]. Microwave irradiation conditions have been widely used for the 1,2,3-triazole ring construction via the Huisgen reaction.

The solvent- and catalyst-free reaction of TMS azide (**67**) with diphenylacetylene (**68**) was more efficient under microwave irradiation in a Biotage Initiator⁺ reactor than under conventional heating (Scheme 6.27) [72]. The yield of 4,5-diphenyl-1,2,3-triazole (**69**) was a few times higher when microwave heating was applied.

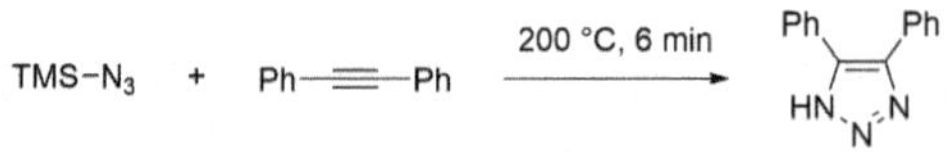

Scheme 6.27 Huisgen reaction under microwave irradiation and conventional heating.

Conditions	Yield, %
Microwave	98
Conventional heating	28

Vinyl acetate (**71**) was effectively used as an alkyne analog in the cycloaddition with diverse azides under microwave irradiation [73]. Heating benzyl azide (**70**) in vinyl acetate (**71**) under microwave irradiation in a Biotage Initiator reactor at 120 °C for five hours afforded 1-benzyl-1,2,3-triazole (**72**) in nearly quantitative yield (Scheme 6.28). The same cycloaddition under conventional heating (reflux in excess of **71**) was substantially less efficient, and **72** was obtained in a 55% yield after a one-week reaction.

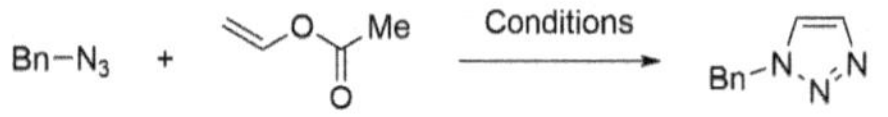

Conditions	Yield, %
Microwave, 120 °C, 5 h	99
Reflux, 7 days	55

Scheme 6.28 Huisgen reaction of benzyl azide (**70**) and vinyl acetate (**71**) under microwave irradiation and conventional heating.

An interesting microwave-assisted intramolecular Huisgen reaction was reported by Balducci et al. [74]. Azido-substituted alkyne **73** was heated under microwave irradiation at 160 °C for three one-hour cycles to produce bicyclic triazole **74**, which was isolated in a 65% yield (Scheme 6.29). Importantly, the conventional heating was inefficient for this reaction and after heating in a sealed vial at the same temperature for 24 hours, no product **74** was formed.

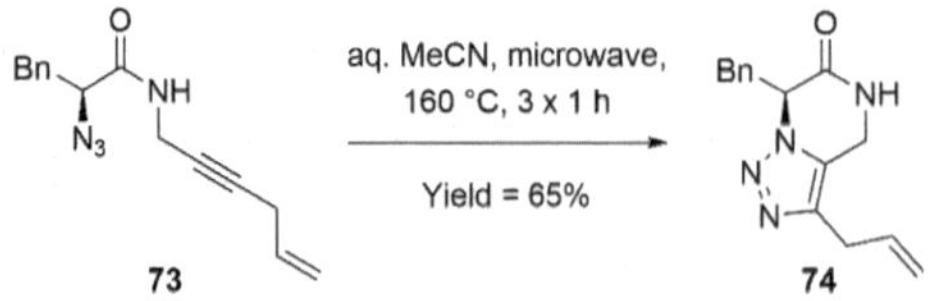

Scheme 6.29 Intramolecular Huisgen reaction under microwave irradiation.

Young's group explored effects of microwave irradiation parameters on the Huisgen cycloaddition in a CEM Discover reactor using a model reaction of benzyl azide (**70**) and phenylacetylene (**75**) [75]. They found that microwave power and the mode in which this power was applied had a tremendous effect on the reaction outcome.

The best yield of triazoles **76** and **77** (formed in equal quantities) was obtained when a high power (300 W) was applied in a pulsed power mode, maintaining temperature at 168 °C with a δT range of 15 °C (Scheme 6.30).

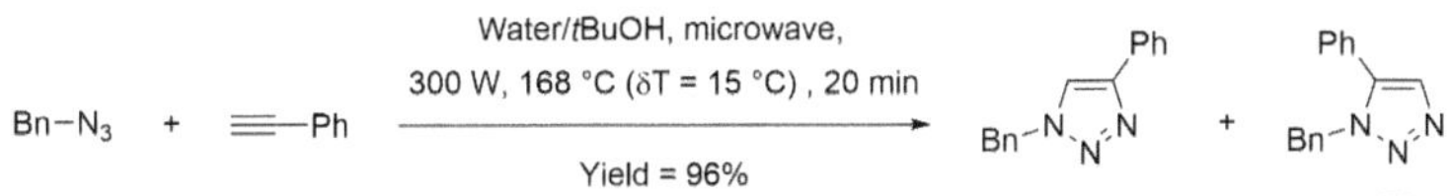

Bn–N$_3$ + ≡–Ph $\xrightarrow[\text{Yield = 96\%}]{\substack{\text{Water/}t\text{BuOH, microwave,} \\ \text{300 W, 168 °C (}\delta T\text{ = 15 °C) , 20 min}}}$ Bn-triazole **76** + Bn-triazole **77**

70   **75**   **76**   **77**

Scheme 6.30 Huisgen reaction benzyl azide (**70**) and phenylacetylene (**75**) under microwave irradiation.

The microwave irradiation with simultaneous cooling of the reaction mixture in a CEM Discovery CoolMate microwave reactor was effective for bioconjugations based on the Huisgen cycloaddition [75]. The reaction between an alkyne-substituted protein and an azide-functionalized fluorophore was performed without substantial protein denaturation applying power up to 200 W and allowing the temperature to rise from −30 to 40 °C.

6.2.21 Finnegan Tetrazole Synthesis

A related cycloaddition of sodium azide (**78**) to a triple bond of nitriles resulted in the formation of tetrazoles and this synthetic approach is known as the Finnegan tetrazole synthesis [76]. This reaction was found to benefit from microwave irradiation, substantially reducing the reaction time and expanding the reaction scope. The cycloaddition of sodium azide (**78**) to benzonitrile (**79**) was effectively performed under microwave irradiation in a Milestone MicroSYNTH Ethos 1600 reactor at 100 °C in the presence of triethylammonium chloride using nitrobenzene as a medium (Scheme 6.31) [77]. The reaction resulted in the formation of 5-phenyltetrazole (**80**), which was isolated in good yield without the need for chromatographic purification.

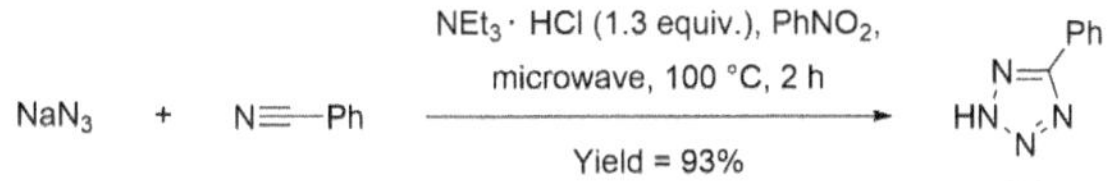

NaN$_3$ + N≡–Ph $\xrightarrow[\text{Yield = 93\%}]{\substack{\text{NEt}_3 \cdot \text{HCl (1.3 equiv.), PhNO}_2, \\ \text{microwave, 100 °C, 2 h}}}$ 5-phenyltetrazole

78   **79**   **80**

Scheme 6.31 Finnegan tetrazole synthesis under microwave irradiation.

Among numerous modifications of the microwave-promoted Finnegan tetrazole synthesis, there are efficient variations with in situ formation of nitriles for the subsequent one-pot reaction with sodium azide (**78**). For example, the reaction of benzaldehyde (**47**) with iodine and aqueous ammonia generated benzonitrile (**79**), which was subjected to the zinc bromide-catalyzed Finnegan reaction with sodium azide (**78**) under microwave irradiation to afford 5-phenyltetrazole (**80**) (Scheme 6.32) [78].

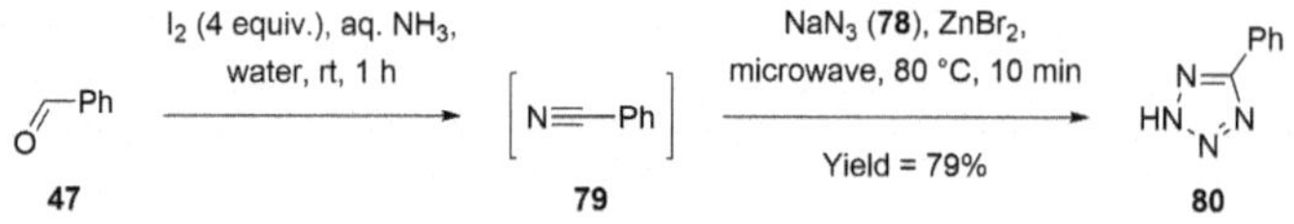

Scheme 6.32 One-pot synthesis of 5-phenyltetrazole (**80**) under microwave irradiation.

6.2.22 Four-component Ugi-azide Reaction

Another tetrazole synthesis is based on the four-component Ugi-azide reaction, which involves an isocyanide, an aldehyde, an amine, and an azide [79]. This reaction was also reported [80] to be efficient under microwave irradiation. For example, the microwave irradiation of cyclohexylcaboxaldehyde (**81**) and tritylamine (**82**) in ethanol resulted in the formation of Schiff base **85**, which was involved in the reaction with trimethylsilyl azide (**83**) and cyclohexyl isocyanide (**84**) to afford tetrazole **86** (Scheme 6.33).

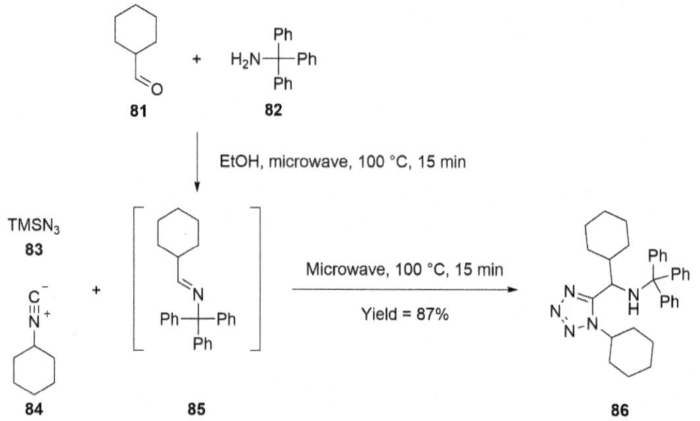

Scheme 6.33 Four-component Ugi-azide reaction under microwave irradiation.

6.2.23 Kröhnke Pyridine Synthesis

The multicomponent synthesis of 2,4,6-triphenylpyridines, using the reaction between β-benzoylstyrene, *N*-phenacetylpyridinium bromide, and ammonium acetate was reported by Fritz Kröhnke in 1961 [81]. In the four-component variation of this synthesis, β-benzoylstyrene is formed in situ from benzaldehyde and acetophenone. Microwave irradiation was successfully utilized to promote the Kröhnke pyridine synthesis in various modifications and for diverse substrates. For example, microwave-assisted Krohnke pyridine synthesis was used for the preparation of modified natural androgen 5α-dihydrotestosterone (DHT) derivatives with a fused pyridine ring in moderate to good yields [82]. For example, pyridine-fused DHT **89** was prepared by reacting steroidal arylideneketone **87** with *N*-phenacetylpyridinium iodide (**88**) and ammonium acetate in a 77% yield after 20 minutes of microwave irradiation at 90 °C using a CEM Discover reactor (Scheme 6.34).

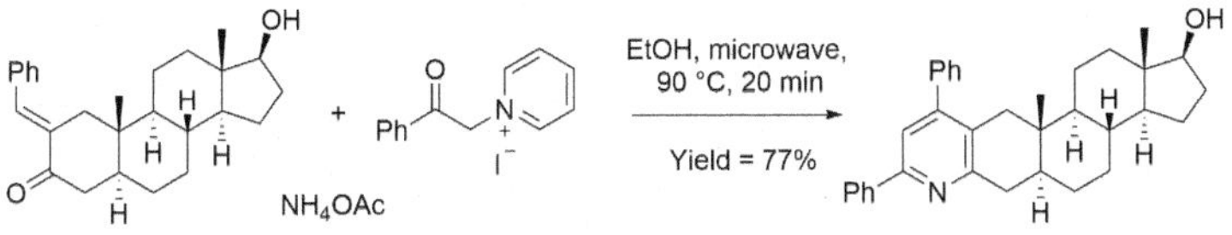

Scheme 6.34 Kröhnke pyridine ring annelation under microwave irradiation.

The four-component modification of the Kröhnke pyridine synthesis between *p*-chlorobenzaldehyde (**90**), α-acetylpyridine (**91**), and ammonium acetate was efficient under microwave irradiation in an Emrys Creator reactor (Scheme 6.35) [83]. The yield of pyridine **92** under conventional heating was lower, while the reaction time was fifteen times longer. The microwave power was found to significantly influence the reaction outcome with the best yield observed at a power of 200 W. The power dependence of the microwave-promoted four-component Kröhnke pyridine synthesis was also observed by other researchers [84].

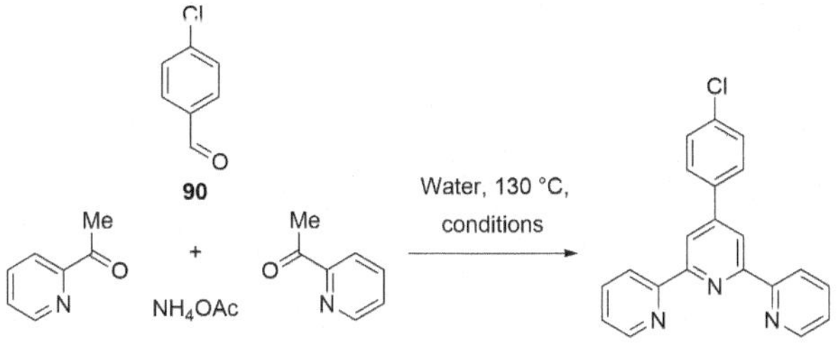

Conditions	Yield, %
Microwave, 200 W, 16 min	92
Conventional heating, 6 h	78

Scheme 6.35 Kröhnke pyridine synthesis under microwave irradiation and conventional heating.

6.2.24 Bohlmann–Rahtz Pyridine Synthesis

In 1957, Ferdinand Bohlmann and Dieter Rahtz reported [85] the synthesis of pyridines using cycloaddition of enamines to α-carbonylacetylenes. Microwave irradiation in a CEM Discover and conventional heating were almost equally efficient in the Bohlmann–Rahtz synthesis of pyridine **95** from enamine **93** and benzoylacetylene (**94**) (Scheme 6.36) [86]. However, microwave irradiation was preferred for other substrates and conditions. For example, the microwave-assisted reaction of enamine **93** with **96** afforded a higher yield of **97** than the same reaction under thermal conditions (Scheme 6.37) [86].

The Bohlmann–Rahtz synthesis of pyridine **95** was also optimized for a microwave flow reactor [87].

Conditions	Yield, %
Microwave	98
Conventional heating	95

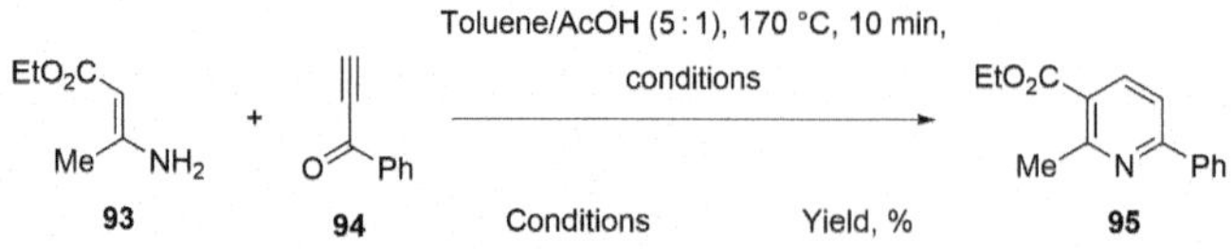

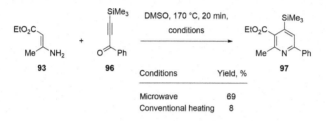

Scheme 6.36 Bohlmann–Rahtz pyridine synthesis under microwave irradiation and conventional heating.

Conditions	Yield, %
Microwave	69
Conventional heating	8

Scheme 6.37 Higher efficiency of the Bohlmann–Rahtz pyridine synthesis under microwave irradiation versus conventional heating.

6.2.25 Boger Reaction

The pyridine ring formation, using the inverse-electron demand Diels–Alder reaction of enamines with 1,2,4-triazines, was developed by Dale Boger [88]. In the synthesis of **100** (a key intermediate for the preparation of the louisianin alkaloid family) using the Boger reaction, microwave irradiation was found to be a convenient alternative to conventional heating (Scheme 6.38) [89]. Despite a lower yield of **100** when cycloaddition of enamine **98** and triazine **99** was performed under microwave irradiation, the reaction time was substantially reduced compared to the synthesis under conventional heating, and the addition of silica to the reaction mixture was also unnecessary.

Conditions	Yield, %
Microwave, toluene, 120 °C, 1 h	80
Reflux, xylene, SiO_2, 18 h	89

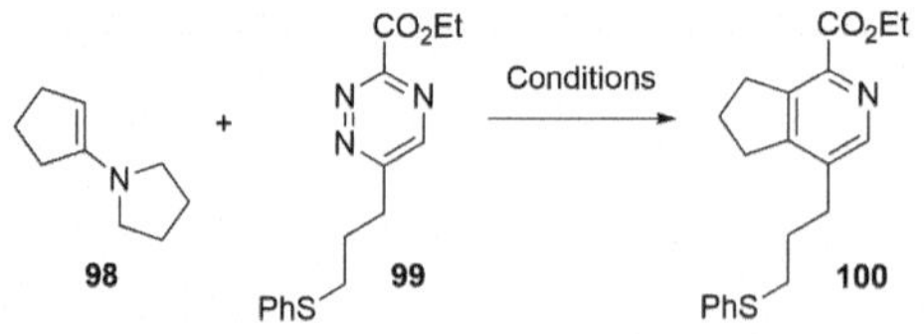

Scheme 6.38 Boger pyridine synthesis under microwave irradiation and conventional heating.

6.2.26 Skraup Reaction

In 1880, Zdenko Hans Skraup reported [90] quinoline synthesis by heating aniline or nitrobenzene with glycerol in the presence of sulfuric acid. The best results were obtained when aniline and nitrobenzene were used together. In this case, nitrobenzene served as an oxidizing agent, while sulfuric acid catalyzed the reaction by dehydrating glycerol to the reactive intermediate, acrolein. The best yield of quinoline achieved under these conditions was 25%.

Microwave irradiation was found to improve the efficiency of the Skraup reaction. Without adding an oxidizing agent, in aqueous media, aniline (**6**) reacted with glycerol (**101**) and was successfully converted into quinoline (**102**) under controlled microwave irradiation in a Monowave 300 reactor (Anton Paar, Austria) at 200 °C for 10 minutes using 36 °C/min heating ramp (Scheme 6.39) [91]. The microwave conditions were demonstrated to be suitable for the synthesis of a variety of substituted quinolines, while reactions of nitroanilines resulted in the formation of phenanthrolines.

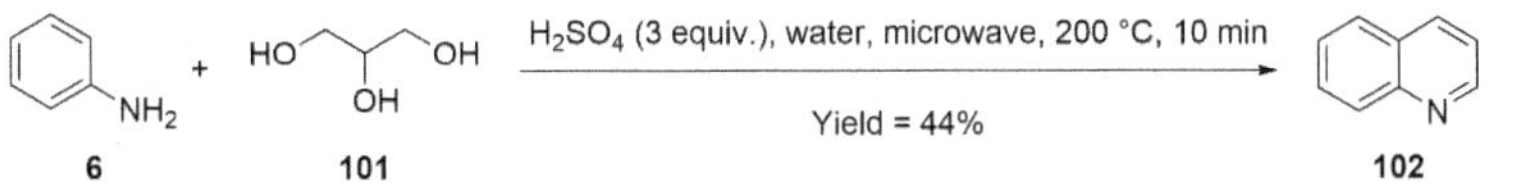

Scheme 6.39 Skraup reaction of aniline (**6**) under microwave irradiation.

The microwave-assisted Skraup reaction, utilizing nitrobenzene (**103**) as a starting material, resulted in the formation of 6-hydroxyquinoline (**104**) (Scheme 6.40) [92]. The presence of sulfuric acid was found to be an important factor governing the reaction regioselectivity. When a mixture of equimolar quantities of aniline (**6**) and nitrobenzene (**103**) was used for the microwave-assisted Skraup reaction under the same conditions, quinoline (**102**) was obtained in a 42% yield with 6-hydroxyquinoline (**104**) in trace amount.

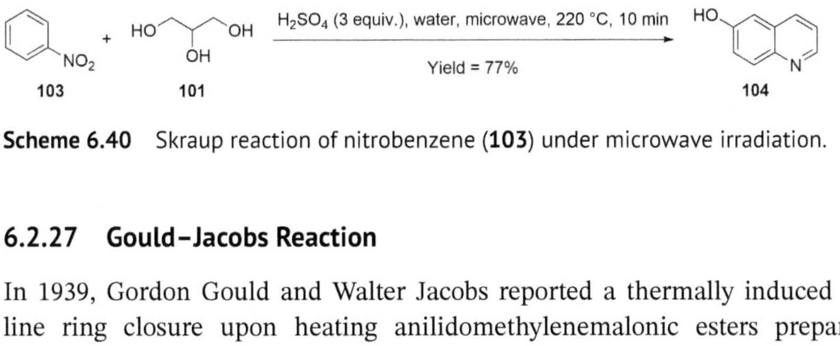

Scheme 6.40 Skraup reaction of nitrobenzene (**103**) under microwave irradiation.

6.2.27 Gould–Jacobs Reaction

In 1939, Gordon Gould and Walter Jacobs reported a thermally induced quinoline ring closure upon heating anilidomethylenemalonic esters prepared by the reaction of aniline with ethoxymethylenemalonic ester [93]. This synthetic approach was later applied to the preparation of related heterocyclic systems. Microwave irradiation was found to facilitate the Gould–Jacobs reaction, effectively combining two steps in one short synthetic process. The Biotage's application note [94] demonstrated that synthesis of quinoline **106** can be achieved in five minutes directly from aniline (**6**) and diethyl ethoxymethylenemalonate (**105**) upon microwave irradiation at 300 °C in a Biotage Initiator⁺ reactor (Scheme 6.41).

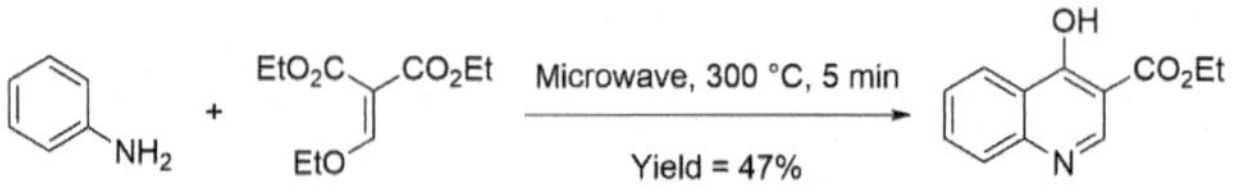

176 | *6 Synthesis of Heterocyclic Compounds Under Microwave Irradiation Using Name Reactions*

Scheme 6.41 Gould–Jacobs reaction under microwave irradiation.

6.2.28 Friedländer Quinoline Synthesis

In 1882, Paul Friedländer reported [95] a synthesis of quinoline by the condensation of *o*-aminobenzaldehyde and acetaldehyde, and in subsequent work [96] expanded the scope of both carbonyl compounds. Microwave irradiation was found to significantly enhance the performance of the Friedländer quinoline synthesis. For example, the reaction time between 2-amino-3-hydroxybenzaldehyde (**107**) and 3-pentanone (**108**) was substantially shorter under microwave irradiation in an Anton Paar Monowave 300 reactor than under conventional heating, while the yield of the resulting quinoline **109** was four times higher for the microwave-mediated protocol (Scheme 6.42) [97].

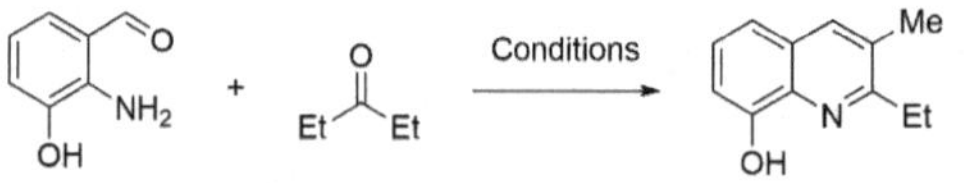

Conditions	Yield, %
Microwave, EtOH, KOH (2.2 equiv.), 130 °C, 40 min	95
Reflux, EtOH, KOH (2.7 equiv.), 24 h	24

Scheme 6.42 Friedländer quinoline synthesis under microwave irradiation and conventional heating.

An effective synthesis of fused quinolines was achieved in the Friedländer reaction of cyclic ketones with 2-aminobenzophenones under microwave irradiation [98]. This can be illustrated by the high-yielding synthesis of 9-phenyl-1,2,3,4-tetrahydroacridine (**111**) from 2-aminobenzophenone (**110**) and cyclohexanone (**20**) under irradiation in a CEM Discover microwave reactor (Scheme 6.43).

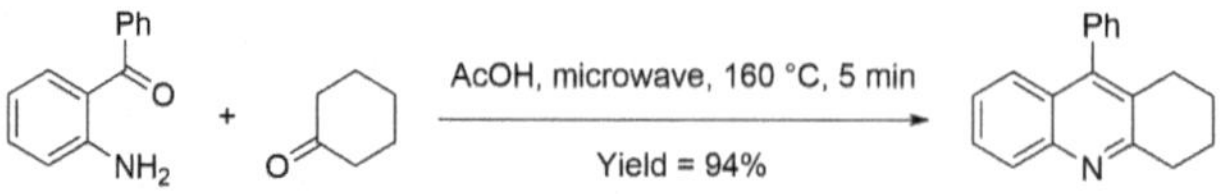

Scheme 6.43 Friedländer reaction under microwave irradiation.

6.2.29 Povarov Reaction

The Povarov reaction was developed in 1962 for the synthesis of tetrahydroquinolines and quinolines [99]. This acid-catalyzed three-component reaction involves aldehydes, anilines, and alkenes or alkynes, and it benefits from microwave irradiation conditions. The quinoline synthesis using the Povarov reaction was performed in a CEM Discover microwave reactor using phenylacetylene (**38**), aniline (**6**), and paraformaldehyde (**112**) under the (±)-camphor-10-sulfonic acid (CSA) catalysis in a 2,2,2-trifluoroethanol (TFE) medium (Scheme 6.44) [100]. Microwave irradiation provided more efficient activation than conventional heating, thus resulting in a higher yield of 4-phenylquinoline (**113**) and tremendous time-saving.

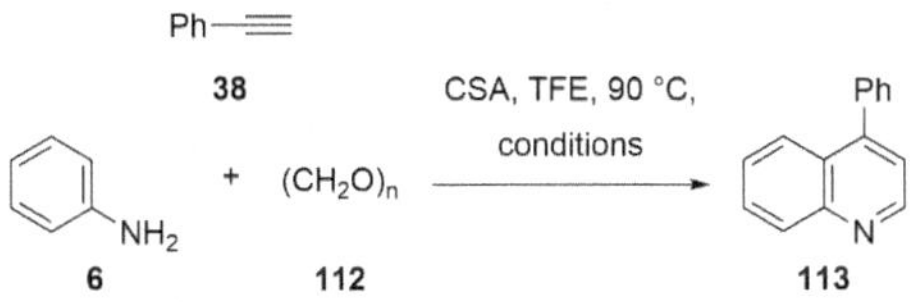

Conditions	Yield, %
Microwave, 20 min	89
Conventional heating, 24 h	46

Scheme 6.44 Povarov reaction under microwave irradiation and conventional heating.

6.3 Conclusion

Technological advances in microwave reactors made applications of microwave irradiation an efficient approach for promoting chemical transformations with the benefits of safety, short reaction times, high yield, and reproducibility. Improvements in energy transfer in reactions under microwave irradiation versus conventional heating typically result in many-fold increases in yields of heterocycles and considerable reaction time shortening (often from hours to minutes). The advantages of microwave-assisted protocols breathe new life into classical name reactions as illustrated above in the synthesis of a wide range of heterocycles: pyrroles, furans, thiophenes, pyrazoles, imidazole, oxazoles, thiazoles, triazoles, tetrazoles, indoles, pyridines, and quinolines.

Acknowledgments

The funding by the Ministry of Higher Education, Malaysia, under the Fundamental Research Grant Scheme (Grant no. FRGS/1/2020/STG04/MUSM/02/2) is acknowledged.

References

1 Gedye, R., Smith, F., Westaway, K. et al. (1986). The use of microwave ovens for rapid organic synthesis. *Tetrahedron Lett.* 27: 279–282. https://doi.org/10.1016/S0040-4039(00)83996-9.

2 Giguere, R.J., Bray, T.L., Duncan, S.M., and Majetich, G. (1986). Application of commercial microwave ovens to organic synthesis. *Tetrahedron Lett.* 27: 4945–4948. https://doi.org/10.1016/S0040-4039(00)85103-5.

3 Adam, D. (2003). Out of the kitchen. *Nature* 421: 571–572. https://doi.org/10.1038/421571a.

4 Kamboj, M., Bajpai, S., Yadav, M., and Singh, S. (2023). Microwave radiations: a green approach to the synthesis of five-membered heterocyclic compounds. *Curr. Green Chem.* 10: 57–72. https://doi.org/10.2174/2213346110666230102095423.

5 Meera, G., Rohit, K.R., Saranya, S., and Anilkumar, G. (2020). Microwave assisted synthesis of five membered nitrogen heterocycles. *RSC Adv.* 10: 36031–36041. https://doi.org/10.1039/D0RA05150K.

6 Sharma, A. and Piplani, P. (2017). Microwave-activated synthesis of pyrroles: a short review. *J. Heterocycl. Chem.* 54: 27–34. https://doi.org/10.1002/jhet.2550.

7 Franco, P.D., Caruso, L., Nadur, F.N. et al. (2021). Recent advances in microwave-assisted synthesis and functionalization of 1,2,3- and 1,2,4-triazoles. *Curr. Org. Chem.* 25: 2815–2839. https://doi.org/10.2174/1385272825666211011111408.

8 Henary, M., Kananda, C., Rotolo, L. et al. (2020). Benefits and applications of microwave-assisted synthesis of nitrogen containing heterocycles in medicinal chemistry. *RSC Adv.* 10: 14170–14197. https://doi.org/10.1039/D0RA01378A.

9 Sahoo, M.B., Banik, K.B., Kumar, V.V.R.B. et al. (2023). Microwave induced green synthesis: sustainable technology for efficient development of bioactive pyrimidine scaffolds. *Curr. Med. Chem.* 30: 1029–1059. https://doi.org/10.2174/0929867329666220622150013.

10 Kumar, K. (2022). Microwave-assisted diversified synthesis of pyrimidines: an overview. *J. Heterocycl. Chem.* 59: 205–238. https://doi.org/10.1002/jhet.4376.

11 Junaid, A. and Dolzhenko, A.V. (2019). Microwave-assisted synthesis of 1,3,5-triazines: efficient approaches to therapeutically valuable scaffold. *Heterocycles* 98: 1678–1706. https://doi.org/10.3987/rev-19-920.

12 Piloty, O. (1910). Synthese von Pyrrolderivaten: Pyrrole aus Succinylobernsteinsäureester, Pyrrole aus Azinen. *Ber. Dtsch. Chem. Ges.* 43: 489–498. https://doi.org/10.1002/cber.19100430182.

13 Robinson, G.M. and Robinson, R. (1918). A new synthesis of tetraphenylpyrrole. *J. Chem. Soc., Trans.* 113: 639–645. https://doi.org/10.1039/CT9181300639.

14 Milgram, B.C., Eskildsen, K., Richter, S.M. et al. (2007). Microwave-assisted Piloty–Robinson synthesis of 3,4-disubstituted pyrroles. *J. Org. Chem.* 72 (10): 3941–3944. https://doi.org/10.1021/jo070389+.

15 Elming, N. and Clauson-Kaas, N. (1952). The preparation of pyrroles from furans. *Acta Chem. Scand.* 6: 867–874. https://doi.org/10.3891/acta.chem.scand.06-0867.

16 Clauson-Kaas, N. and Tyle, Z. (1952). Preparation of cis- and trans 2,5-dimethoxy-2-(acetamidomethyl)-2,5-dihydrofuran, of cis- and trans 2,5-dimethoxy-2-(acetamidomethyl)-tetrahydrofuran and of 1-phenyl-2-(acetamidomethyl)-pyrrole. *Acta Chem. Scand.* 6: 667–670. https://doi.org/10.3891/acta.chem.scand.06-0667.

17 Miles, K.C., Mays, S.M., Southerland, B.K. et al. (2009). The Clauson–Kaas pyrrole synthesis under microwave irradiation. *Arkivoc* xiv: 181–190.

18 Abid, M., Landge, S.M., and Török, B. (2006). An efficient and rapid synthesis of *N*-substituted pyrroles by microwave assisted solid acid catalysis. *Org. Prep. Proced. Int.* 38: 495–500. https://doi.org/10.1080/00304940609356444.

19 Rohit, K.R., Meera, G., and Anilkumar, G. (2022). A solvent-free manganese(II)-catalyzed Clauson–Kaas protocol for the synthesis of *N*-aryl pyrroles under microwave irradiation. *J. Heterocycl. Chem.* 59: 194–200. https://doi.org/10.1002/jhet.4372.

20 Knorr, L. (1884). Synthese von Furfuranderivaten aus dem Diacetbernsteinsäureester. *Ber. Dtsch. Chem. Ges.* 17: 2863–2870. https://doi.org/10.1002/cber.188401702254.

21 Knorr, L. (1885). Einwirkung des Diacetbernsteinsäureesters auf Ammoniak und primäre Aminbasen. *Ber. Dtsch. Chem. Ges.* 18: 299–311. https://doi.org/10.1002/cber.18850180154.

22 Paal, C. (1884). Ueber die Derivate des Acetophenonacetessigesters und des Acetonylacetessigesters. *Ber. Dtsch. Chem. Ges.* 17: 2756–2767. https://doi.org/10.1002/cber.188401702228.

23 Paal, C. (1885). Synthese von Thiophen- und Pyrrolderivaten. *Ber. Dtsch. Chem. Ges.* 18: 367–371. https://doi.org/10.1002/cber.18850180175.

24 Veitch, G.E., Bridgwood, K.L., Rands-Trevor, K., and Ley, S.V. (2008). Magnesium nitride as a convenient source of ammonia: preparation of pyrroles. *Synlett* 2597–2600. https://doi.org/10.1055/s-0028-1083504.

25 Minetto, G., Raveglia, L.F., and Taddei, M. (2004). Microwave-assisted Paal–Knorr reaction. A rapid approach to substituted pyrroles and furans. *Org. Lett.* 6: 389–392. https://doi.org/10.1021/ol0362820.

26 Minetto, G., Raveglia, L.F., Sega, A., and Taddei, M. (2005). Microwave-assisted Paal–Knorr reaction – three-step regiocontrolled synthesis of polysubstituted furans, pyrroles and thiophenes. *Eur. J. Org. Chem.* 5277–5288. https://doi.org/10.1002/ejoc.200500387.

27 Xu, B., Li, G., Li, J., and Shi, Y. (2016). Total synthesis of the proposed structure of Marineosin A. *Org. Lett.* 18: 2028–2031. https://doi.org/10.1021/acs.orglett.6b00632.

28 Gewald, K., Schinke, E., and Böttcher, H. (1966). Heterocyclen aus CH-aciden Nitrilen, VIII. 2-Amino-thiophene aus methylenaktiven Nitrilen, Carbonylverbindungen und Schwefel. *Chem. Ber.* 99: 94–100. https://doi.org/10.1002/cber.19660990116.

29 Revelant, G., Dunand, S., Hesse, S., and Kirsch, G. (2011). Microwave-assisted synthesis of 5-substituted 2-aminothiophenes starting from arylacetaldehydes. *Synthesis* 2935–2940. https://doi.org/10.1055/s-0030-1261032.

30 Sridhar, M., Rao, R.M., Baba, N.H.K., and Kumbhare, R.M. (2007). Microwave accelerated Gewald reaction: synthesis of 2-aminothiophenes. *Tetrahedron Lett.* 48: 3171–3172. https://doi.org/10.1016/j.tetlet.2007.03.052.

31 Fischer, E. and Jourdan, F. (1883). Ueber die hydrazine der brenztraubensäure. *Ber. Dtsch. Chem. Ges.* 16: 2241–2245. https://doi.org/10.1002/cber.188301602141.

32 Fischer, E. and Hess, O. (1884). Synthese von Indolderivaten. *Ber. Dtsch. Chem. Ges.* 17: 559–568. https://doi.org/10.1002/cber.188401701155.

33 Desroses, M., Wieckowski, K., Stevens, M., and Odell, L.R. (2011). A microwave-assisted, propylphosphonic anhydride (T3P®) mediated one-pot Fischer indole synthesis. *Tetrahedron Lett.* 52: 4417–4420. https://doi.org/10.1016/j.tetlet.2011.06.053.

34 Li, B.L., Xu, D.-Q., and Zhong, A.G. (2012). Novel SO_3H-functionalized ionic liquids catalyzed a simple, green and efficient procedure for Fischer indole synthesis in water under microwave irradiation. *J. Fluorine Chem.* 144: 45–50. https://doi.org/10.1016/j.jfluchem.2012.09.010.

35 Hughes-Whiffing, C.A. and Perry, A. (2021). One-pot, three-component Fischer indolisation–*N*-alkylation for rapid synthesis of 1,2,3-trisubstituted indoles. *Org. Biomol. Chem.* 19: 627–634. https://doi.org/10.1039/d0ob02185g.

36 Bischler, A. (1892). Ueber die Entstehung einiger substituirter Indole. *Ber. Dtsch. Chem. Ges.* 25: 2860–2879. https://doi.org/10.1002/cber.189202502123.

37 Vara, Y., Aldaba, E., Arrieta, A. et al. (2008). Regiochemistry of the microwave-assisted reaction between aromatic amines and α-bromoketones to yield substituted 1*H*-indoles. *Org. Biomol. Chem.* 6: 1763–1772. https://doi.org/10.1039/B719641E.

38 La Regina, G., Bai, R., Rensen, W.M. et al. (2013). Toward highly potent cancer agents by modulating the C-2 group of the arylthioindole class of tubulin polymerization inhibitors. *J. Med. Chem.* 56: 123–149. https://doi.org/10.1021/jm3013097.

39 Hemetsberger, H. and Knittel, D. (1972). Synthese und Thermolyse von α-Azidoacrylestern. *Monatsh. Chem.* 103: 194–204. https://doi.org/10.1007/BF00912944.

40 Ranasinghe, N. and Jones, G.B. (2013). Extending the versatility of the Hemetsberger–Knittel indole synthesis through microwave and flow chemistry. *Bioorg. Med. Chem. Lett.* 23: 1740–1742. https://doi.org/10.1016/j.bmcl.2013.01.066.

41 Batcho, A.D. and Leimgruber, W. (1985). Indoles from 2-methylnitrobenzenes by condensation with formamide acetals followed by reduction: 4-benzyloxyindole. *Org. Synth.* 63: 214–220.

42 Siu, J., Baxendale, I.R., and Ley, S.V. (2004). Microwave assisted Leimgruber–Batcho reaction for the preparation of indoles, azaindoles and pyrroylquinolines. *Org. Biomol. Chem.* 2: 160–167. https://doi.org/10.1039/B313012F.

43 Cadogan, J.I.G. (1969). Phosphite-reduction of aromatic nitro-compounds as a route to heterocycles. *Synthesis* 11–17. https://doi.org/10.1055/s-1969-34189.

44 Sundberg, R.J. (1965). Deoxygenation of nitro groups by trivalent phosphorus. Indoles from *o*-nitrostyrenes. *J. Org. Chem.* 30: 3604–3610. https://doi.org/10.1021/jo01022a006.

45 Kaur, M., Mehta, V., Abdullah Wani, A. et al. (2021). Synthesis of 1,4-dihydropyrazolo[4,3-*b*]indoles via intramolecular C(sp_2)-N bond formation involving nitrene insertion, DFT study and their anticancer assessment. *Bioorg. Chem.* 114: 105114. https://doi.org/10.1016/j.bioorg.2021.105114.

46 von Pechmann, H. (1895). Ueber Diazomethan. *Ber. Dtsch. Chem. Ges.* 28: 855–861. https://doi.org/10.1002/cber.189502801189.

47 Mirjana, E.-M., Irena, Z., and Marina, J. (2006). Microwave-assisted synthesis of pyrazoles by 1,3-dipolar cycloaddition of diazo compounds to acetylene derivatives. *Heterocycles* 68 (9): 1961–1967. https://doi.org/10.3987/com-06-10803.

48 Debus, H. (1858). Ueber die Einwirkung des Ammoniaks auf Glyoxal. *Justus Liebigs Ann. Chem.* 107: 199–208. https://doi.org/10.1002/jlac.18581070209.

49 Radziszewski, B. (1882). Ueber die Constitution des Lophins und verwandter Verbindungen. *Ber. Dtsch. Chem. Ges.* 15: 1493–1496.

50 Radzisewski, B. (1882). Ueber Glyoxalin und seine Homologe. *Ber. Dtsch. Chem. Ges.* 15: 2706–2708. https://doi.org/10.1002/cber.188201502245.

51 Drefahl, G. and Herma, H. (1960). Untersuchungen über Stilbene, XXVIII. Stilbenyl-imidazole. *Chem. Ber.* 93: 486–492. https://doi.org/10.1002/cber.19600930232.

52 Wolkenberg, S.E., Wisnoski, D.D., Leister, W.H. et al. (2004). Efficient synthesis of imidazoles from aldehydes and 1,2-diketones using microwave irradiation. *Org. Lett.* 6: 1453–1456. https://doi.org/10.1021/ol049682b.

53 Gelens, E., Kanter, F., Schmitz, R. et al. (2006). Efficient library synthesis of imidazoles using a multicomponent reaction and microwave irradiation. *Mol. Diversity* 10: 17–22.

54 Crouch, R.D., Howard, J.L., Zile, J.L., and Barker, K.H. (2006). Microwave-mediated synthesis of lophine: developing a mechanism to explain a product. *J. Chem. Educ.* 83: 1658. https://doi.org/10.1021/ed083p1658.

55 Van Leusen, A.M., Wildeman, J., and Oldenziel, O.H. (1977). Chemistry of sulfonylmethyl isocyanides. 12. Base-induced cycloaddition of sulfonylmethyl isocyanides to carbon, nitrogen double bonds. Synthesis of 1,5-disubstituted and 1,4,5-trisubstituted imidazoles from aldimines and imidoyl chlorides. *J. Org. Chem.* 42: 1153–1159. https://doi.org/10.1021/jo00427a012.

56 Rashamuse, T.J., Harrison, A.T., Mosebi, S. et al. (2020). Design, synthesis and biological evaluation of imidazole and oxazole fragments as HIV-1 integrase-LEDGF/p75 disruptors and inhibitors of microbial pathogens. *Bioorg. Med. Chem.* 28: 115210. https://doi.org/10.1016/j.bmc.2019.115210.

57 Samanta, S.K., Kylänlahti, I., and Yli-Kauhaluoma, J. (2005). Microwave-assisted synthesis of imidazoles: reaction of *p*-toluenesulfonylmethyl isocyanide and polymer-bound imines. *Bioorg. Med. Chem. Lett.* 15: 3717–3719. https://doi.org/10.1016/j.bmcl.2005.05.066.

58 van Leusen, A.M., Hoogenboom, B.E., and Siderius, H. (1972). A novel and efficient synthesis of oxazoles from tosylmethylisocyanide and carbonyl compounds. *Tetrahedron Lett.* 13: 2369–2372. https://doi.org/10.1016/S0040-4039(01)85305-3.

59 Mukku, N., Madivalappa Davanagere, P., Chanda, K., and Maiti, B. (2020). A facile microwave-assisted synthesis of oxazoles and diastereoselective oxazolines using aryl-aldehydes, *p*-toluenesulfonylmethyl isocyanide under controlled basic conditions. *ACS Omega* 5: 28239–28248. https://doi.org/10.1021/acsomega.0c04130.

60 Robinson, R. (1909). CCXXXII.—a new synthesis of oxazole derivatives. *J. Chem. Soc. Trans.* 95: 2167–2174. https://doi.org/10.1039/CT9099502167.

61 Gabriel, S. (1910). Eine synthese von oxazolen und thiazolen. I. *Ber. Dtsch. Chem. Ges.* 43: 134–138.

62 Brain, C.T. and Paul, J.M. (1999). Rapid synthesis of oxazoles under microwave conditions. *Synlett* 1642–1644. https://doi.org/10.1055/s-1999-2905.

63 Szabó, T., Dancsó, A., Ábrányi-Balogh, P. et al. (2019). First reported propylphosphonic anhydride (T3P®) mediated Robinson–Gabriel cyclization. Synthesis of natural and unnatural 5-(3-indolyl)oxazoles. *Tetrahedron Lett.* 60: 1353–1356. https://doi.org/10.1016/j.tetlet.2019.04.024.

64 Hantzsch, A. and Weber, J.H. (1887). Ueber Verbindungen des Thiazols (Pyridins der Thiophenreihe). *Ber. Dtsch. Chem. Ges.* 20: 3118–3132. https://doi.org/10.1002/cber.188702002200.

65 Gaikwad, S.A., Patil, A.A., and Deshmukh, M.B. (2009). An efficient, uncatalyzed, and rapid synthesis of thiazoles and aminothiazoles under microwave irradiation and investigation of their biological activity. *Phosphorus, Sulfur Silicon Relat. Elem.* 185: 103–109. https://doi.org/10.1080/10426500802715163.

66 Einhorn, A., Bischkopff, E., Szelinski, B. et al. (1905). Ueber die N-Methylolverbindungen der Säureamide. *Justus Liebigs Ann. Chem.* 343: 207–305. https://doi.org/10.1002/jlac.19053430207.

67 Brunner, K. (1914). Eine neue Darstellungsweise von sekundären Säureamiden. *Ber. Dtsch. Chem. Ges.* 47: 2671–2680. https://doi.org/10.1002/cber.19140470351.

68 Lee, J., Hong, M., Jung, Y. et al. (2012). Synthesis of 1,3,5-trisubstituted-1,2,4-triazoles by microwave-assisted N-acylation of amide derivatives and the consecutive reaction with hydrazine hydrochlorides. *Tetrahedron* 68: 2045–2051. https://doi.org/10.1016/j.tet.2012.01.003.

69 Pellizzari, G. (1894). Nuova sintesi del triazolo e dei suoi derivati. *Gazz. Chim. Ital.* 24: 222–229.

70 Bechara, W.S., Khazhieva, I.S., Rodriguez, E., and Charette, A.B. (2015). One-pot synthesis of 3,4,5-trisubstituted 1,2,4-triazoles via the addition of hydrazides to activated secondary amides. *Org. Lett.* 17: 1184–1187. https://doi.org/10.1021/acs.orglett.5b00128.

71 Rostovtsev, V.V., Green, L.G., Fokin, V.V., and Sharpless, K.B. (2002). A stepwise Huisgen cycloaddition process: copper(I)-catalyzed regioselective "ligation" of azides and terminal alkynes. *Angew. Chem. Int. Ed.* 41: 2596–2599. https://doi.org/10.1002/1521-3773(20020715)41:14<2596::AID-ANIE2596>3.0.CO;2-4.

72 Roshandel, S., Suri, S.C., Marcischak, J.C. et al. (2018). Catalyst and solvent free microwave-assisted synthesis of substituted 1,2,3-triazoles. *Green Chem.* 20: 3700–3704. https://doi.org/10.1039/C8GC01516C.

73 Hansen, S.G. and Jensen, H.H. (2009). Microwave irradiation as an effective means of synthesizing unsubstituted N-linked 1,2,3-triazoles from vinyl acetate and azides. *Synlett* 3275–3276. https://doi.org/10.1055/s-0029-1218366.

74 Balducci, E., Bellucci, L., Petricci, E. et al. (2009). Microwave-assisted intramolecular Huisgen cycloaddition of azido alkynes derived from α-amino acids. *J. Org. Chem.* 74: 1314–1321. https://doi.org/10.1021/jo802463r.

75 Chatkewitz, L.E., Halonski, J.F., Padilla, M.S., and Young, D.D. (2018). Investigation of copper-free alkyne/azide 1,3-dipolar cycloadditions using microwave irradiation. *Bioorg. Med. Chem. Lett.* 28: 81–84. https://doi.org/10.1016/j.bmcl.2017.12.007.

76 Finnegan, W.G., Henry, R.A., and Lofquist, R. (1958). An improved synthesis of 5-substituted tetrazoles. *JACS* 80: 3908–3911. https://doi.org/10.1021/ja01548a028.

77 Roh, J., Artamonova, T.V., Vávrová, K. et al. (2009). Practical synthesis of 5-substituted tetrazoles under microwave irradiation. *Synthesis* 2175–2178. https://doi.org/10.1055/s-0029-1216840.

78 Shie, J.-J. and Fang, J.-M. (2007). Microwave-assisted one-pot tandem reactions for direct conversion of primary alcohols and aldehydes to triazines and tetrazoles in aqueous media. *J. Org. Chem.* 72: 3141–3144. https://doi.org/10.1021/jo0625352.

79 Ugi, I. and Steinbrückner, C. (1961). Isonitrile, II. Reaktion von Isonitrilen mit Carbonylverbindungen, Aminen und Stickstoffwasserstoffsäure. *Chem. Ber.* 94: 734–742. https://doi.org/10.1002/cber.19610940323.

80 Zhao, T., Boltjes, A., Herdtweck, E., and Dömling, A. (2013). Tritylamine as an ammonia surrogate in the Ugi tetrazole synthesis. *Org. Lett.* 15: 639–641. https://doi.org/10.1021/ol303348m.

81 Zecher, W. and Kröhnke, F. (1961). Eine neue Synthese substituierter Pyridine, I. Grundzüge der Synthese. *Chem. Ber.* 94: 690–697. https://doi.org/10.1002/cber.19610940317.

82 Kiss, M.A., Peřina, M., Bereczki, L. et al. (2023). Dihydrotestosterone-based A-ring-fused pyridines: microwave-assisted synthesis and biological evaluation in prostate cancer cells compared to structurally related quinolines. *J. Steroid Biochem. Mol. Biol.* 231: 106315. https://doi.org/10.1016/j.jsbmb.2023.106315.

83 Tu, S., Jia, R., Jiang, B. et al. (2007). Kröhnke reaction in aqueous media: one-pot clean synthesis of 4′-aryl-2,2′:6′,2″-terpyridines. *Tetrahedron* 63: 381–388. https://doi.org/10.1016/j.tet.2006.10.069.

84 Zhou, J.-F., Sun, X.-J., Lou, F.-W. et al. (2012). A facile one-pot, three-component synthesis of 3,3′-(4-arylpyridine-2,6-diyl)bis(2H-chromen-2-one) derivatives under microwave irradiation. *Res. Chem. Intermed.* 39: 1401–1408. https://doi.org/10.1007/s11164-012-0696-5.

85 Bohlmann, F. and Rahtz, D. (1957). Über eine neue Pyridinsynthese. *Chem. Ber.* 90: 2265–2272. https://doi.org/10.1002/cber.19570901021.

86 Bagley, M.C., Lunn, R., and Xiong, X. (2002). A new one-step synthesis of pyridines under microwave-assisted conditions. *Tetrahedron Lett.* 43: 8331–8334. https://doi.org/10.1016/S0040-4039(02)01975-5.

87 Bagley, M.C., Fusillo, V., Jenkins, R.L. et al. (2013). One-step synthesis of pyridines and dihydropyridines in a continuous flow microwave reactor. *Beilstein J. Org. Chem.* 9: 1957–1968. https://doi.org/10.3762/bjoc.9.232.

88 Boger, D.L. and Panek, J.S. (1981). Diels–Alder reaction of heterocyclic azadienes. I. Thermal cycloaddition of 1,2,4-triazine with enamines: simple preparation of substituted pyridines. *J. Org. Chem.* 46: 2179–2182. https://doi.org/10.1021/jo00323a044.

89 Catozzi, N., Edwards, M.G., Raw, S.A. et al. (2009). Synthesis of the louisianin alkaloid family via a 1,2,4-triazine inverse-electron-demand Diels–Alder approach. *J. Org. Chem.* 74: 8343–8354. https://doi.org/10.1021/jo901761r.

90 Skraup, Z.H. (1880). Eine Synthese des Chinolins. *Monatsh. Chem. Verw. Teile Anderer Wiss.* 1: 316–318. https://doi.org/10.1007/BF01517073.

91 Saggadi, H., Luart, D., Thiebault, N. et al. (2014). Quinoline and phenanthroline preparation starting from glycerol via improved microwave-assisted modified Skraup reaction. *RSC Adv.* 4: 21456–21464. https://doi.org/10.1039/C4RA00758A.

92 Saggadi, H., Luart, D., Thiebault, N. et al. (2014). Toward the synthesis of 6-hydroxyquinoline starting from glycerol via improved microwave-assisted modified Skraup reaction. *Catal. Commun.* 44: 15–18. https://doi.org/10.1016/j.catcom.2013.07.029.

93 Gould, R.G. and Jacobs, W.A. (1939). The synthesis of certain substituted quinolines and 5,6-benzoquinolines. *JACS* 61: 2890–2895. https://doi.org/10.1021/ja01265a088.

94 Biotage (2012). Gould–Jacobs quinoline-forming reaction: a comparison of heating using microwave irradiation to 250 °C and 300 °C. *Application Note, AN-056.0112*.

95 Friedlaender, P. (1882). Ueber o-Amidobenzaldehyd. *Ber. Dtsch. Chem. Ges.* 15: 2572–2575. https://doi.org/10.1002/cber.188201502219.

96 Friedländer, P. and Gohring, C.F. (1883). Ueber eine Darstellungsmethode im Pyridinkern substituirter Chinolinderivate. *Ber. Dtsch. Chem. Ges.* 16: 1833–1839. https://doi.org/10.1002/cber.18830160265.

97 Garrison, A.T., Abouelhassan, Y., Yang, H. et al. (2017). Microwave-enhanced Friedländer synthesis for the rapid assembly of halogenated quinolines with antibacterial and biofilm eradication activities against drug resistant and tolerant bacteria. *MedChemComm* 8: 720–724. https://doi.org/10.1039/C6MD00381H.

98 Bailey, H.V., Mahon, M.F., Vicker, N., and Potter, B.V.L. (2020). Rapid and efficient microwave-assisted Friedländer quinoline synthesis. *ChemistryOpen* 9: 1113–1122. https://doi.org/10.1002/open.202000247.

99 Povarov, L.S., Grigos, V.I., and Mikhailov, B.M. (1963). Reaction of benzylideneaniline with some unsaturated compounds. *Bull. Acad. Sci. USSR Div. Chem. Sci.* 12: 1878–1880.

100 Chandra, D., Dhiman, A.K., Kumar, R., and Sharma, U. (2019). Microwave-assisted metal-free rapid synthesis of C4-arylated quinolines via Povarov type multicomponent reaction. *Eur. J. Org. Chem.* 2753–2758. https://doi.org/10.1002/ejoc.201900325.

7

Microwave- and Ultrasound-Assisted Enzymatic Reactions

Nafseen Ahmed[1], Chandan K. Mandal[2], Varun Rai[3], Abbul Bashar Khan[1], and Kamalakanta Behera[3]

[1]Jamia Millia Islamia (Central University), Department of Chemistry, Jamia Nagar, Okhla, New Delhi 110025, India
[2]Amity University Gurugram, Department of Applied Chemistry (CBFS - ASAS), Manesar, Panchgaon, Haryana 122413, India
[3]University of Allahabad, Department of Chemistry, Old Katra, Prayagraj (Allahabad), Uttar Pradesh 211002, India

7.1 Introduction

Macromolecular biological catalysts that catalyze or enhance chemical processes are known as enzymes. Enzymes are recognized to catalyze a broad variety of biological events, not only in the food business. Like other catalysts, enzymes enhance the rate of reaction by reducing their activation energy. They are employed in several industrial processes and play significant roles in a variety of industries, including those that deal with biofuels, textiles, dyes, water treatment, medicines, feed, and food [1]. According to the green chemistry tenets, the use of enzymes as catalysts lessen environmental harm by encouraging more selective reactions under more tolerable temperature, pressure, and pH conditions. Enzymes also have the potential to increase conversion yields and lower purification stage costs, making them an intriguing substance from both an economic and environmental perspective. Due to these qualities, the use of enzyme catalysis in industrial processes has increased, particularly in recent years when governmental incentives and international demands have emerged to look for and use techniques that are less damaging to the environment [2]. The chemical industry is using microwave technology more and more since it provides a higher rate of transformation and yield in a small amount of time. Microwaves are electromagnetic radiation along with a frequency range of 300 MHz to 300 GHz, and wavelengths between 1 m and 1 mm [3]. Enzymatic reactions have been the main use of the standard microwave wavelength of 12.24 cm for 2.45 GHz frequency [4, 5]. In the reaction system, microwave energy interacts with the polar molecules to produce vibrations that are

Green Chemical Synthesis with Microwaves and Ultrasound, First Edition.
Edited by Dakeshwar Kumar Verma, Chandrabhan Verma, and Paz Otero Fuertes.
© 2024 WILEY-VCH GmbH. Published 2024 by WILEY-VCH GmbH.

unaffected by the reaction's temperature. Ionic conductance causes instantaneous localized superheating as a result [6, 7]. Enzyme stability is enhanced by microwave radiation because it delays denaturation. Compared to typical heating methods, enzymes such as Candida *antarctica* lipase B have demonstrated enhanced stability [8, 9]. Enzymatic catalysis and microwave irradiation interact to considerably accelerate the reaction rate. According to various publications, several enzymes from bacteria, fungi, and plants are used in the biodegradation of harmful organic contaminants. Microbial enzymes regulate the cost-effective and environmentally friendly biotechnology known as bioremediation. Chemical processes that take various hours under normal circumstances may frequently be completed in a matter of minutes with extremely high yields and reaction selectivities when heated using microwave radiation, which was first recognized as an effective heating source for chemical reactions in the middle of the 1980s [10–12]. Reactions can be carried out in open flasks with the use of solvents with relatively high boiling points, such as acetonitrile and dimethyl formamide, to reduce the hazards of possible explosions. Another benefit of implementing microwave technology is the potential for using gentle, less harmful reagents as well as solvents. There have been numerous studies on the application of microwave irradiation in organic synthesis, including those for the synthesis of heterocycles, organometallic processes, and rearrangement reactions [10, 13–16]. The fundamental theory of microwave dielectric heating, pertinent dielectric parameters, and microwave-assisted organic reactions have all been covered in several review publications [17–21]. Due to its great efficiency and energy savings, low equipment needs, and zero pollutant output, ultrasound is employed for a high number of purposes in a variety of fields. Because of its mechanical and radical impacts, ultrasound is known to cause modifications in product characteristics and speed up some chemical reactions and industrial processes [22–25]. Although some studies claim that ultrasonic does not completely inactivate all enzymes under mild conditions, it has long been used to inactivate enzymes [26]. Enzymatic processes can be increased using ultrasound, which also has favorable impacts on enzyme activity. The use of ultrasound in enzymatic reactions has a lot of possible applications in the industry. The application of ultrasonic therapy throughout the course of enzymatic reactions will be examined in this paper. We will also talk about how ultrasound stimulates enzymatic reactions.

7.2 Influence Microwave Radiation on the Stability and Activity of Enzymes

Microwave radiation has an important effect on enzyme activity and the stability of efficiency in the esterification processes; some of these effects are shown in Table 7.1. The reaction media's polarity and hydration state have the most impact on the enzyme activity in a microwave system. The hydrophobicity of the solvent rises with enzymatic activity. Under microwave heating, enzymes have a propensity to operate differently, becoming activated and energized. Using Novozym 435

Table 7.1 Various applications of ultrasonic-assisted enzymolysis.

Application	Ultrasound conditions	Ultrasound equipment	Major observation	References
Using ultrasound, the antioxidant hydrolysate from peanuts (Arachin conarachin L.) is enzymatically broken down.	Temperature – 62 °C; Power – 150 W; frequency – 28 kHz initial pH – 8.5; incubation time – 25 min;	Kunshan Ultrasonic Instruments' tri-frequency digital fully automated ultrasonic cleaner, model number KQ-300DZ. Co., Ltd. (Shanghai, China)	Peanut hydrolysate has up to 90% more DPPH free radical scavenging capacity.	[27]
Pretreatment with ultrasound on the enzymatic breakdown of protein in tea residue	Temperature – 20, 30, 40 and 50 °C; Power – 377 W/l; frequency – 20 kHz; sonication time – 13 min; initial pH – 8.0; solid-liquid ratio of 51.5 g/l;	Fanbo Biological Engineering Co, Ltd., Wuxi, China; Model FBTQ 2000; Single Frequency Counter Current Ultrasound (SFCCU) with a 2 cm flat tip probe; working at a fixed frequency of 20 kHz, a 300 rpm circulating pump speed with pulse on and off intervals of 3 and 2 s, respectively.	Tea protein hydrolysate conversion rate went from 56% to 61%, and the concentration of polypeptides was increased to 24.12 mg/ml.	[28]
Alkylphenols' enzymatic hydrolysis and the breakdown of 17β-oestradiol glucuronide are both accelerated by ultrasonic technology.	frequency – 20 kHz; Power – 70 W; initial temperature – 20 °C; sonication time – 1–20 minutes; initial pH – 7.4;	Bandelin Sonoplus HD 2070 ultrasound system. manufactured in Berlin, Germany (20 kHz, 70 W).	20-minute decrease in the amount of time needed for enzymatic hydrolysis	[29]
Bagasse hemicellulose enzymatic hydrolysis with ultrasound assistance	Temperature – 45 °C; Power – 300 W; initial pH – 6.0; sonication time – 20 min;	–	6.5% more reducing sugar was produced, and the rate of hydrolysis improved.	[30]

(Continued)

Table 7.1 (Continued)

Application	Ultrasound conditions	Ultrasound equipment	Major observation	References
Protein from sunflower meal enzymes with ultrasound assistance	Temperature – 45 °C; Power density – 220 W/l; pump circulation speed – 100 rpm; time – 15 min; dual frequency – 20/40 kHz; 2 s/5 s pulsed off/on time;	Miebo Biotech Co. in Jiangsu, China, created dual-frequency ultrasound.	Improvement of the protein enzymolysis of sunflower meal with an 11.29% decrease in the Michaelis constant and a 1.96% rise in the breakdown rate	[31]
the impact of ultrasonic pretreatment on the β-Lg's absorption rate, antioxidant capacity, and allergenicity.	Temperature – lower than 15 °C; Power density – 120 W/cm^2; 7 s/3 s pulsed on/off time; time – 15 min; frequency – 20 kHz;	The USA-made Misonix Qsonica Q700 Sonicator has a 3 mm microtip probe and operates at 20 kHz.	Improvement of IgG/IgE's capacity to bind	[32]
Rice protein is broken down by enzymes.	Temperature – 53.46 °C; Power density – 50 W/l; dual frequency – 20/40 kHz; time – 10 min;	Jiangsu University created multi-mode S-type ultrasonic technology with five frequencies at 20, 28, 35, 40, and 50 kHz.	Intensification of ACE inhibitory action	[33]
Camel milk casein and whey protein bioactive characteristics modified by ultrasonication	Power – 400 W; time – 45 min; frequency – 30 kHz	Omni Sonic Ruptor 400, a benchtop homogenizer manufacturer in Kennesaw, Georgia, USA	An increase in ACE-inhibitory activity.	[34]
Proteins from wheat germ and silkworm pupa are converted by enzymes into ACE-inhibitory peptides.	Power density – 410 W/100 ml; time – 32 min; 2 s/2 s pulsed on/off time; frequency – 20 kHz	A 1.5 cm flat tip probe for an ultrasonic cell crusher (JY92-II, Haishukesheng Ultrasonic Equipment Co., Ningbo, China)	Increase in ACE-inhibitory activity of 21.0–40.7%	[35, 36]

Study	Parameters	Instrument	Results	Ref.
Characterization of modified soldier fly protein and hydrolysate produced by multiple-frequency ultrasound	Power: 600 W; pulse duration: 15 s/5 s; frequency: 40 2 kHz	Multi-Frequency Ultrasound (MFU) (Fanbo Bio-Eng. Comp., Wuxi, China)	Surface charge, reconstitutability function, and the thiol value of the sonicated samples all significantly improved. The protein hydrolysates also had an increase in lightness (L^*) of about 7.46%.	[37]
Egg white protein hydrolysates' functional characteristics after ultrasound probe treatment	Time – 2–20 minutes; frequency – 20 ± 0.2 kHz Power amplitude – 40%;	Sonicator with a probe	Enhancements in the egg white protein hydrolysates' solubility, foaming, and emulsifying properties.	[38]
Ultrasonication and controlled enzymatic degradation of sunflower protein	Time – 15 min; Power – 220 W; 5 s/ 2 s pulsed on/off time; dual frequency – 20/40 kHz;	A reaction tank, a control panel, two peristaltic pumps, two generators, two probes, and a reaction vessel were all included in the instrument.	Enhanced emulsification, foaming, and solubility abilities	[39]
Pretreatment of chicken bone protein using ultrasound	Temperature – 25–65 °C; frequency – 40 kHz; Time – 30–150 min; Power – 240–400 W;	An ultrasonic bath is available from Kunshan Ultrasonic Instruments Co., Ltd. in Suzhou, China.	Greater surface hydrophobicity and yield of protein extraction	[40]
Resveratrol was extracted in an aqueous, two-phase process with the use of ultrasound from Polygonum cuspidatum enzymatic hydrolysates.	Temperature – 25 °C; Time – 30 min; Frequency – 53 kHz;	The Kudos Ultrasonic Instrument Co., Ltd., SK5200HP, is an ultrasonic bath.	79.4% more resveratrol was produced.	[41]

and a microwave, the enzyme activity was increased during the production of 2-hydroxypropyl laurate [42] and propyl 3-xobutanoate [43]. Enzymes become inactive at extremely high temperatures because the catalytically active sites are removed. The interactions between enzymes and their surroundings play a major role in determining their stability. Non-polar liquids heated by microwaves improve the structural rigidity and the enzyme's conformation, increasing their stability as a result. Even polar solvents under a microwave can occasionally increase the stability of enzymes. Microwave radiation has a great affinity for polar liquids. Therefore, much less power will be required when using polar solvents than when using nonpolar solvents to achieve the necessary temperature for synthesis. Highly polar solvents are able to combine with microwave energy more when heated by microwaves. In order to sustain a temperature of 100 °C in butanol, a power of 24 W is necessary, but 210 W is required in ethyl butyrate [8]. Recently, it has been reported that Novozym 435 has demonstrated improved stability during microwave heating compared to conventional heating with the help of a highly protic solvent, like butanol. Enzymes that have been immobilized in the reaction medium are more stable and reusable because they are shielded from severe reaction conditions, such as the microwave's temperature as well as the acidity of the medium [44]. The economic viability of enzymatic processes is heavily influenced by the reusability of enzymes. The transesterification of ethyl butyrate and butanol is used for attached lipase from Candida antarctica. Microwave heating consistently showed higher enzyme stability over conventional heating. The biocatalyst demonstrated a greater starting rate and conversion under microwave heating with each run or reuse. The residual activity of the lipase employed in the traditional systems, and the microwave systems, was found to differ by a factor of 2 after a total of 6 consecutive batches. While the biocatalyst was preincubated in ethyl butyrate at 100 °C for 30 minutes during storage, 39% of its activity had been preserved under the traditional approach whereas 45% was preserved under the microwave system [9]. Utilizing Novozym 435, experiments on reusability were conducted in the production of citronellyl acetate. In this research, conversions were examined over the course of three cycles under both microwave and traditional heating conditions. The initial rates and conversions were found to be greater under the microwave system in all cycles (microwave irradiation: initial rate $0.821e^{-0.207(\text{reuse})}$; traditional heating: initial rate $0.631e^{-0.232(\text{reuse})}$) [45]. The biocatalyst was shown to be reusable in both temperature and microwave environments in the transesterification of β-ketoesters that was mediated by microwaves and solvent-free enzymes. It was found that the microwave environment had no effect on the catalyst's stability, and the transesterification processes could be repeated ten times in a row with little activity loss [46].

7.3 Principle of Ultrasonic-Assisted Enzymolysis

One use of contemporary food physical processing technology is ultrasonic-assisted enzymolysis, which is a relatively novel strategy. The acoustic cavitation phenomena of ultrasound technology is the fundamental idea behind ultrasonic-assisted enzymolysis [27, 31, 47–49]. Acoustic cavitation, which is the fast-oscillating movement or collapse of small bubbles in a medium, is an outcome of the sound energy generated during the sonication process and causes mechanical action and also heat and chemical reactions [50]. High pressures and high temperatures are produced quickly by this acoustic cavitation process, which results in shock waves, free radicals, and shearing forces with high intensities [51]. Phase alterations are the cause of the thermal effects of ultrasonography that alter temperature and enthalpy [52]. In order to minimize the Gibbs free energy and the activating energy of reactions like enzymolysis, protein molecular structure could be changed using ultrasound pretreatment [53]. This is accomplished by decreasing the Michaelis constant, which represents the concentration of substrate during the process and assesses the kinetics of enzyme reactions, raising the starting rate of the enzymolysis reaction [54]. When ultrasound and microjet waves combine to produce ultrasonic-assisted enzymolysis, a large mechanical shearing force is created, which leads the bubbles created by the noise-induced cleavage phenomena to burst. As a result, proteins are further denatured to liberate their hydrophilic groups, increasing their solubility, and enabling enzymes to attach to substrates more readily [55]. This improves the efficiency of the enzymolysis process throughout. The shear produced by the ultrasonic waves during enzymolysis improves the mass conveyance of extractible components by producing forces proportionately targeting the specific protein or enzyme in the media [56]. Additionally, the cavitation bubbles' collapse causes chaotic changes in the ultrasonic medium's pressure and airflow velocity, which intensifies the collision of the sample matrix. As a consequence of internal motion and diffusion, such thermodynamic changes will distribute the matrix's minute particle holes, which will result in sample deterioration, dissolution, and surface abrasion [57]. By increasing the exposure of vulnerable peptide bonds, ultrasonic-assisted enzymolysis causes protein unfolding and increases proteolysis [58]. It improves the yield of bioactive peptides from the enzymatic hydrolysis of proteins. In order to boost the degree of protein and protein complex hydrolysis, sonochemistry and ultrasound work together to disrupt the medium matrix, an arrangement of interactional forces, hydrogen bonds, and dipole attractions combining the molecules [59]. The percentage of hydrolysis is frequently used to quantify the degradation of proteins and to compare different protein hydrolysates. To increase the effectiveness, speed, and accuracy of the enzyme extraction procedure, ultrasonic-assisted enzymolysis also employs ultrasound waves to disrupt the alignments and structure of the cell wall and membrane. Thus, to take advantage of these benefits, the cell wall can be broken down via ultrasonic-assisted pretreatment. The combined effects of thermal and mechanical actions caused by acoustic cavitation about ultrasonic waves form the basis of the ultrasonic-assisted enzymolysis principle. They generally aid in improving the movement of mass effectiveness while also improving the diffusion of the enzyme and the substrate, including the ambiguity of the configuration of the substrate.

7.4 Applications of Ultrasonic-Assisted Enzymolysis

7.4.1 Proteins and Other Plant Components Can Be Transformed and Extracted

Protein functionality and enzymatic cleavage efficiency have both been successfully improved by ultrasonic pretreatment [30, 35, 60]. Numerous papers have shown that ultrasonic pretreatment just before enzymolysis would enhance the frequency of substrate and enzyme interaction as well as mass transfer efficiency [59]. In these experiments, plant cell components were modified and made more protein-soluble using ultrasonic as a pretreatment step. In addition to other plant components like peptides and polypeptides, this enhanced the extraction of protein. This is because protein solubility, which influences the characteristics of protein gelation, foaming, and emulsification, whose outcomes all depend on initial protein solubility, is the most straightforward method to determine the denaturation of proteins and aggregation, making it an accurate measure to determine the performance of the protein. In addition, [49] sunflower meal protein and sunflower meal protein hydroxylate were used in Dabbour et al.'s [61] investigation of the effects of ultrasonic-assisted enzymolysis on the efficiency of the enzymatic process. In the study, ultrasonic-assisted enzymolysis, at a density of power of 200 W/l and dual frequency of 20/40 kHz increased the efficiency of sunflower meal protein enzymolysis, lowering the Michaelis constant by 11.29%, raising the breakdown rate by 1.96%, and lowering the activation energy, enthalpy, and entropy by 24.28%, 26.13%, and 9.10%, respectively. Sunflower meal protein hydroxylate, and sunflower meal protein were also subjected to the sonication to modify their physical and structural characteristics, which improved their capacity to absorb and disseminate the oil. Jia et al. [32] also said that ultrasonic pretreatment combined with enzymatic hydrolysis was extremely successful in changing the functioning of proteins. In a different work by Ayim et al. [31], ultrasonic pretreatment was utilized to considerably improve the enzymolysis of tea residue protein, particularly at the early reaction rate, at a power density of 377 W/l as well as the frequency of 20 kHz. In the research, the process of enzymolysis was successfully enhanced overall with the application of ultrasonic pretreatment. Compared to the conventional enzymolysis process without ultrasonic pretreatment (−5 W/l, frequency, −28 kHz, as well as sonication time 25 minutes), Liang et al.'s [47] ultrasonic-assisted enzymolysis of maize protein hydrolysate increased the degree of hydrolysis and protein conversion rate by 5.42% and 11.27%, respectively. This was related to a reduction in particle size and changes to the structure of the protein. Ultrasonic pretreatment enhanced the proteolysis process as evidenced by the results of the degree of hydrolysis, which increased from 2.4% to 6%, oil holding capacity, and emulsifying stability in the research by Resendiz–Vasquez et al. [48] to examine the impact of ultrasonic prior to the treatment on the ability to function as well as enzymolysis effectiveness of the jackfruit seeds protein. Additionally, the general efficiency of the enzymolysis process was enhanced by making the protein more soluble using ultrasonic pretreatment at various ultrasonic powers of 200, 400, and

600 W. Jia et al.'s investigation into the effects of ultrasonic power on the process of extraction of anthocyanin from cherry wine lees discovered that the optimal ultrasonic power of 300 W considerably enhanced both the pace and amount of anthocyanin that was extracted [62]. This is another method for using ultrasound for an enhanced extraction process. This was ascribed to the ultrasound's acoustic impacts and its capacity to break down the cell walls of cherry wine lees, which boosted anthocyanin hydrolysis and, in turn, its extraction rate and efficiency. Similar to this was Tang et al.'s investigation into the potential speed-up effects of acoustic pretreatment on the baggage enzyme's hydrolysis [33]. According to the study, compared to nonultrasonic pretreatment, ultrasonic pretreatment at a power of 300 W increased the rate of baggage hydrolysis by 6.5%. In a different study, Li et al. [59] looked at the structural characteristics and enzymolysis kinetics of rice protein that had undergone pretreatment with energy-gathered ultrasound and ultrasonic-assisted alkali. The study's findings demonstrated that sonication treatments had a positive impact on the protein's functioning, imparting and boosting the enzymatic hydrolysis's effectiveness.

7.4.2 Modification of Protein Functionality

In order to improve protein functioning during enzymolysis, ultrasound can be utilized as a technique. It causes the helical region to unfold in a way that alters protein functioning in a way that is preferable [30, 38, 54]. Mintah et al. [39] examined the structural, physical, and functional alterations generated by multi-frequency (fixed and sweep) ultrasound at 600 W power and 402 kHz frequency to the edible soldier fly protein and hydrolysate. The protein and hydrolysates were structurally and functionally characterized, and it was discovered that ultrasonic treatment ($p < 0.05$) improved the lightness (L^*) of the protein hydrolysates by around 7.46%. Additionally, the sonicated samples' surface charge, reconstitutability function, and thiol value, all significantly increased in comparison to the control group, while their turbidity values and particle sizes decreased. The functionality of protein-hydrolysate from grass carp was examined by Yang et al. [37] using energy-divergent ultrasound, which releases energy from a probe inside the ultrasonic bath, and energy-gathered ultrasound, which uses a number of divergent transducers distributed around the walls of the ultrasonic bath. The ultrasound conditions were 100 W/l power density, 20 minutes of sonication time, and $30 \pm 2\,°C$ temperature. When compared to the control, the Maillard reaction products from the research showed greater absorbance values at 294 and 420 nm, increasing by 21.43% and 61.14% for energy-divergent ultrasound and energy-gathered ultrasound, respectively. This was ascribed to modifications made to the protein molecules' microstructure during ultrasonic pretreatment, which encouraged the protein's enzymatic breakdown and released more peptides and amino acids into the protein-hydrolysate that were essential to the Maillard process. The degree of graft of the products from both types of ultrasound pretreatment methods also increased, with improvements in the random coil content (increment of 115.02% and 14.55%), surface hydrophobicity (increment of 36.85% and

27.46%), truthfulness, and acceptance of Maillard reaction products from both types of ultrasound pretreatment methods being noted. These results suggest that ultrasonic pretreatment improves protein enzymolysis [63, 64] and stimulates the production of protein-hydrolysate for the Maillard reaction, as well as improving the organoleptic substances attributes of reaction products by altering the protein functionality. When rice protein hydrolysates and chitosan were combined to create edible films, applying ultrasound to the process, Wang et al. [40] discovered that it decreased the size of the particles and viscosity of the film-forming solution while improving the amount of elongation at the break of the composite films by 125% and the smoothness of the surface, proving that ultrasound is a useful tool for enhancing the properties of edible composite films. In a different investigation, Stefanovic et al. [65] used ultrasound to enhance the physicochemical and functional characteristics of egg white protein hydrolysates. The study's findings demonstrated that ultrasonic pretreatment affected how the proteins in egg whites were broken down during proteolysis, enhancing the proteins' capacity for solubility, foaming, and emulsification. The following research also used ultrasound to increase protein functionality – [29] controlled enzymatic cleavage of sunflower protein, analysis of the breakdown of proteins kinetics, and structural characterization of sunflower protein by Dabbour et al. [66] as well as anal extraction of functional protein hydrolysates from chicken eggshell membrane [67], enzymatic cleavage of chicken bone protein by Dong et al. [68] and technical cleavage of chicken bone protein by Mintah et al. [69]. All these researches demonstrate that ultrasound is a useful technique for improving protein extraction yield and quality as well as its ability to alter the functioning of isolated proteins to give them desirable properties.

7.4.3 Enhancement of Biological Activity

Several researchers have looked at how to improve biological activity using enzymolysis by ultrasonic assistance. Antioxidant activity has been effectively increased in a variety of procedures by using ultrasound technology. The antioxidant properties of peanut (*Arachin conarachin* L.) and antioxidant hydrolysate were enhanced in the work by Yu et al. [27], using ultrasonic-assisted enzymolysis. The results of the study show that, under the optimal ultrasonic circumstances of 150 W of power, 28 kHz ultrasonic intensity, 62 °C temperature, and an incubation duration of 25 minutes with an initial pH of 8.5, peanut hydrolysate's DPPH free radical scavenging activity may reach up to 90.06%. The control trial (without sonication) showed a DPPH free radical scavenging activity of 58.64%; this is a considerable improvement. It was also shown that the antioxidant hydroxylates from peanuts showed eight new forms of antioxidant action. Liang et al. [47] carried out a similar investigation on the effects of low-frequency ultrasound on the enzymatic hydroxylation of maize protein. They found that the DPPH and hydroxyl (OH) radical elimination capacity were significantly enhanced under the optimal conditions of 28 kHz frequency, 5 W/l power, 2 s/2 s on/off time, 50 °C temperature is recommended, 25 minutes of sonication time, as well as a pH of 9.0. Falleh et al. [34] looked at how ultrasound affected the functions of extracts from Mesembryanthemum edule shoots, and they found that these extracts had higher levels of scavenging activity, reducing power, and inhibitory activity against β-carotene bleaching than extracts obtained without ultrasound pretreatment. The impact of ultrasound on the potential allergenicity of specific materials and numerous other bioactivities has also been researched in relation to antioxidant activity. Shao et al. [36] examined how ultrasonic pretreatment affected the ability of β-Lg to be digested, to act as an antioxidant, and to trigger allergic reactions. Under ultrasound circumstances of 20 kHz frequency, 20 W/cm^2, and a 15-minute ultrasonic exposure time, the binding ability of IgG/IgE was enhanced in the study. This demonstrates how the extremely high pressure and shear produced by ultrasound had an impact on the conformational shape of β-Lg, exposed more IgG and IgE epitopes, and enhanced its binding capacity. According to Lebon et al. [70], liquid deagglomeration and fragmentation are brought on by the high pressure of sonic cavitation. Additionally, the IC50 value of 1.80 mg/ml for the ABTS+ scavenging activity of β-Lg was raised. This development was brought about by the ultrasound-induced unfolding of β-Lg's structure and increase in surface area, which offered more active free radical sites [71]. The ACE inhibitory activity of the protein hydrolysate during GI-simulated digestion was utilized as an indicator in Yang et al.'s [37] inquiry into the effects of multi-mode S-type ultrasonic pretreatment on the enzymolysis of rice protein. According to the research project's findings, all ultrasound performing ways considerably ($p < 0.05$) increased ACE inhibitory activity, with the dual frequency of 20/40 kHz having the biggest effect, increasing ACE inhibitory activity by 27.47% and 38.28%, respectively, in comparison to the control and ultrasound clearing machine, under ultrasound conditions of 50 W/l density of power, 10 minutes of sonication duration, and 53.46 °C.

7.4.4 Ultrasonic-Assisted Acceleration of Hydrolysis Time

Studies have looked at how ultrasonic-assisted enzymolysis affects reaction times and speeds up the enzymolysis procedures in addition to raising and enhancing enzymolysis efficiency. In the experiment performed by Vallejo et al., the rate of enzymatic hydrolysis was significantly slowed down by 20 minutes, utilizing an acoustic probe at a single cycle and 10% amplitude [41], that used ultrasonics to accelerate the enzymatic hydrolysis of alkylphenols and 17β-oestradiol glucuronide in fish bile. This process has been used on additional samples from other sources after being successfully verified in the research. In a study by Abadia–Garcia et al. [60], the time required for the decomposition of protein from whey by vegetable proteases was reduced by 95% as a result of preprocessing with ultrasound. The degree of hydrolysis enhanced between 26% and 63% in the samples that had undergone ultrasound treatment, proving that the procedure was more efficiently completed, especially in terms of speed. Uluko et al. [72] also discovered that ultrasonic pretreatment increased the pace and effectiveness of hydrolysis. Jia et al. [32] reported that using an ultrasonic cell crusher for 210 minutes at 20 kHz and 40 W power boosted the enzymatic hydrolysis of the dehydrated germ of wheat protein. Additionally, after ultrasonic pretreatment in Wang et al.'s study [73] on the effects of ultrasound on the morphology and antioxidant activity of conglycinin (7S) and glycinin (11S), as well as their hydrolysates, the amount of hydrolysis and free SH groups considerably increased. In Table 7.1, many uses of ultrasonic-assisted enzymolysis are listed as follows.

7.5 Enzymatic Reactions Supported by Ultrasound

Although the enzymatic reaction is frequently the most important step throughout operations, it is typically rate- and cost-limited [74]. The degree of hydrolysis is always constrained by conventional enzymatic procedures, which take a very long period. There has been a lot of interest in using various methods to help enzymatic processes. It is known that ultrasound may be utilized to speed up a variety of enzymatic reactions as well as other physical, chemical, and biological processes [75]. The reaction rate is always the primary focus of studies on ultrasound-aided enzymatic processes. Enzymatic processes can be aided by ultrasound in both liquid–liquid phase systems and solid–liquid phase systems (Table 7.2).

7.5.1 Lipase

The food, pharmaceutical, and cosmetics industries frequently use lipase-catalyzed processes for esterification, hydrolysis, glycerolysis, and transesterification [79]. At the boundary between the aqueous phase containing the lipase and the oil phase, lipase catalyzes reactions. The heterogeneous system may hinder the enzymatic process. Throughout the entire process of enzyme catalyzing, ultrasound is used to enhance lipase performance. In an environment free of solvents, Huang et al.'s study [80] examined the impact of ultrasound on the lipase-catalyzed hydrolysis of soy oil. Compared to the vibrating bath method, an ultrasonic bath produced droplets with a smaller size and a greater interfacial area. Additionally, smaller oil drops in water were produced with increased ultrasonic power. The initial hydrolysis rate grew as the interfacial area did. Consequently, the initial reaction rate was improved by the use of ultrasonography as a tool. This form of support has been shown in the treatment of used motor oil [76]. Like this, waste cooking oil hydrolyzed using lipase and ultrasonic in combination was more effective than using normal stirring techniques. In addition to enzyme-catalyzed hydrolysis, transesterification, and esterification are also affected by ultrasound. The synthesis of methyl esters by lipase-catalyzed direct transesterification may benefit from ultrasound [81]. It increased the oil's solubility in the aqueous phase in comparison to not utilizing ultrasonic treatment, and it also yielded more methyl esters in less time.

7.5.2 Protease

Recent research has shown that ultrasonic can speed up the enzymatic hydrolysis of proteins, causing them to degrade, alter in structure, and form functional peptides [82]. Even for complicated materials, ultrasound-assisted enzymatic digestion

Table 7.2 Applications of ultrasound-assisted enzymatic reactions.

Enzyme	Application	System	Type of US device/ optimum conditions	Observations and major results	References
Carboxymethyl-cellulase	Hydrolysis of parthenium hysterophorus	Solid–liquid phase	Ultrasonic bath reactor (25 cm × 15 cm × 10 cm), 35 kHz, 35 W, duty cycle 10%	Enhanced enzyme/substrate affinity and hydrolysis reaction velocity, enhanced enzymatic hydrolysis kinetics by a factor of six, and an increase in net sugar yield of around 20%.	[74]
Immobilized lipase	Discarded cooking oil is hydrolyzed.	Solid–water–oil phase	Ultrasonic bath reactor, 22 kHz, 100 W, duty cycle 50%.	Higher conversion (highest at 75.19%), faster processing time, and milder operating parameters	[76]
Alliinase	Hydrolysis of alliin	Liquid–liquid phase	Ultrasonic bath reactor, 40 kHz, 0.5 W/cm^2.	Improved alliinase activity and thermal stability; altered enzyme structure; improved mass-transfer; and free radical snare by reaction products (allicin).	[77]
Pectinase	Pectin hydrolysis	Liquid–liquid phase	Ultrasonic horn type reactor ($d = 10$ mm), 22 kHz, 4.5 W/ml	32.59% more hydrolysis occurred; pectinase's activity was boosted by modification.	[78]

may be completed in a matter of minutes. When combined with immobilized trypsin, bovine serum albumin, lactalbumin, carbonic anhydrase, and ovalbumin may be broken down from overnight treatment to six minutes [83]. Temperature effects on ultrasonic-assisted tryptic protein digestion were examined, and it was discovered that ultrasound performed best at cooler temperatures. The contribution of ultrasound decreased as the incubation temperature increased [84]. Hydrolysis using enzymes has been utilized frequently to separate precious metal ions from environmental and biological components, however, it is time-consuming. As a result, it has been proposed that ultrasonic might hasten the enzymatic digestion process employing trypsin, pancreatin, and pepsin. Temperatures are lowered via ultrasound. The contribution from ultrasound decreased as the incubation temperature rose [84]. The released element was considerably boosted by aided enzymatic digestion, and the treatment period was reduced to 30 minutes. Ultrasonic cavitation damages the cell membrane, which makes it easier for proteases to assault the cytosolic material and decrease the pretreatment period [85].

7.5.3 Polysaccharide Enzymes

To enhance polysaccharide characteristics and creating useful oligosaccharides, polysaccharide breakdown is a topic of extensive research. The effects of increasing low-intensity ultrasonic irradiation on the cellulase-catalyzed depolymerization of aqueous guar gum solution were examined by Prajapat et al. [86]. Three methods of depolymerizing guar gum were investigated in the study – enzymatic depolymerization with and without ultrasound, and simply ultrasound. In comparison to traditional enzymatic degradation methods, the use of ultrasound and cellulase increases the degree of depolymerization while simultaneously speeding up the process. Ma carefully investigated the synergistic impact of pectinase and ultrasound [78, 87]. The structural features of pectin and its enzymatic breakdown with the aid of ultrasound were investigated. According to the findings, pectin treated with ultrasound and pectinase dissolved much more quickly than the control within 12 minutes. Pectinase inactivation caused the sonoenzymatic hydrolysis rate to fall below the enzymatic hydrolysis rate after 12 minutes. The operating temperature has an impact on the synergistic relationship between pectinase and ultrasonography. Mass transport is a crucial restriction during the entire enzymatic process. Given that the substrate and the enzyme are both macromolecules, there may be serious mass transfer restrictions. Ultrasonic therapy can be used to get around this difficulty. The synergistic interaction of enzymes and ultrasound led to a reduction in the diffusion-limiting barrier between enzymes and substrates, which improved mass transfer inside the reaction vessel [88].

7.6 Biodiesel Production via Ultrasound-Supported Transesterification

Triglycerides (TG) and methanol are normally transesterified conventionally with the goal of creating biodiesel. The interfaces between the two immiscible phases,

TG, and methanol, are the sites of this reaction, which is normally catalyzed by an acid or a base. To boost the overall reaction rate and product conversion, the interfacial surface area between these two immiscible phases must be maximized. Standard mechanical agitation can increase the interfacial surface area and mass transfer characteristics of the process, but advances in ultrasound-assisted technologies have been found to accelerate transesterification processes [89]. These advancements enable the manufacture of biodiesel from a variety of feedstocks using ultrasonic irradiation as a unique, more effective mixing technique [90]. In several applications, ultrasonication has already demonstrated great promise. For instance, ultrasonication has been used in procedures like the extraction of lipids from *Chlorella vulgaris* [91], synthesis of saturated aliphatic esters from aliphatic acids [92], removal of oil from used coffee grounds using a two-phase solvent extraction method [93], and evaluation of oxidation stability [94]. The use of ultrasonication technology in the manufacture of biodiesel is timely and supported by current research as its applications grow.

7.6.1 Homogenous Acid-Catalyzed Ultrasound-Assisted Transesterification

For processing oils or fats with high amounts of free fatty acids (FFAs), acid-catalyzed transesterification is advantageous [95] because acid catalysts can be used for both the esterification and transesterification processes. Despite this, acids are more typically used for esterification than transesterification since the rate of acid-catalyzed transesterification is lower than that of base-catalyzed transesterification. Table 7.3 gathers a number of publications on the use of ultrasound in the homogeneous acid catalyst-based biodiesel manufacturing process, together with the typical operating parameters [96–98, 100, 101]. Sulfuric acid (H_2SO_4) is the acid catalyst that is most frequently utilized, as seen in the table. As an illustration, Hanh et al.'s research [96] on the ultrasound-assisted esterification of oleic acid with short-chain alcohols (ethanol, propanol, and butanol) revealed that 95% of fatty acid ethyl ester conversion was feasible with the use of a catalyst containing 5 wt% H_2SO_4. The ideal esterification conditions were a 3:1 molar ratio of ethanol to oleic acid, 5 wt% of H_2SO_4 at 60 °C, and a two-hour irradiation period. It was discovered that the length of the ultrasonic irradiation duration affected the esters' high conversion and purity.

7.6.2 Transesterification with Ultrasound Assistance and Homogenous Base Catalysis

Alkaline metal hydroxides , KOH and NaOH are the only two homogeneous base catalysts used in ultrasound-assisted transesterification. Most research studies have concentrated on using base catalysts rather than acid catalysts since base-catalyzed transesterification is speedier than acid-catalyzed transesterification. Base catalysts may be administered at low concentrations (0.2 wt%) and yet result in high biodiesel conversion (nearly 100%) in a short amount of time (<30 minutes). NaOH and

Table 7.3 Bibliographic collection of common parameters in homogeneous acid-catalyzed transesterifications with ultrasound assistance.

Ultrasonic reactor type	Frequency (kHz); power (W)	Feedstock	Catalyst type; loading (wt%)	Temperature (°C)	Alcohol type; alcohol to feedstock molar ratio	Time (min)	Biodiesel conversion (%)	References
Indirect sonication (bath)	40; 700	Oleic acid	H_2SO_4; 5	60	Ethanol; 3 : 1	120	>90	[96]
—	22; 120	Palm fatty acid distillate	H_2SO_4; 5	40	Methanol; 7 : 1	200	>90	[97]
—	25; 210	Palm fatty acid distillate	H_2SO_4; 5	60	Isopropanol; 5 : 1	275	75	[98]
—	40; 60	Nile tilapia-free fatty acid	H_2SO_4; 2	30	Methanol; 9 : 1	90	98.2	[99]
—	40; 200	Soybean waste oil	H_2SO_4; 3.5	60	Methanol; 9 : 1	60	99.9	[100]
—	–; 210	Jatropha oil	H_2SO_4; 4a	60	Methanol; 7 : 1	60a	96.4	[101]

KOH are also inexpensive and suitable for usage at low reaction pressures and temperatures [95]. Base catalyst usage also aids in preventing or minimizing corrosion issues [102]. A bibliographic compilation of relevant papers on the use of ultrasound for the homogeneous base catalyst-based biodiesel synthesis process is presented in Table 7.4 [97, 98, 100, 103–122]. As can be shown, all biodiesel conversions catalyzed by homogeneous base catalysts are over 90%, regardless of the choice of feedstock (edible or nonedible), the type of ultrasonic reactor (direct or indirect sonication), or the catalyst type (NaOH or KOH). With reaction times of under 60 minutes, the optimal catalyst concentrations were discovered to be between 0.2 and 2 wt%. In ultrasound-assisted homogeneous base-catalyzed transesterification, many different types of edible and nonedible vegetable oils such as palm [103–105], cottonseed [106, 107], sunflower seed [108], soybean [109–113], rapeseed [114], coconut [115], tung [116–118], canola [119], castor [120], waste cooking oil [121, 122], Jatropha curcas [123], and babassu [124] have been employed as feedstock. The commonality of significantly brief reaction times, typically less than one hour, and in some cases less than 10 minutes to achieve high conversions, seems to be the overarching conclusion in all of these studies. In the presence of a KOH catalyst, Teixeira et al. [125] investigated the utilization of several feedstocks as a source of TG for ultrasound-assisted transesterification. Beef tallow was chosen over vegetable oil because of its historically cheap cost and highly centralized production in slaughter and processing facilities. The results demonstrated that rapid ultrasonication (400 W, 24 kHz) was required to create biodiesel with a conversion rate comparable to that of conventional transesterification.

7.6.3 Heterogeneous Acid-Catalyzed Ultrasound-Assisted Transesterification

Early research mostly focuses on homogeneous catalyzed transesterification in ultrasound-assisted biodiesel synthesis. However, because of the significant limitations of high catalyst usage and challenging product separation, heterogeneous catalysts have recently been developed for use in ultrasound-assisted transesterification. The use of heterogeneous catalysts comes with a variety of difficulties. First off, because TG, the heterogeneous catalyst, and alcohol are immiscible, heterogeneous catalysis often proceeds more slowly. Second, processes that are heterogeneously catalyzed typically require more alcohol. Finally, because of the significant mass transfer restrictions brought on by immiscibility, heterogeneous catalysis is advantageous at higher temperatures and pressures. Regardless of whether acid, base, or enzyme catalysts are used, the pace of the reaction will be accelerated when utilizing ultrasound. Table 7.5 summarizes the bibliographic compilation of published works that employ ultrasonic and heterogeneous acid catalysts to speed up the transesterification of biodiesel [126–130]. For C_8–C_{10} fatty acid odor–cut esterification, Kelkar et al. [126] employed cavitation with a superacid catalyst, chlorosulfonic acid supported on zirconium. The molar ratio of methanol to fatty acids was maintained at 10:1. The conversion was almost 75% in five hours with a catalyst loading of 1 wt%. At 2 wt% catalyst addition, a conversion

Table 7.4 Bibliographic compilation of typical parameters for homogeneous base-catalyzed transesterifications with ultrasound assistance.

Ultrasonic reactor type	Frequency (kHz); power (W)	Feedstock	Catalyst type; loading (wt%)	Temperature (°C)	Alcohol type; alcohol to oil molar ratio	Time (min)	Biodiesel conversion (%)	References
Indirect sonication (bath)	28 & 40; 1200	Commercial edible oil	NaOH; 1/KOH; 0.5	36 ± 2	Methanol; 6 : 1	40/20	98/99	[103]
—	40 kHz; 1200 W	Commercial edible oil, corn, grape seed, canola, palm, sesame, synthetic oils	KOH; 0.5	36 ± 2	Methanol; 6 : 1	30	na	[104]
—	40; 1200	Commercial edible oil, corn, grape seed, canola, palm oils	KOH; 0.5	36 ± 2	Methanol; 6 : 1	60	>94	[105]
Direct sonication (horn)	24; 200	Cotton seed oil	NaOH; 2	60	Methanol; 7 : 1	20	95	[106]
Direct sonication (transducer)	40; —	Crude cottonseed oil	NaOH; 1	25	Methanol; 6.2 : 1	8	98	[107]
Direct sonication (horn)	24; 200	Sunflower oil	NaOH; 2	60	Methanol; 7 : 1	20	95	[108]
Direct sonication (transducer)	19.7; 100	Soybean oil	NaOH; na	45	Methanol; 6 : 1	10–20	100	[109]
Direct sonication (sonotrode)	24; 400	Soybean oil	KOH; 1	89	Methanol; 6 : 1	5	99	[110]
Indirect sonication (bath)	40; 4870 W/m^2	Soybean oil	NaOH; 0.2	29	Methanol; 9 : 1	30	100	[111]

Method	Power (W); Frequency (kHz)	Feedstock	Catalyst; wt %	Temp (°C)	Alcohol; molar ratio	Time (min)	Yield (%)	Ref.
Direct sonication (transducer)	611; 139	Soybean oil	KOH; 0.5	26 ± 1	Methanol; 6 : 1	30	90	[112]
Direct sonication (horn)	20; 2200	Soybean oil	NaOH; 1	40	Methanol; 5 : 1	1.5	96	[113]
—	24; 200	Rapeseed oil	NaOH; 2	60	Methanol; 7 : 1	20	96	[114]
—	24; 200	Coconut oil	KOH; 0.75	—	Ethanol; 6 : 1	7	98	[115]
Direct sonication (probe)	25; 300	Tung oil/blended oil (20% tung, 50% canola, 30% palm)	KOH; 1	20–30	Methanol; 6 : 1	5	87–91/92–94	[116]
—	25; 300	Blended oil (60% tung, 30% canola, 10% palm)	KOH; 2	25	Methanol; 6 : 1	5	98.04	[117]
—	20; 300	Tung oil/blended oil (20% tung, 50% canola, 30% palm)	KOH; 2	25	Methanol; 6 : 1	5	98.33	[118]
Direct sonication (sonotrode)	20; 2000	Canola oil	KOH; 0.7	25 ± 2	Methanol; 5 : 1	50	99	[119]
Indirect sonication (bath)	45; 2400	Castor oil	KOH; 0.7	40	Methanol; 6 : 1	20	92.2	[120]
Direct sonication (sonotrode)	20; 2000	Waste cooking oil	KOH; 1	20–25	Methanol; 4 : 1	0.93	93.8	[121]
Direct sonication (transducer)	20; 200	Waste cooking oil	KOH; 1	45	Methanol; 6 : 1	40	na	[122]
Direct sonication (horn)	24; 200	Jatropha curcas oil	KOH; 0.75	—	Ethanol; 4 : 1	7–8	na	[123]
Indirect sonication (bath)	20; 600	Babassu oil	KOH; 1	30	Ethanol; 6 : 1	10	97	[124]
Direct sonication (sonotrode)	24; 400	Beef tallow	KOH; 0.5	60	Methanol; 6 : 1	1.17	na	[125]

Table 7.5 Bibliographic list of common variables in transesterifications that are aided by ultrasound and heterogeneous acid.

Ultrasonic reactor type	Frequency (kHz); power (W)	Feedstock	Catalyst type; loading (wt%)	Temperature (°C)	Alcohol type; alcohol to oil molar ratio	Time (min)	Biodiesel conversion (%)	References
Indirect sonication (bath)	20; 120	Fatty acid odor cut	Zirconium-supported chlorosulfonic acid; 1	40	Methanol; 10:1	300	75	[126]
Direct sonication (horn)	20; 400 W	Jatropha oil	Activated carbon-supported tungstophosphoric acid; 4.23	65	Methanol; 25 : 1	40	91	[127]
—	20; 400	—	Gamma alumina-supported tungstophosphoric acid; 4.44	65	Methanol; 19 : 1	50	84	[128]
—	20; 400	—	Cesium doped heteropolyacid; 2.9	65	Methanol; 25 : 1	34	90.5	[129]
—	20; 400	—	Activated carbon-supported tungstophosphoric acid; 4	65	Methanol; 20 : 1	40	87.33	[130]

of almost 97% was attained in seven hours. The introduction of a heterogeneous catalyst lengthened the reaction time from 1.5 to 7 hours when compared to H_2SO_4. However, using a heterogeneous acid catalyst increased the biodiesel conversion from 95% to 97% at the same catalytic loading (2 wt%). During the process, the superacid was shown to become inactive, which may have been caused by the water molecule adhering to the active sites. Catalyst deactivation was also brought on by changes in the catalyst's morphology brought on by cavitation effects. The catalyst reusability cycles test showed that the regenerated catalysts had no effect on reaction rates, indicating that the suggested regeneration was ineffective since it was unable to completely remove the water molecules from the active sites. The performance of several heterogeneous acid catalysts has been thoroughly developed and researched by the research team of Badday et al. [127–130].

7.6.4 Heterogeneous Base-Catalyzed Ultrasound-Assisted Transesterification

Heterogeneous base catalysts have received more research attention than heterogeneous acid catalysts for transesterification for a variety of reasons. These drawbacks include slower reaction rates, the potential for unfavorable side reactions, longer reaction times, and higher temperatures for heterogeneous acid catalysts. Furthermore, it has been asserted that heterogeneous base catalysts are more active than heterogeneous acid catalysts due to the metal oxide group they contain [102]. The use of heterogeneous base catalysts in conjunction with ultrasound has recently been the focus of research. The bibliographic listing of publications on the use of ultrasound to accelerate biodiesel transesterification using heterogeneous base catalysts is summarized in Table 7.6, together with the typical operating conditions [131–134]. Biodiesel conversions range from 80% to 98.5%, as seen in Table 7.4. According to Verziu et al. [131], commercial heterogeneous base CaO (CS) and microcrystalline CaO (MS) catalysts were used in conjunction with ultrasonic mixing to transesterify sunflower, soybean, and rapeseed oils. It was demonstrated that microcrystalline CaO from less costly sources, such as limestone and calcium hydroxide, displayed better catalytic activity during the transesterification of sunflower oil as compared to CS CaO. Heat activation and ultrasonic activation were two separate techniques used to prepare MS CaO. Each catalyst was thermally activated at different temperatures (450, 515, and 600 °C) or ultrasonically activated in methanol at room temperature. The outcomes of ultrasonic transesterification, using both thermally and ultrasonically activated CaO (MS), were compared. The MS CaO was ultrasonically activated, and after 45 minutes, there was a conversion of over 92% at 75 °C. In different research using CaO, the production of biodiesel from processed palm oil was investigated using commercial heterogeneous catalysts, namely CaO and potassium phosphate (K_3PO_4). However, CaO (90%) yielded a little higher conversion than K_3PO_4 (80%) (Table 7.6), despite both catalysts producing high biodiesel conversions. In the testing, a conventional mechanically stirred reactor was also employed for comparison. Different mixing qualities in the ultrasonic reactor and the mechanically

Table 7.6 Bibliographic list of typical transesterification parameters using a heterogeneous base as the catalyst.

Ultrasonic reactor type	Frequency (kHz); power (W)	Feedstock	Catalyst type; loading (wt%)	Temperature (°C)	Alcohol type; alcohol to oil molar ratio	Time (min)	Biodiesel conversion (%)	References
Direct sonication (horn)	35	Sunflower oil	CaO	75	Methanol; 4 : 1	45	80	[131]
Indirect sonication (bath)	40; 160	Palm oil	CaO; 3	65	Methanol; 6 : 1	120	90	[132]
—	35; 35	Soybean oil	CaO; 6	62	Methanol; 10 : 1	60	80	[133]
Direct sonication (horn)	20; 750	—	$C_2H_6CaO_2$; 1	65	Methanol; 9 : 1	90	90	[134]

stirred reactor led to different catalytic performances in terms of activation and deactivation. In the catalyst reusability test employing CaO, the conversion from the prior cycle using the ultrasonic reactor was higher compared to that of the mechanically stirred reactor.

7.6.5 Enzyme-Catalyzed Ultrasound-Assisted Transesterification

Although it has also been mentioned in the literature, there is currently little research on enzyme-catalyzed ultrasound-assisted transesterification. This method has a number of benefits, including good selectivity, reduced energy use, less waste products and side products, no soap production, and simple recovery of catalysts and glycerol [95]. As indicated in Table 7.7, Novozym 435, a commercial lipase preparation from Candida antarctica, has been used to investigate enzyme-catalyzed transesterification. For instance, Yu et al. [135] studied the generation of soybean biodiesel utilizing Novozym 435 (Candida antarctica lipase B immobilized on polyacrylic resin). They claimed that, in a comparatively shorter amount of time (from more than 12 hours to only 4 hours), ultrasonic irradiation provided the same 459 biodiesel conversion as vibration mixing. By combining ultrasonic irradiation with vibration mixing, the rate of biodiesel generation was significantly increased, with 96% conversion obtained in four hours and no evident reduction in lipase activity after repeated use for five cycles. While this was going on, Gharat and Rhatod [136] looked into the transesterification of immobilized Novozym 435 using ultrasound in a solvent-free method utilizing dimethyl carbonate to produce biodiesel from used cooking oil. Three distinct experimental setups – ultrasonic irradiation (UI) without stirring, UI combined with stirring, and stirring alone – were run. At four hours, the stirring only produced a 38.69% FAME conversion. In contrast, during the same reaction period, UI without stirring dramatically increased the conversion of enzymatic transesterification up to 57.68%. Under UI in conjunction with stirring, the greatest conversion of 86.61% was attained for the same reaction time. The rate of FAME conversion and enzyme activity gradually dropped after many applications of Novozym 435.

7.7 Conclusions

The significance of microwave and ultrasound technology in enzymatic processes has been covered in this chapter. Studies have shown that microwave heating can increase the enzyme's activity, stability, and selectivity. However, as enzymes are temperature-sensitive macromolecules, the application of microwave irradiation in enzymatic synthesis is still restricted due to the high temperatures associated with microwave heating. To speed up the reaction and increase product yields, ultrasound is frequently employed at various stages of the enzymatic reaction process. The enhancement is caused by the cavitation that ultrasound causes. Cavitation is a dynamic phenomenon, and the operating conditions have a significant impact on how it behaves. Enzymes respond well to low-intensity ultrasound,

Table 7.7 Bibliographic list of common transesterification parameters using ultrasound with enzyme catalysis.

Ultrasonic reactor type	Frequency (kHz); power (W)	Feedstock	Enzyme type; loading (wt%)	Temperature (°C)	Alcohol type; alcohol to oil molar ratio	Time (min)	Biodiesel conversion (%)	References
Indirect sonication (bath)	40; 500	Soybean oil	Novozym 435; 6	40	Methanol; 6 : 1	240	96	[135]
—	25; 200	Waste cooking oil	—	60	Dimethyl carbonate; 6 : 1	240	86.61	[136]
—	28; 100 W	High acid value waste oil	Novozym 435; 8	40–45	Propanol; 3 : 1	50	94.86	[137]

but high-intensity ultrasound has the opposite effect. Designing an effective ultrasound-aided enzymatic hydrolysis system requires research into the impact of operating parameters and the ultrasonic boosting effect on enzyme processes. In the chemical industry, microwave systems still need to be investigated. They also offer a lot of worldwide research and development potential.

Acknowledgments

Dr. K. Behera acknowledges the University of Allahabad and Dr. Abbul Bashar Khan acknowledges Jamia Millia Islamia for the facilities and funding.

References

1 Beilen, J.B.V. and Li, Z. (2002). Enzyme technology: an overview. *Curr. Opin. Biotechnol.* 13 (4): 338–344.
2 Silva, A.P.T., Bredda, E.H., Castro, H.F., and Ros, C.M. (2020). Enzymatic catalysis: an environmentally friendly method to enhance the transesterification of microalgal oil with fusel oil for production of fatty acid esters with potential application as biolubricants. *Fuel* 273: 117786. https://doi.org/10.1016/j.fuel.2020.117786
3 Kappa, C.O. (2004). Controlled microwave heating in modern organic synthesis. *Anew. Chem. Int. Ed.* 43: 6250–6284.
4 Horikoshi, S., Schiffmann, R.F., Fukushima, J., and Serpone, N. (2018). Microwave-assisted chemistry. In: *Microwave Chemical and Materials Processing*, 243–319. Singapore: Springer Singapore. https://doi.org/10.1007/978-981-10-6466-1_9.
5 Mierzwa, D. and Pawlowski, A.D. (2017). Convective drying of potatoes assisted by microwave and infrared radiation – process kinetics and quality aspect. *J. Food Nutr. Res.* 56: 351–361.
6 Mazo, P., Restrepo, G., and Rios, L. (2011). Biodiesel: feedstocks and processing technologies. In: *Alternative Methods for Fatty Acid Alkyl-Esters Production: Microwaves, Radio-Frequency and Ultrasound* (ed. M. Stoytcheva), 269–288. In tech (National Renewable Energy Laboratory).
7 Gedye, R., Smith, F., Westaway, K. et al. (1986). The use of microwave ovens for rapid organic synthesis. *Tetrahedron Lett.* 27: 279–282.
8 Rejasse, B., Lamare, S., Legoy, M.D., and Besson, T. (2004). Stability improvement of immobilized Candida antarctica lipase B in an organic medium under microwave radiation. *Org. Biomol. Chem.* 2: 1086–1089.
9 Rejasse, B., Bessonen, T., Legoy, M.D., and Lamare, S. (2006). Influence of microwave radiation on free Candida antarctica lipase B activity and stability. *Org. Biomol. Chem.* 4: 3703–3707.
10 Lidstrom, P., Tierney, J., Wathey, B., and Westman, J. (2001). Microwave assisted organic synthesis—a review. *Tetrahedron* 57: 9225–9283.

11 Larhed, M. and Hallberg, A. (2001). Microwave-assisted high-speed chemistry: a new technique in drug discovery. *Drug Discovery Today* 6: 406–416.

12 Elander, N., Jones, J.R., Lu, S.Y., and Stone-Elander, S. (2000). Microwave-enhanced radiochemistry. *Chem. Soc. Rev.* 29: 239–249.

13 Suarez, M., Loupy, A., Salfran, E. et al. (1999). Synthesis of decahydroacridines under microwaves using ammonium acetate supported on alumina. *Heterocycles* 51: 21–27.

14 Loupy, A. and Regnier, S. (1999). Solvent-free microwave-assisted beckmann rearrangement of benzaldehyde and 2-hydroxyacetophenone oximes. *Tetrahedron Lett.* 40: 6221–6224.

15 Olofsson, K., Kim, S.Y., Larhed, M. et al. (1999). High-speed, highly fluorous organic reactions. *J. Organomet. Chem.* 64: 4539–4541.

16 Orru, R.V.A. and de Greef, M. (2003). Recent advances in solution phase multicomponent methodology for the synthesis of heterocyclic compounds. *Synth. Stuttg.* 2003: 1471–1499.

17 Nuchter, M., Ondruschka, B., Bonrath, W., and Gum, A. (2004). Microwave assisted synthesis—a critical technology overview. *Green Chem.* 6: 128–141.

18 Gabriel, C., Gabriel, S., Grant, E.H. et al. (1998). Dielectric parameters relevant to microwave dielectric heating. *Chem. Soc. Rev.* 27: 213–223.

19 Gedye, R.N. and Wei, J.B. (1998). Rate enhancement of organic reactions by microwaves at atmospheric pressure. *Can. J. Chem.* 76: 525–532.

20 Langa, F., DelaCruz, P., DelaHoz, A. et al. (1997). Microwave irradiation: more than just a method for accelerating reactions. *Contemp. Org. Synth.* 4: 373–386.

21 Loupy, A., Petit, A., Hamelin, J. et al. (1998). New solvent free organic synthesis using focused microwaves. *Synth. Stuttg.* 09: 1213–1234.

22 Tao, Y. and Sun, D.W. (2015). Enhancement of food processes by ultrasound: a review. *Crit. Rev. Food Sci. Nutr.* 55 (4): 570–594.

23 McClement, D.J. (1995). Advances in the application of ultrasound in food analysis and processing. *Trends Food Sci. Technol.* 6 (9): 293–299.

24 Rastogi, N.K. (2011). Opportunities and challenges in application of ultrasound in food processing. *Crit. Rev. Food Sci. Nutr.* 51 (8): 705–722.

25 Wang, W., Ma, X., Jiang, P. et al. (2016). Characterization of pectin from grapefruit peel: a comparison of ultrasound-assisted and conventional heating extractions. *Food Hydrocolloids* 61: 730–739.

26 Nguyen, T.T.T. and Le, V.V.M. (2013). Effects of ultrasound on cellulolytic activity of cellulase complex. *Int. Food Res. J.* 20 (2): 557–563.

27 Yu, L., Sun, J., Liu, S. et al. (2012). Ultrasonic-assisted enzymolysis to improve the antioxidant activities of peanut (*Arachin conarachin* L.) antioxidant hydrolysate. *Int. J. Mol. Sci.* 13: 9051–9068.

28 Nadar, S.S., Rao, P., and Rathod, V.K. (2018). Enzyme assisted extraction of biomolecules as an approach to novel extraction technology: a review. *Food Res. Int.* 108: 309–330.

29 Dabbour, M., He, R., Mintah, B. et al. (2019). Changes in functionalities, conformational characteristics and antioxidative capacities of sunflower protein by controlled enzymolysis and ultrasonication action. *Ultrason. Sonochem.* 58: 104625.

30 Zhang, Y., Wang, B., Zhou, C. et al. (2016). Surface topography, nano-mechanics and secondary structure of wheat gluten pretreated by alternate dual-frequency ultrasound and the correlation to enzymolysis. *Ultrason. Sonochem.* 31: 267–275.

31 Ayim, I., Ma, H., Alenyorege, E.A. et al. (2018). Influence of ultrasound pretreatment on enzymolysis kinetics and thermodynamics of sodium hydroxide extracted proteins from tea residue. *J. Food Sci. Technol.* 55: 1037–1046.

32 Jia, J., Ma, H., Zhao, W. et al. (2010). The use of ultrasound for enzymatic preparation of ACE-inhibitory peptides from wheat germ protein. *Food Chem.* 119: 336–342.

33 Tang, X., Dai, L., Tang, Y., and Sun, W. (2014). Study on optimization of bagasse hemicellulose enzymolysis with response surface analysis. *J. Sustain. Bioenergy Syst.* 4: 249–259.

34 Falleh, H., Ksouri, R., Lucchessi, M.-E. et al. (2012). Ultrasound-assisted extraction: effect of extraction time and solvent power on the levels of polyphenols and antioxidant activity of *Mesembryanthemum edule* L. aizoaceae shoots. *Trop. J. Pharm. Res.* 11: 243–249.

35 Jin, Y., Liang, R., Liu, J. et al. (2017). Effect of structure changes on hydrolysis degree, moisture state, and thermal denaturation of egg white protein treated by electron beam irradiation. *LWT* 77: 134–141.

36 Shao, Y., Zhang, Y., Liu, J., and Tu, Z. (2020). Influence of ultrasonic pretreatment on the structure, antioxidant and IgG/IgE binding activity of β-lactoglobulin during digestion in vitro. *Food Chem.* 312: 126080.

37 Yang, X., Li, Y., Li, S. et al. (2020). Effects and mechanism of ultrasound pretreatment of protein on the Maillard reaction of protein-hydrolysate from grass carp (*Ctenopharyngodon idella*). *Ultrason. Sonochem.* 64: 104964.

38 Yang, X., Li, Y., Li, S. et al. (2017). Effects of ultrasound pretreatment with different frequencies and working modes on the enzymolysis and the structure characterization of rice protein. *Ultrason. Sonochem.* 38: 19–28.

39 Mintah, B.K., He, R., Dabbour, M. et al. (2020). Characterization of edible soldier fly protein and hydrolysate altered by multiple-frequency ultrasound: structural, physical, and functional attributes. *Process Biochem.* 95: 157–165.

40 Wang, L., Ding, J., Fang, Y. et al. (2020). Effect of ultrasonic power on properties of edible composite films based on rice protein hydrolysates and chitosan. *Ultrason. Sonochem.* 65: 105049.

41 Vallejo, A., Usobiaga, A., Ortiz-Zarragoitia, M. et al. (2010). Focused ultrasound-assisted acceleration of enzymatic hydrolysis of alkylphenols and 17β-oestradiol glucuronide in fish bile. *Anal. Bioanal.Chem.* 398: 2307–2314.

42 Yadav, G.D. and Lathi, P.S. (2006). Intensification of enzymatic synthesis of propylene glycol monolaurate from 1,2-propanediol and lauric acid under microwave irradiation: kinetics of forward and reverse reactions. *Enzyme Microb. Technol.* 38: 814–820.

43 Yadav, G.D. and Lathi, P.S. (2004). Synergism between microwave and enzyme catalysis in intensification of reactions and selectivities: transesterification of methyl acetoacetate with alcohols. *J. Mol. Catal. A: Chem.* 223: 51–56.

44 Bansode, S.R. and Rathod, V.K. (2018). Enzymatic sythesis of Isoamyl butyrate under microwave irradiation. *Chem. Eng. Process. Process Intensif.* 129: 71–76.

45 Yadav, G.D. and Borkar, I.V. (2009). Kinetic and mechanistic investigation of microwave-assisted lipase catalyzed synthesis of citronellyl acetate. *Ind. Eng. Chem. Res.* 48: 7915–7922.

46 Risso, M., Mazzini, M., Kröger, S. et al. (2012). Microwave-assisted solvent-free lipase catalyzed transesterification of β-ketoesters. *Green Chem. Lett. Rev.* 5: 539–543.

47 Liang, Q., Ren, X., Ma, H. et al. (2017). Effect of low-frequency ultrasonic-assisted enzymolysis on the physicochemical and antioxidant properties of corn protein hydrolysates. *J. Food Qual.* 2017: 1–10.

48 Resendiz-Vazquez, J.A., Urías-Silvas, J.E., Ulloa, J.A. et al. (2019). Effect of ultrasound-assisted enzymolysis on jackfruit (*Artocarpus heterophyllus*) seed proteins: structural characteristics, technofunctional properties and the correlation to enzymolysis. *J. Food Process. Technol.* 10 (5). https://doi.org/10.4172/2157-7110.1000796.

49 Dabbour, M., He, R., Mintah, B. et al. (2018). Ultrasound assisted enzymolysis of sunflower meal protein: kinetics and thermodynamics modeling. *J. Food Process Eng* 41: 12865.

50 Fernandes, F.A.N., Linhares, F.E., and Rodrigues, S. (2008). Ultrasound as pre-treatment for drying of pineapple. *Ultrason. Sonochem.* 15: 1049–1054.

51 Suslick, K.S. (1989). The chemical effects of ultrasound. *Sci. Am.* 260: 80–86.

52 Dular, M. and Coutier-Delgosha, O. (2013). Thermodynamic effects during growth and collapse of a single cavitation bubble. *J. Fluid Mech.* 736: 44–66.

53 Vickers, N.J. (2017). Animal communication: when i'm calling you, will you answer too? *Curr. Biol.* 27: 713–715.

54 Jin, J., Ma, H., Wang, K. et al. (2015). Effects of multi-frequency power ultrasound on the enzymolysis and structural characteristics of corn gluten meal. *Ultrason. Sonochem.* 24: 55–64.

55 Patist, A. and Bates, D. (2008). Ultrasonic innovations in the food industry: from the laboratory to commercial production. *Innovative Food Sci. Emerg. Technol.* 9: 147–154.

56 Ji, J., Lu, X., Cai, M., and Xu, Z. (2006). Improvement of leaching process of Geniposide with ultrasound. *Ultrason. Sonochem.* 13: 455–462.

57 Priyadarshi, A., Khavari, M., Subroto, T. et al. (2021). On the governing fragmentation mechanism of primary intermetallics by induced cavitation. *Ultrason. Sonochem.* 70: 105260.

58 Kadam, S.U., Tiwari, B.K., Álvarez, C., and O'Donnell, C.P. (2015). Ultrasound applications for the extraction, identification and delivery of food proteins and bioactive peptides. *Trends Food Sci. Technol.* 46: 60–67.

59 Li, S., Yang, X., Zhang, Y. et al. (2016). Enzymolysis kinetics and structural characteristics of rice protein with energy-gathered ultrasound and ultrasound assisted alkali pretreatments. *Ultrason. Sonochem.* 31: 85–92.

60 Abadía-García, L., Castano-Tostado, E., Ozimek, L. et al. (2016). Impact of ultrasound pretreatment on whey protein hydrolysis by vegetable proteases. *Innovative Food Sci. Emerg. Technol.* 37: 84–90.

61 Dabbour, M., He, R., Mintah, B. et al. (2020). Ultrasound pretreatment of sunflower protein: Impact on enzymolysis, ACE-inhibition activity, and structure characterization. *J. Food Process. Preserv.* 44: e14398.

62 Jia, C., Han, F., Miao, X. et al. (2019). Study on optimization of extraction process of anthocyanin from cherry wine lees. *J. Nutr., Health Food Eng.* 9: 18–27.

63 Zhou, C., Ma, H., Yu, X. et al. (2013). Pretreatment of defatted wheat germ proteins (by-products of flour mill industry) using ultrasonic horn and bath reactors: effect on structure and preparation of ACE-inhibitory peptides. *Ultrason. Sonochem.* 20: 1390–1400.

64 Chandrapala, J., Zisu, B., Palmer, M. et al. (2011). Effects of ultrasound on the thermal and structural characteristics of proteins in reconstituted whey protein concentrate. *Ultrason. Sonochem.* 18: 951–957.

65 Stefanovic, A.B., Jovanovic, J.R., Balanc, B.D. et al. (2018). Influence of ultrasound probe treatment time and protease type on functional and physicochemical characteristics of egg white protein hydrolysates. *Poult. Sci.* 97: 2218–2229.

66 Dabbour, M., Alenyorege, E.A., Mintah, B. et al. (2020). Proteolysis kinetics and structural characterization of ultrasonic pretreated sunflower protein. *Process Biochem.* 94: 198–206.

67 Jain, S. and Anal, A.K. (2016). Optimization of extraction of functional protein hydrolysates from chicken egg shell membrane (ESM) by ultrasonic assisted extraction (UAE) and enzymatic hydrolysis. *LWT Food Sci. Technol.* 69: 295–302.

68 Dong, Z.Y., Li, M.Y., Tian, G. et al. (2019). Effects of ultrasonic pretreatment on the structure and functionality of chicken bone protein prepared by enzymatic method. *Food Chem.* 299: 125103.

69 Mintah, B.K., He, R., Dabbour, M. et al. (2019). Techno-functional attribute and antioxidative capacity of edible insect protein preparations and hydrolysates thereof: effect of multiple mode sonochemical action. *Ultrason. Sonochem.* 58: 104676.

70 Lebon, G.S.B., Tzanakis, I., Pericleous, K., and Eskin, D. (2018). Experimental and numerical investigation of acoustic pressures in different liquids. *Ultrason. Sonochem.* 42: 411–421.

71 Wang, Z., Wang, C., Zhang, C., and Li, W. (2017). Ultrasound-assisted enzyme catalyzed hydrolysis of olive waste and recovery of antioxidant phenolic compounds. *Innovative Food Sci. Emerg. Technol.* 44: 224–234.

72 Uluko, H., Zhang, S., Liu, L. et al. (2014). Pilot-scale membrane fractionation of ACE inhibitory and antioxidative peptides from ultrasound pretreated milk protein concentrate hydrolysates. *J. Funct. Foods* 7: 350–361.

73 Wang, Y., Wang, Z., Handa, C.L., and Xu, J. (2017). Effects of ultrasound pre-treatment on the structure of β-conglycinin and glycinin and the antioxidant activity of their hydrolysates. *Food Chem.* 218: 165–172.

74 Singh, S., Agarwal, M., Bhatt, A. et al. (2015). Ultrasound enhanced enzymatic hydrolysis of *Parthenium hysterophorus*: a mechanistic investigation. *Bioresour. Technol.* 192: 636–645.

75 Mason, T.J. (1997). Ultrasound in synthetic organic chemistry. *Chem. Soc. Rev.* 26: 443.

76 Waghmare, G.V. and Rathod, V.K. (2016). Ultrasound assisted enzyme catalyzed hydrolysis of waste cooking oil under solvent free condition. *Ultrason. Sonochem.* 32: 60–67.

77 Wang, J., Cao, Y., Sun, B. et al. (2011). Effect of ultrasound on the activity of alliinase from fresh garlic. *Ultrason. Sonochem.* 18: 534–540.

78 Ma, X., Zhang, L., Wang, W. et al. (2016). Synergistic effect and mechanisms of combining ultrasound and pectinase on pectin hydrolysis. *Food Bioprocess Technol.* 9: 1249–1257.

79 Lerin, L.A., Loss, R.A., Remonatto, D. et al. (2014). A review on lipase-catalyzed reactions in ultrasound-assisted systems. *Bioprocess. Biosyst. Eng.* 37: 2381–2394.

80 Huang, J., Liu, Y., Song, Z. et al. (2010). Kinetic study on the effect of ultrasound on lipase-catalyzed hydrolysis of soy oil: study of the interfacial area and the initial rates. *Ultrason. Sonochem.* 17: 521–525.

81 Sivaramakrishnan, R. and Incharoensakdi, A. (2017). Direct transesterification of Botryococcus sp. catalysed by immobilized lipase: ultrasound treatment can reduce reaction time with high yield of methyl ester. *Fuel* 191: 363–370.

82 Domínguez-Vega, E., García, M.C., Crego, A.L., and Marina, M.L. (2010). First approach based on direct ultrasonic assisted enzymatic digestion and capillary-high performance liquid chromatography for the peptide mapping of soybean proteins. *J. Chromatogr. A* 1217: 6443–6448.

83 Vale, G., Santos, H.M., Carreira, R.J. et al. (2011). An assessment of the ultrasonic probe-based enhancement of protein cleavage with immobilized trypsin. *PROTEOMICS* 11: 3866–3876.

84 Shin, S., Yang, H.-J., Kim, J., and Kim, J. (2011). Effects of temperature on ultrasound-assisted tryptic protein digestion. *Anal. Biochem.* 414: 125–130.

85 Peña-Farfal, C., Moreda-Piñeiro, A., Bermejo-Barrera, A. et al. (2004). Ultrasound bath-assisted enzymatic hydrolysis procedures as sample pretreatment for the multielement determination in mussels by inductively coupled plasma atomic emission spectrometry. *Anal. Chem.* 76: 3541–3547.

86 Prajapat, A.L., Subhedar, P.B., and Gogate, P.R. (2016). Ultrasound assisted enzymatic depolymerization of aqueous guar gum solution. *Ultrason. Sonochem.* 29: 84–92.

87 Ma, X., Wang, W., Wang, D. et al. (2016). Degradation kinetics and structural characteristics of pectin under simultaneous sonochemical-enzymatic functions. *Carbohydr. Polym.* 154: 176–185.

88 Easson, M.W., Condon, B., Dien, B.S. et al. (2011). The application of ultrasound in the enzymatic hydrolysis of switchgrass. *Appl. Biochem. Biotechnol.* 165: 1322–1331.

89 Refaat, A.A. (2010). Different techniques for the production of biodiesel from waste vegetable oil. *Int. J. Environ. Sci. Technol.* 7: 183–213.

90 Gole, V.L. and Gogate, P.R. (2012). A review on intensification of synthesis of biodiesel from sustainable feed stock using sonochemical reactors. *Chem. Eng. Process. Process Intensif.* 53: 1–9.

91 Kim, Y.-H., Park, S., Kim, M.H. et al. (2013). Ultrasound-assisted extraction of lipids from *Chlorella vulgaris* using [Bmim][MeSO$_4$]. *Biomass Bioenergy* 56: 99–103.

92 Abdullah, M. and Bulent Koc, A. (2013). Oil removal from waste coffee grounds using two-phase solvent extraction enhanced with ultrasonication. *Renew. Energy* 50: 965–970.

93 Hobuss, C.B., Venzke, D., Pacheco, B.S. et al. (2012). Ultrasound-assisted synthesis of aliphatic acid esters at room temperature. *Ultrason. Sonochem.* 19: 387–389.

94 Avila Orozco, F.D., Sousa, A.C., Domini, C.E. et al. (2013). An ultrasonic-accelerated oxidation method for determining the oxidative stability of biodiesel. *Ultrason. Sonochem.* 20: 820–825.

95 Veljkovic, V.B., Avramovic, J.M., and Stamenkovic, O.S. (2012). Biodiesel production by ultrasound-assisted transesterification: State of the art and the perspectives. *Renew. Sustain. Energy Rev.* 16: 1193–1209.

96 Hanh, H.D., Dong, N.T., Okitsu, K. et al. (2009). Biodiesel production by esterification of oleic acid with short-chain alcohols under ultrasonic irradiation condition. *Renew. Energy* 34: 780–783.

97 Deshmane, V.G., Gogate, P.R., and Pandit, A.B. (2009). Ultrasound-assisted synthesis of biodiesel from palm fatty acid distillate. *Ind. Eng. Chem. Res.* 48: 7923–7927.

98 Deshmane, V.G., Gogate, P.R., and Pandit, A.B. (2009). Ultrasound assisted synthesis of isopropyl esters from palm fatty acid distillate. *Ultrason. Sonochem.* 16: 345–350.

99 Santos, F.F.P., Malveira, J.Q., Cruz, M.G.A., and Fernandes, F.A.N. (2010). Production of biodiesel by ultrasound assisted esterification of *Oreochromis niloticus* oil. *Fuel* 89: 275–279.

100 Santos, F.F.P., Matos, L.J.B.L., Rodrigues, S., and Fernandes, F.A.N. (2009). Optimization of the production of methyl esters from soybean waste oil applying ultrasound technology. *Energy Fuels* 23: 4116–4120.

101 Deng, X., Fang, Z., and Liu, Y. (2010). Ultrasonic transesterification of *Jatropha curcas* L. oil to biodiesel by a two-step process. *Energy Convers. Manage.* 51: 2802–2807.

102 Ramachandran, K., Suganya, T., Nagendra Gandhi, N., and Renganathan, S. (2013). Recent developments for biodiesel production by ultrasonic assist transesterification using different heterogeneous catalyst: a review. *Renew. Sustain. Energy Rev.* 22: 410–418.

103 Stavarache, C., Vinatoru, M., Nishimura, R., and Maeda, Y. (2005). Fatty acids methyl esters from vegetable oil by means of ultrasonic energy. *Ultrason. Sonochem.* 12: 367–372.

104 Stavarache, C., Vinatoru, M., and Maeda, Y. (2006). Ultrasonic versus silent methylation of vegetable oils. *Ultrason. Sonochem.* 13: 401–407.

105 Stavarache, C., Vinatoru, M., and Maeda, Y. (2007). Aspects of ultrasonically assisted transesterification of various vegetable oils with methanol. *Ultrason. Sonochem.* 14: 380–386.

106 Georgogianni, K.G., Kontominas, M.G., Pomonis, P.J. et al. (2008). Alkaline conventional and in situ transesterification of cottonseed oil for the production of biodiesel. *Energy Fuels* 22: 2110–2115.

107 Fan, X., Chen, F., and Wang, X. (2010). Ultrasound-assisted synthesis of biodiesel from crude cottonseed oil using response surface methodology. *J. Oleo Sci.* 59: 235–241.

108 Georgogianni, K.G., Kontominas, M.G., Pomonis, P.J. et al. (2008). Conventional and in situ transesterification of sunflower seed oil for the production of biodiesel. *Fuel Process. Technol.* 89: 503–509.

109 Ji, J., Wang, J., Li, Y. et al. (2006). Preparation of biodiesel with the help of ultrasonic and hydrodynamic cavitation. *Ultrasonics* 44: e411–e414.

110 Singh, A.K. and Fernando, S.D. (2006). Base catalyzed fast-transesterification of soybean oil using ultrasonication. In: *2006 ASAE Annual Meeting* Portland, Oregon, July 9-12, 2006. St. Joseph, MI: American Society of Agricultural and Biological Engineers.

111 Santos, F.F.P., Rodrigues, S., and Fernandes, F.A.N. (2009). Optimization of the production of biodiesel from soybean oil by ultrasound assisted methanolysis. *Fuel Process. Technol.* 90: 312–316.

112 Mahamuni, N.N. and Adewuyi, Y.G. (2009). Optimization of the synthesis of biodiesel via ultrasound-enhanced base-catalyzed transesterification of soybean oil using a multifrequency ultrasonic reactor. *Energy Fuels* 23: 2757–2766.

113 Chand, P., Chintareddy, V.R., Verkade, J.G., and Grewell, D. (2010). Enhancing biodiesel production from soybean oil using ultrasonics. *Energy Fuels* 24: 2010–2015.

114 Georgogianni, K.G., Katsoulidis, A.K., Pomonis, P.J. et al. (2009). Transesterification of rapeseed oil for the production of biodiesel using homogeneous and heterogeneous catalysis. *Fuel Process. Technol.* 90: 1016–1022.

115 Kumar, D., Kumar, G., Poonam, and Singh, C.P. (2010). Fast, easy ethanolysis of coconut oil for biodiesel production assisted by ultrasonication. *Ultrason. Sonochem.* 17: 555–559.

116 Van Manh, D., Chen, Y.-H., Chang, C.-C. et al. (2011). Biodiesel production from Tung oil and blended oil via ultrasonic transesterification process. *J. Taiwan Inst. Chem. Eng.* 42: 640–644.

117 Manh, D.-V., Chen, Y.-H., Chang, C.-C. et al. (2012). Effects of blending composition of tung oil and ultrasonic irradiation intensity on the biodiesel production. *Energy* 48: 519–524.

118 Manh, D.-V., Chen, Y.-H., Chang, C.-C. et al. (2012). Parameter evaluation of biodiesel production from unblended and blended Tung oils via ultrasound-assisted process. *J. Taiwan Inst. Chem. Eng.* 43: 368–373.

119 Thanh, L.T., Okitsu, K., Sadanaga, Y. et al. (2010). Ultrasound-assisted production of biodiesel fuel from vegetable oils in a small scale circulation process. *Bioresour. Technol.* 101: 639–645.

120 Encinar, J.M., González, J.F., and Pardal, A. (2012). Transesterification of castor oil under ultrasonic irradiation conditions. Preliminary results. *Fuel Process. Technol.* 103: 9–15.

121 Hingu, S.M., Gogate, P.R., and Rathod, V.K. (2010). Synthesis of biodiesel from waste cooking oil using sonochemical reactors. *Ultrason. Sonochem.* 17: 827–832.

122 Thanh, L.T., Okitsu, K., Sadanaga, Y. et al. (2010). A two-step continuous ultrasound assisted production of biodiesel fuel from waste cooking oils: a practical and economical approach to produce high quality biodiesel fuel. *Bioresour. Technol.* 101: 5394–5401.

123 Kumar, G. and Kumar, D. (2013). Monitoring of base catalyzed ethanolysis of Jatropha curcas oil by reversed phase high performance liquid chromatography assisted by ultrasonication. *J. Environ. Chem. Eng.* 1: 962–966.

124 Paiva, E.J.M., da Silva, M.L.C.P., Barboza, J.C.S. et al. (2013). Non-edible babassu oil as a new source for energy production – a feasibility transesterification survey assisted by ultrasound. *Ultrason. Sonochem.* 20: 833–838.

125 Teixeira, L.S.G., Assis, J.C.R., Mendonça, D.R. et al. (2009). Comparison between conventional and ultrasonic preparation of beef tallow biodiesel. *Fuel Process. Technol.* 90: 1164–1166.

126 Kelkar, M.A., Gogate, P.R., and Pandit, A.B. (2008). Intensification of esterification of acids for synthesis of biodiesel using acoustic and hydrodynamic cavitation. *Ultrason. Sonochem.* 15: 188–194.

127 Badday, A.S., Abdullah, A.Z., and Lee, K.-T. (2013). Optimization of biodiesel production process from Jatropha oil using supported heteropolyacid catalyst and assisted by ultrasonic energy. *Renew. Energy* 50: 427–432.

128 Badday, A.S., Abdullah, A.Z., and Lee, K.-T. (2013). Ultrasound-assisted transesterification of crude Jatropha oil using alumina-supported heteropolyacid catalyst. *Appl. Energy* 105: 380–388.

129 Badday, A.S., Abdullah, A.Z., and Lee, K.-T. (2013). Ultrasound-assisted transesterification of crude Jatropha oil using cesium doped heteropolyacid catalyst: interactions between process variables. *Energy* 60: 283–291.

130 Badday, A.S., Abdullah, A.Z., and Lee, K.-T. (2014). Transesterification of crude Jatropha oil by activated carbon-supported heteropolyacid catalyst in an ultrasound-assisted reactor system. *Renew. Energy* 62: 10–17.

131 Verziu, M., Coman, S.M., Richards, R., and Parvulescu, V.I. (2011). Transesterification of vegetable oils over CaO catalysts. *Catal. Today* 167: 64–70.

132 Choedkiatsakul, I., Ngaosuwan, K., and Assabumrungrat, S. (2013). Application of heterogeneous catalysts for transesterification of refined palm oil in ultrasound-assisted reactor. *Fuel Process. Technol.* 111: 22–28.

133 Choudhury, H.A., Chakma, S., and Moholkar, V.S. (2014). Mechanistic insight into sonochemical biodiesel synthesis using heterogeneous base catalyst. *Ultrason. Sonochem.* 21: 169–181.

134 Deshmane, V.G. and Adewuyi, Y.G. (2013). Synthesis and kinetics of biodiesel formation via calcium methoxide base catalyzed transesterification reaction in the absence and presence of ultrasound. *Fuel* 107: 474–482.

135 Yu, D., Tian, L., Wu, H. et al. (2010). Ultrasonic irradiation with vibration for biodiesel production from soybean oil by Novozym 435. *Process Biochem.* 45: 519–525.

136 Gharat, N. and Rathod, V.K. (2013). Ultrasound assisted enzyme catalyzed transesterification of waste cooking oil with dimethyl carbonate. *Ultrason. Sonochem.* 20: 900–905.

137 Wang, J. (2007). Lipase-catalyzed production of biodiesel from high acid value waste oil using ultrasonic assistant. *Chin. J. Biotechnol.* 23: 1121–1128.

8

Microwave- and Ultrasound-Assisted Synthesis of Polymers

Anupama Singh[1], Sushil K. Sharma[2], and Shobhana Sharma[1]

[1] S.S. Jain Subodh P.G. College, Department of Chemistry, Jaipur, Rajasthan 324005, India
[2] University of Kota, Department of Pure and Applied Chemistry, Kota, Rajasthan 324005, India

8.1 Introduction

Polymers play an important role in our everyday life as they make our routine life more flexible in various ways [1, 2]. Synthetic polymers are highly flexible and possess more strength, inertness, and resistivity [3, 4]. The synthetic route of synthetic polymers is quite difficult and time-consuming, which restricts the use of polymers. Moreover, the nonbiodegradable nature of synthetic polymers makes permanent waste to them [5, 6]. To minimize all these problems, researchers started using microwave (MW) and ultrasound-assisted techniques. The use of microwave- and ultrasound-assisted techniques led to new innovations in polymer chemistry [7]. Recently, researchers used ionization radiation in the polymerization method and a new emerging field of polymer radiation chemistry arose [8]. Ionizing radiation is a feasible technology to synthesize economic, efficient, and safe polymers [9]. Several types of radiation sources are used in radiation techniques for polymerization processes. The different ionizing radiations generate reactive intermediates to initiate polymerization reactions. The various reaction paths were followed by the intermediates, which led to new bond formation and rearrangement. These reactive intermediates formed by irradiation induce various reactions. The ionizing radiation developed cross-linking at room temperature in polymers, and the degree of cross-linking was dependent on radiation dose [10, 11]. Polymer synthesis through microwave irradiation is the simplest and easiest way. It is a well-established technique for promoting and driving various chemical reactions of polymers. Microwave assistance provides clean synthesis and improves reaction rates and the purity of the product [12]. It is economical, highly selective, and provides more yield. Microwave assistance leads to uniform distribution of heat, and so it reduces reaction time [13]. The main limitation of microwave-assisted technique is that microwave irradiation restricts the penetration depth of absorbing substance [14]. The microwave-assisted synthesis is based on aligning dipoles of the substance in an external field. The technique leads to internal heat production for

Green Chemical Synthesis with Microwaves and Ultrasound, First Edition.
Edited by Dakeshwar Kumar Verma, Chandrabhan Verma, and Paz Otero Fuertes.
© 2024 WILEY-VCH GmbH. Published 2024 by WILEY-VCH GmbH.

the required chemical reaction. Ultrasonic irradiation develops mechanical effects at microscopic and macroscopic levels [15]. The ultrasound-assisted technology in the future will lead to remarkable improvements in various industries [16]. Sonochemistry is quite important for mechanistic and synthetic investigations [17]. The main limitation of using sonochemistry is that ultrasound production from electricity is quite difficult. The ultrasound-assisted synthesis basically creates physical and chemical effects [18]. The physical effect is acoustic cavitations which destroy the intermolecular force of attraction of the liquid [19]. The acoustic cavitation causes the production, growth, and crash of bubbles which ultimately creates physical and chemical effects [20, 21]. The microwave heating and ultrasound-assisted techniques lead to a decrease in reaction time and also contribute to green and sustainable chemistry. Both microwave and ultrasonic methods promoted the chemical reaction, but the simultaneous use of microwaves and ultrasound waves proved to be far more effective in reducing the reaction time and increasing the chemical efficiency of the final product. Thus, microwave and ultrasonic radiation are better than the conventional heating method.

In the present article, attempts have been made to summarize the various methods used in the synthesis of different polymers under microwave and ultrasound assistance. The article also explains the variation in quality, and yield of polymers by altering the time of irradiation. This article provides a fruitful pathway for researchers working on microwave- and ultrasound-assisted polymer synthesis. This article shows that the use of these techniques is far better than old conventional methods.

8.2 Microwave-Assisted Synthesis of Polymers

Microwave technologies are more useful as compared to the conventional method. It increases the desired temperature in a short span, shortens the reaction time, and the product obtained is of the highest purity [22]. It also increases the molecular mass of the polymer with its yield [23]. Various types of polymerization reactions can be carried out in MW oven, whether it is traditional or modern, as shown in Figure 8.1. In MW heating, materials are exposed to electromagnetic radiations between 300 MHz and 300 GHz, having wavelengths between 1 m and 1 mm. As per their dielectric properties, certain organic materials, including polymers, interact with microwaves, and it results in a rise in the temperature [24]. In microwave-assisted synthesis, there are no chances of overheating the product and the reaction is eco-friendly [25]. Zhang et al. synthesized a biodegradable, edible polymer that is starch-based and can absorb a lot of water. This type of material is used for various applications as it has gel-like properties. For e.g. starch-grafted PGA can be used to control the release of lipophilic and hydrophilic bioactive agents in the GI tract [26]. Tally et al. synthesized a pH-sensitive superabsorbent hydrogel composite, sodium alginate-g-poly(acrylic acid-*co*-acrylamide) by using polyvinyl pyrrolidone (PVP) by microwave technique. Here, potassium peroxodisulfate (KPS) was used as an initiator [27]. Novel soil conditioners called SAP are synthesized

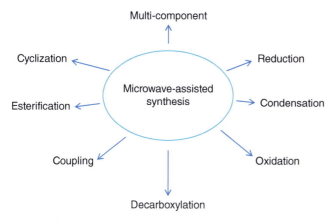

Figure 8.1 Various types of microwave-assisted synthesis of polymer.

by this technique. Sapare cross-linked polymerscan absorb large amounts of water and can release them to the soil slowly [28, 29]. Recently, various nano-composites associated with metal particles have also been synthesized by this technique [30, 31]. These types of polymers have vast applications ranging from optical fibers to sensors. These types of materials are very sensitive for DNA screening because of their flexibility and redox properties [32, 33]. MW-assisted reactions also depend on the ability of the reaction mixture to absorb MW energy and also upon the type of solvent used. The ability of a specific solvent to convert MW energy into heat is determined by the loss of tangents (δ). The higher the tan δ value, the better the solvent for MW-assisted synthesis, and for heating efficiency. The solvents employed for MW-assisted chemistry are classified as high (tan δ > 0.5), medium (0.1 < tan δ < 0.5), and low microwave absorbing (tan δ < 0.1), as shown in Table 8.1.

The main topics of investigations can be divided into four areas. In this book chapter, we will divide the polymerization reactions into four basic parts: radical

Table 8.1 Loss tangents of frequently used solvents in microwave-assisted reactions (2.45 GHz, 20 °C).

Solvent	tan δ	Solvent	tan δ
Ethylene glycol (EG)	1.350	1,2-Dichloroethane	0.127
Ethanol	0.941	Water	0.123
Dimethyl sulfoxide (DMSO)	0.825	Chloroform	0.091
Methanol	0.659	Acetonitrile	0.062
1,2-Dichlorobenzene	0.280	Tetrahydrofuran	0.047
Ethylene glycol (EG)	1.350	1,2-Dichloroethane	0.127
Ethanol	0.941	Water	0.123
Dimethyl sulfoxide (DMSO)	0.825	Chloroform	0.091

polymerizations, step-growth polymerizations, ring-opening polymerizations, and polymer modifications.

(a) **Radical polymerizations**: Free radical polymerization is an important technique used in industrial polymerization which produces polymers of high molecular weight. Polymers like polyethylene, polystyrene, poly(vinyl chloride), poly(vinyl acetate), poly(acrylonitrile), poly(methyl methacrylate) (PMMA), and poly(tetrafluoroethylene) can be produced by radical polymerization reaction [34–38]. Porto et al. reported the free radical polymerization reaction of vinyl acetate, styrene, and acrylonitrile in a domestic oven [35]. Radical polymerization can be subdivided into four important parts: free radical copolymerization, emulsion polymerizations, preparation of composite materials, and controlled radical polymerizations. Free radical polymerization is a versatile route for synthesizing a vast variety of different important polymer composites. Goretzki et al. reported the direct solvent-free amidation of methyl acrylic acid with aliphatic and aromatic amines in the year 2004 [36]. This was done by irradiating the reactants for 30 minutes and approximately 96% of the yield was obtained. In the same way, (meth)acrylic acid and (R)-1-phenylethylamine on MW irradiation give chiral (meth)acrylamides [37]. Stange et al. reported the copolymerization of styrene and MMA with MW irradiation by using various organic peroxides as initiators [39]. He reported an enhanced monomer conversion. Various groups of scientists have reported the preparation of graft copolymers. Singh et al. reported the polymerization of MMA, acrylamide, and acrylonitrile in a domestic MW [40–42]. MW technique was intensively studied on styrene and MMA under the technique of emulsion polymerization. Wu and Guo studied the emulsion polymerization of styrene with sodium dodecyl sulfate and potassium persulfate in a domestic MW oven [43].

(b) **Step-growth polymerizations**: In this type of polymerization reaction, bifunctional or multifunctional monomers react to form dimmers, then to trimers, long oligomers, and finally get converted to long-chain polymers. In recent years MW assisted esterification, amidation, imidation, and polymers from metal-catalyzed cross-coupling reactions have been widely studied and reported. As per Velmathi et al., polymerization of poly(butylenes succinate) was done from butane-1,4-diol and succinic acid in the presence of 1,3-dichloro1,1,3,3-tetrabutyldistannoxane by irradiating in MW for 20 minutes, and later on the yield and molecular weight was compared from traditional methods [44]. Huang et al. prepared polyaspartic acid from aspartic acid under solvent-free and catalyst-free conditions by irradiating with MW radiations for a few minutes. Ammonia with maleic anhydride was added to the mixture, and then the whole mixture was hydrolyzed under basic conditions to polyaspartic acid [45].

(c) **Ring-opening polymerizations**: It is a form of chain growth polymerization in which a polymer terminus attacks cyclic monomer to form a long polymer. Reactants can be radical, anionic, or cationic. It is the most widely adopted method to produce biopolymers. Leuchs synthesized polypeptide in 1906 and this is the oldest reported history of ring-opening polymerization [46]. A lot of research is carried out on cyclic ethers, acetals, amides (lactams), esters (lactones), and siloxanes by MW and it is of commercial interest too. Lactides are being used to prepare polyesters by ROP in MW-assisted synthesis and it is the most attractive research because of their biocompatibility and thermal properties [47–49]. In polymer chemistry, a metal-catalyzed cross-coupling reaction is the most favorable method to produce carbon-to-carbon and carbon-to-heteroatom bonds. This method is mostly used to synthesize conjugated polymers which are very important as they are used in electronic devices. Nehls et al. reported the first microwave-assisted synthesis of semiconducting polymer [50]. They used palladium-catalyzed Suzuki and Stille reactions. Khan and Hect reported the dependency of solvent in MW-assisted synthesis of polycondensation of poly(m-phenylene ethylene)s [51].

(d) **Polymer modifications**: Polymer modification plays an important role in polymer chemistry as it is used to produce the polymer with desired properties [52, 53]. Polymer modifications are performed to imbibe desired properties like thermal stability, biological resistance, flexibility, degradability, and compatibility with the modified material [54, 55]. Azides and alkynes are reported to produce polymers of high efficiency by this method.

Kretschmann et al. mentioned the synthesis of hydrophobically modified gelatinous, poly(acrylic acid) by irradiating poly(acrylic acid) and aminoalkyl-functionalized adamantanes by MW radiations for 20 minutes [56]. Metallic and bimetallic cross-linked polyvinyl alcohol nanocomposites were also prepared by MW irradiation [57].

8.3 Ultrasound-Assisted Synthesis of Polymers

There are numerous research papers reported on ultrasound assisted synthesis of polymers. Table 8.2 contains the various reported ultrasound-assisted polymer synthetic methods. Bhanvase et al. reported a novel ultrasound-assisted in situ semi-batch emulsion polymerization method for the synthesis of PANI/ZnMoO$_4$ nanocomposite [58]. The ultrasonic irradiation enhances the rate of diffusion of aniline, and the formation of radicals due to this the rate of polymerization increases. The ultrasound generates a cavitation jet which allows myristic acid functionalized ZnMoO$_4$ nanocomposite to mix with aniline. Further, ultrasound irradiations help in the formation of core–shell structure of PANI/ZM nanocomposite. The rate of polymerization improves with an increase in ultrasonic irradiation time. At low temperatures, intense cavity collapse and more free radicals form. A novel 3D supramolecular metal–organic coordination framework was suggested by Mirtamizdoust et al. [59]. The zig-zag one-dimensional polymer [Pb(1-2-pinh)N$_3$H$_2$O]$_n$, by weak interaction with neighboring atoms, creates a 3D supramolecular structure. The thermolysis of this one-dimensional polymer at 180 °C temperature converts nano-cauliflower-shaped lead(II) polymer into oxide. The polymer synthesizes through ultrasonic assistance from Schiff-base ligand and

Table 8.2 The use of various ultrasound-assisted polymer synthetic methods for the preparation of different types of polymers under variable conditions.

S. No.	Process	Reported by	Polymer	Precursor/Monomer	Conditions
1.	Ultrasound-assisted in situ semi-batch emulsion polymerization	Bhanvase et al. (2015)	PANI/ZnMoO$_4$	Myristic acid functionalized ZnMoO$_4$ and polyaniline	Placed in an ultrasonicator for 70 min at 4 °C
2.	Ultrasonic method	Mirtamizdoust et al. (2016)	[Pb(l-2-pinh)N$_3$H$_2$O]$_n$	2-Pyridine carbaldehyde isonicotinoyl hydrazone, lead(II) acetate and sodium azide	Ultrasonic probes operate at 20 kHz and 600 W
3.	Piezochemical activated polymerization	Mohapatra et al. (2016)	Poly(n-butyl acrylate)	Acrylate monomers, Cu(II) based catalytic precursor	Alternating 4 s applied ultrasound irradiation at a temperature of 15–25 °C
4.	Ultrasound-assisted in situ emulsion polymerization	Poddar et al. (2016)	PMMA/ZnO nanocomposites	Methyl methacrylate, zinc acetate dehydrate, sodium hydroxide	Ultrasound processor having a frequency of 20 kHz and 200 W power, reaction time 1 h
5.	In situ sonochemical oxidative polymerization	Hemmati et al. (2017)	PPy-DBSNa/ZnO nanocomposite, …	ZnO-NPs, Pyrrole, sodium dodecylbenzene sulfonate (DBSNa), anhydrous iron(III) chloride (FeCl$_3$)	500 rpm stirring rate for 2 h
6.	Esterification reaction	Li et al. (2017)	PVA/PA polymer	Polyvinyl alcohol and phytic acid	3 h of ultrasound irradiation
7.	One-pot ultrasonic irradiation	Huang et al. (2018)	PEG-functionalized FCNs	PEG-SH and FCNs	Ultrasonically heating for 30 min
8.	In situ polymerization	Cherifi et al. (2018)	Poly(GMA)/maghnite	GMA, organophilic clay	Ultrasound for 1 h

No.	Method	Reference	Material	Components	Conditions
9.	High high-energy ultrasound probe method	Lahcen et al. (2019)	(Mag-MIPs) and US-magMIPs	Sulfamethoxazole, methacrylic acid, ethylene glycol dimethacrylate	Ultrasonicate for 2 h at 65 °C under nitrogen
10.	Ultrasound initiated reversible addition-fragmentation chain transfer dispersion polymerization	Wan et al. (2020)	Poly(ethylene glycol) trithiocarbonate (PEG113-CDTPA)macro-CTA	Poly(ethylene glycol), 4-cyano-4 (((dodecylthio)carbonothioyl)thio) pentanoic acid (CDTPA)	Ultrasonic irradiation for 1 h
11.	Ultrasound-assisted solution casting technique	Swaminathan et al. (2020)	PVA/PAM blend polymer	Poly(vinyl alcohol), poly(acrylamide), ammonium thiocyanate	Ultrasound frequency of about 20 kHz required for 20 min
12.	Sonochemical method	Dadashi et al. (2021)	Polymer, $[Pb(L)_2(CH_3COO)]_n$	Pyridine-4-carbaldehyde thiosemicarbazone and lead(II) acetate	High-density ultrasonic probe used
13.	In situ-sono-PISA self-assembly	Wan et al. (2021)	Copolymer nano-spheres PEG_{113}-b-$PDMAEMA_{24}$-b-$PHPMA_n$	DMAEMA	Sono-PISA requires room temperature and a 990 kHz ultrasonic reactor
14.	Ultrasonic method	Cherednichenko et al. (2022)	NFC/PS composites	Styrene monomer and nanofibrillar cellulose	Ultrasonicate for several minutes and hours
15.	Ultrasonic method	Afzal et al. (2022)	Ultrasound gel	Carbopol 980, methylparaben, propylparaben	Probe having a 4.5 MHz frequency
16.	Ultrasound probe-assisted synthesis	Lamaoui et al. (2022)	magNIP	Bisphenol-A	Ultrasonication for 9 min
17.	One-pot sono-RAFT process	Padmakumar et al. (2023)	CCS polymer	N-acryloyl morpholine and N,N_0-dimethylacrylamide	Ultrasonic vessel for 4 h at 840 kHz/80 W

sodium azide. The IR spectra and single-crystal X-ray elucidate the structure of the polymer. The theoretical studies determine the electronic properties and the structure of the polymer. Mohapatra et al. described a method in which mechanical energy of ultrasonic agitation was used in the polymerization of acrylate monomers [60]. $BaTiO_3$ nanoparticles act as piezoelectric transducer which is required to control ultrasound-mediated radical polymerization. This is a piezochemical-activated polymerization that applies the conversion of mechanical energy into polymer synthesis. The activator of polymerization is generated through piezocatalytic reduction of Cu(II)-based catalytic precursor.

The two-step in situ ultrasonic method was analyzed by Poddar et al. for the preparation of PMMA/ZnO nanocomposites [61]. In the first step, zinc acetate is converted into nanocomposites of ZnO through an in situ ultrasound-assisted hydrolysis process. The in situ ultrasound-assisted emulsion polymerization process leads to the preparation of PMMA/ZnO nanocomposites from ZnO nanocomposites in the second step. The main advantage of this method is that nanoscale ZnO particles are uniformly encapsulated and dispersed in the matrix of polymethyl methacrylate. The sonication provides fine emulsification and intense micromixing due to which kinetics of polymerization are enhanced at moderate temperature. Hemmati et al. prepared PPy-DBSNa/ZnO (polypyrrole-sodium dodecylbenzene-sulfonate/zinc oxide) nanocomposite through in situ sonochemical oxidative polymerization, in which ZnO nanoparticle cores possess the effective coating of pyrrole [62]. This nanocomposite acts as an efficient nanosorbent which is used in ultrasound meliorated dispersive micro solid-phase extraction of anti-hypertensive drugs with high-performance liquid chromatography ultraviolet detection. This technique is helpful in pre-concentrating the three anti-hypertensive drugs present in human plasma and urine matrices. The polyvinyl alcohol/phytic acid (PVA/PA) composite was synthesized by Li et al. from monomeric units of polyvinyl alcohol and phytic acid through ultrasound-assisted esterification reaction [63]. The alteration in mechanical properties of PVA/PA polymeric film is observed on changing sonication time. The hydroxyl group of polyvinyl alcohol and the phosphonic group of phytic acid combine to form a covalent phosphonate bond. The solidification processes lead to the formation of PVA/PA polymer film. The formation of ether bonds, phosphonate bonds, and hydrogen bonds enhances the thermal stability and ductility of polymeric film whereas the surface resistivity of polymeric film is remarkably reduced. Huang et al. demonstrated a highly efficient technique for the preparation of PEG-functionalized FCN polymer [64]. This polymer was synthesized from fluorescent carbon nanoparticles (FCNS) obtained from cigarette ash and thiol group-containing poly(ethylene oxide) (PEG-SH), having a thiol group through one-pot ultrasonic treatment. The FCNS composite exhibited cytotoxicity against L929 cells. The interesting feature of this research is the use of cigarette ash which acts as an economic and renewable carbon source. Moreover, organoclay nanocomposites are synthesized and characterized by Cherifi et al. through two different ultrasonic-assisted methods [65]. In the first method, GMA and the organoclay in situ polymerize, and in the second method, poly(GMA) blend with the help of ultrasound assistance. The poly(GMA) nanocomposites show improvement in quality, yield, and thermal properties in comparison to the pure blend poly(GMA). Another simple and fast high-energy ultrasound probe method was investigated by Lahcen et al. for the synthesis of magnetic molecularly imprinted and non-imprinted polymers (Mag-MIPs and US-magMIPs) [66]. These polymers were developed from high-energy ultrasound probes. US-magMIPs show higher selectivity for sulfamethoxazole templates than others like sulfacetamide, sulfadiazine, and sulfamerazine. This method is a time as well as an energy saver as compared to the conventional method. The several morphologies of block copolymer were investigated by Wan et al. through green synthesis [67]. They developed a block copolymer by ultrasound-initiated reversible addition-fragmentation chain transfer dispersion polymerization. In this process, water was used as a solvent and source of initiator. The ultrasound-assisted solution casting technique was reported by Swaminathan et al. for the structural evaluation of PVA/PAM blend polymer electrolytes which contain different ratios [68]. The alteration in electrical, structural, and electrochemical properties was observed with the addition of variable weight percentage of ionic dopant (ammonium thiocyanate) in PVA/PAM (50 : 50) blend film in the presence of ultrasound. In the presence of ultrasound-assistance PVA/PAM, the blend polymer formed shows high conductivity with those that form in the absence of ultrasound irradiation. The sonochemical method discussed by Dadashi et al. involves the generation of 1D polymer, $[Pb(L)_2(CH_3COO)]_n$, in which lead(II) acetate interacts with pyridine-4-carbaldehyde thiosemicarbazone [69]. The ultrasound technique is employed to investigate the nanostructure of the polymer in an aqueous solution, and the single crystal X-ray analysis of this polymeric structure was evaluated through a heat gradient. The branched-tube method decides the value of the heat gradient in an aqueous solution. Further, PbO clusters were achieved from the polymer through the thermolysis method at 180 °C in which oleic acid acts as a surfactant. Polymer–metal nanocomposites exhibited applications in various fields. Wan et al. reported a polymer–metal nanocomposites with high catalytic activity through in situ sono-PISA self-assembly [70]. This method contains ultrasound as the reducing source and initiator. The macro-RAFT agent having poly[2-(dimethylamino)ethyl methacrylate] (DMAEMA), under ultrasound-assisted self-assembly generates nanotriblock copolymer. The triblock copolymer nanocomposite contains tertiary amine groups. The polymer acts as a scaffold for the in situ reduction of Au and Pd ions. The polymer generates radicals through the sonolysis of water in the absence of any reducing agents. The pickering emulsion polymerization method was proposed by Cherednichenko et al. to prepare nanofibrillar cellulose/polystyrene (NFC/PS) composites [71]. Two methods were adopted and compared for the synthesis of NFC/PS composites in which sonication time varied. The sonication increases the stability of styrene-water emulsion and enhances the rate of polymerization which decreases the reaction time. The composite possesses improved thermal and mechanical properties. Apart from this, the polymer also shows high wetting performance and biodegradability. Ultrasound imaging is a very important technique available in hospitals and diagnostic centers. During ultrasound imaging, a gel is used which acts as a coupling medium in the whole procedure and replaces air from the transducer and the skin.

It is difficult for ultrasound waves to travel through the air and gel facilitates this traveling. Afzal et al. performed a comparative study on commercially available ultrasound gel and formulated gel in which commercially available gel is employed as a control [72]. The different concentrations of methylparaben and carbopol 980 led to the nine different samples of formulated gel. The economic formulate gel provides the alternative to costly commercial ultrasound gel followed in hospitals and diagnostic centers. On the basis of computational studies, Lamaoui et al. evaluated the appropriate monomer and porogen solvent for designing molecularly imprinted polymer (MIP) of bisphenol-A [73]. They developed a very easy and rapid technique based on a high-power ultrasound probe for the synthesis of magMIP. The theoretical studies predict that methyl methacrylate and acetone were appropriate to synthesize molecularly imprinted polymer. The synthesized molecularly imprinted polymer possesses rapid and high adsorptive power with good selectivity. Sips and pseudo-second-order models were employed to study the adsorption isotherm and kinetics of magMIP. A core-crosslinked star (CCS) polymer having a tri-block arm was synthesized by Padmakumar et al. which shows high arm-to-star conversion [74]. The one-pot sono-RAFT process is used in two-step to prepare CCS polymer in which N,N_0-dimethylacrylamide forms the arms and N,N_0-methylenebis(acrylamide) acts as a crosslinking agent. In the sono-RAFT method, the polymerization is studied under certain variable factors like power supply, and crosslinker-to-arm ratios. The variable applied power decreases the reaction time. A core-crosslinked star polymer develops from the triblock arm and evaluates a very high arm-to-star conversion ratio.

8.4 Conclusion

In this book chapter, the recent developments of microwave- and ultrasound-assisted synthesis of polymers are outlined. By controlling certain specific parameters like pressure, temperature, and choice of solvent, a vast variety of polymers with good yield can be synthesized and it can prove to be a remarkable eco-friendly technique in the polymer industry. A lot of research is being carried out in this field and that shows that nearly all types of polymerization reactions can be carried out under them with some exceptions. A comparative study between the traditional conventional method shows better results in MW and ultrasonic method as these reactions take less time, require lesser amount of reactant, give good yield, provide solvent-free reaction, and are eco-friendly too. One of the biggest challenges in this field is the regulation and calibration of temperature. Optical fibers and certain specifically designed thermometers are used to control the temperature, but there is a need to develop more precise equipment for temperature control and calibration during microwave heating and especially during polymer synthesis. Using the ultrasound method of polymer synthesis improves the overall reaction rate and also provides a higher degree of polymerization. Sonochemistry is essentially used in the case of chain growth polymerization reactions including reversible-deactivation radical polymerization (RDRP) techniques and novel applications. This technique does not involve any toxic chemical initiators and also has complete control of the polymerization process. It is a novel, clean, and green technology that needs further research by coupling with various thermo-, mechano-, and photochemical stimuli. It has a vast potential to be applied in industrial setups.

References

1 He, X., Wang, W., Yang, S. et al. (2023). Adhesive tapes: from daily necessities to flexible smart electronics. *Appl. Phys. Rev.* 10 (1). https://doi.org/10.1063/5.0107318.

2 Baruah, R.K., Yoo, H., and Lee, E.K. (2023). Interconnection technologies for flexible electronics: materials, fabrications, and applications. *Micromachines* 14 (6): 1131. https://doi.org/10.3390/mi14061131.

3 Javid-Naderi, M.J., Behravan, J., Karimi-Hajishohreh, N., and Toosi, S. (2023). Synthetic polymers as bone engineering scaffold. *Polym. Adv. Technol.* https://doi.org/10.1002/pat.6046.

4 Yuce-Erarslan, E., Domb, A.A.J., Kasem, H. et al. (2023). Intrinsically disordered synthetic polymers in biomedical applications. *Polymers* 15 (10). 2406. https://doi.org/10.3390/polym15102406.

5 Mangal, M., Rao, C.V., and Banerjee, T. Bioplastic: an eco-friendly alternative of non-biodegradable plastic. *Polym. Int.* https://doi.org/10.1002/pi.6555.

6 Kumar, R., Sadeghi, K., Jang, J., and Seo, J. (2023). Mechanical, chemical, and bio-recycling of biodegradable plastic: a review. *Sci. Total Environ.* 163446. https://doi.org/10.1016/j.scitotenv.2023.163446.

7 Pandey, T., Sandhu, A., Sharma, A., and Ansari, M.J. (2023). Recent advances in applications of sonication and microwave. In: *Ultrasound and Microwave for Food Processing* (ed. G.A. Nayik, M. Ranjha, X.A. Zeng, et al.), 441–470. Elsevier. https://doi.org/10.1016/B978-0-323-95991-9.00003-5.

8 Pal, T.S. and Singha, N.K. (2022). Radiation-induced polymer modification and polymerization. In: *Radiation Technologies and Applications in Materials Science* (ed. S.R. Chowdhury), 99–122. CRC Press.

9 Salmieri, S., Leila, B., and Monique, L. (2022). Using ionizing technologies on natural compounds and wastes for the development of advanced polymers and active packaging materials. In: *Ionizing Radiation Technologies: Managing and Extracting Value from Wastes* (ed. S. Shayanfar and S.D. Pillai), 180–209. Wiley. https://doi.org/10.1002/9781119488583.ch11.

10 Kolhe, A., Chauhan, A., and Dongre, A. (2022). A review on various methods for the cross-linking of polymers. *Res. J. Pharm. Dosage Forms Technol.* 183–188. http://dx.doi.org/10.52711/0975-4377.2022.00003.

11 Bai, Y., Yang, C., Yuan, B. et al. (2023). A UV cross-linked gel polymer electrolyte enabling high-rate and high voltage window for quasi-solid-state supercapacitors. *J. Energy Chem.* 76: 41–50. https://doi.org/10.1016/j.jechem.2022.09.015.

12 Swamy, K.M.K., Yeh, W.B., Lin, M.J., and Sun, C.M. (2003). Microwave-assisted polymer-supported combinatorial synthesis. *Curr. Med. Chem.* 10 (22): 2403–2423. https://doi.org/10.2174/0929867033456594.

13 Bao, C., Serrano-Lotina, A., Niu, M. et al. (2023). Microwave-associated chemistry in environmental catalysis for air pollution remediation: a review. *Chem. Eng. J.* 466: 142902. https://doi.org/10.1016/j.cej.2023.142902.

14 Kalinke, I., Kubbutat, P., Taghian Dinani, S. et al. (2022). Critical assessment of methods for measurement of temperature profiles and heat load history in microwave heating processes—a review. *Compr. Rev. Food Sci. Food Saf.* 21 (3): 2118–2148. https://doi.org/10.1111/1541-4337.12940.

15 Chen, R., Zheng, D., Ma, T. et al. (2017). Effects and mechanism of ultrasonic irradiation on solidification microstructure and mechanical properties of binary TiAl alloys. *Ultrason. Sonochem.* 38: 120–133. https://doi.org/10.1016/j.ultsonch.2017.03.006.

16 Kiss, A.A., Geertman, R., Wierschem, M. et al. (2018). Ultrasound-assisted emerging technologies for chemical processes. *J. Chem. Technol. Biotechnol.* 93 (5): 1219–1227. https://doi.org/10.1002/jctb.5555.

17 Qi, K., Zhuang, C., Zhang, M. et al. (2022). Sonochemical synthesis of photocatalysts and their applications. *J. Mater. Sci. Technol.* 123: 243–256. https://doi.org/10.1016/j.jmst.2022.02.019.

18 Ranjan, A., Singh, S., Malani, R.S., and Moholkar, V.S. (2016). Ultrasound-assisted bioalcohol synthesis: review and analysis. *RSC Adv.* 6 (70): 65541–65562. https://doi.org/10.1039/C6RA11580B.

19 Kalva, A., Sivasankar, T., and Moholkar, V.S. (2009). Physical mechanism of ultrasound-assisted synthesis of biodiesel. *Ind. Eng. Chem. Res.* 48 (1): 534–544. https://doi.org/10.1021/ie800269g.

20 Luo, J., Fang, Z., Smith, R.L., and Qi, X. (2015). Fundamentals of acoustic cavitation in sonochemistry. In: *Production of Biofuels and Chemicals with Ultrasound* (ed. Z. Fang, R.L. Smith, and X. Qi), 3–33. Springer.

21 Hasaounia, I., Mazouz, D., and Kerboua, K. (2022). Physical effects and associated energy release. In: *Energy Aspects of Acoustic Cavitation and Sonochemistry* (ed. O. Hamdaoui and K. Kerboua), 35–49. Elsevier. https://doi.org/10.1016/B978-0-323-91937-1.00018-9.

22 Loupy, A. (2006). *Microwaves in Organic Synthesis*, 2e. Weinheim: Wiley-VCH.

23 Kappe, C.O. and Stadler, A. (2005). *Microwaves in Organic and Medicinal Chemistry*, 1e. Weinheim: Wiley-VCH.

24 Kappe, C.O. and Dallinger, D. (2006). The impact of microwave synthesis on drug discovery. *Nat. Rev. Drug Discovery* 5: 51. https://doi.org/10.1038/NRD1926.

25 Tselinskii, I.V., Brykov, A.S., and Astrat'ev, A.A. (1996). Effect of microwave heating on organic reactions of different types. *Russ. J. Gen. Chem.* 66: 1653.

26 Zhang, Z., Zhang, R., Chen, L. et al. (2015). Designing hydrogel particles for controlled or targeted release of lipophilic bioactive agents in the gastrointestinal tract. *Eur. Polym. J.* 72: 698–716. https://www.researchgate.net/publication/276263513.

27 Tally, M. and Atassi, Y. (2015). Optimized synthesis and swelling properties of a pH-sensitive semi-IPN superabsorbent polymer based on sodium alginate-g-poly(acrylic acid-*co*-acrylamide) and polyvinylpyrrolidone and obtained via microwave irradiation. *J. Polym. Res.* 22: 181. https://doi.org/10.1007/s10965-015-0822-3.

28 Brandon-Peppas, L. and Harland, R.S. (1990). *Absorbent Polymer Technology*. Amsterdam: Elsevier.

29 Pó, R. (1994). Water-absorbent polymers: a patent survey. *J. Macromol. Sci., Part C: Polym. Rev.* 34: 607.

30 Kampf, G., Dietze, B., Große-Siestrup, C. et al. (1998). Microbicidal activity of a new silver-containing polymer, SPI-ARGENT II. *Antimicrob. Agents Chemother.* 42: 2440–2442.

31 Fang, N., Lee, H., Sun, C., and Zhang, X. (2005). Sub-diffraction-limited optical imaging with a silver superlens. *Science* 308: 534–537.

32 Sun, Y., Gates, B., Mayers, B., and Xia, Y. (2002). Crystalline silver nanowires by soft solution processing. *Nano Lett.* 2: 165–168.

33 Natan, M.J. (2001). Submicrometer metallic barcodes. *Science* 294: 137–141.

34 Gizdavic-Nikolaidis, M.R., Pupe, J.M., Jose, A. et al. (2023). Eco-friendly enhanced microwave synthesis of polyaniline/chitosan-AgNP composites, their physical characterisation and antibacterial properties. *Synth. Met.* 293: 117273. https://doi.org/10.1016/j.synthmet.2022.117273.

35 Porto, A.F., Sadicoff, B.L., Amorim, M.C.V., and de Mattos, M.C.S. (2002). Microwave-assisted free radical bulk-polyaddition reactions in a domestic microwave oven. *Polym. Test.* 21: 145. https://doi.org/10.1016/S0142-9418(01)00061-7.

36 Goretzki, C., Krlej, A., Steffens, C., and Ritter, H. (2004). Green polymer chemistry: microwave-assisted single-step synthesis of various (meth)acrylamides and poly(meth)acrylamides directly from (meth)acrylic acid and amines. *Macromol. Rapid Commun.* 25: 513. https://doi.org/10.1002/MARC.200300154.

37 Iannelli, M. and Ritter, H. (2005). Microwave-assisted direct synthesis and polymerization of chiral acrylamide. *Macromol. Chem. Phys.* 206: 349. https://doi.org/10.1002/MACP.200400422.

38 Iannelli, M., Alupei, V., and Ritter, H. (2005). Selective microwave-accelerated synthesis and polymerization of chiral methacrylamide directly from methacrylic acid and (*R*)-1-phenyl-ethylamine. *Tetrahedron* 61: 1509. https://doi.org/10.1016/J.TET.2004.11.068.

39 Stange, H. and Greiner, A. (2007). Microwave-assisted free radical copolymerizations of styrene and methyl methacrylate. *Macromol. Rapid Commun.* 28 (4): 504–508. https://doi.org/10.1002/marc.200600841.

40 Singh, V., Tripathi, D.N., Tiwari, A., and Sanghi, R. (2006). Microwave synthesized chitosan-graft-poly(methyl methacrylate): an efficient Zn^{2+} ion binder. *Carbohydr. Polym.* 65: 35. https://doi.org/10.1016/J.CARBPOL.2005.12.002.

41 Singh, V., Tiwari, A., Tripathi, D.N., and Sanghi, R. (2006). Microwave enhanced synthesis of chitosan-graft-polyacrylamide. *Polymer* 47: 254. https://doi.org/10.1016/J.POLYMER.2005.10.101.

42 Singh, V., Tripathi, D.N., Tiwari, A., and Sanghi, R.S. (2005). Microwave promoted synthesis of chitosan-graft-poly(acrylonitrile). *J. Appl. Polym. Sci.* 95: 820. https://doi.org/10.1002/APP.21245.

43 Gao, J. and Wu, C. (2005). Modified structural model for predicting particle size in the microemulsion and emulsion polymerization of styrene under microwave irradiation. *Langmuir* 21: 782. https://doi.org/10.1021/LA048972Y.

44 Velmathi, S., Nagahata, R., Sugiyama, J.I., and Takeuchi, K. (2005). A rapid eco-friendly synthesis of poly (butylene succinate) by a direct polyesterification under microwave irradiation. *Macromol. Rapid Commun.* 26 (14): 1163–1167. https://doi.org/10.1002/marc.200500176.

45 Huang, J.L., Zhang, Y.L., Cheng, Z.H., and Tao, H.C. (2007). Microwave-assisted synthesis of polyaspartic acid and its effect on calcium carbonate precipitate. *J. Appl. Polym. Sci.* 103 (1): 358–364.

46 Leuchs, H. (1906). "Glycine-carbonic acid". *Ber. Dtsch. Chem. Ges.* 39: 857. https://doi.org/10.1002/cber.190603901133

47 Dao, B.N., Groth, A.M., and Hodgkin, J.H. (2007). Microwave-assisted aqueous polyimidization using high-throughput techniques. *Macromol. Rapid Commun.* 28: 604. https://doi.org/10.1002/MARC.200600747.

48 Pitt, C.G., Chasalow, F.I., Hibionada, Y.M. et al. (1981). Aliphatic polyesters. I. The degradation of poly(ε-caprolactone) *in vivo*. *J. Appl. Polym. Sci.* 26: 3779. https://doi.org/10.1002/APP.1981.070261124.

49 Yasin, M. and Tighe, B.J. (1992). Polymers for biodegradable medical devices: VIII. Hydroxybutyrate-hydroxyvalerate copolymers: physical and degradative properties of blends with polycaprolactone. *Biomaterials* 13: 9. https://doi.org/10.1016/0142-9612(92)90087-5.

50 Nehls, B.S., Asawapirom, U., Fuldner, S. et al. (2004). Semiconducting polymers via microwave-assisted Suzuki and Stille cross-coupling reactions. *Adv. Funct. Mater.* 14: 352. https://doi.org/10.1002/ADFM.200400010.

51 Khan, A. and Hecht, S. (2004). Microwave-accelerated synthesis of lengthy and defect-free poly(*m*-phenyleneethynylene)s via AB′ and A_2 + BB′ polycondensation routes. *Chem. Commun.* 300. https://doi.org/10.1039/B312762A.

52 Laurichesse, S. and Avérous, L. (2014). Chemical modification of lignins: towards biobased polymers. *Prog. Polym. Sci.* 39 (7): 1266–1290.

53 Nemani, S.K., Annavarapu, R.K., Mohammadian, B. et al. (2018). Surface modification of polymers: methods and applications. *Adv. Mater. Interfaces* 5 (24): 1801247. https://doi.org/10.1002/admi.201801247.

54 Youssef, A.M. (2013). Polymer nanocomposites as a new trend for packaging applications. *Polym. Plast. Technol. Eng.* 52 (7): 635–660. https://doi.org/10.1080/03602559.2012.762673.

55 Jasso-Gastinel, C.F., Soltero-Martínez, J.F.A., and Mendizábal, E. (2017). Introduction: modifiable characteristics and applications. In: *Modification of Polymer Properties* (ed. C.F. Jasso-Gastinel and J.M. Kenny), 1–21. William Andrew Publishing. https://doi.org/10.1016/B978-0-323-44353-1.00001-4.

56 Kretschmann, O., Schmitz, S., and Ritter, H. (2007). Microwave-assisted synthesis of associative hydrogels. *Macromol. Rapid Commun.* 28: 1265. https://doi.org/10.1002/MARC.200700117.

57 Nadagouda, M.N. and Varma, R.S. (2007). Preparation of novel metallic and bimetallic cross-linked poly(vinyl alcohol) nanocomposites under microwave irradiation. *Macromol. Rapid Commun.* 28: 465. https://doi.org/10.1002/MARC.200600735.

58 Bhanvase, B.A., Darda, N.S., Veerkar, N.C. et al. (2015). Ultrasound assisted synthesis of PANI/ZnMoO$_4$ nanocomposite for simultaneous improvement in anticorrosion, physico-chemical properties and its application in gas sensing. *Ultrason. Sonochem.* 24: 87–97. https://doi.org/10.1016/j.ultsonch.2014.11.009.

59 Mirtamizdoust, B., Bieńko, D.C., Hanifehpour, Y. et al. (2016). Preparation of a novel nano-scale lead(II) zig-zag metal–organic coordination polymer with ultrasonic assistance: synthesis, crystal structure, thermal properties, and NBO analysis of [Pb(μ-2-pinh)N$_3$H$_2$O]$_n$. *J. Inorg. Organomet. Polym. Mater.* 26: 819–828. https://doi.org/10.1007/s10904-016-0385-8.

60 Mohapatra, H., Kleiman, M., and Esser-Kahn, A.P. (2017). Mechanically controlled radical polymerization initiated by ultrasound. *Nat. Chem.* 9 (2): 135–139. https://doi.org/10.1038/nchem.2633.

61 Poddar, M.K., Sharma, S., and Moholkar, V.S. (2016). Investigations in two-step ultrasonic synthesis of PMMA/ZnO nanocomposites by in-situ emulsion polymerization. *Polymer* 99: 453–469. https://doi.org/10.1016/j.polymer.2016.07.052.

62 Hemmati, M., Rajabi, M., and Asghari, A. (2017). Ultrasound-promoted dispersive micro solid-phase extraction of trace anti-hypertensive drugs from biological matrices using a sonochemically synthesized conductive polymer nanocomposite. *Ultrason. Sonochem.* 39: 12–24. https://doi.org/10.1016/j.ultsonch.2017.03.024.

63 Li, J., Li, Y., Song, Y. et al. (2017). Ultrasonic-assisted synthesis of polyvinyl alcohol/phytic acid polymer film and its thermal stability, mechanical properties and surface resistivity. *Ultrason. Sonochem.* 39: 853–862. https://doi.org/10.1016/j.ultsonch.2017.06.017.

64 Huang, H., Cui, Y., Liu, M. et al. (2018). A one-step ultrasonic irradiation assisted strategy for the preparation of polymer-functionalized carbon quantum dots and their biological imaging. *J. Colloid Interface Sci.* 532: 767–773. https://doi.org/10.1016/j.jcis.2018.07.099.

65 Cherifi, Z., Boukoussa, B., Zaoui, A. et al. (2018). Structural, morphological and thermal properties of nanocomposites poly(GMA)/clay prepared by ultrasound and in situ polymerization. *Ultrason. Sonochem.* 48: 188–198. https://doi.org/10.1016/j.ultsonch.2018.05.027.

66 Lahcen, A.A., García-Guzmán, J.J., Palacios-Santander, J.M. et al. (2019). Fast route for the synthesis of decorated nanostructured magnetic molecularly imprinted polymers using an ultrasound probe. *Ultrason. Sonochem.* 53: 226–236. https://doi.org/10.1016/j.ultsonch.2019.01.008.

67 Wan, J., Fan, B., Liu, Y. et al. (2020). Room temperature synthesis of block copolymer nano-objects with different morphologies via ultrasound initiated

RAFT polymerization-induced self-assembly (sono-RAFT-PISA). *Polym. Chem.* 11 (21): 3564–3572. https://doi.org/10.1039/D0PY00461H.

68 Swaminathan, A., Ravi, R., Sasikumar, M. et al. (2020). Preparation and characterization of PVA/PAM/NH$_4$ SCN polymer film by ultrasound-assisted solution casting method for application in electric double layer capacitor. *Ionics* 26: 4113–4128. https://doi.org/10.1007/s11581-020-03542-4.

69 Dadashi, J., Hanifehpour, Y., Mirtamizdoust, B. et al. (2021). Ultrasound-assisted synthesis and DFT calculations of the novel 1D Pb(II) coordination polymer with thiosemicarbazone derivative ligand and its use for preparation of PbO clusters. *Crystals* 11 (6): 682. https://doi.org/10.3390/cryst11060682.

70 Wan, J., Fan, B., and Thang, S.H. (2021). Sonochemical preparation of polymer–metal nanocomposites with catalytic and plasmonic properties. *Nanoscale Adv.* 3 (11): 3306–3315. https://doi.org/10.1039/D1NA00120E.

71 Cherednichenko, K.A., Sayfutdinova, A.R., Kraynov, A. et al. (2022). A rapid synthesis of nanofibrillar cellulose/polystyrene composite via ultrasonic treatment. *Ultrason. Sonochem.* 90: 106180. https://doi.org/10.1016/j.ultsonch.2022.106180.

72 Afzal, S., Zahid, M., Rehan, Z.A. et al. (2022). Preparation and evaluation of polymer-based ultrasound gel and its application in ultrasonography. *Gels* 8 (1): 42. https://doi.org/10.3390/gels8010042.

73 Lamaoui, A., Palacios-Santander, J.M., Amine, A., and Cubillana-Aguilera, L. (2022). Computational approach and ultrasound probe-assisted synthesis of magnetic molecularly imprinted polymer for the electrochemical detection of bisphenol A. *Mater. Sci. Eng. B* 277: 115568. https://doi.org/10.1016/j.mseb.2021.115568.

74 Padmakumar, A.K., Jafari, V.F., Singha, N.K. et al. (2023). Synthesis of star polymers using ultrasound-induced RAFT polymerization. *J. Polym. Sci.* https://doi.org/10.1002/pol.20230077.

9

Synthesis of Nanomaterials Under Microwave and Ultrasound Irradiation

Ahmed A. Mohamed

University of Sharjah, Department of Chemistry, Sharjah 27272, UAE

9.1 Introduction

Nanomaterials have entrenched an inevitable role in many applications over the last few decades. Synthesis of nanomaterials using green routes finds ubiquitous biomedical applications in nanomedicine engineering. Conjugation of nanoparticles with biocompatible functionalities leads to acting as reducing, capping, and stabilizing agents. Nanoparticles are synthesized using chemical and physical reduction methods among other procedures. The most contemporary approach is the chemical reduction of salts [1–3]. Although not specifically focusing on the different synthesis approaches, some recent review articles put high prominence on microwave and ultrasonication irradiation as the most preferred techniques due to the green approach. Some synthesis approaches, such as chemical precipitation and hydrothermal process, constructed robust nanoparticles; however, their investigation is still limited due to the need for reducing agents [4, 5].

The recent approaches for nanoparticle synthesis are microwave and ultrasonication irradiation. The approaches are emerging for metal, metalloid, and nonmetal nanoparticle synthesis and modification. Although not a major area of study in many laboratories, and though it has just emerged, this research area aims to develop synthetic methods to stabilize this emerging class of nanoparticles for further applications [6, 7]. Still, achieving efficient methods for the synthesis of ultrapure nanomaterials is a challenge for many applications, especially in the field of the biomedical industry due to the lack of biocompatibility, weak mechanical stability, and short-term stability. The most common method for preventing aggregation and achieving pure nanomaterials is through irradiation techniques. Surface properties like charge, energy, wettability, and texture can be controlled using these routes.

In this chapter, we discuss the synthesis of nanomaterials using microwave and ultrasonication irradiation. The two approaches have merit in their synthesis of pure nanomaterials due to the possibility of carrying the reactions without supporting agents or reducing agents. It is desirable to develop biocompatible and catalytic

Green Chemical Synthesis with Microwaves and Ultrasound, First Edition.
Edited by Dakeshwar Kumar Verma, Chandrabhan Verma, and Paz Otero Fuertes.
© 2024 WILEY-VCH GmbH. Published 2024 by WILEY-VCH GmbH.

materials that have strong diverse physicochemical properties. Metal-polymer nanocomposites are superior in their physicochemical properties compared to their unmodified counterparts. They are prepared by mixing the nanoparticles with the polymer matrixes and fabricated or engineered using various techniques such as microwave and sonochemical irradiation. Their fabrication strategies have attained wide attention due to the possibilities of tailoring both the physical properties and surface modification. Thus, nano-texturing of polymers using irradiation routes can lead to the development of surfaces that have interesting properties.

9.2 Synthesis of Metal Nanoparticles

Gold nanoparticles are the most studied nanomaterials because of their wide applications in environmental remediation, biomedical, and energy. Microwave, biological, and physical approaches were used in their fabrication. Typically, they are fabricated using the commercially available $HAuCl_4$ acid which is highly soluble in water. They can be fabricated in acidic, neutral, and basic media in the presence or absence of a surfactant. The wide range of obtained morphologies made this material class very interesting in biomedical applications, particularly in cancer diagnosis and treatment. Gold in the oxidation state (III) has been the main starting material for the reduction reactions for the formation of gold nanomaterials. This is due to the facile reduction of gold(III) compared with gold(I) using green routes.

A few gold nanomaterials were fabricated using a microwave procedure. For example, hexagonal-shaped gold nanoplates were fabricated from $HAuCl_4$ acid using the microwave-hydrothermal method in the presence of sodium citrate [8]. The addition of PVP in different molar ratios aided in controlling the size of the isolated nanoparticles [9]. Folic acid, $HAuCl_4$, and sodium hydroxide were mixed, and using the microwave-hydrothermal procedure under 100 °C, 15 ± 5 nm size nanoparticles were obtained [10]. The black seed extract was used to synthesize the gold triangular and hexagonal morphologies obtained [11].

Ultrasonic irradiation was used in controlling the size of gold nanoparticles under irradiation powers of 60, 150, and 210 W [12]. $HAuCl_4$/trisodium citrate salt mixture was used in the synthesis. Gold nanoparticles size of 16 nm were obtained from the samples under 60 and 150 W; however, a 12 nm size was obtained using 210 W. The obtained gold nanoparticles and the mechanism of the reduction of gold(III) starting material are shown in Figure 9.1. Spherical alkyne-functionalized silica nanoparticles with a porous structure were loaded with gold nanoparticles [13]. The gold acid was spontaneously reduced on the alkynyl carbamate functionalities present on silica to produce 11 nm-sized gold-supported hybrid nanoparticles. The reaction occurred without the addition of a reducing agent and stabilizer.

The green synthesis of silver nanoparticles has been of great interest using microwave procedures for applications as antibacterial agents. Silver nanoparticles were synthesized as a single metal or as alloys with gold. The main variation in the synthesis is the green reducing agents and the templates. The typical starting salt in the synthesis of silver nanoparticles is silver nitrate. The green routes avoid the use

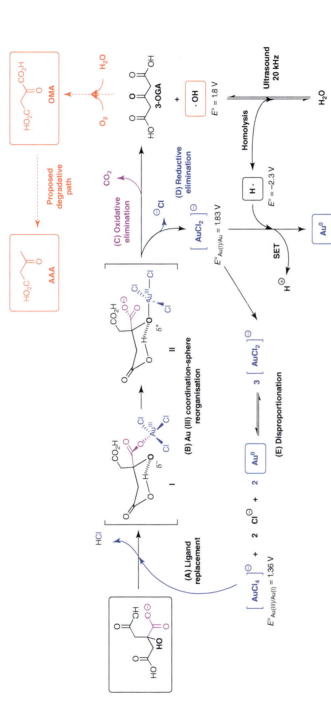

Figure 9.1 Chemical and ultrasonic reduction of chloroauric acid. (A) Ligand replacement. (B) Gold(III) coordination sphere reorganization. (C) Reductive elimination. (D) Oxidative elimination. Homolysis: homolytic cleavage of solvent generates a local accumulation of high-energy species such as hydroxyl and hydrogen radicals. (E) Disproportionation. Source: Reproduced with permission from Fuentes-García et al. [12]/Elsevier.

of chemicals which minimizes the need for extra steps for purification. Microwave heating was used for the synthesis of silver nanoparticles in the presence of beet juice [14], grape pomace extract [15], sugar cane [16], and guava leaf [17]. Additional routes involved the use of green reducing agents which also acted as temples such as starch, sucrose, glucose, glutathione, carboxymethyl cellulose sodium, polymethacrylic acid sodium salt, poly(propyleneimine) dendrimer [18–24].

The sonochemical method was used in the synthesis of silver nanoparticles at a faster rate than other methods such as hydrothermal under the same conditions Figure 9.2 [25]. Interestingly, this procedure produced monodispersed and smaller-sized nanoparticles of 8 nm compared with the hydrothermal approach. *Metha aquatica* leaf extract was used as the reducing and capping agent.

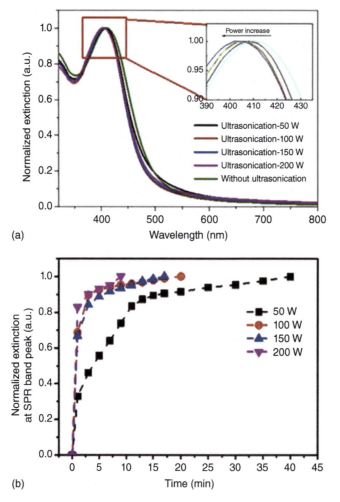

Figure 9.2 (a) Effect of ultrasonic power on the synthesis of AgNPs. (b) Temporal evolution of the extinction at SPR band of the AgNPs prepared under different ultrasonic powers. Source: Reproduced with permission from Nouri et al. [25]/Elsevier.

Ultrasonic power from 50 to 200 W reduced the reduction time. It was noticed that increasing ultrasound power resulted in a smaller wavelength due to the formation of silver nanoparticles with a smaller size with excellent monodispersity.

9.3 Synthesis of Carbon Dots

Carbon dots are important luminescent materials and can be synthesized from large molecules such as carbohydrates, PEG, and PVA in the presence of a small amount of inorganic salts and acids. Ultrasmall sizes of carbon dots are usually obtained. For example, 1–3 nm-sized carbon dots were prepared from PEG and PVA in the presence of phosphoric acid under microwave irradiation [26]. PEG was used in the synthesis of carbon dots of 4.5 nm size [27]. Another example of the synthesis of carbon dots involves the microwave pyrolysis of dextrin in the presence of sulfuric acid [28]. In another synthesis, 2–7 nm size carbon dots were synthesized from glucose [29].

There are a few factors that affect the efficiency of ultrasonication irradiation as a tool in the synthesis of carbon dots which are the solvent of choice, ultrasound frequency, power of ultrasound, sonication time, doping material, and catalyst [30]. More focus has been given to the choice of the starting material in the synthesis of carbon dots. They can be molecular compounds such as ammonium citrate, natural sources as simple as milk or sugar, and organic molecules of small size such as amino acids.

A few metals were deposited on carbon dots for the targeted applications. An example is Sn@carbon dots, Figure 9.3. Fabrication of carbon dots was carried out using ultrasound irradiation of polyethylene glycol. Tin was added and heated at the melting point followed by ultrasonication [31].

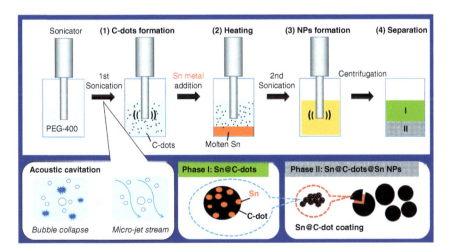

Figure 9.3 Sonochemical synthesis of C-dots, Sn@C-dots, and Sn@C-dots@Sn NPs. Source: Reproduced with permission from Kumar et al. [31]/Royal Society of Chemistry.

9.4 Synthesis of Metal Oxides

Iron oxide nanoparticles were synthesized using a microwave procedure. Two iron oxide nanoparticles are discussed in this part which are α-Fe_2O_3 and magnetic Fe_3O_4. Both oxides were synthesized using iron(III) nitrate, iron(III) chloride, potassium hexacyanoferrate(III), and iron(II) sulfate salts. The synthesis steps were conducted in the presence of polymers such as polyethylene glycol (PEG) and polyvinylpyrrolidone (PVP). In a few synthesis steps, hydrazine, ammonia, and urea-reducing agents were added. α-Fe_2O_3 nanoparticles were synthesized using a microwave-assisted hydrothermal approach at 100 °C using 0.02 M $Fe(NO_3)_3 \cdot 9H_2O$ [32]. $FeCl_3 \cdot 6H_2O$ in the presence of PVP was used in the synthesis of α-Fe_2O_3 nanorods [33]. Magnetic Fe_3O_4 can be also synthesized using a $FeCl_3$/hydrazine/PEG mixture in an aqueous solution under microwave heating at 100 °C [34]. Another approach involved mixing iron(III) chloride with iron(II) sulfate salts in NH_3 to form Fe_3O_4 NPs [35]. Iron salts resulted in Fe_3O_4 NPs synthesis in the presence of a reducing agent. However, another approach was carried out in the presence of a basic medium. For example, hexagonal Fe_3O_4 was prepared using a microwave, using iron(II) sulfate in the presence of a mixture of sodium acetate, citric acid, a base such as sodium hydroxide, and monosodium phosphate [36].

Recently, we synthesized SnO_2 in the AuNPs-SnO_2 core-shell arrangement [37]. The synthesis of SnO_2 involved the hydrolysis and oxidation of divalent tin(II) chloride to tetravalent SnO_2. The procedure involved mixing aryldiazonium gold salt with $SnCl_2$ in water. The reaction occurred spontaneously at room temperature. The literature showed a few attempts to utilize microwave heating and microwave-assisted hydrothermal processes for the synthesis of SnO_2 NPs using $SnCl_4 \cdot 5H_2O$, $SnCl_4$/urea, $SnCl_4$/HCl, $SnCl_4$/glucose, $SnCl_4$/HCl/urea, $SnCl_4$/HCl/ammonia starting materials. A few morphologies were obtained such as nanoparticles, porous nanotubes, and nanotubes Figure 9.4 [38, 39].

The fabrication of TiO_2 nanomaterials was achieved using microwave-assisted approaches such as hydrolysis and hydrothermal. For example, the hydrolysis of $TiCl_4$ in dilute sulfuric acid resulted in anatase TiO_2 NPs using a microwave-assisted hydrolysis procedure. Using a higher concentration of 1% sulfuric acid resulted in the aggregation of the nanoparticles [40]. Microwave-assisted hydrothermal process of $TiCl_4$ in 0.1% sulfuric acid in the presence of PVP resulted in the formation of needle-shaped crystals [41]. Using the same procedure, ultra-long wires of rutile TiO_2 were fabricated using $TiCl_3$ starting material at 200 °C in water [42]. The synthesis of TiO_2 nanotubes of 8–12 diameter using microwave heating was carried out from rutile and anatase phases in the presence of sodium hydroxide [43].

Copper oxide nanoparticles were prepared using microwave and microwave-assisted methods. Several starting materials, reaction conditions, heating temperatures, and reaction times were used. Results obtained showed the sensitivity of the copper oxide morphology to the reaction conditions. For example, a microwave-assisted hydrothermal reaction of copper acetate salt in the presence of urea at 150 °C for 30 minutes resulted in CuO NPs [44]. By changing the anion from

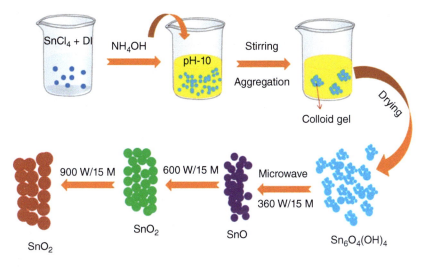

Figure 9.4 Formation of SnO_2 NPs under varying microwave conditions. Source: Reproduced with permission from Parthibavarman et al. [38]/Springer Nature.

acetate to nitrate, the morphology of CuO changed from NPs to nanorods, which underwent further assembly to flower-like. Nanosheets of CuO were constructed when copper acetate was used in the presence of 0.1 M NaOH [45]. Moreover, hollow nanostructures of CuO were hierarchical when urea was added to the reaction mixture [46]. Copper(II) chloride was used also as a starting salt which showed a dramatic effect on the CuO morphology. For example, CuO nanoplates were fabricated using a microwave-hydrothermal assisted procedure using $CuCl_2 \cdot 2H_2O$ in the presence of sodium hydroxide [47]. However, by using sodium carbonate instead of sodium hydroxide, urchin-like CuO was formed [48]. A minor change in the experimental conditions such as the reducing agent resulted in copper nanofibers instead of CuO. For example, single-crystalline copper nanowires were obtained using $CuCl_2$ in the presence of a reducing agent such as ascorbic acid and hexadecyl amine [49].

Three different salts of zinc nitrate, acetate, and chloride were used for the synthesis of ZnO NPs in the presence of the anionic surfactant sodium di-2-ethylhexyl-sulfosuccinate. The different morphologies are shown in Figure 9.5 [50]. Zinc nitrate was used for the synthesis of ZnO rods in the presence of hexamethylenetetramine using microwave heating at 90 °C [51]. Zinc nitrate was also used in the synthesis of ZnO nanotubes, in the presence of urea under microwave irradiation [52]. Zinc acetate, in the presence of a base such as ammonia, was used in the synthesis of ZnO nanorods using microwave irradiation at 90 °C [53]. ZnO nanorods were synthesized using zinc acetate starting material in the presence of potassium hydroxide under microwave irradiation for 20 minutes [54]. ZnO was prepared as flowers, nanospheres, and nanorods morphology using zinc acetate/NaOH/guanidium carbonate in an aqueous solution [55]. Lastly, zinc chloride/ammonia was used as a starting material for the synthesis of ZnO using a microwave-assisted hydrothermal procedure. Controlling the reaction time resulted in different morphologies [56].

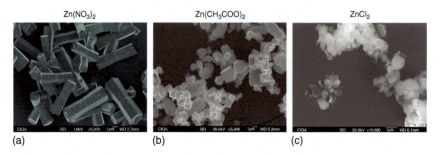

Figure 9.5 Field-emission FESEM images of ZnO nanostructures produced via MW-assisted synthesis at 140 °C and 600 W. The image shows the effect that using different precursor salts (a) $Zn(NO_3)_2$, (b) $Zn(CH_3COO)_2$, and (c) $ZnCl_2$, has on the morphological development of ZnO during MW synthesis. The scale bar is 1 µm. Source: Reproduced with permission from Barreto et al. [50]/Multidisciplinary Digital Publishing Institute.

Micro flowers fabrication of ZnO in the hexagonal prisms with planar and pyramid morphology was explained. The reaction progress is dependent on the pH and the chemical state of divalent zinc in the starting salt in the presence of sodium hydroxide. The reactions (9.1)–(9.4) are as follows [57, 58]:

$$Zn(NO_3)_2 \cdot 6H_2O \leftrightarrow Zn^{2+} + 2NO_3^- \tag{9.1}$$

$$NaOH \leftrightarrow Na^+ + OH^- \tag{9.2}$$

$$Zn^{2+} + 4OH^- \leftrightarrow Zn(OH)_4^{2-} \tag{9.3}$$

$$Zn(OH)_4^{2-} \rightarrow ZnO \downarrow + H_2O + 2OH^- \tag{9.4}$$

Ultrasound irradiation was used in the synthesis of various metal oxides. Here, we report a few examples to describe the facile synthesis without the need for a supporting electrolyte. For example, ultrasonic irradiation was used in the synthesis of

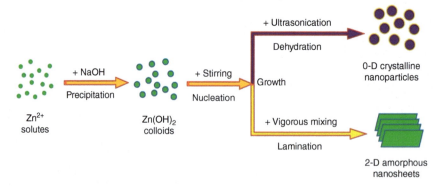

Figure 9.6 Fabrication of ZnO using chemical precipitation and ultrasonic irradiation. Source: Reproduced with permission from Huy et al. [61]/Elsevier.

Fe$_3$O$_4$ nanoparticles using various iron salts under high temperatures and pressure. Capping agents can be used such as dextran and graphene oxide [59, 60]. The synthesis of ZnO using ultrasonic irradiation was reported [61]. Zinc nitrate salt was used in the presence of sodium hydroxide Figure 9.6. It is interesting to note the catalytic and adsorption merit of the nanoparticles synthesized using ultrasonic irradiation over the chemical precipitation procedure.

The modification and stabilization of 20–30 nm-sized aluminum nanoparticles were effectively accomplished using alkyl-substituted epoxide to yield nanoparticles of very low Al$_2$O$_3$ content. The sonochemical technique produced 5 nm-sized air-stable aluminum nanoparticles capped with oleic acid attached via an Al—O—C bond with the carbon atom formerly involved in the carboxylic acid functional group [62].

9.5 Synthesis of Silicon Dioxide

Microwave-assisted synthesis of silica nanoparticles has advanced for their applications in electronics [63]. Porous and non-porous silica nanoparticle synthesis is of great interest. SiO$_x$/C materials are very interesting, for example, in silica material applications. Hence, we selected this class of nono-porous silica for discussion which exists in a core-shell arrangement. Typically, they are prepared using tetramethyl orthosilicate in the microwave at 70 °C [64]. The presence of a surfactant such as CTAB and a base such as ammonia is critical in the progress of the synthesis. The obtained particles were then carbonized at 1200 °C under a nitrogen atmosphere to form SiO$_x$@carbon. An example of porous silica is included, which is SBA-15 (Santa Barbara Amorphous-15). It is a highly stable mesoporous silica sieve developed by researchers at the University of California at Santa Barbara. The silica framework Ti-SBA-15 was synthesized using TEOS, titanium isopropoxide, P123, and ethanol. The synthesis using a microwave-hydrothermal method at 90 °C occurred, which was followed by titanium inclusion [65].

9.6 Conclusion

The applications of microwave and ultrasonication irradiation synthesis have shown excellent success in the fabrication of metal nanoparticles, semiconductors, metal oxides, carbon nanomaterials, and metal sulfides. Further, the two approaches resulted in various arrangements such as core-shell. The fabrication of different morphologies such as nanotubes, nanoparticles, and nano urchins was achieved. Ultrasmall-size nanoparticles can be also synthesized. The most important feature of the two approaches is the facile fabrication of different nanostructures using multiple starting salts under acidic, neutral, and basic conditions. Green synthesis using plant extracts and bacteria can be carried out using microwave and ultrasonic methods.

References

1 Mirzaei, A. and Neri, G. (2016). Microwave-assisted synthesis of metal oxide nanostructures for gas sensing application: a review. *Sens. Actuators B* 237: 749–775.

2 Kumar, A., Kuang, Y., Liang, Z., and Sun, X. (2020). Microwave chemistry, recent advancements, and eco-friendly microwave-assisted synthesis of nanoarchitectures and their applications: a review. *Mater. Today Nano* 11: 100076.

3 Chauhan, D.S., Gopal, C., Kumar, D. et al. (2018). Microwave induced facile synthesis and characterization of ZnO nanoparticles as efficient antibacterial agents. *Mater. Discover.* 11: 19–25.

4 Yang, G. and Park, S.J. (2019). Conventional and microwave hydrothermal synthesis and application of functional materials: a review. *Materials* 12: 1177.

5 Savun-Hekimoğlu, B. (2020). A review on sonochemistry and its environmental applications. *Acoustics* 2: 766–775.

6 Asgharzadehahmadi, S., Abdul Raman, A.A., Parthasarathy, R., and Sajjadi, B. (2016). Sonochemical reactors: review on features, advantages and limitations. *Renewable Sustainable Energy Rev.* 63: 302–314.

7 Li, Z., Zhuang, T., Dong, J. et al. (2021). Sonochemical fabrication of inorganic nanoparticles for applications in catalysis. *Ultrason. Sonochem.* 71: 105384.

8 Wang, J. and Wang, Z.X. (2007). Rapid synthesis of hexagon-shaped gold nanoplates by microwave assistant method. *Mater. Lett.* 61: 4149.

9 Kundu, S., Wang, K., and Liang, H. (2009). Size-selective synthesis and catalytic application of polyelectrolyte encapsulated gold nanoparticles using microwave irradiation. *J. Phys. Chem. C* 113: 5157.

10 Zhang, Z.W., Jia, J., Ma, Y.Y. et al. (2011). Microwave-assisted one-step rapid synthesis of folic acid modified gold nanoparticles for cancer cell targeting and detection. *MedChemComm* 2: 1079.

11 Fragoon, A., Li, J.J., Zhu, J., and Zhao, J.W. (2012). Biosynthesis of controllable size and shape gold nanoparticles by black seed (*Nigella sativa*) extract. *J. Nanosci. Nanotechnol.* 12: 2337.

12 Fuentes-García, J., Santoyo-Salzar, J., Rangel-Cortes, E. et al. (2021). Effect of ultrasonic irradiation power on sonochemical synthesis of gold nanoparticles. *Ultrason. Sonochem.* 70: 105274.

13 Fazzini, S., Cassani, M.C., Ballarin, B. et al. (2014). Novel synthesis of gold nanoparticles supported on alkyne-functionalized nanosilica. *J. Phys. Chem. C* 118: 24538.

14 Kou, J.H. and Varma, R.S. (2012). Beet juice utilization: expeditious green synthesis of noble metal nanoparticles (Ag, Au, Pt, and Pd) using microwaves. *RSC Adv.* 2: 10283.

15 Baruwati, B. and Varma, R.S. (2009). High value products from waste: grape pomace extract—a three-in-one package for the synthesis of metal nanoparticles. *ChemSusChem* 2: 1041.

16 Chaudhari, P.R., Masurkar, S.A., Shidore, V.B., and Kamble, S.P. (2012). Biosynthesis of silver nanoparticles using *Saccharum officinarum* and its antimicrobial activity. *Micro Nano Lett.* 7: 646.

17 Raghunandan, D., Mahesh, B.D., Basavaraja, S. et al. (2011). Microwave-assisted rapid extracellular synthesis of stable bio-functionalized silver nanoparticles from guava (*Psidium guajava*) leaf extract. *J. Nanopart. Res.* 13: 2021.

18 Sreeram, K.J., Nidhin, M., and Nair, B.U. (2008). Microwave assisted template synthesis of silver nanoparticles. *Bull. Mater. Sci.* 31: 937.

19 Luo, Y.L. and Sun, X.P. (2007). Rapid, single-step preparation of dendrimer-protected silver nanoparticles through a microwave-based thermal process. *Mater. Lett.* 61: 1622.

20 Chen, J., Wang, J., Zhang, X., and Jin, Y.L. (2008). Microwave-assisted green synthesis of silver nanoparticles by carboxymethyl cellulose sodium and silver nitrate. *Mater. Chem. Phys.* 108: 421.

21 Filippo, E., Serra, A., and Manno, D. (2009). Self-assembly and branching of sucrose stabilized silver nanoparticles by microwave assisted synthesis: from nanoparticles to branched nanowires structures. *Colloids Surf.* 348: 205.

22 Baruwati, B., Polshettiwar, V., and Varma, R.S. (2009). Glutathione promoted expeditious green synthesis of silver nanoparticles in water using microwaves. *Green Chem.* 11: 926.

23 Liu, S.H., Lu, F., and Zhu, J.J. (2011). Highly fluorescent Ag nanoclusters: microwave-assisted green synthesis and Cr^{3+} sensing. *Chem. Commun.* 47: 2661.

24 Filippo, E., Manno, D., Buccolieri, A. et al. (2010). Shape-dependent plasmon resonances of Ag nanostructures. *Superlattices Microstruct.* 47: 66.

25 Nouri, A., Tavakkoli Yaraki, M., Lajevardi, A. et al. (2020). Ultrasonic-assisted green synthesis of silver nanoparticles using *Mentha aquatica* leaf extract for enhanced antibacterial properties and catalytic activity. *Colloid Interface Sci. Commun.* 35: 100252.

26 Mitra, S., Chandra, S., Patra, P. et al. (2011). Rapid microwave synthesis of fluorescent hydrophobic carbon dots. *J. Mater. Chem.* 21: 17638.

27 Jaiswal, A., Ghosh, S.S., and Chattopadhyay, A. (2012). One step synthesis of C-dots by microwave mediated caramelization of poly(ethylene glycol). *Chem. Commun.* 48: 407.

28 Puvvada, N., Kumar, B.N., Konar, S. et al. (2012). Synthesis of biocompatible multicolor luminescent carbon dots for bioimaging applications. *Sci. Technol. Adv. Mater.* 13: 045008.

29 Song, Y.C., Shi, W., Chen, W. et al. (2012). Fluorescent carbon nanodots conjugated with folic acid for distinguishing folate-receptor-positive cancer cells from normal cells. *J. Mater. Chem.* 22: 12568.

30 Kumar, R., Kumar, V.B., and Gedanken, A. (2020). Sonochemical synthesis of carbon dots, mechanism, effect of parameters, and catalytic, energy, biomedical and tissue engineering applications. *Ultrason. Sonochem.* 64: 105009.

31 Kumar, V.B., Tang, J., Lee, K.J. et al. (2016). *In situ* sonochemical synthesis of luminescent Sn@C-dots and a hybrid Sn@C-dots@Sn anode for lithium-ion batteries. *RSC Adv.* 6: 66256–66265.

32 Fernández-Barahona, I., Muñoz-Hernando, M., and Herranz, F. (2019). Microwave-driven synthesis of iron-oxide nanoparticles for molecular imaging. *Molecules* 24: 1224.

33 Zhang, X.J. and Li, Q.L. (2008). Microwave assisted hydrothermal synthesis and magnetic property of hematite nanorods. *Mater. Lett.* 62: 988.

34 Qiu, G.H., Huang, H., Genuino, H. et al. (2011). Microwave-assisted hydrothermal synthesis of nanosized α-Fe_2O_3 for catalysts and adsorbents. *J. Phys. Chem. C* 115: 19626.

35 Hong, R.Y., Pan, T.T., and Li, H.Z. (2006). Microwave synthesis of magnetic Fe_3O_4 nanoparticles used as a precursor of nanocomposites and ferrofluids. *J. Magn. Magn. Mater.* 303: 60.

36 Zhou, H.F., Yi, R., Li, J.H. et al. (2010). Microwave-assisted synthesis and characterization of hexagonal Fe_3O_4 nanoplates. *Solid State Sci.* 12: 99.

37 Abla, F., Kanan, S.M., Park, Y. et al. (2021). Exceptionally redox-active precursors in the synthesis of gold core-tin oxide shell nanostructures. *Colloids Surf.* 616: 126266.

38 Parthibavarman, M., Sathishkumar, S., and Prabhakaran, S. (2018). Enhanced visible light photocatalytic activity of tin oxide nanoparticles synthesized by different microwave optimum conditions. *J. Mater. Sci.: Mater. Electron.* 29: 2341–2350.

39 Dheyab, M.A., Aziz, A.A., Jameel, M.S., and Oladzadabbasabadi, N. (2022). Recent advances in synthesis, modification, and potential application of tin oxide nanoparticles. *Surf. Interfaces* 28: 101677.

40 Baldassari, S., Komarneni, S., Mariani, E., and Villa, C. (2005). Rapid microwave–hydrothermal synthesis of anatase form of titanium dioxide. *J. Am. Ceram. Soc.* 88: 3238.

41 Baldassari, S., Komarneni, S., Mariani, E., and Villa, C. (2005). Microwave-hydrothermal process for the synthesis of rutile. *Mater. Res. Bull.* 40: 2014.

42 Zhang, D.Q., Li, G.S., Wang, F., and Yu, J.C. (2010). Green synthesis of a self-assembled rutile mesocrystalline photocatalyst. *CrystEngComm* 12: 1759.

43 Wu, X., Jiang, Q.Z., Ma, Z.F. et al. (2005). Synthesis of titania nanotubes by microwave irradiation. *Solid State Commun.* 136: 513.

44 Qiu, G.H., Dharmarathna, S., Zhang, Y.S. et al. (2012). Facile microwave-assisted hydrothermal synthesis of CuO nanomaterials and their catalytic and electrochemical properties. *J. Phys. Chem. C* 116: 468.

45 Liang, Z.H. and Zhu, Y. (2005). Single-crystalline CuO nanosheets synthesized from a layered precursor. *J. Chem. Lett.* 34: 214.

46 Deng, C.H., Hu, H.M., Zhu, W.L. et al. (2011). Green and facile synthesis of hierarchical cocoon shaped CuO hollow architectures. *Mater. Lett.* 65: 575.

47 Moura, A.P., Cavalcante, L.S., Sczancoski, J.C. et al. (2010). Structure and growth mechanism of CuO plates obtained by microwave-hydrothermal without surfactants. *Adv. Powder Technol.* 21: 197.

48 Chen, G., Zhou, H.F., Ma, W. et al. (2011). Microwave-assisted synthesis and electrochemical properties of urchin-like CuO micro-crystals. *Solid State Sci.* 13: 2137.

49 Liu, Y.Q., Zhang, M., Wang, F.X., and Pan, G.B. (2012). Facile microwave-assisted synthesis of uniform single-crystal copper nanowires with excellent electrical conductivity. *RSC Adv.* 2: 11235.

50 Barreto, G.P., Morales, G., and Quintanilla, M.L.L. (2013). Microwave assisted synthesis of ZnO nanoparticles: effect of precursor reagents, temperature, irradiation time, and additives on nano-ZnO morphology development. *J. Mater.* 2013: 1–11.

51 Hu, X.L., Zhu, Y.J., and Wang, S.W. (2004). Sonochemical and microwave-assisted synthesis of linked single-crystalline ZnO rods. *Mater. Chem. Phys.* 88: 421.

52 Kong, X.R., Duan, Y.Q., Peng, P. et al. (2007). A novel route to prepare ZnO nanotubes by using microwave irradiation method. *Chem. Lett.* 36: 428.

53 Cho, S., Kim, S., Kim, N.H. et al. (2008). In situ fabrication of density-controlled ZnO nanorod arrays on a flexible substrate using inductively coupled plasma etching and microwave irradiation. *J. Phys. Chem. C* 112: 17760.

54 Phuruangrat, A., Thongtem, T., and Thongtem, S. (2009). Microwave-assisted synthesis of ZnO nanostructure flowers. *Mater. Lett.* 63: 1224.

55 Hamedani, N.F., Mahjoub, A.R., Khodadadi, A.A., and Mortazavi, Y. (2011). Microwave assisted fast synthesis of various ZnO morphologies for selective detection of CO, CH_4 and ethanol. *Sens. Actuators, B* 156: 737.

56 Wu, W.T., Shi, L., Zhu, Q.R. et al. (2008). Rapid synthesis of ZnO micro/nanostructures in large scale. *Mater. Lett.* 62: 159.

57 Wojnarowicz, J., Chudoba, T., and Lojkowski, W. (2020). A review of microwave synthesis of zinc oxide nanomaterials: reactants, process parameters and morphologies. *Nanomaterials* 10: 1086.

58 Phuruangrat, A., Thongtem, T., and Thongtem, S. (2014). Controlling morphologies and growth mechanism of hexagonal prisms with planar and pyramid tips of ZnO microflowers by microwave radiation. *Ceram. Int.* 40: 9069–9076.

59 Lüdtke-Buzug, K. and Penxová, Z. (2019). Superparamagnetic iron oxide nanoparticles: an evaluation of the sonochemical synthesis process. *Curr. Dir. Biomed. Eng.* 5: 307–309.

60 Sriram, B., Govindasamy, M., Wanga, S.F. et al. (2019). Novel sonochemical synthesis of Fe_3O_4 nanospheres decorated on highly active reduced graphene oxide nanosheets for sensitive detection of uric acid in biological samples. *Ultrason. Sonochem.* 58: 104618.

61 Huy, N.N., Thanh Thuy, V.T., Thang, N.H. et al. (2019). Facile one-step synthesis of zinc oxide nanoparticles by ultrasonic-assisted precipitation method and its application for H_2S adsorption in air. *J. Phys. Chem. Solids* 132: 99–103.

62 Lewis, W.K., Rosenberger, A.T., Gord, J.R. et al. (2010). Multispectroscopic (FTIR, XPS, and TOFMS−TPD) investigation of the core–shell bonding in

sonochemically prepared aluminum nanoparticles capped with oleic acid. *J. Phys. Chem. C* 114: 6377.

63 Díaz de Greñu, B., de los Reyes, R., Costero, A.M. et al. (2020). Recent progress of microwave-assisted synthesis of silica materials. *Nanomaterials* 10: 1092.

64 Anh Cao, K.L., Arif, A.F., Kamikubo, K. et al. (2019). Controllable synthesis of carbon-coated SiO_x particles through a simultaneous reaction between the hydrolysis–condensation of tetramethyl orthosilicate and the polymerization of 3-aminophenol. *Langmuir* 35: 13681–13692.

65 Nguyen, T.T. and Qian, E.W. (2018). Synthesis of mesoporous Ti-inserted SBA-15 and CoMo/Ti-SBA-15 catalyst for hydrodesulfurization and hydrodearomatization. *Microporous Mesoporous Mater.* 265: 1–7.

10

Microwave- and Ultrasound-Assisted Synthesis of Metal-Organic Frameworks (MOF) and Covalent Organic Frameworks (COF)

Sanjit Gaikwad and Sangil Han

Changwon National University, Department of Chemical Engineering, Changwon-Si, Gyeongsangnam-do 51140, South Korea

10.1 Introduction

MOFs and COFs have gained significant attention in recent years due to their versatile and tunable structures, high surface area, and potential applications in gas storage [1, 2], separation [3, 4], catalysis [5, 6], and sensing [1, 7]. MOFs are a class of porous materials consisting of metal ions or clusters bridged by organic ligands to form a three-dimensional (3D) network, while COFs are a subclass of MOFs composed of organic building blocks linked by covalent bonds to form extended two-dimensional (2D) or 3D structures. Both MOFs and COFs offer advantages over traditional porous materials, such as activated carbon, zeolites, and mesoporous silica due to their high surface area, tunable pore size, and chemical functionality [8, 9].

MOFs and COFs are typically synthesized using solvothermal methods that require high temperature, pressure, and extended reaction times [10, 11]. These conventional approaches suffer from poor reproducibility, limited scalability, low yields, and the use of hazardous solvents. Solvothermal synthesis involves heating a mixture of metal salts and organic ligands in a solvent, at high temperature and pressure [12], while hydrothermal synthesis utilizes water as the solvent and lower temperature and pressure [13, 14]. Both techniques demand careful control of reaction parameters, such as temperature, pressure, time, and pH to achieve the desired products [15, 16]. As a result, alternative synthetic approaches that can ensure sustainable and efficient production of MOFs and COFs have been explored.

Microwave- and ultrasound-assisted synthesis have emerged as promising alternatives due to their unique advantages. These techniques significantly improve the production of MOFs and COFs by reducing reaction times, increasing yields, and improving purity [17–20]. They also offer precise control over the material size, composition, and morphology [16, 21–23]. Microwave-assisted synthesis involves the efficient heating of the reaction mixture through electromagnetic waves, leading

Green Chemical Synthesis with Microwaves and Ultrasound, First Edition.
Edited by Dakeshwar Kumar Verma, Chandrabhan Verma, and Paz Otero Fuertes.
© 2024 WILEY-VCH GmbH. Published 2024 by WILEY-VCH GmbH.

to fast and uniform heating, selective activation of reactants, and reduced reaction time. Ultrasound-assisted synthesis, on the other hand, applies high-frequency sound waves to the reaction mixture, inducing cavitation and acoustic streaming that improve mass transfer, homogenization, and nucleation [24, 25]. This approach has shown improved purity, higher yield, and enhanced mechanical and thermal stability. Furthermore, both microwave- and ultrasound-assisted synthesis offer the benefits of green chemistry, including reduced waste amount, energy consumption, and solvent use. Overall, these alternative synthetic techniques show great promise for the sustainable and efficient production of MOFs and COFs.

In this chapter, we provide a comprehensive review of the microwave- and ultrasound-assisted synthesis of MOFs and COFs which highlights the recent advances, challenges, and opportunities. We start by briefly introducing the definition and importance of MOFs and COFs, and discussing the traditional synthesis methods, followed by a detailed description of the principles and mechanisms of microwave- and ultrasound-assisted synthesis. Then, we present the recent examples of microwave- and ultrasound-assisted synthesis of MOFs and COFs, including their advantages and limitations. Finally, we conclude by discussing the future directions and potential applications of microwave- and ultrasound-assisted synthesis in the field of MOFs and COFs.

10.2 Principles

10.2.1 Principles of Microwave Heating

Microwave heating is a process that uses electromagnetic radiation in the microwave frequency range to heat materials. When microwaves pass through a material, they cause the molecules to rotate and create friction, which results in the generation of heat. The heat is produced uniformly throughout the material due to the even distribution of the microwaves (Figure 10.1). The heating effect is most pronounced in materials with high dielectric constants, such as water, which can effectively absorb and convert the energy of the microwaves into heat. Microwave heating is a fast, efficient, and selective method that can be used in different applications, including food processing, material synthesis, and chemical reactions.

10.2.2 Principle of Ultrasound-Assisted Techniques

Ultrasound-assisted synthesis is a technique that uses high-frequency sound waves to enhance chemical reactions. The principle behind this technique is the phenomenon of cavitation, where the sound waves create tiny bubbles in the reaction mixture. These bubbles rapidly expand and collapse, generating high temperatures and pressures that facilitate the formation of chemical bonds and accelerate the reaction kinetics (Figure 10.2). Additionally, the mechanical forces generated by the collapse of these bubbles can promote the mixing and dispersion of reactants, resulting in more uniform and efficient reactions. Many different applications have

Figure 10.1 Mechanism of microwave heating. Source: From [26], Springer Nature, licensed under CC BY 4.0.

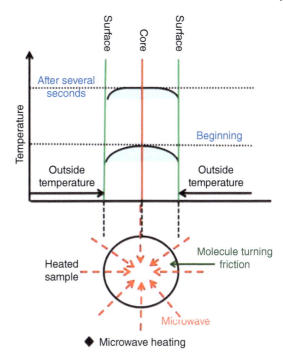

Figure 10.2 Cavitation phenomenon under ultrasound. Source: Reproduced with permission from Chatel [27].

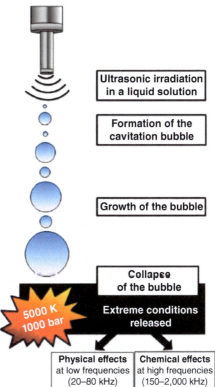

made use of ultrasound-assisted synthesis, including the synthesis of nanoparticles, organic molecules, and polymers as well as the preparation of materials for biomedical and environmental applications.

10.2.3 Advantages and Disadvantages of Microwave- and Ultrasound-Assisted Techniques

Microwave- and ultrasound-assisted synthesis are emerging as promising alternatives to traditional methods due to their numerous advantages. These techniques offer faster reaction times, higher yields, and better control over the size, morphology, and composition of the materials. They can also be more environmentally friendly since they reduce energy consumption and utilize nontoxic solvents.

Microwave-assisted synthesis is particularly effective at promoting rapid and efficient reactions. It works by selectively heating the reactants using electromagnetic radiation, resulting in higher yields, shorter reaction times, and improved product purity compared to traditional methods. In addition, microwave-assisted synthesis reduces the amount of solvent required, which can benefit the environment and economy. However, nonuniform heating can be a limitation of this technique, leading to localized hotspots and uneven product distribution. The use of microwave radiation can also be hazardous and requires careful safety measures to prevent accidents.

Ultrasound-assisted synthesis, on the other hand, utilizes acoustic waves to promote reactions by creating high-pressure and high-temperature zones within the reaction mixture. This method can enhance mass transport and accelerate reactions, resulting in faster reaction times and higher yields. Ultrasound-assisted synthesis can also improve the homogeneity of the reaction mixture and enhance the crystallinity and morphology of the resulting materials. However, the formation of unwanted by-products can be a limitation due to the high-energy conditions created by the acoustic waves. Additionally, the equipment required for ultrasound-assisted synthesis can be expensive and may require specialized expertise.

In summary, while microwave- and ultrasound-assisted synthesis offers many benefits over traditional methods, it is important to consider the specific advantages and disadvantages of each technique for a given application. Careful consideration of these factors can ensure the successful and sustainable production of MOF and COF.

10.3 MOF Synthesis by Microwave and Ultrasound Method

A class of porous materials known as MOFs have extremely flexible chemical and structural properties [28]. They consist of secondary building blocks (SBUs), such as metal ions, metal clusters, or metal–oxygen clusters, which are connected by organic ligands to create 3D lattices [29]. MOFs are promising materials for several uses, including gas storage, catalysis, drug delivery, and sensing [30, 31]. This is because of their high porosity, large interior surface area, and unique atomic-level structures. Metal clusters can be added to MOFs to boost stability

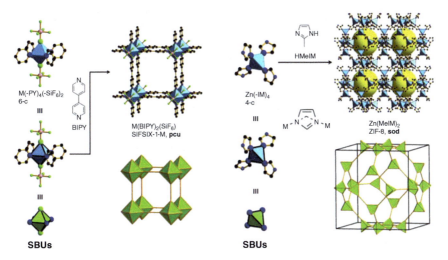

Figure 10.3 Different MOFs synthesized by using different SBUs. Source: Reproduced with permission from Schoedel [34]/Elsevier.

and porosity. The coordination bonds and building blocks containing metal are typically produced in situ in a single pot [32]. The SBUs, which are essential in determining the topology of MOFs, are required for the construction of potentially permeable periodic networks [33, 34] (Figure 10.3). Because of their great porosity and meticulously designed atomic-level structures and components, MOFs have long been considered one of the most intriguing families of materials from a scientific and technical perspective [35].

10.3.1 Microwave-Assisted Synthesis of MOF

The use of microwave heating for the synthesis of MOFs is a common practice in material chemistry. Gaikwad et al. demonstrated the synthesis of Bimetallic UTSA-16-Zn [36] and seeded growth PAN/UTSA-16-Zn/Co mat for CO_2 capture using MW irradiation [37] (Figure 10.4). MW heating offers the advantage of uniform heating of the entire sample, and it eliminates the need for heat transfer within the mixture. Additionally, the scalability of the process is less complicated since the heating process is independent of the reaction mixture volume.

The synthesis of MIL-100-(Cr) was carried out for the first time using MW heating at 220 °C for four hours, and then it was optimized to reduce the time to one hour while maintaining similar physicochemical properties [38]. Several other MOFs were developed using transition metals (Cu, Co, Ni, and Fe) with improved surface area and crystal homogeneity via the MW method [39]. Additionally, a crystalline Zr-MOF with highly defined crystals and an octahedral shape with sharp edges was synthesized using MW [40]. MOF-199 was synthesized within 30 minutes with a yield of 77%, showing crystalline structures with a high surface area ($1636\,m^2/g$) due to decreased agglomeration [41]. MW was also used to synthesize an efficient zirconium-based MOF (UiO-66) using acetic acid as an additive and

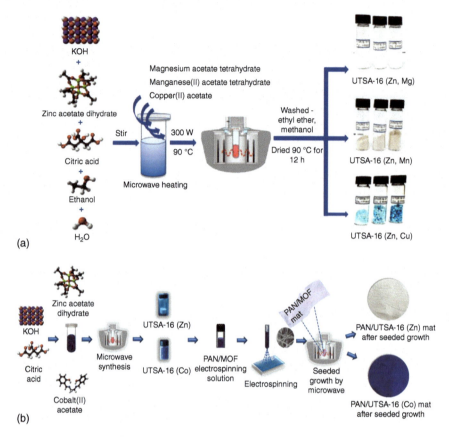

Figure 10.4 Schematic representations of the microwave-assisted synthesis of (a) bimetallic UTSA-16-(Zn, X) and (b) seeded growth PAN/UTSA-16 (Co/Zn) fiber mats. Source: Reproduced with permission from Gaikwad et al. [36, 37]/Elsevier.

trifluoroacetic acid as a modulator, resulting in the creation of catalytic metal sites after removing them from the metal clusters [42]. Gaikwad et al. developed UTSA-16-(Co) using the microwave method, reducing the synthesis time from 48 to 4 hours, with improved crystallinity and CO_2 adsorption capacity (5.5 mmol/g at 298 K, 1 bar) [18]. Additionally, Gusain and Bux synthesized Mg(BTC) MOF using the microwave method in one hour [43]. Liang and D'Alessandro improved the crystallinity and purity of zirconium-bearing MOFs (MIL-140A and MIL-140B) using the microwave method at 220 °C in 15 minutes [44]. Wang et al. synthesized MOF-23 by microwave method with improved crystallinity and yield compared to the solvothermal method [45]. Chen et. al [46] synthesized MOF-74-Ni for CO_2 capture application by reducing reaction time from seven to one hour at 125 °C with improved CO_2 adsorption capacity 3.47 mmol/g (298 K, 0.15 bar) compared to the hydrothermal (HT) (2.44 mmol/g) and condensation reflux (CE) (3.04 mmol/g) methods for the same condition (Figure 10.5). Different MOFs synthesized by the microwave method with different reaction condition shown in Table 10.1.

10.3 MOF Synthesis by Microwave and Ultrasound Method | 255

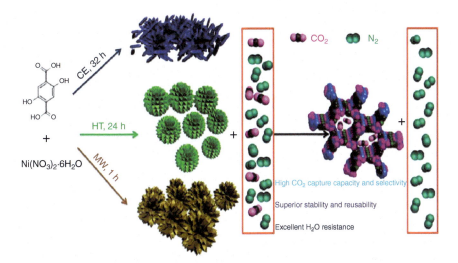

Figure 10.5 Schematic representation of MOF-74-Ni synthesized by the microwave method. Source: Reproduced with permission from Chen et al. [46]/American Chemical Society.

Table 10.1 MOFs synthesized by the microwave technique at different reaction conditions.

MOF	Temperature (°C)	Power (W)	Reaction time (min)	References
MIL-100	220	—	240	[38]
MIL-53 (Fe)	100	300	60	[47]
MIL-140a	220	—	1	[48]
MIL-53 (Cr)	210	—	120	[49]
MIL-53 (Al)	200	—	150	[49]
MIL-147 (V)	175	—	120	[49]
MIL-88B	150	200	15	[50]
UiO-66	120	—	15	[42]
MOF-74(Co)	130	300	60	[51]
NH2-MIL-88B	150	200	15	[50]
NH2-MIL-101	150	—	15	[52]
MOF-5	95–135	—	10–60	[53]
MIL-101	220	300	15	[54]
CPO-27	70	400	44–180	[55]
HKUST-1	150–220	300	1–240	[56]
UTSA-16 (Co)	90	300	30–480	[18]
UTSA-16 (Zn)	90	300	60–360	[17]
UTSA-16 (Zn, X)	90	300	60	[36]
ZIF-8	243	80	180	[57]

10.3.2 Ultrasound-Assisted Synthesis of MOFs

Ultrasound synthesis technique has emerged as a green and efficient method for the preparation of MOFs, as it allows faster reaction rate, reduced reaction time, lower temperature, and the use of less solvent [58] under mild conditions. Huh et al. [59] developed [M(bpydc)(H_2O)H_2O]$_n$ MOFs using three different metals (M = Zn, Co, and Ni) with a reduced reaction time from 48 to 2 hours. Ni-MOFs, made by the ultrasonication method, showed the highest surface area (2021 m^2/g) and pore volume 0.882 cm^3/g) with 81% crystallinity and smaller particle size of 3 μm compared to similar MOFs made by the other methods with surface area (~1500 m^2/g), pore volume (0.5 cm^3/g), and crystallinity (67%) [60–63].

Abbasi et al. [65] demonstrated that 3D cobalt(II) MOF was synthesized in 15 minutes (80 W) using ultrasound technology, surpassing the conventional method which typically takes 72 hours at 180 °C with improved heavy metal ion removal performance by 20%. Yu et al. [64] synthesized MOF-525 and MOF-545 in 2.5 hours and 0.5 hours, respectively, via hydrolysis of dimethyl-4-nitrophenyl phosphate (DMNP) and adsorption of bisphenol-A (BPA) (Figure 10.6). Haque et al. used MIL-53 (Fe) to investigate the reaction rates for the different synthesis methods: conventional, microwave, and ultrasound [66]. They revealed that the time required for the MOF synthesis is in the order of decreasing time – conventional method (1.5–3 days at 70–80 °C), microwave method (1.5–2.5 hours at 60–70 °C), and ultrasound method (0.5–1 hour at 50–70 °C). Increased reaction rates for the microwave and ultrasound methods are due to the uniform heating and cavitation effect [67, 68]. Furthermore, the ultrasound method exhibited crystal and nucleation growth rates of 9.89×10^{-2} and 1.67×10^{-1} min^{-1}, which are higher than those of the conventional methods [66] (Figure 10.7). Different MOFs synthesized by the ultrasound method are shown with reaction conditions of temperature, power, and reaction time in Table 10.2.

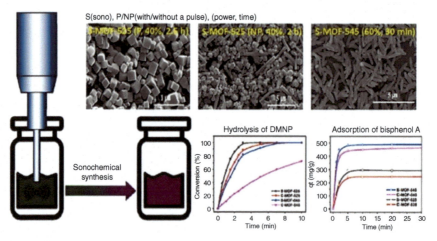

Figure 10.6 Schematic representation of MOF-525 and MOF-545 synthesized using ultrasound method via hydrolysis of DMNP and adsorption of BPA data. Source: Reproduced with permission from Yu et al. [64]/Elsevier.

Figure 10.7 SEM images of MIL-53 (Fe) synthesized at 70 °C using (a) conventional (3 days), (b) microwave (2 hours), and (c) ultrasound (35 minutes). Source: Reproduced with permission from Haque et al. [66]/John Wiley & Sons.

Table 10.2 Different MOFs synthesized by ultrasound at different reaction conditions and time.

MOF	Temperature (°C)	Power (W)	Reaction time (min)	References
HKUST-1	—	60	30–60	[69]
MOF-177	—	300	40	[70]
MOF-5	155	—	10	[71]
MIL-53 (Cr)	210	—	120	[49]
MIL-53 (Fe)	—	300	15	[72]
Zn-TMU-34		300	160	[73]
CPO-27-Co	70	262.5	75	[55]
Cd-TMU-7	RT	—	90	[74]
MIL-88A(Fe)	50	—	15	[15]
Zn-MOF	40	190	20	[75]
Ni/Zn MOF	30	240	35	[75]
Cu-MOF	30	190	20	[75]
TMU-50	RT	305	60	[76]
TMU-51	RT	305	60	[76]
NH2-MIL-53(Fe)	80	305	60	[77]
Ta-MOF	50, 70, 90	210, 250, 290	12, 20, 28	[78]
Sn-MOF	25	155	5	[79]
Ni-MOF	25	370	20	[80]
Ca(II)-MOF	—	12	60	[81]

RT: room temperature

10.4 Factors That Affect MOF Synthesis

10.4.1 Solvent

The solvent system used in MOF synthesis has a significant impact on the morphology of MOFs [82]. Solvents not only act as space-filling molecules but also

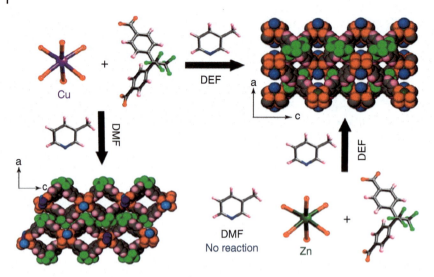

Figure 10.8 Schematic details and scheme of synthesis of the F-MOFs. Source: Reproduced with permission from Pachfule et al. [84].

serve as structure-directing agents [72] and coordinating agents with metal ions. High-boiling solvents maintain a constant temperature, and prevent solvent evaporation, which can lead to the formation of small, irregular MOF crystals and polar solvents with high dielectric constants stabilize metal ions in solution and promote the nucleation and growth of MOF crystals [83] such as dimethylformamide (DMF), dimethylformamide (DEF), dimethylsulfoxide (DMSO), dimethylacetamide (DMA), alcohols, acetone, and acetonitrile. Solvent mixtures may also be used depending on the solubility of starting materials. The MOF synthesis is significantly influenced by the solvent's polarity, organic linker's solubility, and protolysis properties. Different solvents under identical reaction conditions result in MOFs of different morphologies, which could be attributed to differences in the degree of deprotonation of the organic linker in different solvent systems. For instance, Pachfule et al. [84] synthesized fluorinated MOFs such as F-MOF-4, Cu-MOF-4B, and Zn-FMOF-4B, which exhibited 2D structures with or without interdigitation, depending on the solvent used (Figure 10.8). Li et al. [85] synthesized MOFs from biphenyl tricarboxylic acid (H_3BPT) and $Cd(NO_3)_2 \cdot 4H_2O$ using different solvents (DMF, DMA, and DEF) and found that the resulting coordination structures are different depending on the solvent as a coordinating agent.

10.4.2 Temperature and pH

The synthesis of MOFs is significantly impacted by both the temperature and pH of the reaction medium. With different pH ranges, linkers can adopt different coordination modes. For example, increasing pH results in a higher degree of linker deprotonation. Al^{3+} ion can coordinate with four, six, and eight carboxyl O-atoms at different pH levels, leading to the formation of MIL-121 (pH = 1.4), MIL-118 (pH = 2), and MIL-120 (pH = 12.2), respectively [86]. Higher pH values promote the formation of interpenetrated networks while lower pH values favor the formation of non-interpenetrated networks [87]. Moreover, the color of MOFs can also vary depending on the pH of the reaction medium. Luo et al. [88] conducted experiments on three Co–MOF complexes and found that altering the pH values can produce MOFs with different structures, colors, and adsorption capabilities.

MOFs with diverse architectures were synthesized at different temperatures. MOFs 5 and 7, which were synthesized at a lower temperature of 140 °C, revealed 2D networks with unique topologies, whereas MOFs 6 and 8, which were synthesized at a higher temperature of 180 °C, showed two-fold interpenetrated 3D frameworks (Figure 10.9). This suggests that temperature was important in shaping the structure of the MOFs. When the reaction temperature was raised from 140 to 180 °C, the MOFs 5 and 7 transitioned from non-interpenetrated 2D networks to interpenetrated 3D frameworks. Higher temperatures may have developed the construction of multidimensional frameworks with greater voids, resulting in the formation of interpenetrated structures [89, 90].

Alongside compositional parameters such as solvents, pH, and the molar ratio of starting materials, the process parameters including temperature, pressure, and time require careful analysis. High temperature is favorable for MOF crystallization due

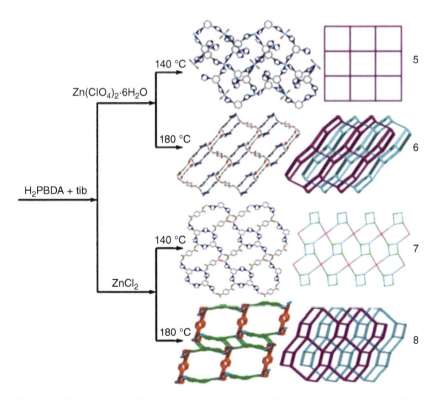

Figure 10.9 Schematic diagram of the synthesis conditions and topologies for MOFs 5 – 8 [89].

to the high solubility of reactants, leading to the formation of large and high-quality crystals [91, 92]. The temperature of the reaction mixture also affects the nucleation and crystal growth rates as well as the morphology of the synthesized MOFs. The hydrothermal method at a higher temperature leads to denser, less hydrated, and higher-dimensional MOFs with higher thermal stability. Bernini et al. [93] showed that two Ho-succinate MOFs differ in thermal stability and the MOF synthesized at a higher temperature was more stable than the one prepared at room temperature.

10.5 Application of MOF

MOFs have demonstrated great selectivity and efficiency in the removal of impurities from water such as heavy metals, dyes, and organic pollutants (Table 10.3). MIL-101 is a highly porous MOF that has been demonstrated to successfully remove heavy metals such as lead and cadmium from contaminated water [112]. Other MOFs, such as UiO-66 and UiO-66-NH$_2$ have shown outstanding dye and organic pollutant adsorption [113]. Furthermore, researchers have investigated the addition of functional groups such as carboxyl, amino, and sulfonic acid groups to MOFs to improve their adsorption properties of heavy metals, dyes, and organic pollutants.

MOFs have been studied for gas adsorption applications such as CO_2, methane, hydrogen, and other hazardous chemicals due to the uniform porous structure of MOFs. Several variables, such as open metal sites, surface area, pore size, and MOF activation might affect the adsorption capacity of these materials [114]. It is possible to manipulate the adsorption capabilities of MOFs to suit particular energy and environmental applications by modifying their unique properties of surface area, pore size, and surface functionality [115].

Due to their exceptional tunability at the nanometer scale, MOFs are becoming an increasingly attractive candidate as solid support in heterogeneous catalysis [116]. MOFs are suitable for various ranges of catalytic applications, including H_2 production [99], dye degradation [100], and electrocatalytic water oxidation [117]. MOF-derived materials also enable precise control over the design of catalyst systems for organic reactions.

MOFs have demonstrated considerable promising sensing technologies. MOF-based sensors are used for different applications such as food safety and preservation, gas sensing, and water purification. MOF-based sensors were applied to detect contaminants such as thiram, glyphosate, and inorganic fluorides for food safety and preservation [118]. MOFs have been employed in the development of luminous sensors for detecting gases and volatile organic compounds (VOCs) based on chemiresistive, magnetic, ferroelectric, and colorimetric mechanisms [119]. MOFs have also been combined with other materials such as TiO_2 [106], Fe_3O_4 [108], and PDMS [120] to improve sensor performance such as electrical, mechanical, electrochemical, and photochemical properties.

MOFs have emerged as promising candidates for drug delivery due to their exclusive properties, such as their nontoxic nature, ability to load multiple drugs, uniform porous structure, and directed or stimulus-based delivery methods. These properties

Table 10.3 Applications of MOFs with name and synthesis method.

Application	Specific application	MOF	Synthesis method	References
Antibiotics adsorption	Oxytetracycline chloride and tetracycline from water	ZIF-8	String method, RT	[94]
	Adsorption of Metronidazole (MNZ) from aqueous solution	$[Cu_3(\mu_6\text{-tma})_2]_n$ (HKUST-1)	Reflux condensation, 100 °C	[95, 96]
Toxic pollutant removal	Arsenic Adsorption from groundwater	MOF-74	Solvothermal method, 120 °C	[97]
	Benzothiophene Adsorption from liquid fuel	Zr(BTC); HPW (1.5)/Zr(BTC)	Solvothermal method, 120 °C	[98]
Gas adsorption	CO_2 adsorption	UTSA-16-Co	Microwave method 90 °C, 300 W	[18]
	CO_2 adsorption	UTSA-16-Zn	Microwave method 90 °C, 300 W	[17]
Photocatalyst	H_2 production	NH_2-MIL-125	Solvothermal method 150 °C	[99]
	Degradation rhodamine B	Zn–Zr MOFs	Solvothermal method 150 °C	[100]
Battery	Additive Lithium–battery	UiO-66	Solvothermal method 120 °C	[101, 102]
	Sodium-ion battery	MOF-5 DC	Solvothermal method 150 °C	[103]
	Lithium–selenium battery	ZIF-67@Se@MnO_2	Solvothermal method 90 °C	[104]
Supercapacitor	Supercapacitors	CoCuNi-bdc	Solvothermal method 120 °C	[105]
Electrochemical sensor	Detection of glucose	MIL-125/TiO_2	hydrothermal method 150 °C	[105–107]
	Electrochemical sensor for chlorogenic acid	Fe_3O_4@MIL-100(Fe)	hydrothermal method 160 °C	[108, 109]
Drug delivery	Insulin delivery	UiO-68-NH_2	Solvothermal method 90 °C	[110]
	Ibuprofen delivery	Cu-MOF/IBU@GM	String method 125 °C	[111]

have contributed to the growing interest in MOFs as an effective drug delivery system [121]. For various illnesses, including infections diabetic mellitus, ophthalmic conditions, and malignancies, MOFs are being investigated as a promising drug delivery system (DDS). Zr-MOFs are employed to treat cancer because of their exceptional stability and large surface area [122]. MOFs have shown significant potential in magnetic drug delivery, allowing targeted delivery of medications to specific organs or areas of the body without being affected by surrounding cells [123].

10.6 COF Synthesis by Microwave and Ultrasound Method

COFs are porous and crystalline materials that can be produced using a variety of organic precursors. Depending on the organic precursors employed, many forms of COFs can be produced. Organic precursors typically utilized in COF synthesis include boroxine, imine, imide, boronic ester, hydrazone, azine, and triazine [124]. There are various synthesis methods for COFs including sonochemical synthesis [125], solution-based synthesis [126], and reflux condensation synthesis [127]. To achieve a high degree of crystallinity and porosity, it is necessary to develop specialized reaction methods and post-synthesis treatments.

10.6.1 Ultrasound-Assisted Synthesis of COFs

Ultrasound-assisted synthesis is a technique that employs ultrasound waves to improve the reaction rate and yield of COFs, which can expedite the crystallization rate by generating highly localized heat and pressure (~5000 K and ~1000 bar) owing to acoustic cavitation in solutions, resulting in ultrafast heating and cooling. This process is cost-effective, energy-efficient, and has a shorter synthesis period as it bypasses the induction phase. In 2012, Ahn and his colleagues were the first to pioneer ultrasound synthesis of boron-based COFs (COF-1 and COF-5) under ultrasonication in one hour, resulting in a high Brunauer–Emmett–Teller (BET) surface area of 2122 m^2/g and a 100-fold decrease in crystal size (250 nm) compared to the conventional solvothermal methods (Figure 10.10) [128]. Sonochemical synthesis enabled a large-scale COF-5 synthesis, showing approximately nine times higher space–time yield (45 kg/m^3/day) than the COF analogs synthesized under solvothermal conditions. Shim et al. fabricated a core–shell structure by synthesizing COF-5 on the surface of carbon nanotubes and graphene and evaluated its potential for CO_2 capture applications [129]. Although the use of ultrasound waves has been found to improve the reaction rate and yield of COF synthesis, the full potential of this technique has not been realized yet. Therefore, further research is needed to fully understand the effectiveness of ultrasound-assisted synthesis of COFs.

10.6.2 Microwave-Assisted Synthesis of COF

Microwave-assisted synthesis has emerged as a promising method for the rapid and efficient synthesis of COFs. The use of microwave radiation as a heating source allows for faster and more efficient energy transfer, resulting in reduced reaction times and increased yields. In the case of COFs, the rapid heating and cooling supplied by microwave radiation can lead to the formation of highly crystalline and porous materials. For the synthesis of COFs, numerous methods have been reported (Table 10.4). Yaghi and colleagues used this solvothermal method for the synthesis of the first two COFs [140]. However, due to the long synthesis time, the solvothermal method is ineffective for completing the reaction.

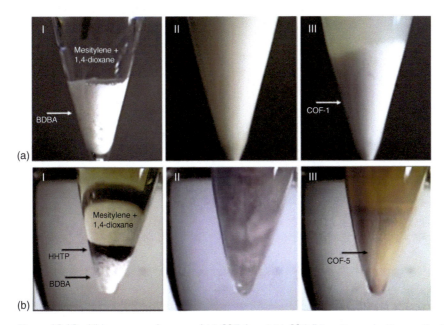

Figure 10.10 Video capture images of (a) COF-1 and (b) COF-5 by ultrasonication method; (I) setup, (II) applying sonication, and (III) after the reaction. Source: Reproduced with permission from Yang et al. [128].

Zhao et al. synthesized a hexagonal structured LZU-1 2D COF using 1,3,5-Triformylbenzene (TFB), 4-(tert-butoxycarbonylamino)-aniline (NBPDA), and poly(N-vinylpyrrolidone) (PVP) under microwave conditions at 120 °C for 30 minutes [141]. Similarly, Das et al. developed TAPB-MPPD and TAB-DFP COFs under microwave conditions for two hours with a yield of 90% for oil capture applications [142]. COF-5 was synthesized using microwave conditions for 20–70 minutes at 100 °C under stirring, and compared to the solvothermal method, the microwave synthesis was 200 times faster and resulted in improved porosity [143]. Yang et al. also utilized the microwave technique to synthesize AEM-COF-2 in 40 minutes at 120 °C, while the solvothermal method required eight days [144]. Wei et al. synthesized a COF using the microwave method in one hour at 100 °C while the solvothermal method took three days at 120 °C. The microwave method showed a high bond formation rate and produced a more crystalline structure compared to the solvothermal method [11]. Using the microwave method, Zhang et al. synthesized a 2D COF with 1,3,5-tris(4-aminophenyl)benzene (TAPB) and 1,1′-bis(2,4-dinitrophenyl)-[4,4′-bipyridine]-1,1′-diium dichloride (BDB). The resulting COF was then used for the "bottom-up" growth approach to create 2D-COF-based mixed matrix membranes for CO_2 separation applications [134].

10.6.3 Structure of COF (2D and 3D)

COFs have 2D or 3D frameworks based on structural and geometrical characteristics [145]. 2D COFs are the most widely documented, with over 200 structures reported.

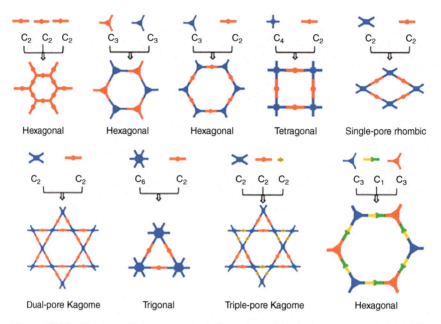

Figure 10.11 Topology diagrams representing a general basis for the construction of 2D COFs. Source: Reproduced with permission from Abuzeid et al. [147]/Elsevier.

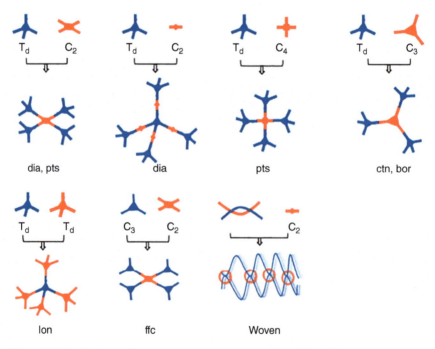

Figure 10.12 Topology diagrams representing a general basis for COF design and construction of 3D COFs. Source: Reproduced with permission from Abuzeid et al. [147]/Elsevier.

Table 10.4 Different 2D-COFs synthesized by microwave method under different reaction conditions and times.

COF	Temperature (°C)	Power (W)	Reaction time (min)	References
LZU-PEI	120	—	30	[130]
NiCOF	200	—	180	[131]
TAPB-TDA	—	200	60	[132]
IrNCOF	185–200	—	12–120	[131]
CTF-0	110	300	30	[133]
2D-COF	100	-	120	[134]
TH-COF	70	100	30	[135]
Pd@CTF	250	100	20	[136]
MA-PMDA	200	—	120	[137]
MA-NTDA	200	—	120	[137]
CTF-DCB	—	800	20	[138]
PtNCOF	200	—	210	[131]
CTF-BPDCN	—	220	180	[138]
PI-COF	—	200	120	[137]
IrO$_x$@CTF	250	100	3–10	[136]
DPCOF	200	—	180	[78]
LZU-Cys	120	—	30	[79]
Au@LZU1	120	—	20	[139]
LZU-1	120	—	30	[130]

These COFs are made up of building blocks that are joined together by covalent bonds, resulting in the formation of polymeric layers [146]. The figures presented in Figures 10.11 and 10.12 illustrate a topology diagram that provides a systematic and well-structured method for creating polygons by integrating building units to form extended lattices in 2D and 3D COFs. The diagrams provide a visual representation of the underlying structures and patterns necessary for creating these complex crystalline networks. Numerous techniques have been developed to create COFs with a diverse range of topologies using various building units. In general, the size and shape of the construction unit define its topology. When compared to other porous materials, COFs' ability to predict the skeleton structure is unique, allowing for the building of complex designs [148, 149].

3D COFs possess advantages such as high porosity and low density, which result from the linkages extending into the three dimensions of space. The development of 3D COFs has been hindered by the limited availability of tetrahedral building blocks and other related synthetic difficulties [146], which has resulted in a relatively low number of reported instances to date. Furthermore, the number of publications focused on 3D COF synthesis using microwave irradiation is significantly lower than

Figure 10.13 Scanning electron micrograph of COF-102 synthesized in a sealed microwave tube. Source: Reproduced with permission from Campbell et al. [143].

those for 2D COFs. COF-102 was the first 3D COF synthesized in a microwave oven in 2009 in 20 minutes at 100 °C, which is much faster than the conventional method for four days at 85 °C (Figure 10.13) [143]. Hei et al. made NO_2-PAF-1 COF by using the microwave method with a much faster time (30 minutes, 150 W) than the conventional method (48 hours, 175 °C) and used it for CO_2 capture [150].

10.7 Factors Affecting the COF Synthesis

The synthesis of COFs is a complex process involving many variables that can affect their properties such as purity, porosity, and structure of COFs [151]. The purity of COFs can be affected by the presence of impurities in the starting materials, as well as the synthesis conditions [152]. The porosity of COFs can be controlled by adjusting reaction conditions, such as the type and quantity of solvent employed, the reaction time and temperature, and the choice of organic building blocks [153]. Some of the most critical parameters that can influence COF synthesis include precursor molecules, solvent system, reaction temperature, and time. The precursor molecules utilized in the synthesis of COFs can have a substantial impact on the material's structure and properties. The precursor molecules can influence the porosity, stability, and usefulness of the COFs. The size, shape, and rigidity of the precursor molecules can also influence COF production. By supplying reactive sites for covalent bonding, the functional groups on the precursor molecules can also play an important role in the creation of COFs. The solvent system has a significant impact on the reaction rate, the solubility of the precursor molecules, shape, and crystallinity, and the stability of the resultant COFs.

In the synthesis of COFs, reaction temperature and time are key parameters that affect the rate of the reaction, the stability of the precursor molecules, and the morphology of COFs. High temperatures can accelerate reactions, but they can also result in the formation of amorphous or weakly crystalline COFs. Longer reaction periods can result in higher yields, but they can also produce contaminants or side products.

10.8 Applications of COFs

COFs are a significant and versatile type of porous organic materials that find diverse applications (Table 10.5). Triphenylamine (TPA) or triphenyl triazine (TPT)-based COFs have been synthesized, and the degree of crystallinity and surface area is significantly dependent on the symmetry and planarity of the monomers [184]. Higher N-content triaryl triazine and more planar units were discovered to improve their interactions with CO_2. 3D COFs have also been used as scaffolds in carbon dioxide adsorption for COF-102 and COF-103, which showed exceptional carbon dioxide uptake of up to 1200 and 1190 mg/g at 50 bar and 298 K, respectively [160]. COFs have recently emerged as promising candidates for efficient and robust catalysis due to their catalytic sites and resistance to heat, water, and organic solvents. COF-LZU1 was incorporated with a metal ion through post-treatment to form Pd/COF-LZU1, which was used as a catalyst in the Suzuki–Miyaura coupling reaction. The resulting catalyst demonstrated exceptional catalytic activity, producing yields of 96–98% [173].

The platforms for many applications including gas storage and separation, sensing, energy storage, catalysis, water purification, and electrical double-layer supercapacitors (EDLCs) require porous materials with high surface areas such as COFs. The DAAQ-TFP COF was the first COF known to have stable capacitance at 40 F/g across 5000 charge/discharge cycles [185]. The capacitance of the electrode was increased by 400% when an orientated thin film of 2D COF was connected by β-ketoenamine [186]. COFs with various redox-active moieties and N-atom-rich groups have recently been found to possess good electrochemical properties because of their large surface areas, organized porosities, and functional units. COFs can be formed using organic chromophores as building blocks for solar energy conversion applications. For example, organic chromophores are included in COF frameworks to create electroactive COFs as active layers for bulk heterojunction solar cells.

Table 10.5 Different applications of COFs in different fields.

Application	Specific application	COF	References
Toxic pollutant removal	Hg(II) removal from Water	COF-SH	[154]
	Remove Cu(II) from sewage	COF@PDA	[155]
	Determination of trace PCAs in plant-derived food	TAPT-DHTA	[156]
	Removal of acid blue 9	MIL-101-NH$_2$@COF	[157]
	Removal of Hg^{+2} and Hg^0	COF-S-SH	[158]
Gas storage	CO$_2$ adsorption	PCTF5	[159]
	H$_2$, CH$_4$ and CO$_2$ adsorption	COF-1, COF-5, COF-6, COF-8, COF-10, COF-102, COF-103	[160]
	H$_2$ adsorption	COF-11Å, COF-18Å, COF-14Å, CTC-COF, COF-16Å	
Electrochemical sensor	Detection of the explosive picric acid	COF-Cage 4	[161]
	Detection of ammonia	TPE-Ph COF	[162]
	Detection and Facile Removal of Hg^{+2}	COF-LZU8	[163]
	Detection of nitroaromatic explosives	COF-3 and COP-4	[164]
	Detection of Gallic acid and uric acid	ACOF-TaTp	[165, 166]
Filler in polymer membranes	BSA removal	TpEB	[167]
	Dye removal	GO-CTN	[168]
	Organic foulant removal	COF- (TpPa-2)	[169]
	CO$_2$/CH$_4$ separation	COF-MMM	[170]
	H$_2$O/ethanol separation	COF-TpHz	[171]
	Desalination	PA-rCON	[172]
Catalyst	Suzuki–Miyaura coupling reaction	Pd/COF-LZU1	[173]
	Suzuki–Miyaura coupling reaction	Pd/H$_2$P-Bph-COF	[174]
	Henry reactions	M/Salen-COFs	[175]
	Michael-addition	COF-SQ	[176]
Drug delivery	Delivery of 5-fluorouracil, captopril, ibuprofen	PI-3-COF, PI-2-COF	[177, 178]
	Delivery of Quercetin (3,3′,4′,5,7pentahydroxyflavone)	TTI-COF	[179]
	Delivery of anticancer 5-Fu	TpASH	[180]
Biosensing	Sensing-EGFR specific aptasensor	p-COF	[181]
	Sensing-Single strand DNA	TPA-COF	[182]
	Sensing-Aptasensor for detection of enrofloxacin and ampicillin	Py-M-COF	[183]

10.9 Future Predictions

MOFs and COFs exhibit tremendous potential as materials with diverse applications in areas such as energy storage, sensing, drug delivery, gas storage and separation, and catalysis. We can expect increased commercialization of MOFs and COFs in the near future as research continues to expand and novel applications are discovered, resulting in a widespread availability for industrial usage and greater integration into our daily lives. Overcoming stability issues such as breakdown under extreme conditions, like high temperatures or moisture exposure, and the development of a mass production process will be the major focus of future research. Additionally, there will be a drive to tailor the properties of MOFs and COFs for specific applications such as achieving higher selectivity, flexibility and tunability, chemical stability, mechanical strength, biocompatibility, porosity, and conductivity. Electrical conductivity, in particular, can facilitate electron transmission, making these materials appropriate for use as electrodes in energy storage devices such as batteries and supercapacitors. Applications of MOFs and COFs at the nanoscale are of tremendous interest, and we may see further progress in this area when researchers develop new techniques to manipulate these materials at the nanoscale level. As environmental concerns continue to intensify, sustainable synthesis techniques for MOFs and COFs will receive more attention with a focus on developing more ecologically friendly synthesis techniques such as the utilization of renewable energy sources and reducing the use of organic solvents. MOFs and COFs hold great promise, and as more research is conducted and new applications are discovered, we can anticipate these materials to become increasingly significant in various industries.

10.10 Summary

In conclusion, the microwave- and ultrasound-assisted synthesis of MOFs and COFs offer several advantages including faster reaction rates, higher yields, and the ability to control crystal size and morphology. The use of these methods also reduces environmental concerns, as they require minimal solvents and less heating time. Furthermore, MOFs and COFs obtained through these methods exhibit improved properties, which makes them promising candidates for various industrial applications in the future. The use of microwave- and ultrasound-assisted synthesis has also shown potential for post-synthesis modification and the fabrication of films and membranes using mild conditions that preserve the framework's integrity. While limitations still exist, such as scale-up development for mass production, further exploration of MOFs and COFs using these methods could lead to the development of novel porous crystalline materials with cost-efficiency and superior performance.

Acknowledgments

This research was supported by the Basic Science Research Program through the National Research Foundation of Korea (NRF) grant funded by the Korean government (MSIT) (NRF-2021R1F1A1045789, 2021K2A9A1A06093672), the Cooperative R&D between Industry, Academy, and Research Institute funded Korea Ministry of SMEs and Startups (S3310832).

References

1 Lahcen, A.A., Surya, S.G., Beduk, T. et al. (2022). Metal–organic frameworks meet molecularly imprinted polymers: insights and prospects for sensor applications. *ACS Appl. Mater. Interfaces* 14 (44): 49399–49424.
2 Mason, J.A., Veenstra, M., and Long, J.R. (2014). Evaluating metal–organic frameworks for natural gas storage. *Chem. Sci.* 5 (1): 32–51.
3 Wang, S., Yang, Y., Liang, X. et al. (2023). Ultrathin ionic COF membrane via polyelectrolyte-mediated assembly for efficient CO_2 separation. *Adv. Funct. Mater.* 2300386.
4 Qian, Q., Asinger, P.A., Lee, M.J. et al. (2020). MOF-based membranes for gas separations. *Chem. Rev.* 120 (16): 8161–8266.
5 Du, Y.-X., Zhou, Y.-T., and Zhu, M.-Z. (2023). Co-based MOF derived metal catalysts: from nano-level to atom-level. *Tungsten* 1–16.
6 Sun, M., Liu, Z., Wu, L. et al. (2023). Bioorthogonal-activated in situ vaccine mediated by a COF-based catalytic platform for potent cancer immunotherapy. *J. Am. Chem. Soc.* 145: 5330–5341.
7 Liu, X., Huang, D., Lai, C. et al. (2019). Recent advances in covalent organic frameworks (COFs) as a smart sensing material. *Chem. Soc. Rev.* 48 (20): 5266–5302.
8 Liu, Y., Chen, L., Yang, L. et al. (2023). Porous framework materials for energy & environment relevant applications: a systematic review. *Green Energy Environ.* 9: 217–310.
9 Wu, Y. and Weckhuysen, B.M. (2021). Separation and purification of hydrocarbons with porous materials. *Angew. Chem. Int. Ed.* 60 (35): 18930–18949.
10 Zhang, S., Wang, J., Zhang, Y. et al. (2021). Applications of water-stable metal-organic frameworks in the removal of water pollutants: a review. *Environ. Pollut.* 291: 118076.
11 Wei, H., Chai, S., Hu, N. et al. (2015). The microwave-assisted solvothermal synthesis of a crystalline two-dimensional covalent organic framework with high CO_2 capacity. *Chem. Commun.* 51 (61): 12178–12181.
12 Kamal, K., Bustam, M.A., Ismail, M. et al. (2020). Optimization of washing processes in solvothermal synthesis of nickel-based MOF-74. *Materials* 13 (12): 2741.
13 Feng, S.H. and Li, G.H. (2017). Chapter 4 - Hydrothermal and solvothermal syntheses. In: *Modern Inorganic Synthetic Chemistry*, 2e (ed. R. Xu and Y. Xu), 73–104. Amsterdam: Elsevier.
14 Chen, W., Du, L., and Wu, C. (2020). Hydrothermal synthesis of MOFs. In: *Metal-Organic Frameworks for Biomedical Applications*, 141–157. Elsevier.

15 Chalati, T., Horcajada, P., Gref, R. et al. (2011). Optimisation of the synthesis of MOF nanoparticles made of flexible porous iron fumarate MIL-88A. *J. Mater. Chem.* 21 (7): 2220–2227.

16 Łuczak, J., Kroczewska, M., Baluk, M. et al. (2023). Morphology control through the synthesis of metal-organic frameworks. *Adv. Colloid Interface Sci.* 102864.

17 Gaikwad, S., Kim, S.-J., and Han, S. (2020). Novel metal–organic framework of UTSA-16 (Zn) synthesized by a microwave method: outstanding performance for CO_2 capture with improved stability to acid gases. *J. Ind. Eng. Chem.* 87: 250–263.

18 Gaikwad, S. and Han, S. (2019). A microwave method for the rapid crystallization of UTSA-16 with improved performance for CO_2 capture. *Chem. Eng. J.* 371: 813–820.

19 Khan, N.A. and Jhung, S.H. (2015). Synthesis of metal-organic frameworks (MOFs) with microwave or ultrasound: rapid reaction, phase-selectivity, and size reduction. *Coord. Chem. Rev.* 285: 11–23.

20 Mu, X., Zhan, J., Feng, X. et al. (2018). Exfoliation and modification of covalent organic frameworks by a green one-step strategy: Enhanced thermal, mechanical and flame retardant performances of biopolymer nanocomposite film. *Composites Part A* 110: 162–171.

21 Mao, H., Li, S.-H., Zhang, A.-S. et al. (2021). Furfural separation from aqueous solution by pervaporation membrane mixed with metal organic framework MIL-53 (Al) synthesized via high efficiency solvent-controlled microwave. *Sep. Purif. Technol.* 272: 118813.

22 Vaitsis, C., Sourkouni, G., and Argirusis, C. (2019). Metal organic frameworks (MOFs) and ultrasound: a review. *Ultrason. Sonochem.* 52: 106–119.

23 Masoomi, M.Y., Bagheri, M., and Morsali, A. (2016). High adsorption capacity of two Zn-based metal–organic frameworks by ultrasound assisted synthesis. *Ultrason. Sonochem.* 33: 54–60.

24 Yao, Y., Pan, Y., and Liu, S. (2020). Power ultrasound and its applications: a state-of-the-art review. *Ultrason. Sonochem.* 62: 104722.

25 Pollet, B.G. and Ashokkumar, M. (2019). *Introduction to Ultrasound, Sonochemistry and Sonoelectrochemistry*. Springer Nature.

26 Zhang, F., Zhou, T., Liu, Y., and Leng, J. (2015). Microwave synthesis and actuation of shape memory polycaprolactone foams with high speed. *Sci. Rep.* 5 (1): 1–12.

27 Chatel, G. (2019). Sonochemistry in nanocatalysis: the use of ultrasound from the catalyst synthesis to the catalytic reaction. *Curr. Opin. Green Sustainable Chem.* 15: 1–6.

28 Baumann, A.E., Burns, D.A., Liu, B., and Thoi, V.S. (2019). Metal-organic framework functionalization and design strategies for advanced electrochemical energy storage devices. *Commun. Chem.* 2 (1): 86.

29 Li, G., Xia, L., Dong, J. et al. (2020). Chapter 10 - Metal-organic frameworks. In: *Solid-Phase Extraction* (ed. C.F. Poole), 285–309. Elsevier.

30 Zhou, H.-C., Long, J.R., and Yaghi, O.M. (2012). *Introduction to Metal–Organic Frameworks*, 673–674. ACS Publications.

31 MacGillivray, L.R. (2010). *Metal-organic Frameworks: Design and Application*. Wiley.

32 Kitagawa, S. (2014). Metal–organic frameworks (MOFs). *Chem. Soc. Rev.* 43 (16): 5415–5418.

33 Ha, J., Lee, J.H., and Moon, H.R. (2020). Alterations to secondary building units of metal–organic frameworks for the development of new functions. *Inorg. Chem. Front.* 7 (1): 12–27.

34 Schoedel, A. (2020). Secondary building units of MOFs. In: *Metal-Organic Frameworks for Biomedical Applications*, 11–44. Elsevier.

35 Cai, G., Yan, P., Zhang, L. et al. (2021). Metal–organic framework-based hierarchically porous materials: synthesis and applications. *Chem. Rev.* 121 (20): 12278–12326.

36 Gaikwad, R., Gaikwad, S., and Han, S. (2022). Bimetallic UTSA-16 (Zn, X; X= Mg, Mn, Cu) metal organic framework developed by a microwave method with improved CO_2 capture performances. *J. Ind. Eng. Chem.* 111: 346–355.

37 Gaikwad, R., Gaikwad, S., Kim, Y., and Han, S. (2021). Electrospun fiber mats with multistep seeded growth of UTSA-16 metal organic frameworks by microwave reaction with excellent CO_2 capture performance. *Microporous Mesoporous Mater.* 323: 111233.

38 Jhung, S.-H., Lee, J.-H., and Chang, J.-S. (2005). Microwave synthesis of a nanoporous hybrid material, chromium trimesate. *Bull. Korean Chem. Soc.* 26 (6): 880–881.

39 Aguiar, L.W., da Silva, C.T.P., de Lima, H.H.C. et al. (2018). Evaluation of the synthetic methods for preparing metal organic frameworks with transition metals. *AIMS Mater. Sci.* 5 (3): 467–478.

40 Ren, J., Segakweng, T., Langmi, H.W. et al. (2014). Microwave-assisted modulated synthesis of zirconium-based metal–organic framework (Zr-MOF) for hydrogen storage applications. *Int. J. Mater. Res.* 105 (5): 516–519.

41 Minh, T.T. and Thien, T.V. (2017). Synthesis of metal-organic framework-199: comparison of microwave process and solvothermal process. *Hue Univ. J. Sci. Nat. Sci.* 126 (1C): 107–116.

42 Taddei, M., Dau, P.V., Cohen, S.M. et al. (2015). Efficient microwave assisted synthesis of metal–organic framework UiO-66: optimization and scale up. *Dalton Trans.* 44 (31): 14019–14026.

43 Gusain, D. and Bux, F. (2019). Synthesis of magnesium based metal organic framework by microwave hydrothermal process. *Inorg. Chem. Commun.* 101: 172–176.

44 Liang, W. and D'Alessandro, D.M. (2013). Microwave-assisted solvothermal synthesis of zirconium oxide based metal–organic frameworks. *Chem. Commun.* 49 (35): 3706–3708.

45 Wang, X.-F., Zhang, Y.-B., Huang, H. et al. (2008). Microwave-assisted solvothermal synthesis of a dynamic porous metal-carboxylate framework. *Cryst. Growth Des.* 8 (12): 4559–4563.

46 Chen, C., Feng, X., Zhu, Q. et al. (2019). Microwave-assisted rapid synthesis of well-shaped MOF-74 (Ni) for CO_2 efficient capture. *Inorg. Chem.* 58 (4): 2717–2728.

47 Jia, J., Xu, F., Long, Z. et al. (2013). Metal–organic framework MIL-53 (Fe) for highly selective and ultrasensitive direct sensing of $MeHg^+$. *Chem. Commun.* 49 (41): 4670–4672.

48 Liang, W., Babarao, R., and D'Alessandro, D.M. (2013). Microwave-assisted solvothermal synthesis and optical properties of tagged MIL-140A metal–organic frameworks. *Inorg. Chem.* 52 (22): 12878–12880.

49 Khan, N.A., Jun, J.W., Jeong, J.H., and Jhung, S.H. (2011). Remarkable adsorptive performance of a metal–organic framework, vanadium-benzenedicarboxylate (MIL-47), for benzothiophene. *Chem. Commun.* 47 (4): 1306–1308.

50 Ma, M., Bétard, A., Weber, I. et al. (2013). Iron-based metal–organic frameworks MIL-88B and NH2-MIL-88B: high quality microwave synthesis and solvent-induced lattice "breathing". *Crystal Growth Des.* 13 (6): 2286–2291.

51 Cho, H.-Y., Yang, D.-A., Kim, J. et al. (2012). CO_2 adsorption and catalytic application of Co-MOF-74 synthesized by microwave heating. *Catal. Today* 185 (1): 35–40.

52 Taylor-Pashow, K.M.L., Della Rocca, J., Xie, Z. et al. (2009). Postsynthetic modifications of iron-carboxylate nanoscale metal–organic frameworks for imaging and drug delivery. *J. Am. Chem. Soc.* 131 (40): 14261–14263.

53 Choi, J.-S., Son, W.-J., Kim, J., and Ahn, W.-S. (2008). Metal–organic framework MOF-5 prepared by microwave heating: factors to be considered. *Microporous Mesoporous Mater.* 116 (1–3): 727–731.

54 Bromberg, L., Diao, Y., Wu, H. et al. (2012). Chromium(III) terephthalate metal organic framework (MIL-101): HF-free synthesis, structure, polyoxometalate composites, and catalytic properties. *Chem. Mater.* 24 (9): 1664–1675.

55 Haque, E. and Jhung, S.H. (2011). Synthesis of isostructural metal–organic frameworks, CPO-27s, with ultrasound, microwave, and conventional heating: Effect of synthesis methods and metal ions. *Chem. Eng. J.* 173 (3): 866–872.

56 Khan, N.A. and Jhung, S.-H. (2009). Facile syntheses of metal-organic framework Cu_3 $(BTC)_2$ $(H_2O)_3$ under ultrasound. *Bull. Korean Chem. Soc.* 30 (12): 2921–2926.

57 Lee, Y.-R., Jang, M.-S., Cho, H.-Y. et al. (2015). ZIF-8: a comparison of synthesis methods. *Chem. Eng. J.* 271: 276–280.

58 Vaitsis, C., Sourkouni, G., and Argirusis, C. (2020). Chapter 11 - Sonochemical synthesis of MOFs. In: *Metal-Organic Frameworks for Biomedical Applications* (ed. M. Mozafari), 223–244. Woodhead Publishing.

59 Huh, S., Jung, S., Kim, Y. et al. (2010). Two-dimensional metal–organic frameworks with blue luminescence. *Dalton Trans.* 39 (5): 1261–1265.

60 Sargazi, G., Afzali, D., Daldosso, N. et al. (2015). A systematic study on the use of ultrasound energy for the synthesis of nickel–metal organic framework compounds. *Ultrason. Sonochem.* 27: 395–402.

61 Liu, J., Wang, Y., Benin, A.I. et al. (2010). CO_2/H_2O adsorption equilibrium and rates on metal–organic frameworks: HKUST-1 and Ni/DOBDC. *Langmuir* 26 (17): 14301–14307.

62 Peralta, D., Chaplais, G., Simon-Masseron, A. et al. (2012). Metal–organic framework materials for desulfurization by adsorption. *Energy Fuels* 26 (8): 4953–4960.

63 Vitillo, J.G., Regli, L., Chavan, S. et al. (2008). Role of exposed metal sites in hydrogen storage in MOFs. *J. Am. Chem. Soc.* 130 (26): 8386–8396.

64 Yu, K., Lee, Y.-R., Seo, J.Y. et al. (2021). Sonochemical synthesis of Zr-based porphyrinic MOF-525 and MOF-545: Enhancement in catalytic and adsorption properties. *Microporous Mesoporous Mater.* 316: 110985.

65 Abbasi, A., Moradpour, T., and Van Hecke, K. (2015). A new 3D cobalt(II) metal–organic framework nanostructure for heavy metal adsorption. *Inorg. Chim. Acta* 430: 261–267.

66 Haque, E., Khan, N.A., Park, J.H., and Jhung, S.H. (2010). Synthesis of a metal–organic framework material, iron terephthalate, by ultrasound, microwave, and conventional electric heating: a kinetic study. *Chemistry* 16 (3): 1046–1052.

67 Tompsett, G.A., Conner, W.C., and Yngvesson, K.S. (2006). Microwave synthesis of nanoporous materials. *Chemphyschem* 7 (2): 296–319.

68 Conner, W.C., Tompsett, G., Lee, K.-H., and Yngvesson, K.S. (2004). Microwave synthesis of zeolites: 1. Reactor engineering. *J. Phys. Chem. B* 108 (37): 13913–13920.

69 Li, Z.-Q., Qiu, L.-G., Xu, T. et al. (2009). Ultrasonic synthesis of the microporous metal–organic framework $Cu_3(BTC)_2$ at ambient temperature and pressure: an efficient and environmentally friendly method. *Mater. Lett.* 63 (1): 78–80.

70 Jung, D.-W., Yang, D.-A., Kim, J. et al. (2010). Facile synthesis of MOF-177 by a sonochemical method using 1-methyl-2-pyrrolidinone as a solvent. *Dalton Trans.* 39 (11): 2883–2887.

71 Son, W.-J., Kim, J., Kim, J., and Ahn, W.-S. (2008). Sonochemical synthesis of MOF-5. *Chem. Commun.* 47: 6336–6338.

72 Gordon, J., Kazemian, H., and Rohani, S. (2012). Rapid and efficient crystallization of MIL-53 (Fe) by ultrasound and microwave irradiation. *Microporous Mesoporous Mater.* 162: 36–43.

73 Razavi, S.A.A., Masoomi, M.Y., and Morsali, A. (2018). Morphology-dependent sensing performance of dihydro-tetrazine functionalized MOF toward Al(III). *Ultrason. Sonochem.* 41: 17–26.

74 Masoomi, M.Y., Bagheri, M., and Morsali, A. (2017). Porosity and dye adsorption enhancement by ultrasonic synthesized Cd(II) based metal-organic framework. *Ultrason. Sonochem.* 37: 244–250.

75 Abdieva, G.A., Patra, I., Al-Qargholi, B. et al. (2022). An efficient ultrasound-assisted synthesis of Cu/Zn hybrid MOF nanostructures with high microbial strain performance. *Front. Bioeng. Biotechnol.* 10: 861580.

76 Yan, X.-W., Gharib, M., Esrafili, L. et al. (2022). Ultrasound irradiation assisted synthesis of luminescent nano amide-functionalized metal-organic frameworks; application toward phenol derivatives sensing. *Front. Chem.* 10: 855886.

77 Xue-Mina, W.U., Long-Xuea, L.I.U., Linga, L.I.U. et al. Ultrasound assisted synthesis of nanoscale NH2-MIL-53 (Fe) for the adsorption of dye. *Jiegou Huaxue* 40: 42–46.

78 Sargazi, G., Afzali, D., Mostafavi, A., and Ebrahimipour, S.Y. (2017). Ultrasound-assisted facile synthesis of a new tantalum (V) metal-organic framework nanostructure: design, characterization, systematic study, and CO_2 adsorption performance. *J. Solid State Chem.* 250: 32–48.

79 Brainer, N.S., dos Santos, T.V., Barbosa, C.D.E.S., and Meneghetti, S.M.P. (2020). Simple and fast ultrasound-assisted synthesis of Sn-MOFs and obtention of SnO_2. *Mater. Lett.* 280: 128512.

80 Shahriari, T., Zeng, Q., Ebrahimi, A. et al. (2022). An efficient ultrasound assisted electrospinning synthesis of a biodegradable polymeric Ni-MOF supported by PVA-fibrous network as a novel CH_4 adsorbent. *Appl. Phys. A* 128 (5): 446.

81 Li, D., Li, L.-F., Zhang, Z.-F. et al. (2020). Ultrasound-assisted synthesis of a new nanostructured Ca(II)-MOF as 5-FU delivery system to inhibit human lung cancer cell proliferation, migration, invasion and induce cell apoptosis. *J. Coord. Chem.* 73 (2): 266–281.

82 Akhbari, K. and Morsali, A. (2011). Effect of the guest solvent molecules on preparation of different morphologies of ZnO nanomaterials from the $[Zn_2(1,4-bdc)_2(dabco)]$ metal-organic framework. *J. Coord. Chem.* 64 (20): 3521–3530.

83 Domingues, N.P., Moosavi, S.M., Talirz, L. et al. (2022). Using genetic algorithms to systematically improve the synthesis conditions of Al-PMOF. *Commun. Chem.* 5 (1): 170.

84 Pachfule, P., Das, R., Poddar, P., and Banerjee, R. (2011). Solvothermal synthesis, structure, and properties of metal organic framework isomers derived from a partially fluorinated link. *Crystal Growth Des.* 11 (4): 1215–1222.

85 Li, L., Wang, S., Chen, T. et al. (2012). Solvent-dependent formation of Cd(II) coordination polymers based on a C2-symmetric tricarboxylate linker. *Crystal Growth Des.* 12 (8): 4109–4115.

86 Volkringer, C., Loiseau, T., Guillou, N. et al. (2010). High-throughput aided synthesis of the porous metal–organic framework-type aluminum pyromellitate, MIL-121, with extra carboxylic acid functionalization. *Inorg. Chem.* 49 (21): 9852–9862.

87 Yuan, F., Xie, J., Hu, H.-M. et al. (2013). Effect of pH/metal ion on the structure of metal–organic frameworks based on novel bifunctionalized ligand 4′-carboxy-4,2′ : 6′,4″-terpyridine. *CrystEngComm* 15 (7): 1460–1467.

88 Luo, L., Lv, G.-C., Wang, P. et al. (2013). pH-Dependent cobalt(II) frameworks with mixed 3,3′,5,5′-tetra(1H-imidazol-1-yl)-1,1′-biphenyl and 1,3,5-benzenetricarboxylate ligands: synthesis, structure and sorption property. *CrystEngComm* 15 (45): 9537–9543.

89 Sun, Y.-X. and Sun, W.-Y. (2014). Influence of temperature on metal-organic frameworks. *Chin. Chem. Lett.* 25 (6): 823–828.

90 Su, Z., Fan, J., Okamura, T.-a. et al. (2010). Interpenetrating and self-penetrating zinc(II) complexes with rigid tripodal imidazole-containing ligand and benzenedicarboxylate. *Crystal Growth Des.* 10 (4): 1911–1922.

91 Zhang, C.Y., Wang, M.Y., Li, Q.T. et al. (2013). Hydrothermal synthesis, crystal structure, and luminescent properties of two zinc(II) and cadmium(II) 3D metal-organic frameworks. *Z. Anorg. Allg. Chem.* 639 (5): 826–831.

92 Yang, L.-T., Qiu, L.-G., Hu, S.-M. et al. (2013). Rapid hydrothermal synthesis of MIL-101 (Cr) metal–organic framework nanocrystals using expanded graphite as a structure-directing template. *Inorg. Chem. Commun.* 35: 265–267.

93 Bernini, M.C., Brusau, E.V., Narda, G.E. et al. (2007). *The Effect of Hydrothermal and Non-Hydrothermal Synthesis on the Formation of Holmium(III) Succinate Hydrate Frameworks*. Wiley Online Library.

94 Li, N., Zhou, L., Jin, X. et al. (2019). Simultaneous removal of tetracycline and oxytetracycline antibiotics from wastewater using a ZIF-8 metal organic-framework. *J. Hazard. Mater.* 366: 563–572.

95 Kalhorizadeh, T., Dahrazma, B., Zarghami, R. et al. (2022). Quick removal of metronidazole from aqueous solutions using metal–organic frameworks. *New J. Chem.* 46 (19): 9440–9450.

96 Wang, X., Zhang, W., Zhang, X. et al. (2022). A novel design of self-assembled metal-organic frameworks MIL-53 (Fe) modified resin as a catalyst for catalytic degradation of tetracycline. *J. Cleaner Prod.* 348: 131385.

97 Sun, J., Zhang, X., Zhang, A., and Liao, C. (2019). Preparation of Fe-Co based MOF-74 and its effective adsorption of arsenic from aqueous solution. *J. Environ. Sci.* 80: 197–207.

98 Ullah, L., Zhao, G., Hedin, N. et al. (2019). Highly efficient adsorption of benzothiophene from model fuel on a metal-organic framework modified with dodeca-tungstophosphoric acid. *Chem. Eng. J.* 362: 30–40.

99 Fu, B., Sun, H., Liu, J. et al. (2022). Construction of MIL-125-NH_2@ $BiVO_4$ composites for efficient photocatalytic dye degradation. *ACS Omega* 7 (30): 26201–26210.

100 Zhang, X., Yu, R., Wang, D. et al. (2022). Green photocatalysis of organic pollutants by bimetallic Zn-Zr metal-organic framework catalyst. *Front. Chem.* 10: 918941.

101 Chu, F., Hu, J., Wu, C. et al. (2018). Metal–organic frameworks as electrolyte additives to enable ultrastable plating/stripping of Li anode with dendrite inhibition. *ACS Appl. Mater. Interfaces* 11 (4): 3869–3879.

102 Yang, X., Zhu, P., Ren, J. et al. (2019). Surfactant-assisted synthesis and electrochemical properties of an unprecedented polyoxometalate-based metal–organic nanocaged framework. *Chem. Commun.* 55 (9): 1201–1204.

103 Ingersoll, N., Karimi, Z., Patel, D. et al. (2019). Metal organic framework-derived carbon structures for sodium-ion battery anodes. *Electrochim. Acta* 297: 129–136.

104 Ye, W., Li, W., Wang, K. et al. (2018). ZIF-67@ Se@ MnO_2: a novel Co-MOF-based composite cathode for lithium–selenium batteries. *J. Phys. Chem. C* 123 (4): 2048–2055.

105 Mohd Zain, N.K., Vijayan, B.L., Misnon, I.I. et al. (2018). Direct growth of triple cation metal–organic framework on a metal substrate for electrochemical energy storage. *Ind. Eng. Chem. Res.* 58 (2): 665–674.

106 Zhang, D., Chen, H., Li, P. et al. (2019). Humidity sensing properties of metal organic framework-derived hollow ball-like TiO_2 coated QCM sensor. *IEEE Sens. J.* 19 (8): 2909–2915.

107 Udourioh, G.A., Solomon, M., Matthews-Amune, C.O. et al. (2023). Current trends in the synthesis, characterization and application of metal organic frameworks. *React. Chem. Eng.*

108 Chen, Y., Huang, W., Chen, K. et al. (2019). A novel electrochemical sensor based on core-shell-structured metal-organic frameworks: the outstanding analytical performance towards chlorogenic acid. *Talanta* 196: 85–91.

109 Ru, C., Gu, Y., Li, Z. et al. (2019). Effective enhancement on humidity sensing characteristics of sulfonated poly (ether ketone) via incorporating a novel bifunctional metal–organic–framework. *J. Electroanal. Chem.* 833: 418–426.

110 Zou, J.-J., Wei, G., Xiong, C. et al. (2022). Efficient oral insulin delivery enabled by transferrin-coated acid-resistant metal-organic framework nanoparticles. *Sci. Adv.* 8 (8): eabm4677.

111 Javanbakht, S., Nezhad-Mokhtari, P., Shaabani, A. et al. (2019). Incorporating Cu-based metal-organic framework/drug nanohybrids into gelatin microsphere for ibuprofen oral delivery. *Mater. Sci. Eng., C* 96: 302–309.

112 Zhang, H., Hu, X., Li, T. et al. (2022). MIL series of metal organic frameworks (MOFs) as novel adsorbents for heavy metals in water: a review. *J. Hazard. Mater.* 128271.

113 Cheng, S., Xie, P., Yu, Z. et al. (2022). Enhanced adsorption performance of UiO-66 via modification with functional groups and integration into hydrogels. *Environ. Res.* 212: 113354.

114 Gheytanzadeh, M., Baghban, A., Habibzadeh, S. et al. (2021). Towards estimation of CO_2 adsorption on highly porous MOF-based adsorbents using gaussian process regression approach. *Sci. Rep.* 11 (1): 1–13.

115 Anderson, R., Biong, A., and Gómez-Gualdrón, D.A. (2020). Adsorption isotherm predictions for multiple molecules in MOFs using the same deep learning model. *J. Chem. Theory Comput.* 16 (2): 1271–1283.

116 Pascanu, V., González Miera, G., Inge, A.K., and Martín-Matute, B. (2019). Metal–organic frameworks as catalysts for organic synthesis: a critical perspective. *J. Am. Chem. Soc.* 141 (18): 7223–7234.

117 Huang, Y., Jiang, L.W., Shi, B.Y. et al. (2021). Highly efficient oxygen evolution reaction enabled by phosphorus doping of the Fe electronic structure in iron–nickel selenide nanosheets. *Adv. Sci.* 8 (18): 2101775.

118 Cheng, W., Tang, X., Zhang, Y. et al. (2021). Applications of metal-organic framework (MOF)-based sensors for food safety: enhancing mechanisms and recent advances. *Trends Food Sci. Technol.* 112: 268–282.

119 Li, H.-Y., Zhao, S.-N., Zang, S.-Q., and Li, J. (2020). Functional metal–organic frameworks as effective sensors of gases and volatile compounds. *Chem. Soc. Rev.* 49 (17): 6364–6401.

120 Shu, Y., Shang, Z., Su, T. et al. (2022). A highly flexible Ni–Co MOF nanosheet coated Au/PDMS film based wearable electrochemical sensor for continuous human sweat glucose monitoring. *Analyst* 147 (7): 1440–1448.

121 Maranescu, B. and Visa, A. (2022). Applications of metal-organic frameworks as drug delivery systems. *Int. J. Mol. Sci.* 23 (8): 4458.

122 He, S., Wu, L., Li, X. et al. (2021). Metal-organic frameworks for advanced drug delivery. *Acta Pharm. Sin. B* 11 (8): 2362–2395.

123 Mallakpour, S., Nikkhoo, E., and Hussain, C.M. (2022). Application of MOF materials as drug delivery systems for cancer therapy and dermal treatment. *Coord. Chem. Rev.* 451: 214262.

124 Ahmed, I. and Jhung, S.H. (2021). Covalent organic framework-based materials: synthesis, modification, and application in environmental remediation. *Coord. Chem. Rev.* 441: 213989.

125 Zhao, W., Yan, P., Yang, H. et al. (2022). Using sound to synthesize covalent organic frameworks in water. *Nat. Synth.* 1 (1): 87–95.

126 Liu, M., Liu, Y., Dong, J. et al. (2022). Two-dimensional covalent organic framework films prepared on various substrates through vapor induced conversion. *Nat. Commun.* 13 (1): 1411.

127 Pp, R., Mondal, P.K., and Chopra, D. (2018). Synthesis and characterization of a 2D covalent organic framework (COF) of hexagonal topology using boronate linkages. *J. Chem. Sci.* 130: 1–7.

128 Yang, S.-T., Kim, J., Cho, H.-Y. et al. (2012). Facile synthesis of covalent organic frameworks COF-1 and COF-5 by sonochemical method. *RSC Adv.* 2 (27): 10179–10181.

129 Yoo, J., Lee, S., Hirata, S. et al. (2015). In situ synthesis of covalent organic frameworks (COFs) on carbon nanotubes and graphenes by sonochemical reaction for CO_2 adsorbents. *Chem. Lett.* 44 (4): 560–562.

130 Hao, K., Guo, Z., Lin, L. et al. (2021). Covalent organic framework nanoparticles for anti-tumor gene therapy. *Sci. China Chem.* 64 (7): 1235–1241.

131 Spaulding, V., Zosel, K., Duong, P.H.H. et al. (2021). A self-assembling, biporous, metal-binding covalent organic framework and its application for gas separation. *Mater. Adv.* 2 (10): 3362–3369.

132 Chen, L., Du, J., Zhou, W. et al. (2020). Microwave-assisted solvothermal synthesis of covalent organic frameworks (COFs) with stable superhydrophobicity for oil/water separation. *Chemistry* 15 (21): 3421–3427.

133 Kong, D., Han, X., Xie, J. et al. (2019). Tunable covalent triazine-based frameworks (CTF-0) for visible-light-driven hydrogen and oxygen generation from water splitting. *ACS Catal.* 9 (9): 7697–7707.

134 Zhang, Y., Ma, L., Lv, Y., and Tan, T. (2022). Facile manufacture of COF-based mixed matrix membranes for efficient CO_2 separation. *Chem. Eng. J.* 430: 133001.

135 Ji, W., Guo, Y.-S., Xie, H.-M. et al. (2020). Rapid microwave synthesis of dioxin-linked covalent organic framework for efficient micro-extraction of perfluorinated alkyl substances from water. *J. Hazard. Mater.* 397: 122793.

136 Rademacher, L., Beglau, T.H.Y., Heinen, T. et al. (2022). Microwave-assisted synthesis of iridium oxide and palladium nanoparticles supported on a nitrogen-rich covalent triazine framework as superior electrocatalysts for the hydrogen evolution and oxygen reduction reaction. *Front. Chem.* 10: 945261.

137 Kuehl, V.A., Wenzel, M.J., Parkinson, B.A. et al. (2021). Pitfalls in the synthesis of polyimide-linked two-dimensional covalent organic frameworks. *J. Mater. Chem. A* 9 (27): 15301–15309.

138 Sun, T., Liang, Y., and Xu, Y. (2022). Rapid, ordered polymerization of crystalline semiconducting covalent triazine frameworks. *Angew. Chem. Int. Ed.* 61 (4): e202113926.

139 Guntern, Y.T., Vavra, J., Karve, V.V. et al. (2021). Synthetic tunability of colloidal covalent organic framework/nanocrystal hybrids. *Chem. Mater.* 33 (7): 2646–2654.

140 Cote, A.P., Benin, A.I., Ockwig, N.W. et al. (2005). Porous, crystalline, covalent organic frameworks. *Science* 310 (5751): 1166–1170.

141 Zhao, Y., Guo, L., Gándara, F. et al. (2017). A synthetic route for crystals of woven structures, uniform nanocrystals, and thin films of imine covalent organic frameworks. *J. Am. Chem. Soc.* 139 (37): 13166–13172.

142 Chen, L., Zhou, C., Tan, L. et al. (2022). Enhancement of compatibility between covalent organic framework and polyamide membrane via an interfacial bridging method: toward highly efficient water purification. *J. Membr. Sci.* 656: 120590.

143 Campbell, N.L., Clowes, R., Ritchie, L.K., and Cooper, A.I. (2009). Rapid microwave synthesis and purification of porous covalent organic frameworks. *Chem. Mater.* 21 (2): 204–206.

144 Yang, H., Du, Y., Wan, S. et al. (2015). Mesoporous 2D covalent organic frameworks based on shape-persistent arylene-ethynylene macrocycles. *Chem. Sci.* 6 (7): 4049–4053.

145 Peng, P., Shi, L., Huo, F. et al. (2019). In situ charge exfoliated soluble covalent organic framework directly used for Zn–air flow battery. *ACS Nano* 13 (1): 878–884.

146 Zhang, Y., Duan, J., Ma, D. et al. (2017). Three-dimensional anionic cyclodextrin-based covalent organic frameworks. *Angew. Chem. Int. Ed.* 56 (51): 16313–16317.

147 Abuzeid, H.R., El-Mahdy, A.F.M., and Kuo, S.-W. (2021). Covalent organic frameworks: design principles, synthetic strategies, and diverse applications. *Giant* 6: 100054.

148 Huang, N., Wang, P., and Jiang, D. (2016). Covalent organic frameworks: a materials platform for structural and functional designs. *Nat. Rev. Mater.* 1 (10): 1–19.

149 Chen, X., Huang, N., Gao, J. et al. (2014). Towards covalent organic frameworks with predesignable and aligned open docking sites. *Chem. Commun.* 50 (46): 6161–6163.

150 Hei, Z.-H., Huang, M.-H., Luo, Y., and Wang, Y. (2016). A well-defined nitro-functionalized aromatic framework (NO_2-PAF-1) with high CO_2

adsorption: synthesis via the copper-mediated Ullmann homo-coupling polymerization of a nitro-containing monomer. *Polym. Chem.* 7 (4): 770–774.

151 Li, X., Yang, C., Sun, B. et al. (2020). Expeditious synthesis of covalent organic frameworks: a review. *J. Mater. Chem. A* 8 (32): 16045–16060.

152 Zhang, T., Zhang, G., and Chen, L. (2022). 2D conjugated covalent organic frameworks: defined synthesis and tailor-made functions. *Acc. Chem. Res.* 55 (6): 795–808.

153 Sharma, R.K., Yadav, P., Yadav, M. et al. (2020). Recent development of covalent organic frameworks (COFs): synthesis and catalytic (organic-electro-photo) applications. *Mater. Horiz.* 7 (2): 411–454.

154 Ma, Z., Liu, F., Liu, N. et al. (2021). Facile synthesis of sulfhydryl modified covalent organic frameworks for high efficient Hg(II) removal from water. *J. Hazard. Mater.* 405: 124190.

155 Xiao, Y., Ma, C., Jin, Z. et al. (2021). Functional covalent organic framework illuminate rapid and efficient capture of Cu(II) and reutilization to reduce fire hazards of epoxy resin. *Sep. Purif. Technol.* 259: 118119.

156 Xu, G., Hou, L., Li, B. et al. (2021). Facile preparation of hydroxyl bearing covalent organic frameworks for analysis of phenoxy carboxylic acid pesticide residue in plant-derived food. *Food Chem.* 345: 128749.

157 Dinari, M. and Jamshidian, F. (2021). Preparation of MIL-101-NH2 MOF/triazine based covalent organic framework hybrid and its application in acid blue 9 removals. *Polymer* 215: 123383.

158 Sun, Q., Aguila, B., Perman, J. et al. (2017). Postsynthetically modified covalent organic frameworks for efficient and effective mercury removal. *J. Am. Chem. Soc.* 139 (7): 2786–2793.

159 Bhunia, A., Boldog, I., Möller, A., and Janiak, C. (2013). Highly stable nanoporous covalent triazine-based frameworks with an adamantane core for carbon dioxide sorption and separation. *J. Mater. Chem. A* 1 (47): 14990–14999.

160 Furukawa, H. and Yaghi, O.M. (2009). Storage of hydrogen, methane, and carbon dioxide in highly porous covalent organic frameworks for clean energy applications. *J. Am. Chem. Soc.* 131 (25): 8875–8883.

161 Acharyya, K. and Mukherjee, P.S. (2014). A fluorescent organic cage for picric acid detection. *Chem. Commun.* 50 (99): 15788–15791.

162 Dalapati, S., Jin, E., Addicoat, M. et al. (2016). Highly emissive covalent organic frameworks. *J. Am. Chem. Soc.* 138 (18): 5797–5800.

163 Ding, S.-Y., Dong, M., Wang, Y.-W. et al. (2016). Thioether-based fluorescent covalent organic framework for selective detection and facile removal of mercury(II). *J. Am. Chem. Soc.* 138 (9): 3031–3037.

164 Xiang, Z. and Cao, D. (2012). Synthesis of luminescent covalent–organic polymers for detecting nitroaromatic explosives and small organic molecules. *Macromol. Rapid Commun.* 33 (14): 1184–1190.

165 Lin, X., Deng, Y., He, Y. et al. (2021). Construction of hydrophilic N, O-rich carboxylated triazine-covalent organic frameworks for the application in selective simultaneous electrochemical detection. *Appl. Surf. Sci.* 545: 149047.

166 Guan, Q.L., Sun, Y., Huo, R. et al. (2021). Cu-MOF material constructed with a triazine polycarboxylate skeleton: multifunctional identify and microdetecting of the aromatic diamine family (o, m, p-phenylenediamine) based on the luminescent response. *Inorg. Chem.* 60 (4): 2829–2838.

167 Wang, X., Shi, X., and Wang, Y. (2020). In situ growth of cationic covalent organic frameworks (COFs) for mixed matrix membranes with enhanced performances. *Langmuir* 36 (37): 10970–10978.

168 Khan, N.A., Yuan, J., Wu, H. et al. (2019). Mixed nanosheet membranes assembled from chemically grafted graphene oxide and covalent organic frameworks for ultra-high water flux. *ACS Appl. Mater. Interfaces* 11 (32): 28978–28986.

169 Xu, L., Xu, J., Shan, B. et al. (2017). TpPa-2-incorporated mixed matrix membranes for efficient water purification. *J. Membr. Sci.* 526: 355–366.

170 Liu, Y., Wu, H., Wu, S. et al. (2021). Multifunctional covalent organic framework (COF)-Based mixed matrix membranes for enhanced CO_2 separation. *J. Membr. Sci.* 618: 118693.

171 Yang, H., Wu, H., Yao, Z. et al. (2018). Functionally graded membranes from nanoporous covalent organic frameworks for highly selective water permeation. *J. Mater. Chem. A* 6 (2): 583–591.

172 Khan, N.A., Yuan, J., Wu, H. et al. (2020). Covalent organic framework nanosheets as reactive fillers to fabricate free-standing polyamide membranes for efficient desalination. *ACS Appl. Mater. Interfaces* 12 (24): 27777–27785.

173 Ding, S.-Y., Gao, J., Wang, Q. et al. (2011). Construction of covalent organic framework for catalysis: Pd/COF-LZU1 in Suzuki–Miyaura coupling reaction. *J. Am. Chem. Soc.* 133 (49): 19816–19822.

174 Hou, Y., Zhang, X., Sun, J. et al. (2015). Good Suzuki-coupling reaction performance of Pd immobilized at the metal-free porphyrin-based covalent organic framework. *Microporous Mesoporous Mater.* 214: 108–114.

175 Li, L.-H., Feng, X.-L., Cui, X.-H. et al. (2017). Salen-based covalent organic framework. *J. Am. Chem. Soc.* 139 (17): 6042–6045.

176 Li, X., Wang, Z., Sun, J. et al. (2019). Squaramide-decorated covalent organic framework as a new platform for biomimetic hydrogen-bonding organocatalysis. *Chem. Commun.* 55 (38): 5423–5426.

177 Bai, L., Phua, S.Z.F., Lim, W.Q. et al. (2016). Nanoscale covalent organic frameworks as smart carriers for drug delivery. *Chem. Commun.* 52 (22): 4128–4131.

178 Fu, Y., Zhu, X., Huang, L. et al. (2018). Azine-based covalent organic frameworks as metal-free visible light photocatalysts for CO_2 reduction with H_2O. *Appl. Catal., B* 239: 46–51.

179 Vyas, V.S., Vishwakarma, M., Moudrakovski, I. et al. (2016). Exploiting noncovalent interactions in an imine-based covalent organic framework for quercetin delivery. *Adv. Mater.* 28 (39): 8749–8754.

180 Mitra, S., Sasmal, H.S., Kundu, T. et al. (2017). Targeted drug delivery in covalent organic nanosheets (CONs) via sequential postsynthetic modification. *J. Am. Chem. Soc.* 139 (12): 4513–4520.

181 Yan, X., Song, Y., Liu, J. et al. (2019). Two-dimensional porphyrin-based covalent organic framework: a novel platform for sensitive epidermal growth factor receptor and living cancer cell detection. *Biosens. Bioelectron.* 126: 734–742.

182 Peng, Y., Huang, Y., Zhu, Y. et al. (2017). Ultrathin two-dimensional covalent organic framework nanosheets: preparation and application in highly sensitive and selective DNA detection. *J. Am. Chem. Soc.* 139 (25): 8698–8704.

183 Wang, M., Hu, M., Liu, J. et al. (2019). Covalent organic framework-based electrochemical aptasensors for the ultrasensitive detection of antibiotics. *Biosens. Bioelectron.* 132: 8–16.

184 El-Mahdy, A.F.M., Kuo, C.-H., Alshehri, A. et al. (2018). Strategic design of triphenylamine-and triphenyltriazine-based two-dimensional covalent organic frameworks for CO_2 uptake and energy storage. *J. Mater. Chem. A* 6 (40): 19532–19541.

185 Spitler, E.L., Koo, B.T., Novotney, J.L. et al. (2011). A 2D covalent organic framework with 4.7-nm pores and insight into its interlayer stacking. *J. Am. Chem. Soc.* 133 (48): 19416–19421.

186 DeBlase, C.R., Hernández-Burgos, K., Silberstein, K.E. et al. (2015). Rapid and efficient redox processes within 2D covalent organic framework thin films. *ACS Nano* 9 (3): 3178–3183.

11

Solid Phase Synthesis Catalyzed by Microwave and Ultrasound Irradiation

R.M. Abdel Hameed[1], Amal Amr[2], Amina Emad[2], Fatma Yasser[2], Haneen Abdullah[2], Mariam Nabil[2], Nada Hazem[2], Sara Saad[2], and Yousef Mohamed[2]

[1] *Cairo University, Faculty of Science, Chemistry Department, El Gamaa Street, Giza 12613, Egypt*
[2] *Cairo University, Faculty of Science, Biotechnology Department, El Gamaa Street, Giza 12613, Egypt*

11.1 Introduction

Microwave radiation is a band inside the whole electromagnetic spectrum with a frequency range of $3 \times 10^8 – 3 \times 10^{11}$ Hz [1]. When compared to traditional heating methods [2–9], many additional characteristics could be gained by applying microwave radiation for heating purposes. It displayed outstanding reaction kinetics [10–12], enhanced energy output, lowered activation energy [13, 14], and decreased instrument size [15]. Besides, specific volumetric heating features with superior controllability could be presented by employing microwave irradiation tools [16].

The microwave irradiation method is a simple and fast fabrication route for metallic nanoparticles [17–19]. Heat could be easily transferred to the reactants using a transparent reaction container that was assisted by microwaves. Shorter crystallization periods and uniform nucleation processes for the catalyst nanoparticles could be achieved when applying microwave irradiation in relation to commonly applied heating ways. When the temperature gradients could not be neglected, the catalyst nanoparticles were irregularly deposited with a wide size distribution histograms and much reduced product yields would be attained. Therefore, great concern was paid for facilitating many chemical and analytical processes by subjecting their systems to microwaves. These applications include polymerization reactions [20], organic and inorganic fabrication routes [21, 22], extraction processes [23, 24], degrading macromolecules [25, 26], reducing air pollution [27, 28], and contamination of soil [29, 30]. Here, in the present chapter, the authors will discuss the activity of microwave-irradiated nanocomposites for water pollution, biodiesel production, fuel cells, and ORR fields.

Water pollution is one of the main problems that gradually increases due to continued industrialization. Many pollutants could be identified in contaminated water samples, including pesticides, pharmaceuticals, dyes, and organic compounds

Green Chemical Synthesis with Microwaves and Ultrasound, First Edition.
Edited by Dakeshwar Kumar Verma, Chandrabhan Verma, and Paz Otero Fuertes.
© 2024 WILEY-VCH GmbH. Published 2024 by WILEY-VCH GmbH.

[31–35]. Oxidizing these pollutants is considered an efficient solution for their mineralization to reach good water quality restrictions [36–39]. Wastewater toxicity is expected to be reduced by lowering the organic species amount by promoting their biodegradability and destroying their structure [40–43].

Broad attention has been focused on developing direct alcohol fuel cells as suitable energy devices with numerous advantageous features including reduced price, increased energy efficiency, enhanced current density, and facile storage [44]. Since the alcohol oxidation process takes place at the anode counterpart, the choice of its material represents the main challenge to achieving satisfactory energy output. Pt-based anodes were widely studied in alcohol fuel cells and exhibited outstanding oxidation, current densities, and good stability during long operations [45]. However, their increased cost and rapid surface poisoning by accumulated reaction byproducts could limit the commercial application of Pt-containing nanomaterials [46]. Therefore, many research groups are directed to find out reasonable solutions for these drawbacks by lowering the added Pt content in the prepared anode nanocatalyst and promoting its catalytic efficiency. Incorporating transition metals and their oxide species into the formed nanomaterial structure could modify its reactivity and participate in scavenging the surface from any adsorbed reaction poisons via their oxidation [47]. Introducing carbonaceous-based metal oxides (NiO, CoO, MnO_2, ZnO, CeO_2) could effectively solve these problems [48–60].

The outstanding advantages of biodiesel as eco-friendly with increased flash points suggest its promising usage as a safe alternative to diesel. Additionally, biodiesel shows favored combustion and engine behavior with easier handling and transportation when related to diesel. Reduced CO output, unburned hydrocarbons, and particulate species could greatly improve the exhaust emission properties of biodiesel [61–64].

11.2 Wastewater Treatment

Wastewater treatment is an important process because the precipitated species in their samples show great harmful effects on human well-being when present in appreciable amounts. Therefore, extracting these dangerous chemicals is a highly significant issue that directs many scientists to develop simple and safe protocols to have pure drinking water [65–68]. Conventional wastewater treatment generally consists of three stages as primary, secondary, and tertiary parts. The sedimentation process represents the primary step which is essential to getting rid of large-sized sewage particles. In the secondary stage, water samples are biologically treated through the application of activated sludge sequence and operating microbial metabolism to build up active biofilms. Certain physicochemical methods are then required as a tertiary stage involving absorption, membrane separation, and chemical oxidation.

Microwave-assisted preparation routes have been extensively applied in the last few years for many nanomaterials as a result of their advantageous formation of homogenous nuclei within shorter crystallization periods in relation to those obtained by traditional heating ways [69, 70]. Numerous nanocomposites were examined for treating industrial wastewater samples after their contamination by organic dyes [71–74]. For instance, Adekunle et al. [75] fabricated Fe_2O_3 nanoparticles using microwave irradiation and studied their photocatalytic performance for degrading eriochrome black-T (EBT) and murexide in wastewater. They measured the maximum degradation efficiency of these two dyes at iron oxide species of 25 mg loading, after exposing the investigated samples to microwave irradiation within 40 minutes. About 96% of EBT and 98% of murexide content were decomposed. Al-Shehri et al. [76] synthesized Ag_2S nanoparticles by adding varied weights of sodium dodecyl sulfate (SDS), and the obtained suspensions were subjected to microwave irradiation at 700 W for 10 minutes. SEM pictures of these deposits are shown in Figure 11.1. Spherical nanoparticles were regularly distributed in all samples and some clusters were also observed. Larger-sized nanoparticles were formed when SDS was absent from the deposition solution (see Figure 11.1a), displaying diameters of 40–60 and 100–130 nm for nanoparticles and nanoclusters, respectively. However, introducing SDS in increased amounts during the fabrication routes gradually decreased the obtained nanoparticles diameters to have

Figure 11.1 SEM images of Ag_2S nanoparticles after their preparation using microwave irradiation in presence of (a) 0, (b) 0.1, (c) 0.25, (d) 0.5, and (e) 1.0 g SDS. Source: Al-Shehri et al. [76]. © 2020, Reproduced with permission from Elsevier.

ranges of 37–56 nm (for 0.1 g SDS containing sample, see Figure 11.1b), 35–52 nm (for 0.25 g SDS containing sample, see Figure 11.1c) and 30–47 nm (for 0.5 and 1.0 g SDS containing sample, see Figure 11.1d,e). This observation might explain the outperformed activity of fabricated Ag_2S samples in the presence of increased SDS content during the deposition process, as a result of their increased surface area. They could efficiently decolorize wastewater samples containing methyl green dye when subjected to illumination by UV/visible light. Qureshi et al. [77] have fabricated CNTs inside a suitable microwave-assisted reactor using H_2 and acetylene as precursor gases. Crystal violet as a cationic dye in wastewater samples of the textile industry could be significantly adsorbed at these CNTs through the Langmuir model with a maximum capacity of 2.615 mg/g. Adjusting the studied solution pH at 7 with a starting concentration of 10 mg/l was enough to have reasonable dye molecule adsorption within 25 minutes. The surface morphology of microwave-treated CNTs was investigated before and after dye adsorption in Figure 11.2. The smoothed CNT surface before adsorption (see Figure 11.2a) was appreciably altered by opened tubes with some attached functional groups during the adsorption process (see Figure 11.2b).

Many industrial processes including mining, electroplating, and battery manufacture left behind increased volumes of contaminated effluents with heavy metallic species like Cd, Pb, and Ni. They seriously cause fatal threats when accumulated in appreciable concentrations inside the human body [78]. Pb(II) ions could result in anemia, brain damage, and anorexia [79]. Some additional toxicity influences might be observed when extra doses of Cd(II) species were introduced such as nervous system damage, lung injury, and renal dysfunction [80]. Chronic bronchitis, skin dermatitis, and gastrointestinal disorders were the main diseases that accompanied the uptake of polluted water samples with Ni(II) ions [81]. Therefore, many studies were concerned with treating heavy metal pollution problems by developing efficient technologies for the purification of water [82]. Microwave irradiation is a significant tool for wastewater treatment [83–85]. Many materials with increased active surface area and broad pore size distribution histograms could appreciably absorb microwaves [86–96]. For instance, Tripathi et al. [97] have investigated the activity of biomass-based char for heavy metals removal

Figure 11.2 SEM pictures of CNTs before (a) and after (b) adsorbing crystal violet dye. Source: Qureshi et al. [77]. © 2022, Reproduced with permission from Elsevier.

from wastewater samples. Tea waste char was efficiently experienced for getting rid of Hg and Cd. This was explained by the increased BET surface area of this biomass waste char. Employing the microwave pyrolysis method for synthesizing this char resulted in the distribution of an increased number of pores throughout its structure. The removal capacity values of Cd and Hg were estimated as 44.8% and 50.2%, respectively. Qu et al. [98] have grafted rice husk-based cellulose with β-cyclodextrin using epichlorohydrin and glutaraldehyde. This functionalization step was activated through microwave radiation within 17 minutes. Superior capacity values of Pb(II) ions monolayer adsorption could be achieved at this modified cellulose with epichlorohydrin and glutaraldehyde as 216.06 and 279.08 mg/g, respectively. These obtained membranes could be applied to have a purified effluent into the acid batteries from Pb(II) ions with increased selectivity and good stability over four cycles of adsorption–desorption processes.

Treating organic wastewater with microwave irradiation could appreciably contribute to removing various types of pollutants like dyes [99–101], pharmaceuticals [102], phenols [103], and petroleum resides [104]. Microwaves could also be applied during sewage sludge purification in order to generate biogas and promote the anaerobic digestion step [83]. For instance, the reduced biodegradability and increased stability of cutting oil in water emulsion samples represent a serious environmental problem. Physical protocols were supposed to remove these pollutants from wastewater tanks by their transfer into different phases. The wet peroxide oxidation process was applied to degrade cutting oil pollutants with the aid of microwave irradiation. Pulses of microwave power of 800 W were operated through the investigated system resulting in complete demulsification, with a total organic carbon removal percentage of 82 within 10 minutes only [105]. Modified ceramic membranes with catalytic deposits could be efficiently exploited to remove pollutants such as 1,4-dioxane and reduce the fouling problems. Certain radicals could be continuously formed at the hotspot positions at the membrane surface, after its irradiation with microwave to degrade the adsorbed pollutants and dissipate the produced energy to impregnated catalyst nanoparticles [106]. Wang et al. [107] impregnated coal fly ash in 0.5 M H_2SO_4 solution at 303 K for 60 minutes. It could be successfully applied for removing p-nitrophenol from wastewater samples under the action of microwaves with an efficiency value of 90%. Microwave-assisted chelation of beta zeolite helped in its dealumination with subsequent hydrophobicity increase. Its increased surface area enhanced the adsorption ability of beta zeolite for benzene molecules. Moreover, multi-cycle regeneration/adsorption experiments along with TGA analysis revealed facile regeneration of modified zeolite support with unaffected physicochemical and adsorption features [108]. Wang et al. [109] have concluded that microwave irradiated MWCNTs-nickel foam nanocomposite displayed a total organic carbon removal efficiency value of 97% when treating wastewater samples. However, bare MWCNTs and nickel foam showed only 65.2% and 79.3%, respectively. This might be attributed to the creation of focal channels for strong absorption of microwaves resulting in a pronounced heating response.

Activated carbon could be successfully fabricated by heating primary paper mill sludge at a pyrolysis temperature of 800 °C for 20 minutes using KOH as an activating agent in a 20% ratio relative to that of the waste sludge. Microwave irradiation helped in this preparation plan to get support with an enlarged surface area of 1196 m^2/g, which, in turn, facilitated the adsorption of two antibiotics from water samples as amoxicillin and sulfamethoxazole with the greatest capacity values of 204 ± 5 and 217 ± 8 mg/g, respectively [110]. Fe nanopowders were synthesized using NaBH$_4$ under the action of microwave irradiation and examined for treating real wastewater samples containing certain drugs. Adding 360 mg/l Fe nanopowder and subjecting the system to microwave power of 780 W for 60 minutes were sufficient to achieve a chemical oxygen demand percentage of 96.8 with complete organic carbon removal. Contaminated water samples with diclofenac and ibuprofen were efficiently treated after their complete degradation within 35 minutes. The variation of (C/C_0) with time for diclofenac and ibuprofen was represented in Figure 11.3. Linear plots were observed, revealing that the degradation reaction for these pharmaceutics obeys zero-order kinetics. The corresponding reaction rate constants were estimated as 0.0283 and 0.0294 mg/(l min) [111]. The Fenton-like reaction was promoted by passing microwave irradiation through the examined system to remove increased concentrations of pharmaceutics inside wastewater samples. The COD percentage reached 57.53 after irradiating microwave power of 300 W for six minutes. Macromolecular species greatly settled down by forming flocs from aggregated particles. Microwave irradiation treatment could improve the degradation efficiency with enhanced settling of accumulated sludge when compared to traditional heating sequences [112]. Poly(N-isopropylacrylamide) (PNIPAm) nanogel was prepared by microwave irradiation of ammonium peroxodisulfate for 20 minutes. This heating protocol greatly reduced the required reaction time and resulted in a 65% yield. The fabricated microwave-assisted

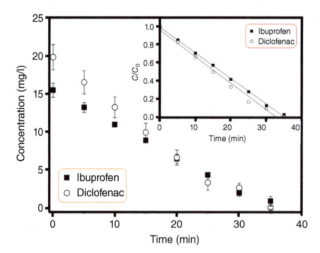

Figure 11.3 Plots of decayed concentration of ibuprofen and diclofenac inside wastewater samples with time at Fe nanopowder surface. The related C/C_0 variation as a function of time was also shown in the inset figure. Source: Vieira et al. [111]. © 2021, Reproduced with permission from Elsevier.

Table 11.1 A list of the detected concentration of removed pollutants, the employed microwave irradiation conditions (time and power), solution pH, and pollutant removal rate percentage for a number of water contaminants.

Pollutant	Concentration (volume)	Microwave irradiation Time (min)	Power (W)	pH	Pollutant removal rate %	References
Volatile fatty acids	2140 mg/l	8	625	7.2	30.7	[114]
Pyridine	100 ml	5	750	9	98	[115]
Nitroglycerin	50 mg/l	8	225	9	55	[116]
Ammonia nitrogen	5 g/l	10	750	11	93	[83]
Carbofuran	800 mg/l	30	750	8	100	[117]
Amoxicillin	6 g/l	7	480	9	96.4	[118]
p-Nitrophenol	6 g/l	8	500	—	99	[119]
Methyl orange	—	8	539	—	98.3	[120]
Congo red	2 g/l	1.5	800	8	89.2	[121]
Sodium dodecyl benzene sulfonate	2.632 g/l	3	750	7	100	[122]
Crystal violet	0.8 mg/l	5	750	9	97	[96]
Phenol	66.7 mg/l	5	750	7	98	[123]
Perchloroethylene	12 g/l	10 s	700	—	95	[124]
Remazol golden yellow	80 mg/l	1.5	400	7	94	[92]
Imidacloprid	26.52 g/l	1	119	6	60	[125]

nanogel could be exploited to efficiently adsorb paraquat molecules from industrial wastewater samples with an efficiency percentage of 60–70 and pesticide release percentage of 35–45 [113]. Extra examples of applying microwave irradiation for treating wastewater samples are presented in Table 11.1. Here, the detected concentration of removed pollutants, the employed microwave irradiation conditions (time and power), solution pH, and pollutant removal rate percentage for a number of water contaminants are listed [83, 92, 96, 114–125].

11.3 Biodiesel Production

Biodiesel is a monoalkylester of fatty acids that are extracted using animal fats and vegetable oils [126–128]. Biodiesel possesses promising features resembling that of petroleum-derived diesel [129–131]. It is a biodegradable and safe fuel with much decreased emitted byproducts since it is free from sulfur and hydrocarbons [132]. Edible vegetable oils, including palm oils, soybeans, and sunflower, could be processed to produce first-generation biodiesels. However, the continuous usability of these oils to supply energy could result in global fluctuation of their price as the main source of foodstuff [133–135]. Therefore, some nonedible plant-based oils were examined as biofuels like *Pongamia pinnata* [136], *Calophyllum inophyllum* [137, 138], *Schleichera triguga* [139], and *Ceiba pentandra* [140]. These second-generation biodiesels could display excellent oxidation stability, desirable cold flow characteristics, and outstanding biodiesel yields.

A number of protocols were planned for producing biodiesel involving micro-emulsion, pyrolysis, and transesterification [141–143]. The formation of methyl or ethyl ester by reacting the respective alcohol with triglycerides in animal fats or vegetable oils was commonly applied [144–148]. The addition of a catalyst to the esterification system was favored [149–151]. Enzymatic, homogeneous, and heterogeneous types were examined. In spite of the increased activity of enzymes at reduced temperatures, and their inert performance toward free fatty acids, the high price of enzymes hindered their commercialization [152, 153]. On the other hand, introducing homogenous catalysts to the reaction mixture displayed a number of difficulties that concerned the system corrosion, the separation steps, and increased sensitivity to the examined fatty acids [154, 155]. Investigating the feasibility of incorporating heterogeneous ingredients to catalyze the esterification process was tried by many research work groups as a result of their tendency to recover the observed difficulties in other catalyst types. Here, the possibility of forming soaps could be efficiently retarded as well as the applied catalyst could be simply separated when the ester formation reaction was completed [156–159]. Different classes of heterogeneous catalysts were investigated during biodiesel formation, such as single and mixed metal oxides, hydrotalcite, zeolites, loaded alkali earth metals onto suitable supports, and ion exchange resins. Metal oxide-based catalysts are favored for biodiesel conversion reactions as a result of their availability, low-cost price, good stability, lowered toxicity, and ability to be recycled [160].

TiO_2 nanoparticles were homogeneously dispersed onto a graphene oxide (GO) surface using a two-step hydrothermal process with the help of microwave irradiation. Different Ti:GO ratios were introduced as 0.1:1, 0.3:1, 0.5:1, and 1:1. Post-sulfonation reaction was then operated to prepare highly stable SO_3H-GO@TiO_2 nanocatalyst over continuous 10 esterification cycles. The crystalline structure of GO@TiO_2 did not become deformed after the sulfonation treatment exhibiting increased surface area and enlarged pore diameter values that accounted for its promoted catalytic performance [161]. Bekhradinassab et al. [162] doped TiO_2 with 2 wt% of both Mn and Fe through a combustion process in the presence of hybridized microwave plasma treatment. A macroporous structure was formed and the walls between pores were gradually destroyed under the action of plasma to have increased diameters allowing for macromolecule diffusion. This synthesized MnFeTiO_2 nanocomposite showed promising performance for biodiesel production with conversion loss % of 2.17 and 0.21 when plasma untreated and treated samples were respectively investigated. This improved behavior of MnFeTiO_2 nanocomposite after plasma irradiation treatment was explained by the increased bond strength between different components inside this nanocomposite that, in turn, efficiently reduced their leaching.

Yang et al. [163] have supported SrO–ZnO nanocatalysts in a metal–organic framework. It offered a significant specific surface area increase by 3.7 folds with superior magnetic properties by introducing Fe_3O_4 nanoparticles. The transesterification reaction time was greatly reduced from 20 to 5 minutes by applying microwave heating at 80 °C to achieve a biodiesel yield of 99%. PEG/MgO/ZSM-5@Fe_3O_4 nanocatalyst was employed in the microwave-assisted transesterification of spirulina platensis using ethanol. The maximum biodiesel yield was 95.8% at the nanocatalyst containing a load of 15 wt% PEG, possessing increased stability when reused for six cycles [164]. Amani et al. [165] have synthesized spinel MgO/$MgFe_2O_4$ nanocatalysts using microwave irradiation. Three urea fuel ratios were introduced during the preparation of mixtures as 1, 1.5, and 2. Based on BET results, the fabricated sample in the presence of a fuel ratio of 1.5 displayed the highest active surface area and pore diameters among the studied mixtures as 50 m^2/g and 4.6 nm, respectively. The reusability experiments also confirmed the ability of MgO/$MgFe_2O_4$ nanocatalysts to produce biodiesel by 78.8% after operation for five cycles. This might be attributed to the advantageous features of applying microwave irradiation for fabricating nanoparticles as a simple, inexpensive, and rapid technique. The formed nanoparticles were also homogeneously distributed as revealed from EDX analysis results in Figure 11.4. Mg, Fe, and O were well dispersed with an increased degree of uniformity. Dutta et al. [166] synthesized green ZnO particles using banana corm extract in the presence of microwave irradiation. SEM image of the ZnO catalyst displayed spherical particles with an appreciable degree of agglomeration. Dynamic light scattering spectroscopy revealed the formation of varied sizes of ZnO particles with the most predominant diameter of 372 nm. A 2.5 wt% ZnO catalyst was able to convert fish lipids to biodiesel with a maximum efficiency of 89.86%. Gas chromatography–mass spectroscopy showed the basic components of fatty acids methyl ester in the formed fish lipid biodiesel.

Zhang et al. [167] synthesized a green bio-catalyst of CaO nanoparticles based on the reduction of their extract from the chicken eggshell by tea decoction. This nanocatalyst was exploited to convert the chicken feather oil into biodiesel when inserted as 1%, together with an increased methanol content by eight times as related to this oil molarity. The transesterification reaction lasted for five minutes, resulting in an optimum conversion of 95%. The biodiesel yield was affected by the microwave irradiation conditions including power, time as well as methanol:oil ratio, and the incorporated catalyst amount. Increasing the power of irradiated microwaves from 200 to 500 W gradually enhanced the conversion rate resulting in recording a maximum yield at 500 W; however, beyond this power value, the obtained yield continuously decreased (see Figure 11.5a). Moreover, the methanol:oil ratio varied in different operating mixtures from 2 up to 14. Increasing the alcohol content was sufficient to achieve an increased biodiesel yield up to 8:1 ratio. An unfavorable reaction rate was then expected when this ratio increased (see Figure 11.5b). The transesterification reaction also required enough time to get significant biodiesel product as revealed from the results in Figure 11.5c, where operating the conversion

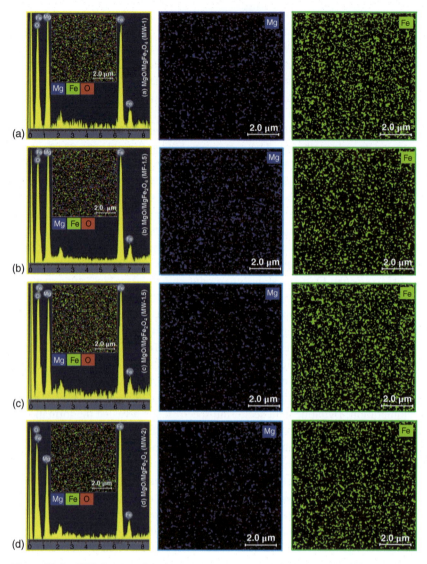

Figure 11.4 EDX charts and their corresponding mapping pictures of MgO/MgFe$_2$O$_4$ nanopowder after its fabrication using microwave irradiation in presence of urea fuel ratios of (a) 1, (c) 1.5, and (d) 2. The obtained analysis results of the heated sample using muffle furnace were also compared in section (b). Source: Amani et al. [165]. © 2019, Reproduced with permission from Elsevier.

process for five minutes could get the optimum yield; while prolonged periods might cause the alcohol evaporation to decay the reaction rate. Additionally, extra reaction time could give the chance for the formation of undesirable byproducts with the possibility of reversing the direction of this reversible reaction. On the other hand, Hsiao et al. [168] prepared CaO nanoparticles and chemically modified them with bromooctane using microwave irradiation. This heating technique greatly reduced the

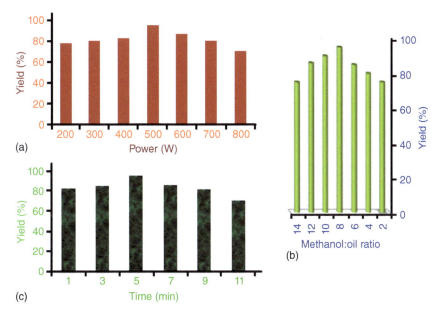

Figure 11.5 Effect of varying the transesterification reaction parameters of chicken feather meal oil using bio-nano CaO derived from chicken egg shell on the obtained yield including (a) microwave power, (b) methanol:oil ratio, and (c) the reaction time. Source: Zhang et al. [167]. © 2022, Reproduced with permission from Elsevier.

preparation time to one-twelfth of that needed by conventional heating methods as well as increased the thermo stability of formed CaO nanospecies. Athar et al. [169] produced biodiesel from increased free fatty acid Jatropha oil by adding methanol in H_2SO_4 solution under the action of microwave irradiation. Using a 3 wt% catalyst and methanol:oil molar ratio of 10:1, 54.07% of biodiesel yield could be attained after 40 minutes. The maximum biodiesel output was 61.1% from a microwave heating system containing a 2 wt% catalyst and methanol:oil ratio of 11:7 after 90 minutes. On the other hand, applying the conventional heating routes requires the addition of increased methanol and catalyst concentrations to the reaction system to produce a reduced yield of 32.31% within a much-prolonged period of 20 hours.

Li et al. [170] have successfully prepared Ca-Benzenetricarboxylic acid (Ca-BTC) and calcined in N_2 atmosphere at 800 °C. The obtained Ca-800 N displayed an outstanding surface area, enabling it to absorb microwaves with increased capacity as indicated by the measured dielectric properties (dielectric constant and dielectric dissipation factor of 33.11 and 3.64 F/m). SEM image of Ca-BTC in Figure 11.6a demonstrated its hexagonal prism skeleton. After its carbonization in the N_2 atmosphere at 800 °C, a multilayered nanosheet structure of Ca-800N was observed besides some uniform nanoparticles that were anchored onto the formed surface, as seen in the respective pictures in Figure 11.6b–d. The elemental components of Ca-800 N, including calcium, oxygen, and carbon were regularly distributed on the nanocatalyst surface in wt% values of 18.62, 27.09, and 54.29 (see Figure 11.6e–g). Hong et al. [171] have exploited waste cooking oil (WCO) to prepare biodiesel

Figure 11.6 SEM pictures of (a) Ca-BTC and (b–d) Ca-800N nanocatalysts. The elemental mapping images of carbon, oxygen and calcium were also presented in sections (e–g), respectively. Source: Li et al. [170]. © 2022, Reproduced with permission from Elsevier.

through a transesterification reaction under the action of microwave irradiation. The low acid value group (LAVG) and high acid value group (HAVG) could be measured when 1.0 and 1.2 wt% catalysts were respectively introduced to the reaction system. This required the application of microwave power of 500 and 600 W for a reaction period of six minutes with a methanol/WCO molar ratio of eight to finally have biodiesel with a domestic acid value below 2.5 mg KOH/g. Ali et al. [172] have demonstrated that WCO could be efficiently converted into biodiesel using an activated limestone-based catalyst inside microwave-assisted reactor. This conversion reaction was greatly influenced by its period, the introduced catalyst content, and the molar ratio of alcohol to oil. About 96.65% biodiesel yield could be formed when the conversion reaction was continued for 55.26 minutes, in the presence of a loaded catalyst % of 5.47, and increased methanol content over oil by 12.21 folds. Employing a microwave-based reactor for this conversion reaction was beneficial in reducing the reaction time by 77% when related to that needed inside a conventional reactor. Decreased emission content of NO_x and particulates during biodiesel combustion was measured in comparison to that observed by petrodiesel.

Palm kernel oil (PKO) is considered as good feedstock for biodiesel production due to its increased oil content. The transesterification reaction of this oil was studied using conventional and microwave-assisted heating routes. By conventional method, methanol:oil molar ratio of 6 with 1 wt% NaOH as the catalyst was introduced to the heating system at 60 °C. After 90 minutes, 96.4% of biodiesel yield was formed. This prolonged reaction time could be greatly reduced to 2.5 minutes when the system was subjected to microwave irradiation by curtailing the separation process from four hours to four minutes, having an increased biodiesel yield of 97.6% [173]. Milano et al. [174] have prepared a blend of WCO with *C. inophyllum* oil (W70CI30) in a 7:3 volume ratio. This blending process appreciably promoted the oxidation stability of WCO (18.03 hours) and its cold flow characteristics were improved recording both pour and cloud points of 2 °C and a cold filter plugging point of 1 °C. Moreover, the microwave treatment of this reaction system greatly reduced its operation time to 9.15 minutes, compared to 82.15 minutes for conventional transesterification conversion. Yadav et al. [175] have exploited *O. sativa* husk as an agricultural waste to prepare some carbonaceous catalysts via carbonization and sulfonation processes. These nanocatalysts were applied to activate the esterification reaction of oleic acid in the presence of H_2SO_4. The maximum biodiesel yield of 99.6% was attained with an activation energy of 76.11 kJ/mol and a pre-exponential factor of 1.4×10^{10} min^{-1}. The increased S content and the total acid density of this biomass nanocatalyst enabled its reusability with good stability over seven cycles. Wahidin et al. [176] have studied the effect of applying 1-ethyl-3-methylimmidazolium methyl sulfate as an ionic liquid during microwave heating for the transesterification system of wet *Nannochloropsis* sp. biomass to produce biodiesel. When the reaction was operated for 25 minutes in the presence of wet algae:methanol ratio of 1:4, and methanol:ionic liquid ratio of 1:0.5, a 40.9% yield of biodiesel was obtained. Phosphomolybdic acid/chitosan was employed as an active catalyst for the transesterification reaction of pomegranate oil. 1.25 wt% catalyst content was added to this reaction system in the presence of a methanol:oil molar ratio of 6:1. The reaction lasted for 74 minutes to get a biodiesel yield of 90%. The first-order kinetics were deduced with an activation energy of 50 kJ/min and frequency factor of 16.47×10^7 min^{-1}. Moreover, the reusability of phosphomolybdic acid/chitosan was examined over six cycles, showing a slight decrease in biodiesel yield from 90% to 71% with an average value of 81.66% (see Figure 11.7) [177].

Gouda et al. [178, 179] examined the activity of the UiO-66-SO$_3$H catalyst to convert oleic acid to methyl oleate. When 20 folds of methanol relative to oleic acid were incorporated in the reaction system, together with a loaded catalyst in 8 wt %, the reaction was heated for one hour at 100 °C under the influence of microwave irradiation. An efficient conversion reaction was operated with a biodiesel yield of 98.3 ± 0.8% and reusability % of 82.12 ± 0.6 up on five consecutive runs. The variation of the natural logarithm of the rate constant of esterification reaction with the reciprocal of absolute temperature exhibited a linear relationship to ascertain the predominance of first-order kinetics with activation energy and pre-exponential factor values of 35.33 kJ/mol and 2.4×10^8 min^{-1}, respectively (see Figure 11.8). Pham et al. [180] have studied the conversion reaction of Basa

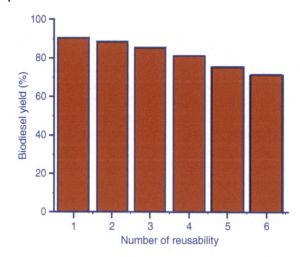

Figure 11.7 The durability test of phosphomolybdic acid/chitosan. Source: Helmi et al. [177]. © 2022, Reproduced with permission from Elsevier.

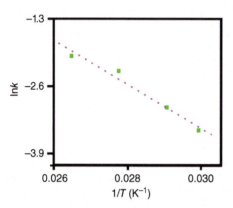

Figure 11.8 Arrhenius plot of oleic acid transesterification reaction using sulfonic acid functionalized UiO-66 nanocatalyst. Source: Gouda et al. [178]. © 2022, Reproduced with permission from Elsevier.

and Tra catfish oils into biodiesel. Different variables could affect the obtained yield including the catalyst concentration, methanol:oil molar ratio, the used co-solvent, the reaction temperature, and time. Increased biodiesel yield within a shorter time was measured using microwave-assisted esterification protocol when compared to those shown by the traditional mechanical stirring method. Lin et al. [181] have investigated the preparation of biodiesel from the esterification reaction of WCO in the presence of waste oyster shells-based catalyst via a microwave heating system. Adding 6 wt% catalyst to the reaction mixture as well as adjusting the methanol:oil ratio at 9 : 1 could participate in optimizing the experimental conditions of this conversion process. Irradiating the reaction mixture using microwaves with a power of 800 W at 65 °C for 180 minutes efficiently catalyzed its rate with the highest yield of 87.3%. Ergan et al. [182] have exploited a dolomite catalyst to activate the conversion of canola oil to biodiesel using conventional and microwave heating systems. Applying microwaves during the transesterification process significantly reduced the needed cost and energy by 58.2% when compared to the traditional heating methods. Furthermore, a faster rate of ester formation by 30.4% was also recorded. A list of some employed catalysts for the transesterification reaction of the

Table 11.2 A list of some employed catalysts for the transesterification reaction of the most used feedstock oils as well as this process conditions (microwave irradiation power, methanol:oil ratio, temperature, and elapsed time), and the obtained yield.

Feedstock	Catalyst	Microwave irradiation power (W)	Transesterification reaction conditions			Yield %	References
			Methanol: oil ratio	T (°C)	Time (min)		
Palm oil	Commercial CaO (5 wt%)	150	9:1	65	60	82.90	[183]
WCO	Bromooctane modified CaO (4 wt%)	300	8:1	65	75	98.20	[168]
WCO	Activated limestone-based catalyst (5.47 wt%)	—	12.21:1	60	55.26	96.65	[172]
WCO	Extracted lignin from sugarcane (15 wt%)	—	18:1	60	15	89.19	[184]
WCO	NaOH (0.8 wt%)	—	12:1	65	2	98.20	[185]
Chicken feather meal oil	Bio CaO nanoparticles extracted from chicken eggshell (1 wt%)	500	8:1	40–60	5	95.00	[167]
Jatropha curcas oil	CaO-based nanoparticles extracted from waste oyster shells	800	9:1	65	180	91.10	[186]
Palm fatty acid distillate	SO_3H-GO@TiO_2 (3 wt%)	—	9:1	70	36	96.70	[161]
Ceiba pentandra oil	KOH (0.84% (w/w))	—	60%	100	388 s	95.42	[187]
Dairy scum oil	KOH (1 wt%)	—	7:1	60	5	93.47	[188]
Argemone oil	NaOH (1.03 wt%)	—	9.5:1	58.8	3.56	99.03	[189]
Oleic acid	Biochar (8 wt%)	50	24:1	80	60	99.60	[175]
Oleic acid	CaO/C (8 wt%)	400	9:1	60	30	97.37	[170]
Soybean oil	SrO–ZnO (8 wt%)	—	11:1	80	120	95.80	[163]
Waste lard	CaO/zeolite (8 wt%)	595	30:1	65	75	90.89	[190]

most used feedstock oils as well as this process conditions (microwave irradiation power, methanol:oil ratio, temperature, and elapsed time) and the obtained yield was presented in Table 11.2 [161, 163, 167, 168, 170, 172, 175, 183–190].

11.4 Oxygen Reduction Reaction

The urgent need to replace fossil fuels with cleaner energy sources with much reduced environmental pollutants and cost-effective production is one of the main concerns for many researchers [191]. ORR could be considered as valuable sustainable energy supply [192]. It takes place inside metal–air batteries and fuel cells, resulting in energy conversion from a chemical form into a renewable electrical one [193]. A number of ideas were planned to achieve promoted ORR activity at the cathode material [194–196]. Platinum is the basic constituent in many cathode nanocatalysts [197]; however, its wide application was limited by the increased cost and fast decay after operation for long periods [198, 199]. These problems could be solved by alloying Pt with suitable metals to increase the exposed surface area to the studied reaction and reduce the required Pt content in the synthesized nanocatalyst as well as enhancing its observed activity [200, 201]. In this respect, 40 wt% Pt nanoparticles were loaded onto MWCNTs using the microwave-assisted impregnation method. ECSA value of this Pt/MWCNTs nanocomposite was measured as 29.50 m^2/g recording 1.56 times enhancement when related to that of comparable Pt/C (E TEK). Rotating disk electrode measurements during ORR demonstrated outperformed mass activity of Pt/MWCNTs at 500 mV [202]. Yin et al. [203] have supported Pt nanoparticles onto CNTs using intermittent microwave heating technique after treating with HF (Pt/CNTs-HF), H_2O_2 (Pt/CNTs-H_2O_2), and HF/H_2O_2 (Pt/CNTs-HF/H_2O_2). TEM images demonstrated a uniform dispersion of Pt nanoparticles onto CNTs surface with average sizes of 2.8, 2.9, 3.3, and 4.0 nm for Pt/CNTs-HF/H_2O_2, Pt/CNTs-HF, Pt/CNTs-H_2O_2, and pristine Pt/CNTs, respectively. Their related utilization efficiencies were also calculated as 86.3%, 69.6%, 46.8%, and 38.9%. The treated Pt/CNTs nanocatalysts with HF, H_2O_2, and HF/H_2O_2 displayed more positive E_{onset} and $E_{1/2}$ values during the ORR study in H_2SO_4 solution when compared to the reported results at untreated Pt/CNTs. The increased conductivity of CNTs as well as good dispersion of Pt nanoparticles motivated the utilization of an increased number of available active sites at Pt/CNTs surface. Moreover, the application of microwave irradiation, after treatment with HF and H_2O_2, increased the functionalization of fabricated CNTs with their subsequent influence in modifying the physical and chemical features of the formed carbonaceous surface resulting in a superior electroactivity. Sahin et al. [204] have examined the electrochemical behavior of supported Pt and Pt–Cr onto carbon for ORR. Rotating ring disk electrode experiments showed that two Tafel slope values were measured for Pt/C and PtCr/C as 66/73 and 116/118 mV/dec in low (Temkin isotherms) and high (Langmuir isotherms) current density regions, respectively. The related exchange current density values were $3.34 \times 10^{-4}/1.62 \times 10^{-3}$ and $2.71 \times 10^{-2}/5.02 \times 10^{-2}$ mA/cm^2 for Pt/C and PtCr/C to illustrate the enhanced oxygen reduction activity at PtCr/C. Sandström et al. [205] have successfully fabricated Pt$_x$Y$_y$ nanoparticles using microwave irradiation. The mass activity of the prepared nanocomposite in Pt:Y molar ratio of 3 : 1 toward ORR was twice of that measured at Pt/carbon Vulcan. Density functional theory studies predicted the migration of Y to the nanocatalyst surface with further oxidation to form Y_2O_3 species, as evidenced by X-ray photoelectron spectroscopy. These metallic oxide species could create numerous active sites at the Pt surface to promote the reaction kinetics and stabilize the overall nanocomposite performance. Alloying Pt with Cu in PtCu/C electrocatalysts in different proportions was found to enhance ORR.

Applying microwave irradiation during the chemical reduction of Pt and Cu ionic species participated in altering the crystal structure of Pt. The lattice diameter was reduced resulting in smaller Pt nanoparticles with better dispersion and increased electroactivity. Oxygen molecules could be adsorbed on the catalytic sites with facile bond breaking, while the formed intermediate byproducts exhibited decreased adsorption capacity to improve the measured electrocatalyst stability. Incorporating Cu nanospecies with a Pt:Cu molar ratio of 2:1 significantly shifted the observed potential toward more positive values with increased limiting current densities in relation to those attained at commercial Pt/C [206]. Pt/Cu nanoparticles were supported onto CNTs using a microwave-assisted polyol process. The synthesis factors, including the heating temperature, solution pH, and polyvinylpyrrolidone (PVP) addition as a stabilizer were significantly controlling the morphology and nanocatalyst loading that, in turn, affected its activity for ORR. Heating the formed PtCu/CNTs-PVP nanocatalyst at varied temperatures such as 400, 500, and 600 °C did not affect its dispersion, where homogeneously distributed nanoparticles at CNTs surface were observed irrespective of the fabrication temperature (see the respective TEM images in Figure 11.9a–c). However, increasing the solution pH in the absence of PVP appreciably increased the agglomeration of deposited nanoparticles as revealed by the TEM image in Figure 11.9d. An increased Cu content was measured for precipitated nanocatalyst particles from solutions with increased pH values, while the absence of PVP from this synthesis solution at reduced pH value would participate in forming nanocatalyst particles with increased Pt loading. The maximum specific activity for ORR was attained at the fabricated nanocatalyst sample from a synthesis solution containing PVP with pH = 11. It was 4.27 times higher than that achieved at the prepared one in the absence of PVP [207]. PdW nanoparticles onto sulfur-doped graphene were synthesized using a microwave-assisted chemical reduction process. Surprising electroactivity for ORR in KOH solution was shown with an E_{onset} value of −20 mV (Ag/AgCl) as 100 mV more positive than the measured one at supported Pd nanoparticles onto graphene (Pd/G) and doped graphene with sulfur (Pd/SG). Koutecky–Levich plots were studied to elucidate the number of included electrons during ORR as 3.94, 2.85, and 2.78 for PdW/SG, Pd/SG, and Pd/G, respectively [208].

Doped carbon supports with heteroatoms could display increased activity toward ORR in acidic and basic solutions, to resemble the measured performance at Pt-based nanomaterials [209–215]. The presence of numerous defects as edges and holes in these doped supports could appreciably increase their active surface area which, in turn, may motivate the fabricated nanocomposite performance [216, 217]. Many routes of introducing nitrogen, sulfur, and boron into carbonaceous nanocatalyst structure were examined, including plasma treatment [218], hydroxide corrosion [219], and template process [220, 221]. Natural biomass sources could be employed to prepare various carbon supports such as lotus root [222], water hyacinth [223], peanut [224], and dandelion seed [225]. For example, Kim et al. [226] have doped graphene support with nitrogen and boron, and the obtained GO powder was reduced through a simple hydrothermal process in the presence of microwave irradiation. Enhanced performance toward ORR was measured at

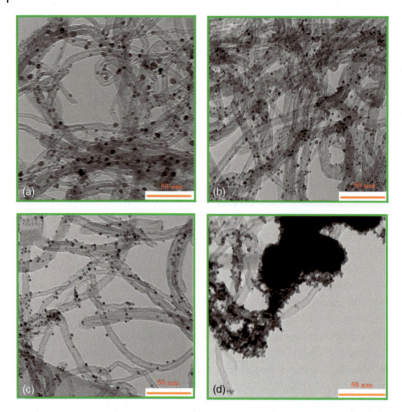

Figure 11.9 TEM pictures of PtCu/PVP nanocatalysts after their heat treatment at (a) 600, (b) 500, and (c) 400 °C. The obtained TEM image for prepared PtCu sample in absence of PVP was also presented in section (d). Source: El-Deeb and Bron [207]. © 2015, Reproduced with permission from Elsevier.

doped graphene in an alkaline solution. Linear sweep voltammograms of this nanocatalyst were studied at varied rotating speeds in Figure 11.10a. Constant E_{onset} values were nearly observed compared to increased current densities as the rotating rate increased. The related Koutecky–Levich curves were drawn in Figure 11.10b at different potentials. The slopes of these linear plots were the basis for calculating the number of transferred electrons during the reduction process. It ranged between 3.05 and 3.84, while the potential was extending from 0.665 down to 0.265 V (see the inset figure). Dai et al. [227] have doped hollow graphite matrix with N and Fe. A superior specific surface area of 1487 m^2/g was measured for this treated graphite support as a result of carbon oxidation in the air by the action of microwave heating. The presence of Fe could activate the formation of a porous support structure with increased graphitization degree that, in turn, facilitated the oxygen reduction process at a doped graphite surface with outstanding stability in both acidic and alkaline electrolytes. Xu et al. [228] derived Fe–N–C nanocatalyst from Fe-doped ZIF-8 and examined its activity for ORR in 0.1 M HClO$_4$ solution. Conventional fabrication routes of this nanomaterial relied on adding certain organic solvents that increased the cost as well as consumed more energy and

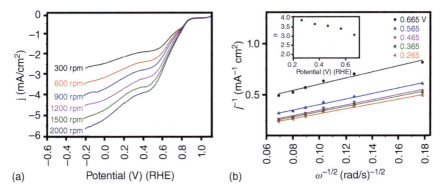

Figure 11.10 (a) Rotating disk electrode measurements of doped graphene support with nitrogen and boron at rotation speeds ranging between 300 and 2000 rpm. (b) The related Koutecky–Levich curves at varied potentials. Source: Kim et al. [226]. © 2016, Reproduced with permission from Elsevier.

time. Instead, a solvent-free protocol was employed for irradiating the preparation system with microwaves in conjunction with ball milling steps. The obtained nanopowder was then subjected to the pyrolysis process at 1000 °C to finally have a hierarchically porous surface with highly dispersed FeN_4 sites. The detailed synthesis sequences are illustrated in Figure 11.11. Excellent activity toward reducing oxygen molecules was measured at this nanocatalyst with $E_{1/2}$ and limiting current density values of 782 mV (RHE) and 6.14 mA/cm², respectively. This result resembled the reported activity of commercial Pt/C (823 mV and 6.04 mA/cm²). Moreover, facilitated reaction kinetics were attained at Fe–N–C nanomaterial as elucidated from its Tafel slope value of 58.3 mV/dec. Increased stability with extraordinary methanol tolerance was also endowed by this nanocatalyst when assembled inside direct methanol fuel cells. Kuo et al. [229] treated cellulose and lignin using the hydrothermal carbonization technique with microwave activation in the NH_3 atmosphere to fabricate N-doped graphitic biochars. They consisted of iron–nitrogen moieties in the core and nitrogen-doped graphitic carbon in the shell. The electrocatalytic activity of these doped biochars with varied Fe percentages was investigated for ORR in alkaline electrolytes. Promoted results at cellulose-derived graphite supports were measured when related to those at lignin-based surfaces. BET-specific surface area of cellulose-derived biochars gradually increased with increased Fe% in the prepared nanomaterial showing the highest one at the doped sample with 1% Fe. However, decreased areas were then recorded for those with Fe content >1%. This might account for the extraordinarily limiting current density of the sample with 1% Fe (3.2 mA/cm²) in relation to those with increased iron concentration during the ORR study.

Due to the observed problems when noble-metal-containing nanomaterials were investigated for ORR, transition metals with their enhanced activity and good stability were applied [230–233]. Binary systems such as Fe–Co, Cu–Co, Mn–Co, and Co–Zn exhibited excellent performance when related to their individual counterparts [234–236] as a result of the synergistic influence between these mixed

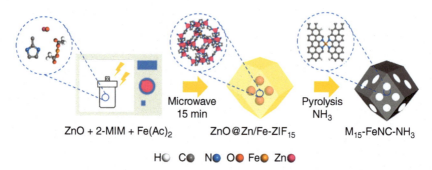

Figure 11.11 Steps of preparing ZnO@Zn/Fe-ZIF15 and M_{15}-FeNC-NH_3. Source: Xu et al. [228]. © 2021, Reproduced with permission from Elsevier.

metals and carbonaceous supports with excellent conductivity and numerous reaction active sites [237–240]. Li et al. [241] fabricated $CoSe_2$/C nanomaterials in different Se:Co molar ratios using microwave irradiation. By introducing Se:Co ratios of 2–4, average crystal diameters of 12.9–15.4 nm were formed. The best oxygen reduction activity was measured at the nanocatalyst containing a Se:Co ratio of three with an overpotential value of 705 mV, and an electron transfer number of four. The presence of a slightly increased content of Se oxide in this nanocomposite structure contributed to the formation of smaller-sized particles, while their increased agglomeration could be observed for the samples containing too many oxide species, resulting in unfavored activity toward the O_2 reduction process. Kang et al. [242] fabricated a CuCo-N/C nanocatalyst using a microwave-assisted hydrothermal process based on the detailed steps in Figure 11.12. Extraordinary oxygen reduction performance was observed at this nanocomposite with an $E_{1/2}$ value of 850 mV (RHE) and a limiting current density of 5.61 mA/cm^2. A Zn-air battery could be constructed with CuCo-N/C to attain a power density of 66.9 mW/cm^2 and V_{oc} of 1468 mV. On discharging for 12 hours, this battery displayed a voltage decay of only 1.4% to outperform that one with a Pt/C electrode. CoFeRu/C nanocatalyst was fabricated by applying a microwave heating step in a mixed solution of ethylene glycol and water, at a radiation power of 600 W for 30 minutes at 220 °C. The needed time during the synthesis routes was significantly reduced by radiating the reaction system with microwaves when related to that observed for conventional thermolysis processes (longer than five hours). This prepared catalyst displayed an enhanced activity for ORR with a charge transfer coefficient value of 0.306 and exchange current density of 2.49×10^{-7} A/cm^2 compared to 7.22×10^{-9} A/cm^2 at Pt/C [243]. WS_2 nanosheets were fabricated using microwave irradiation in the absence of any template. They exhibited outstanding activity for reducing O_2 molecules in 0.1 M KOH solution at an E_{onset} value of 820 mV (Ag/AgCl). Rotating disk measurements at ranged speeds between 100 and 2500 rpm revealed current density rise by shortening the diffusion distance at increased rotating rates as reported in Figure 11.13a. Koutecky–Levich relations in Figure 11.13b suggested that the examined reaction obeyed first-order kinetics with an electron transfer number of 3.29 [244].

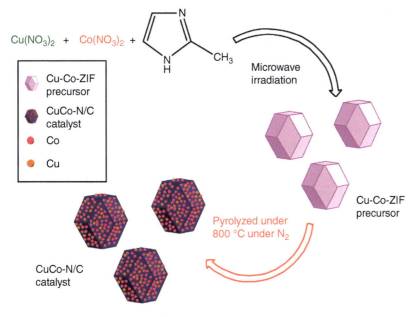

Figure 11.12 Fabrication steps of CuCo-N/C nanocatalyst. Source: Kang et al. [242]. © 2019, Reproduced with permission from Elsevier.

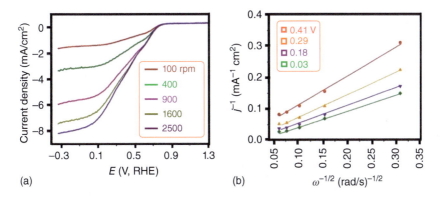

Figure 11.13 (a) Linear sweep voltammograms of modified WS_2 electrode using different rotation speeds in 0.1 M KOH solution after its saturation with O_2 and (b) the related Koutecky–Levich plots. Source: Gnanaprakasam et al. [244]. © 2022, Reproduced with permission from Elsevier.

Semiconductor transition metal oxides, including WO_3, MoO_3, ZnO, MnO_2, Cr_2O_3, MoO_3, V_2O_5, NiO, and CuO showed superior electrical conductivity, better resistance to poison accumulation at their surfaces, and in turn improved their durability during the studied reaction over long operation periods as well as increased thermal stability with elevated melting points. These fascinating features could contribute to motivating their electrochemical behavior for ORR [245–248]. From numerous research reports, Zhang et al. [249] applied the hydrothermal protocol in the presence of microwaves to fabricate MnO_2 nanospecies through the chemical reduction of $KMnO_4$ using different additions of HCl solution. Different morphologies of MnO_2 were obtained as δ-microspheres, α-nanorods, γ-nanosheets, and β-octahedrons when 0.05, 0.10, 0.25, and 0.75 ml HCl solutions were added, respectively. Cyclic voltammetry study indicated that all MnO_2 structures could activate the oxygen reduction process in KOH solution. The unique morphology and crystal phase of α-MnO_2 nanorods enabled a suitable adsorption mode of oxygen molecules, that, in turn, appreciably increased the measured current density (59.1 μA) when related to other prepared MnO_2 forms. Cabello et al. [250] have employed microwave-hydrothermal protocol to fabricate anatase TiO_2 nanoparticles within 20 seconds. This short preparation period could save the needed energy and cost, and restricted particle growth was observed with highly uniform dispersion onto the graphite surface. A good electrocatalytic performance of TiO_2/graphite nanocomposite toward ORR could be reported. The presence of increased TiO_2 content in the examined nanocatalyst was sufficient to attain enhanced electrode activity; however, thick oxide layers could hinder the charge transfer step to decrease the reduction reaction rate. The two-electron pathway was elucidated through H_2O_2 molecule formation that was then reduced to H_2O. Kiran et al. [251] have synthesized $ZnCo_2O_4$ onto GO support with the help of microwave irradiation. The specific mandarin-flower-like structure of this nanocatalyst appreciably increased its measured active surface area. The incorporation of GO into this nanocomposite chemical architecture participated in promoting its electrical conductivity, resulting in improved ORR kinetics. This could explain the reduced overpotential of $ZnCo_2O_4$/GO (300 mV) at −3 mA in relation to that at $ZnCo_2O_4$ (450 mV). $MnFe_2O_4$/C nanocatalyst was prepared using a microwave sintering process. It exhibited outstanding stability when examined for ORR in 6 M KOH solution at −200 mV (Hg/HgO). The almost constant current density of 98 mA/cm^2 was measured within the whole chronoamperometry period. Much reduced resistance value was also observed after doping XC-72R carbon support with this mixed spinel oxide as inferred from the electrochemical impedance spectroscopy (EIS) measurements to ascertain its enhanced activity for the O_2 reduction step. Moreover, when $MnFe_2O_4$/C is assembled inside a primary zinc–air battery, its discharge voltage would gradually increase up to 0.816 V at 80 mA/cm^2 [252]. Applying microwave irradiation during the preparation of spinel $MnCo_2O_4$ resulted in the formation of smaller particles with increased specific surface areas. This could participate in improving this nanocatalyst's performance when examined for ORR [253].

Zhang et al. [254] fabricated Fe_3O_4/Fe/C nanocatalyst using microwave sintering method and the mechanism of ORR was studied at its surface. EIS plots displayed a linear trend in the low-frequency region to demonstrate the predominance of the semi-infinite diffusion process during this reaction. Liu et al. [255] have designated a specific carbon bath for fabricating Fe/Fe_3C nanocatalyst onto doped carbon sheets with nitrogen (Fe/Fe_3C@NC) through irradiation with microwaves within 35 minutes. This effective heating protocol resulted in the formation of a mesoporous nanocatalyst structure with increased specific surface area and numerous active sites for ORR. These synergistic characteristics of Fe/Fe_3C@NC could explain its surprising E_{onset} and $E_{1/2}$ values of 980 and 870 mV (RHE) as 50 mV more positive than the respective ones at 20% Pt/C. Fe/Fe$_3$C@NC also showed a promoted tolerance toward methanol molecules when ORR kinetics were measured in alkaline solution. Tafel slope value of 80 mV/dec was derived at Fe/Fe$_3$C@NC compared to 95 mV/dec at 20%Pt/C to clarify the improved reduction reaction kinetics at this Fe-based nanocatalyst. The chronoamperometry test demonstrated increased stability by 43.7% over 30 000 seconds at Fe/Fe$_3$C@NC when contrasted to that at Pt/C. Ajmal et al. [256] prepared Fe$_3$O$_4$ nanoparticles onto a CNT surface using a microwave-assisted procedure and their activity for ORR was examined. This Fe$_3$O$_4$@CNTs nanomaterial displayed a reduced Tafel slope value of 83 mV/dec than those reported at unsupported Fe$_3$O$_4$ nanoparticles (230 mV/dec) (see Figure 11.14a). The nanocomposite stability was also studied by repeated cyclic voltammetry in 0.1 M KOH solution for 5000 cycles in an O$_2$-saturated atmosphere. The obtained polarization curves in Figure 11.14b demonstrated a small loss in $E_{1/2}$ value (20 mV) and current density (220 µA/cm^2) to ascertain the increased durability of supported Fe$_3$O$_4$ nanoparticles onto CNTs. Mohamed [257] has exploited microwave irradiation to fabricate FeWO$_4$/Fe$_3$O$_4$ nanoparticles onto N-doped rGO sheets (FeWO$_4$/Fe$_3$O$_4$@NrGO). The preparation steps were followed as described in detail in Figure 11.15. Here, metallic salts were dissolved in ammonia solution. A defined content of GO was dispersed in this mixture by adding 1% urea. It was then continuously stirred for 2 hours and microwave radiation was passed through the reaction system for 10 minutes at 350 W. The formed nanocatalyst was then collected and carefully washed. A drying step was sufficient at 80 °C in a vacuum atmosphere for 12 hours. This FeWO$_4$/Fe$_3$O$_4$@NrGO could efficiently catalyze ORR with a limiting current density of 5.1 mA/cm^2 at an E_{onset} value of 930 mV. Synergistic features of combined Fe$_3$O$_4$, FeWO$_4$, and NrGO and the related increase in this nanocomposite active sites could explain its outstanding behavior toward ORR when compared to that of its individual counterparts.

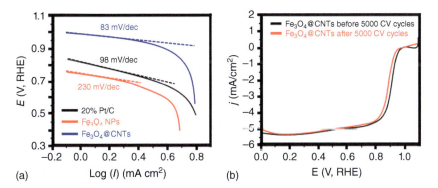

Figure 11.14 (a) Tafel plots of Fe$_3$O$_4$@CNTs nanomaterial and (b) its linear sweep voltammograms before and after cyclization for 5000 runs in soaked 0.1 M KOH solution with O$_2$ at 5 mV/s and rotating speed of 1600 rpm. Source: Ajmal et al. [256]. © 2023, Reproduced with permission from Elsevier.

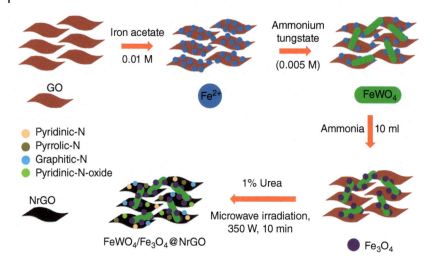

Figure 11.15 Synthesis steps of FeWO$_4$/Fe$_3$O$_4$@NrGO nanocatalyst. Source: Mohamed [257]. © 2021, Reproduced with permission from Elsevier.

11.5 Alcoholic Fuel Cells

Fossil fuels as the main source for most energy demands display a number of dangerous drawbacks to our environment due to the emitted gases in the atmosphere with their subsequent greenhouse phenomenon and serious influence on the health of mankind [258]. Therefore, scientists make an effort to find effective alternatives to these energy sources with satisfied outputs and the least harmful by-products. Fuel cells represent a promising solution to meet this continuous need for valuable energy supplies since they simply produce sufficient amounts of electric energy from the stored chemical energy into some reactants [259, 260]. Alcohol fuel cells exhibited improved kinetics during ORR with reduced levels of corrosion [261–263]. Platinum has been recognized as the most investigated metallic catalyst in different types of fuel cells. Xin et al. [264] have doped graphene with 5.04 wt% nitrogen content through its heating inside the NH$_3$ atmosphere using microwave irradiation. Pt nanocatalyst onto N-doped graphene (Pt/NG) was fabricated by chemical reduction using ethylene glycol with heating inside a microwave oven via pulse mode (10 seconds on/20 seconds off) for 10 runs. This Pt/NG exhibited extraordinary activity for oxidizing methanol molecules in 0.5 M H$_2$SO$_4$ solution with a peak current density of 24.94 mA/cm^2, about two folds higher than that shown at deposited Pt nanoparticles onto undoped graphene support. The electronic structure of graphene was significantly affected by the doping process, resulting in a strong interaction between its surface and Pt species. Moreover, Pt nanoparticles were highly dispersed to finally have numerous active sites for the investigated reaction as revealed from ECSA values of Pt/G and Pt/NG as 37.62 and 80.45 m^2/g, respectively. Digesting MWCNTs in 5 M HNO$_3$ solution using microwave irradiation could effectively open the ends of these tubes, providing an increased surface area for further deposition of platinum nanoparticles. Surface area values of 5.6, 20.1, and 23.9 cm^2 were measured for pristine MWCNTs, treated MWCNTs with HNO$_3$, and digested MWCNTs by microwaves, respectively. This surprising surface area of microwave-treated MWCNTs could account for their improved activity toward methanol oxidation reaction with a mass activity of 253 mA/mg$_{Pt}$ as 1.03 and 1.62 folds higher than the respective values at treated MWCNTs with HNO$_3$ and pristine MWCNTs [265]. Sakthivel et al. [266] have studied the effect of varying the heating time and temperature of microwave irradiation on the formed Pt loading at MWCNTs by polyol reduction protocol. Increasing the heating time decreased the loaded Pt nanoparticles on carbon support by increasing their size to 1.8, 2.0, and 2.4 nm for those heated within 50, 100, and 150 seconds. Raising the heating temperature from 100 to 140 °C gradually increased the obtained Pt loading, reaching 60 wt% for those formed at 140 °C. Further temperature increases resulted in decreased Pt content at MWCNTs. The surfactant:Pt precursor salt ratio also controlled the loaded Pt amount at the carbon surface. A maximum Pt loading of 60 wt% was attained when the surfactant:precursor salt ratio was 2 : 1, while the absence of surfactant during the nanocatalyst preparation greatly reduced this loading value to 18 wt%. For a single-cell assembly with very low loading of examined Pt/MWCNTs as the cathode counterpart, the measured power density was 74 mW/cm^2 compared to 30 mW/cm^2 for the cell that contained the commercial catalyst. Functionalized CNTs were treated with microwave irradiation at 190 °C for 10 minutes for depositing Pt or Pd nanoparticles. Lowered Tafel slopes and activation energy values were attained when ethanol molecules were oxidized at Pd-based nanocomposites, giving a chance for their wide application, especially with their reduced cost in relation to those of Pt nanohybrids [267]. A series of Pt–Sn–Rh nanospecies was fabricated by the chemical reduction route using ethylene glycol and the preparation system was subjected to microwave irradiation. Pt$_{4.5}$Sn$_{1.5}$Rh$_1$/C nanocatalyst exhibited outstanding electroactivity for methanol and ethanol oxidation reactions with mass activity values of 1.70 and 2.18 A/mg$_{Pt}$, respectively as 3.7 and 5.74 folds greater than the measured performance at Pt/C. The durability of Pt$_{4.5}$Sn$_{1.5}$Rh$_1$/C, Pt$_3$Sn$_1$Rh$_1$/C, and Pt$_{4.5}$Sn$_1$Rh$_{1.5}$/C nanocomposites was investigated by repeated cyclization up to 1000 sweep runs as indicated in Figure 11.16. The mass activities of these nanomaterials decayed by 56.88%, 61.25%, and 65% in relation to their initial values in the first cycle. Moreover, the TEM study demonstrated that all synthesized Pt–Sn–Rh-based nanocatalysts retained their main morphology after this stability test. However, a serious agglomeration was observed for Pt/C [268]. Burhan et al. [269] have supported alloyed PtCo nanocatalysts onto hybrid carbon surfaces including (activated carbon-Vulcan carbon) (PtCo@AC-VC) and (reduced GO-Vulcan carbon) (PtCo@rGO-VC), and their electroactivity was compared against PtCo@rGO. After 1000 s from the start of the chronoamperometry experiment in (1 M methanol + 1 M KOH) solution, PtCo@rGO, PtCo@rGO-VC, and PtCo@AC-VC nanocatalysts showed stable residual current densities of 10.16, 14.85, and 23.24 mA/cm^2, respectively. The application of hybrid carbon supports was aiming to improve the characteristics of fabricated nanocatalysts. Proper mixing of AC and VC in PtCo@AC-VC nanocatalyst significantly motivated the proton

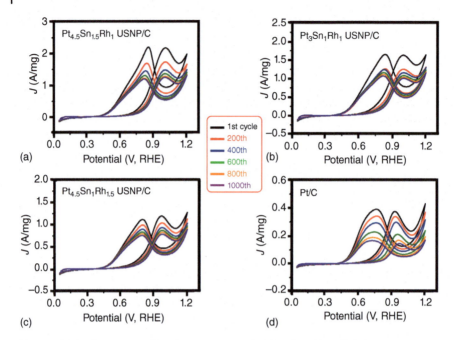

Figure 11.16 Repeated cyclic voltammograms of (a) $Pt_{4.5}Sn_{1.5}Rh1$USNP/C, (b) $Pt_3Sn_1Rh_1$USNP/C, (c) $Pt_{4.5}Sn_1Rh_{1.5}$USNP/C, and (d) commercial Pt/C nanocatalysts for 1000 runs during ethanol oxidation reaction. Source: Hu et al. [268]. © 2020, Reproduced with permission from Elsevier.

transfer process at its surface with better tolerance to the reaction intermediates by accelerating their oxidation rate.

In spite of the increased activity of Pt-based nanocomposites, their expensive cost and lowered kinetics when employed for alcohol's electro-oxidation process might hinder the wide application of these nanomaterials. Palladium is more abundant than platinum and shows faster reaction kinetics at its fabricated nanocatalyst surfaces [270–274]. Depositing Pd nanoparticles onto varied carbon supports could participate in promoting their electroactivity as a result of increasing the active sites for the examined processes. Alloying this noble metal with cheaper and active transition metals [275–280] or metal oxides [281–284] is an additional strategy for outstanding electrode reactivity. Doping rGO with nitrogen and employing microwave irradiation contributed to intercalating Pd nanoparticles inside their sheets, and were attached at the edges as regularly distributed spheres (see cross-sectional SEM pictures of this nanomaterial in Figure 11.17). Improved performance of palladium nanoparticles onto nitrogen doped rGO (Pd/N-rGO) toward ethanol electrooxidation was observed. Mass activities of 32.76 and 42.16 mA/mg_{Pd} were measured at Pd/rGO and Pd/N-rGO, respectively, with a negative shift in the oxidation peak potential by 62 mV [285]. Sikeyi et al. [286] have prepared carbon nano onions by flame pyrolysis of olive oil (p-CNO). Nitrogen (N-CNO) and oxygen (ox-CNO) functionalized carbon supports were then obtained through chemical vapor deposition and reflux protocols, respectively. Pd nanoparticles were then

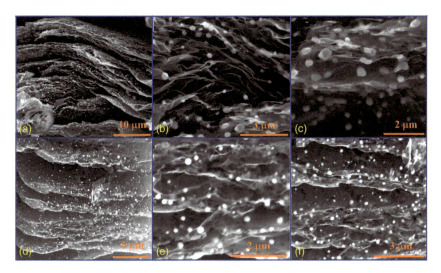

Figure 11.17 SEM pictures of Pd-NrGO nanocatalyst using various magnifications. Source: Kumar et al. [285]. © 2018, Reproduced with permission from Elsevier.

deposited onto these different carbon supports through the chemical reduction method by $NaBH_4$ with the assistance of microwave irradiation. The electroactivity of Pd nanodeposits onto these CNO surfaces toward oxidizing ethanol molecules in alkaline media was examined. Chronoamperometry curves before and after introducing the alcohol molecules to the solution displayed stable, steady-state current densities as shown in Figure 11.18a and b, respectively. P-CNO, ox-CNO, and N-CNO nanocatalysts gradually decayed to reach 0.47, 0.75, and 1.19 mA/cm^2 after 2000 seconds, respectively, in (1 M ethanol + 1 M KOH) solution. Incorporating Pd nanoparticles was beneficial in enhancing the catalytic performance of the formed nanocomposites having current densities of 4.3, 5.1, and 6.9 mA/cm^2

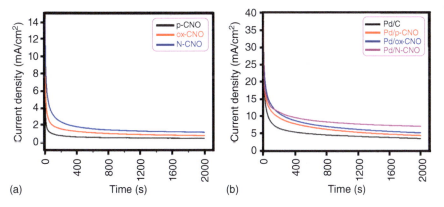

Figure 11.18 Chronoamperograms of (a) p-CNO, ox-CNO and N–CNO nanopowders as well as (b) Pd/C, Pd/p-CNO, Pd/ox-CNO and Pd/N–CNO nanocatalysts in (1 M ethanol + 1 M KOH) solution at −250 mV. Source: Sikeyi et al. [286]. © 2021, Reproduced with permission from Elsevier.

at the end of the chronoamperometry test. The measured resistances by the EIS technique supported these outstanding results where an appreciable decrease was continuously observed after adding alcohol molecules (commercial Pd/C (396.5 Ω) > Pd/p-CNO (195.3 Ω) > Pd/ox-CNO (109.8 Ω) > Pd/N-CNO (97.9 Ω)). The increased electric conductivity and active surface sites in Pd/N-CNO nanocatalyst as well as the good dispersion of Pd nanoparticles onto this carbon surface could explain its surprising activity. Mphahlele et al. [287] have synthesized Pd/C and PdFe/C nanocatalysts using the chemical reduction method in the presence of $NaBH_4$. The application of microwave irradiation during the preparation route appreciably influenced the physical and electrochemical characteristics of formed nanocomposites. Increased crystallinity was observed for the treated samples with microwaves (Pd/C-MW and PdFe/C-MW) when compared to the conventionally fabricated ones (Pd/C-con and PdFe/C-con). However, they exhibited larger diameters (5.3–7.4 nm) in relation to those at Pd/C-con and PdFe/C-con (2.48–3.02 nm). XPS and Raman spectra showed that more defect sites were observed for Pd/C-MW and PdFe/C-MW which in turn increased their electrocatalytic activity toward glycerol molecules with promoted resistance to accumulated poisons. The presence of different metal oxides in the Pd/C chemical structure was responsible for increasing its available surface-active sites for electrocatalytic purposes. Adding MnO_2, SnO_2, and NiO to Pd/C significantly enlarged its ECSA value by 2.85, 6.50, and 10.90 folds, respectively. This might explain the corresponding enhancement in their I_f/I_b and steady-state current densities during ethanol oxidation reaction in NaOH solution [288]. Supported Cu@Pd/SnO_2 nanoparticles onto graphene were attained using a core-shell structured preparation route as assisted by microwave irradiation. Varying the introduced content of SnO_2 in this nanomaterial significantly affected its performance for oxidizing ethanol molecules in basic electrolytes. The presence of 5 or 20 wt% SnO_2 was sufficient to achieve enhanced oxidation current density. Highly stable steady-state behavior was also observed for SnO_2-containing nanocatalysts with a 50.72–72.55% decrease in relation to the start value of the chronoamperometry test. Lowered charge transfer resistances were also shown to reveal the feasibility of the alcohol oxidation process at Cu@Pd/SnO_2 surfaces [289].

Instead of applying noble-metal-based nanomaterials with their increased cost and limited stability, many researchers have studied the activity of transition metals including macromolecules, complexes, and oxides [290–300]. Mahalingam et al. [301] have examined the activity of decorated rGO nanosheets with CuS nanoparticles and ZnS nanorods by microwave irradiation to act as a cathode material in microbial fuel cells. They exhibited an outstanding power density of (1692 ± 15) mW/m^2 with an E_{oc} value of (761 ± 9) mV. This nanocomposite could reach 83% of Pt/C performance. Mehdinia et al. [302] have anchored tin oxide onto rGO/carbon cloth (rGO-SnO_2/CC) using hydrothermal and microwave-assisted preparation steps. This nanomaterial was assembled as an anode inside a microbial fuel cell system to deliver an optimum power density of 1624 mW/cm^2, as 2.8 and 4.8 folds higher than those at respective rGO/CC and bare CC. The increased conductivity and surface area of synthesized nanocomposite participated in facilitating the electron transfer and motivating the formation of bacterial biofilm.

Figure 11.19 Nyquist plots of (a) bare CC, rGO/CC and rGO-SnO$_2$/CC nanocatalysts in (100 mM K$_3$[Fe(CN)$_6$] + 100 mM phosphate buffer) solution as well as (b) rGO-SnO$_2$/CC in (10 g/l glucose + 100 mM phosphate buffer) solution with and without *E. coli*. Source: Mehdinia et al. [302]. © 2014, Reproduced with permission from Elsevier.

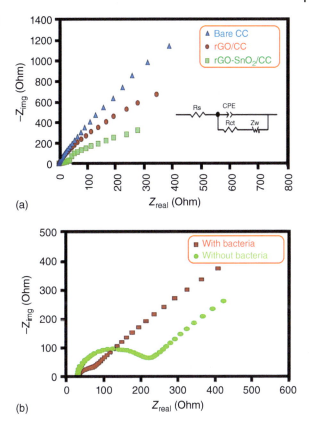

This was illustrated by EIS measurements in Figure 11.19. Nyquist plots of bare CC, rGO/CC, and rGO-SnO$_2$/CC in a phosphate buffer solution in the presence of K$_3$[Fe(CN)$_6$] revealed charge transfer resistance values of 187, 101, and 35 Ω, respectively (see Figure 11.19a). This much lowered resistance at rGO-SnO$_2$/CC confirmed its superior activity. On the other hand, 10 g/l glucose was added to the buffer solution and the related Nyquist plots in the absence and presence of bacteria were studied in Figure 11.19b. Respective resistance values of 180 and 62 Ω were elucidated from these curves to ascertain the biocompatibility of this nanocomposite surface and its ability to form biofilms. Microwave irradiation was exploited to facilitate the deposition of nickel nanoparticles onto the MnO$_x$/C surface by adding NaBH$_4$ as the reducing agent. The activity of fabricated Ni-MnO$_x$/C nanocomposite was studied to oxidize methanol molecules in a KOH solution. Adding 7.5 wt% MnO$_x$ to Ni/C was sufficient to promote the alcohol oxidation current density by 1.43 times with a related catalytic rate constant value of 3.26×10^3 cm^3/(mol s). EIS measurements revealed the increased resistance of Ni-MnO$_x$/C nanocatalyst after long operation at the oxidation peak potential as a result of poison accumulation. However, lowered impedance values were still expected at MnO$_x$ containing nanomaterial when compared to that at Ni/C [51]. Additional studies about the electroactivity of some related nanocatalysts for alcohol electro-oxidation reaction were summarized in Table 11.3, involving their preparation method and ECSA

Table 11.3 A list of microwave irradiated nanocatalysts for fuel cell application with their related preparation method and ECSA values. The oxidized fuel and the measured peak potential, current density, and stability behavior were also reported.

Electrocatalyst	Preparation method	ECSA (m^2/g)	Fuel	Oxidation peak potential (mV) (Ag/AgCl)	Oxidation peak current density (mA/cm^2)	Steady-state current density (mA/cm^2)	References
Pt-CeO$_2$/G[a]	Chemical reduction using ethylene glycol	16.3	Methanol	230	25.5	1.14	[303]
Pt/C	Core-shell method using ethylene glycol	5.65	Ethanol	−206	0.95	0.23	[57]
Ni@Pt/C-0.25		10.00		−226	2.27	1.80	
Ni@Pt/C-0.50		10.65		−226	4.40	4.82	
Ni@Pt/C-1		13.25		−221	6.37	5.52	
Ni@Pt/C-2		18.89		−199	14.23	4.80	
Ni@Pt/C-3		24.05		−217	15.75	14.30	
Pd/C	Chemical reduction using ethylene glycol and NaBH$_4$	9.88	Ethanol	−90	8.55	4.20	[304]
Pd-MnO$_2$/C		32.26		−69	15.92	12.80	
Pd-V$_2$O$_5$/C		55.97		−110	17.26	11.10	
Pd-RuO$_2$/C		171.19		−125	19.26	11.25	
Pd-SnO$_2$/C		229.14		−115	21.27	14.40	
Zn@Pt/AC[b]	Core-shell method using ethylene glycol	37.06	Methanol	785	18.14	0.558	[305]
Mn@Pt/AC		20.37		854	22.22	1.808	
Sn@Pt/AC		53.49		656	44.12	19.777	
Pd/G	Core-shell method using ethylene glycol	54.65	Ethanol	40	10.94	7.89 mA/mg$_{metal}$	[306]
Pd/SnO$_2$–G		94.81		86	30.10	28.85 mA/mg$_{metal}$	
Co@Pd/SnO$_2$–G		217.28		55	39.48	47.78 mA/mg$_{metal}$	
Ni@Pd/SnO$_2$–G		240.33		68	36.51	55.77 mA/mg$_{metal}$	
Cu@Pd/SnO$_2$–G		467.49		99	63.17	84.96 mA/mg$_{metal}$	

a) Pt-CeO$_2$/G: Pt-CeO$_2$/graphene.
b) Zn@Pt/AC: Zn@Pt/activated Vulcan XC-72R carbon black.

values as well as the oxidized fuel, the obtained oxidation peak potential, current density, and steady-state behavior [57, 303–306].

11.6 Conclusion and Future Plans

This book chapter has dealt with applying microwave irradiation to activate the fabrication of many nanomaterials that could be further utilized for electrocatalysis purposes in numerous fields like water treatment, biodiesel production, and fuel cells. The advantageous features of microwave heating including increased uniformity, selectivity, and rapid operation enabled it to outperform the conventional methods.

In spite of the growing need to employ microwave irradiation as a promising technology for nanocatalyst preparation, a number of points that are concerned with the working mechanism of microwave heating technology are still under investigation. Additionally, the coming bottleneck ideas should be taken into account:

- The specific superheating influences of microwaves that are generated at the hotspots with increased feasibility to activate many chemical reaction rates have not been extensively studied yet.
- The modeling design of molecular dynamic reactions in nanosize, with their sufficient characterization, has not been fully examined yet. For instance, many operando analysis tools could be exploited to monitor the rates of microwave-irradiated plasma operations involving color ratio pyrometry and optical emission spectroscopy. This could facilitate collecting sufficient information about the limited space in the studied reaction systems.
- The batch mode of microwave systems displays low-cost operation; however, they need a long time to produce efficient chemical reaction rates with appreciable capacity. Accordingly, much attention should be paid to developing continuous-mode reactors with increased reproducibility and fruitful yields.
- Large-sized microwave heating systems in the industry suffer from the lowered uniformity of produced energy within the subjected absorbers' surfaces. Adjusting the frequency of the microwave irradiation instrument and its pathway depth will promote electromagnetic wave distribution and the obtained product will be improved to a great extent.

References

1 Wang, N., Zheng, T., Jiang, J. et al. (2014). Pilot-scale treatment of *p*-nitrophenol wastewater by microwave-enhanced Fenton oxidation process: effects of system parameters and kinetics study. *Chem. Eng. J.* 239: 351–359.
2 Lai, T.-L., Lee, C.-C., Wu, K.-S. et al. (2006). Microwave-enhanced catalytic degradation of phenol over nickel oxide. *Appl. Catal., B* 68 (3–4): 147–153.

3 Eskicioglu, C., Terzian, N., Kennedy, K.J. et al. (2007). Athermal microwave effects for enhancing digestibility of waste activated sludge. *Water Res.* 41 (11): 2457–2466.

4 Outirite, M., Lebrini, M., Lagrenee, M., and Bentiss, F. (2010). Synthesis of crown compounds containing a 1,3,4-oxadiazole moiety: microwave-assisted synthesis. *J. Heterocycl. Chem.* 47: 555–557.

5 Mokhtar, N.M., Omar, R., and Idris, A. (2012). Microwave pyrolysis for conversion of materials to energy: a brief review. *Energy Sources A* 34 (22): 2104–2122.

6 Sapawe, N., Jalil, A.A., Triwahyono, S. et al. (2013). Cost-effective microwave rapid synthesis of zeolite NaA for removal of methylene blue. *Chem. Eng. J.* 229: 388–398.

7 Mehdizadeh, S.N., Eskicioglu, C., Bobowski, J., and Johnson, T. (2013). Conductive heating and microwave hydrolysis under identical heating profiles for advanced anaerobic digestion of municipal sludge. *Water Res.* 47 (14): 5040–5051.

8 Pawar, R.R., Jadhav, S.V., and Bajaj, H.C. (2014). Microwave-assisted rapid valorization of glycerol towards acetals and ketals. *Chem. Eng. J.* 235: 61–66.

9 Qi, C., Liu, X., Lin, C. et al. (2014). Degradation of sulfamethoxazole by microwave-activated persulfate: kinetics, mechanism and acute toxicity. *Chem. Eng. J.* 249: 6–14.

10 Lin, L., Yuan, S., Chen, J. et al. (2010). Treatment of chloramphenicol-contaminated soil by microwave radiation. *Chemosphere* 78 (1): 66–71.

11 Remya, N. and Lin, J.-G. (2011). Current status of microwave application in wastewater treatment-a review. *Chem. Eng. J.* 166 (3): 797–813.

12 Ju, Y., Fang, J., Liu, X. et al. (2011). Photodegradation of crystal violet in TiO_2 suspensions using UV-vis irradiation from two microwave-powered electrodeless discharge lamps (EDL_{-2}): products, mechanism and feasibility. *J. Hazard. Mater.* 185 (2–3): 1489–1498.

13 Kingman, S.W. and Rowson, N.A. (1998). Microwave treatment of minerals-a review. *Miner. Eng.* 11 (11): 1081–1087.

14 Jones, D.A., Lelyveld, T.P., Mavrofidis, S.D. et al. (2002). Microwave heating applications in environmental engineering – a review. *Resour. Conserv. Recycl.* 34 (2): 75–90.

15 Zhang, X., Wang, Y., Li, G., and Qu, J. (2006). Oxidative decomposition of azo dye C.I. Acid Orange 7 (AO7) under microwave electrodeless lamp irradiation in the presence of H_2O_2. *J. Hazard. Mater.* 134 (1–3): 183–189.

16 Wang, J., Guo, B., Zhang, X. et al. (2005). Sonocatalytic degradation of methyl orange in the presence of TiO_2 catalysts and catalytic activity comparison of rutile and anatase. *Ultrason. Sonochem.* 12 (5): 331–337.

17 Shen, P.K. and Tian, Z. (2004). Performance of highly dispersed Pt/C catalysts for low temperature fuel cells. *Electrochim. Acta* 49 (19): 3107–3111.

18 Xu, C. and Shen, P.K. (2004). Novel $Pt/CeO_2/C$ catalysts for electrooxidation of alcohols in alkaline media. *Chem. Commun.* 19: 2238–2239.

19 Meng, H. and Shen, P.K. (2005). The beneficial effect of the addition of tungsten carbides to Pt catalysts on the oxygen electroreduction. *Chem. Commun.* 35: 4408–4410.

20 Correa, R., Gonzalez, G., and Dougar, V. (1998). Emulsion polymerization in a microwave reactor. *Polymer* 39 (6–7): 1471–1474.

21 Bond, G., Moyes, R.B., and Whan, D.A. (1993). Recent applications of microwave heating in catalysis. *Catal. Today* 17 (3): 427–437.

22 Zhang, Y.M., Wang, P., Han, N., and Lei, H.F. (2007). Microwave irradiation: a novel method for rapid synthesis of D,LL-lactide. *Macromol. Rapid Commun.* 28 (4): 417–421.

23 Prevot, A.B., Gulmini, M., Zelano, V., and Pramauro, E. (2001). Microwave-assisted extraction of polycyclic aromatic hydrocarbons from marine sediments using nonionic surfactant solutions. *Anal. Chem.* 73 (15): 3790–3795.

24 Srogi, K. (2006). A review: application of microwave techniques for environmental analytical chemistry. *Anal. Lett.* 39 (7): 1261–1288.

25 Horikoshi, S., Hidaka, H., and Serpone, N. (2003). Hydroxyl radicals in microwave photocatalysis. Enhanced formation of OH radicals probed by ESR techniques in microwave-assisted photocatalysis in aqueous TiO_2 dispersions. *Chem. Phys. Lett.* 376 (3–4): 475–480.

26 Zhihui, A., Peng, Y., and Xiaohua, L. (2005). Degradation of 4-chlorophenol by microwave irradiation enhanced advanced oxidation processes. *Chemosphere* 60 (6): 824–827.

27 Cha, C.Y. and Kong, Y. (1995). Enhancement of NO_x adsorption capacity and rate of char by microwaves. *Carbon* 33 (8): 1141–1146.

28 Liu, X., Quan, X., Bo, L. et al. (2004). Simultaneous pentachlorophenol decomposition and granular activated carbon regeneration assisted by microwave irradiation. *Carbon* 42 (2): 415–422.

29 Yuan, S., Tian, M., and Lu, X. (2006). Microwave remediation of soil contaminated with hexachlorobenzene. *J. Hazard. Mater.* 137 (2): 878–885.

30 Liu, X. and Yu, G. (2006). Combined effect of microwave and activated carbon on the remediation of polychlorinated biphenyl-contaminated soil. *Chemosphere* 63 (2): 228–235.

31 Malaj, E., von der Ohe, P.C., Grote, M. et al. (2014). Organic chemicals jeopardize the health of freshwater ecosystems on the continental scale. *Proc. Natl. Acad. Sci. U.S.A.* 111 (26): 9549–9554.

32 Le, T.X.H., Nguyen, T.V., Yacouba, Z.A. et al. (2016). Toxicity removal assessments related to degradation pathways of azo dyes: toward an optimization of electro-Fenton treatment. *Chemosphere* 161: 308–318.

33 Danner, M.-C., Robertson, A., Behrends, V., and Reiss, J. (2019). Antibiotic pollution in surface fresh waters: occurrence and effects. *Sci. Total Environ.* 664: 793–804.

34 Charuaud, L., Jarde, E., Jaffrezic, A. et al. (2019). Veterinary pharmaceutical residues from natural water to tap water: sales, occurrence and fate. *J. Hazard. Mater.* 361: 169–186.

35 Liu, N., Jin, X., Feng, C. et al. (2020). Ecological risk assessment of fifty pharmaceuticals and personal care products (PPCPs) in Chinese surface waters: a proposed multiple-level system. *Environ. Int.* 136: 105454.

36 Deng, Y. and Zhao, R. (2015). Advanced oxidation processes (AOPs) in wastewater treatment. *Curr. Pollut. Rep.* 1 (3): 167–176.

37 Dhaka, S., Kumar, R., Lee, S.-H. et al. (2018). Degradation of ethyl paraben in aqueous medium using advanced oxidation processes: efficiency evaluation of UV-C supported oxidants. *J. Cleaner Prod.* 180: 505–513.

38 Zhang, N., Chen, J., Fang, Z., and Tsang, E.P. (2019). Ceria accelerated nanoscale zerovalent iron assisted heterogeneous Fenton oxidation of tetracycline. *Chem. Eng. J.* 369: 588–599.

39 Yang, J., Zhu, M., and Dionysiou, D.D. (2021). What is the role of light in persulfate-based advanced oxidation for water treatment? *Water Res.* 189: 116627.

40 Babu, D.S., Srivastava, V., Nidheesh, P.V., and Kumar, M.S. (2019). Detoxification of water and wastewater by advanced oxidation processes. *Sci. Total Environ.* 696: 133961.

41 Mei, Q., Sun, J., Han, D. et al. (2019). Sulfate and hydroxyl radicals-initiated degradation reaction on phenolic contaminants in the aqueous phase: mechanisms, kinetics and toxicity assessment. *Chem. Eng. J.* 373: 668–676.

42 Wang, Y., Liu, Y., Wu, B. et al. (2020). Comparison of toxicity induced by EDTA-Cu after UV/H_2O_2 and UV/persulfate treatment: species-specific and technology-dependent toxicity. *Chemosphere* 240: 124942.

43 Yu, M.-D., Xi, B.-D., Zhu, Z.-Q. et al. (2020). Fate and removal of aromatic organic matter upon a combined leachate treatment process. *Chem. Eng. J.* 401: 126157.

44 Li, J., Chang, Y., Li, D. et al. (2021). Efficient synergism of V_2O_5 and Pd for alkaline methanol electrooxidation. *Chem. Commun.* 57 (57): 4035–7038.

45 Chen, G., Dai, Z., Sun, L. et al. (2019). Synergistic effects of platinum-cerium carbonate hydroxides–reduced graphene oxide on enhanced durability for methanol electro-oxidation. *J. Mater. Chem. A* 7 (11): 6562–6571.

46 Cui, R., Liu, S., Guo, X. et al. (2020). N-doping holey graphene TiO_2-Pt composite as efficient electrocatalyst for methanol oxidation. *ACS Appl. Energy Mater.* 3 (3): 2665–2673.

47 Liu, X.-J., Sun, Y.-D., Yin, X. et al. (2020). Enhanced methanol electrooxidation over defect-rich Pt-M (M = Fe, Co, Ni) ultrathin nanowires. *Energy Fuels* 34 (8): 10078–10086.

48 Amin, R.S., Abdel Hameed, R.M., El-Khatib, K.M. et al. (2012). Pt–NiO/C anode electrocatalysts for direct methanol fuel cells. *Electrochim. Acta* 59: 499–508.

49 Amin, R.S., Abdel Hameed, R.M., El-Khatib, K.M. et al. (2012). Effect of preparation conditions on the performance of nano Pt–CuO/C electrocatalysts for methanol electro-oxidation. *Int. J. Hydrogen Energy* 37 (24): 18870–18881.

50 Abdel Hameed, R.M. (2015). Optimization of manganese oxide amount on Vulcan XC-72R carbon black as a promising support of Ni nanoparticles for methanol electro-oxidation reaction. *Int. J. Hydrogen Energy* 40 (40): 13979–13993.

51 Abdel Hameed, R.M. (2015). Microwave irradiated Ni–MnO$_x$/C as an electrocatalyst for methanol oxidation in KOH solution for fuel cell application. *Appl. Surf. Sci.* 357 (Part A): 417–428.

52 Abdel Hameed, R.M., Fetohi, A.E., Amin, R.S., and El-Khatib, K.M. (2015). Promotion effect of manganese oxide on the electrocatalytic activity of Pt/C for methanol oxidation in acid medium. *Appl. Surf. Sci.* 359: 651–663.

53 Abdel Hameed, R.M. and El-Sherif, R.M. (2015). Ni-P-SnO$_2$/C composite: synthesis, characterization and electrocatalytic activity for methanol oxidation in KOH solution. *Int. J. Hydrogen Energy* 40 (32): 10262–10273.

54 Abdel Hameed, R.M., Amin, R.S., El-Khatib, K.M., and Fetohi, A.E. (2016). Preparation and characterization of Pt–CeO$_2$/C and Pt–TiO$_2$/C electrocatalysts with improved electrocatalytic activity for methanol oxidation. *Appl. Surf. Sci.* 367: 382–390.

55 Amin, R.S., Fetohi, A.E., Abdel Hameed, R.M., and El-Khatib, K.M. (2016). Electrocatalytic activity of Pt-ZrO$_2$ supported on different carbon materials for methanol oxidation in H$_2$SO$_4$ solution. *Int. J. Hydrogen Energy* 41 (3): 1846–1858.

56 Abdel Hameed, R.M., Amin, R.S., El-Khatib, K.M., and Fetohi, A.E. (2016). Influence of metal oxides on platinum activity towards methanol oxidation in H$_2$SO$_4$ solution. *ChemPhysChem* 17 (7): 1054–1061.

57 Fetohi, A.E., Amin, R.S., Abdel Hameed, R.M., and El-Khatib, K.M. (2017). Effect of nickel loading in Ni@Pt/C electrocatalysts on their activity for ethanol oxidation in alkaline medium. *Electrochim. Acta* 242: 187–201.

58 El-Khatib, K.M., Abdel Hameed, R.M., Amin, R.S., and Fetohi, A.E. (2017). Core-shell structured Cu@Pt nanoparticles as effective electrocatalyst for ethanol oxidation in alkaline medium. *Int. J. Hydrogen Energy* 42 (21): 14680–14696.

59 Zhang, M., Song, Z., Wang, Z. et al. (2021). Platinum quantum dots enhance electrocatalytic activity of bamboo-like nitrogen doped carbon nanotubes embedding Co-MnO nanoparticles for methanol/ethanol oxidation. *J. Colloid Interface Sci.* 590: 164–174.

60 Zhao, W., Ma, L., Gan, M. et al. (2021). Engineering intermetallic-metal oxide interface with low platinum loading for efficient methanol electrooxidation. *J. Colloid Interface Sci.* 604: 52–60.

61 Shahabuddin, M., Liaquat, A.M., Masjuki, H.H. et al. (2013). Ignition delay, combustion and emission characteristics of diesel engine fueled with biodiesel. *Renewable Sustainable Energy Rev.* 21: 623–632.

62 Ghazali, W.N.M.W., Mamat, R., Masjuki, H.H., and Najafi, G. (2015). Effects of biodiesel from different feedstocks on engine performance and emissions: a review. *Renewable Sustainable Energy Rev.* 51: 585–602.

63 Dharma, S., Ong, H.C., Masjuki, H.H. et al. (2016). An overview of engine durability and compatibility using biodiesel-bioethanol-diesel blends in compression-ignition engines. *Energy Convers. Manage.* 128: 66–81.

64 Faried, M., Samer, M., Abdelsalam, E. et al. (2017). Biodiesel production from microalgae: processes, technologies and recent advancements. *Renewable Sustainable Energy Rev.* 79: 893–913.

65 Thakur, V.K. and Thakur, M.K. (2015). Recent advances in green hydrogels from lignin: a review. *Int. J. Biol. Macromol.* 72: 834–847.

66 Thakur, V.K. and Voicu, S.I. (2016). Recent advances in cellulose and chitosan based membranes for water purification: a concise review. *Carbohydr. Polym.* 146: 148–165.

67 Ates, B., Koytepe, S., Ulu, A. et al. (2020). Chemistry, structures, and advanced applications of nanocomposites from biorenewable resources. *Chem. Rev.* 120 (17): 9304–9362.

68 Odoemena, K.I., Rowshon, K.M.D., and Binti, C.M.H. (2020). Advances in utilization of wastewater in agricultural practice: a technical note. *Irrig. Drain.* 69 (1): 149–163.

69 Parsons, J.G., Luna, C., Botez, C.E. et al. (2009). Microwave-assisted synthesis of iron(III) oxyhydroxides/oxides characterized using transmission electron microscopy, X-ray diffraction, and X-ray absorption spectroscopy. *J. Phys. Chem. Solids* 70 (3–4): 555–560.

70 Kijima, N., Yoshinaga, M., Awaka, J., and Akimoto, J. (2011). Microwave synthesis, characterization, and electrochemical properties of α-Fe_2O_3 nanoparticles. *Solid State Ionics* 192 (1): 293–297.

71 Kansal, S.K., Sood, S., Umar, A., and Mehta, S.K. (2013). Photocatalytic degradation of Eriochrome Black T dye using well-crystalline anatase TiO_2 nanoparticles. *J. Alloys Compd.* 581: 392–397.

72 Srivastava, N. and Mukhopadhyay, M. (2014). Biosynthesis of SnO_2 nanoparticles using bacterium *Erwinia herbicola* and their photocatalytic activity for degradation of dyes. *Ind. Eng. Chem. Res.* 53 (36): 13971–13979.

73 Shanker, U., Jassal, V., and Rani, M. (2016). Catalytic removal of organic colorants from water using some transition metal oxide nanoparticles synthesized under sunlight. *RSC Adv.* 6 (97): 94989–94999.

74 Rostami-Vartooni, A. (2017). Green synthesis of CuO nanoparticles loaded on the seashell surface using *Rumex crispus* seeds extract and its catalytic applications for reduction of dyes. *IET Nanobiotechnol.* 11 (4): 349–359.

75 Adekunle, A.S., Oyekunle, J.A.O., Durosinmi, L.M. et al. (2021). Comparative photocatalytic degradation of dyes in wastewater using solar enhanced iron oxide (Fe_2O_3) nanocatalysts prepared by chemical and microwave methods. *Nano-Struct. Nano-Objects* 28: 100804.

76 Al-Shehri, B.M., Shkir, M., Bawazeer, T.M. et al. (2020). A rapid microwave synthesis of Ag_2S nanoparticles and their photocatalytic performance under UV and visible light illumination for water treatment applications. *Phys. E Low Dimens. Syst. Nanostruct.* 121: 114060.

77 Qureshi, S.S., Shah, V., Nizamuddin, S. et al. (2022). Microwave-assisted synthesis of carbon nanotubes for the removal of toxic cationic dyes from textile wastewater. *J. Mol. Liq.* 356: 119045.

78 Sun, H., Shi, B., Yang, F., and Wang, D. (2017). Effects of sulfate on heavy metal release from iron corrosion scales in drinking water distribution system. *Water Res.* 114: 69–77.

79 Pan, W., Pan, C., Bae, Y., and Giammar, D. (2019). Role of manganese in accelerating the oxidation of Pb(II) carbonate solids to Pb(IV) oxide at drinking water conditions. *Environ. Sci. Technol.* 53 (12): 6699–6707.

80 Zeng, G., Liu, Y., Tang, L. et al. (2015). Enhancement of Cd(II) adsorption by polyacrylic acid modified magnetic mesoporous carbon. *Chem. Eng. J.* 259: 153–160.

81 Markou, G., Mitrogiannis, D., Çelekli, A. et al. (2015). Biosorption of Cu^{2+} and Ni^{2+} by *Arthrospira platensis* with different biochemical compositions. *Chem. Eng. J.* 259: 806–813.

82 Guo, X., Yang, Z., Dong, H. et al. (2016). Simple combination of oxidants with zero-valent-iron (ZVI) achieved very rapid and highly efficient removal of heavy metals from water. *Water Res.* 88: 671–680.

83 Lin, L., Yuan, S., Chen, J. et al. (2009). Removal of ammonia nitrogen in wastewater by microwave radiation. *J. Hazard. Mater.* 161 (2–3): 1063–1068.

84 Lin, L., Chen, J., Xu, Z. et al. (2009). Removal of ammonia nitrogen in wastewater by microwave radiation: a pilot-scale study. *J. Hazard. Mater.* 168 (2–3): 862–867.

85 Cravotto, G., Binello, A., Carlo, S.D. et al. (2010). Oxidative degradation of chlorophenol derivatives promoted by microwaves or power ultrasound: a mechanism investigation. *Environ. Sci. Pollut. Res.* 17 (3): 674–687.

86 Jou, C.-J. (2008). Degradation of pentachlorophenol with zero-valence iron coupled with microwave energy. *J. Hazard. Mater.* 152 (2): 699–702.

87 Sun, Y., Zhang, Y., and Quan, X. (2008). Treatment of petroleum refinery wastewater by microwave-assisted catalytic wet air oxidation under low temperature and low pressure. *Sep. Purif. Technol.* 62 (3): 565–570.

88 Bo, L.L., Zhang, Y.B., Quan, X., and Zhao, B. (2008). Microwave assisted catalytic oxidation of *p*-nitrophenol in aqueous solution using carbon-supported copper catalyst. *J. Hazard. Mater.* 153 (3): 1201–1206.

89 Zhang, L., Su, M., and Guo, X. (2008). Studies on the treatment of brilliant green solution by combination microwave induced oxidation with $CoFe_2O_4$. *Sep. Purif. Technol.* 62 (2): 458–468.

90 Bo, L., Quan, X., Wang, X., and Chen, S. (2008). Preparation and characteristics of carbon-supported platinum catalyst and its application in the removal of phenolic pollutants in aqueous solution by microwave-assisted catalytic oxidation. *J. Hazard. Mater.* 157 (1): 179–186.

91 Lai, T.-L., Liu, J.-Y., Yong, K.-F. et al. (2008). Microwave-enhanced catalytic degradation of 4-chlorophenol over nickel oxides under low temperature. *J. Hazard. Mater.* 157 (2–3): 496–502.

92 Bi, X., Wang, P., Jiao, C., and Cao, H. (2009). Degradation of remazol golden yellow dye wastewater in microwave enhanced ClO_2 catalytic oxidation process. *J. Hazard. Mater.* 168 (2–3): 895–900.

93 Lv, S., Chen, X., Ye, Y. et al. (2009). Rice hull/MnFe$_2$O$_4$ composite: preparation, characterization and its rapid microwave-assisted COD removal for organic wastewater. *J. Hazard. Mater.* 171 (1–3): 634–639.

94 Gao, J., Zhao, G., Shi, W., and Li, D. (2009). Microwave activated electrochemical degradation of 2,4-dichlorophenoxyacetic acid at boron-doped diamond electrode. *Chemosphere* 75 (4): 519–525.

95 Zhao, G., Gao, J., Shi, W. et al. (2009). Electrochemical incineration of high concentration azo dye wastewater on the in situ activated platinum electrode with sustained microwave radiation. *Chemosphere* 77 (2): 188–193.

96 He, H., Yang, S., Yu, K. et al. (2010). Microwave induced catalytic degradation of crystal violet in nano-nickel dioxide suspensions. *J. Hazard. Mater.* 173 (1–3): 393–400.

97 Tripathi, M., Agarwal, A., and Mubarak, N.M. (2023). Heavy metal removal from the wastewater using the tea waste derived bio char synthesized through microwave pyrolysis. *Materials Today: Proceedings*.

98 Qu, J., Meng, Q., Lin, X. et al. (2021). Microwave-assisted synthesis of β-cyclodextrin functionalized celluloses for enhanced removal of Pb(II) from water: adsorptive performance and mechanism exploration. *Sci. Total Environ.* 752: 141854.

99 Quan, X., Liu, X., Bo, L. et al. (2004). Regeneration of acid orange 7-exhausted granular activated carbons with microwave irradiation. *Water Res.* 38 (20): 4484–4490.

100 Wang, J., Peng, X., Luan, Z., and Zhao, C. (2010). Regeneration of carbon nanotubes exhausted with dye reactive red 3BS using microwave irradiation. *J. Hazard. Mater.* 178 (1–3): 1125–1127.

101 Zhang, Z., Xu, Y., Ma, X. et al. (2012). Microwave degradation of methyl orange dye in aqueous solution in the presence of nano-TiO$_2$-supported activated carbon (supported-TiO$_2$/AC/MW). *J. Hazard. Mater.* 209–210: 271–277.

102 Qi, X. and Li, Z. (2016). Efficiency optimization of a microwave-assisted Fenton-like process for the pretreatment of chemical synthetic pharmaceutical wastewater. *Desalin. Water Treat.* 57 (25): 11756–11764.

103 Tai, H.-S. and Jou, C.-J.G. (1999). Application of granular activated carbon packed-bed reactor in microwave radiation field to treat phenol. *Chemosphere* 38 (11): 2667–2680.

104 Xia, L., Lu, S., and Cao, G. (2003). Demulsification of emulsions exploited by enhanced oil recovery system. *Sep. Sci. Technol.* 38 (16): 4079–4094.

105 Garcia-Costa, A.L., Luengo, A., Zazo, J.A., and Casas, J.A. (2021). Cutting oil-water emulsion wastewater treatment by microwave assisted catalytic wet peroxide oxidation. *Sep. Purif. Technol.* 257: 117940.

106 Fu, W. and Zhang, W. (2018). Microwave-enhanced membrane filtration for water treatment. *J. Membr. Sci.* 568: 97–104.

107 Wang, N., Jin, L., Li, C. et al. (2022). Preparation of coal fly ash-based Fenton-like catalyst and its application for the treatment of organic wastewater under microwave assistance. *J. Cleaner Prod.* 342: 130926.

108 Qie, Z., Ji, Z., Xiang, H. et al. (2023). Microwave-assisted post treatment to make hydrophobic zeolite Beta for aromatics adsorption in aqueous systems. *Sep. Purif. Technol.* 320: 124148.

109 Wang, W., Li, Z., Zhang, M., and Sun, C. (2020). Preparation of 3D network CNTs-modified nickel foam with enhanced microwave absorptivity and application potential in wastewater treatment. *Sci. Total Environ.* 702: 135006.

110 Sousa, É., Rocha, L., Jaria, G. et al. (2021). Optimizing microwave-assisted production of waste-based activated carbons for the removal of antibiotics from water. *Sci. Total Environ.* 752: 141662.

111 Vieira, Y., Pereira, H.A., Leichtweis, J. et al. (2021). Effective treatment of hospital wastewater with high-concentration diclofenac and ibuprofen using a promising technology based on degradation reaction catalyzed by Fe^0 under microwave irradiation. *Sci. Total Environ.* 783: 146991.

112 Yang, Y., Wang, P., Shi, S., and Liu, Y. (2009). Microwave enhanced Fenton-like process for the treatment of high concentration pharmaceutical wastewater. *J. Hazard. Mater.* 168 (1): 238–245.

113 Kajornprai, T., Seejuntuek, A., Suppakarn, N., Prayoonpokarach, S., and Trongsatitkul, T. (2023). Synthesis of thermoresponsive PNIPAm nanogel adsorbent by microwave-assisted polymerization for wastewater treatment application. *Materials Today: Proceedings*.

114 Yang, L., Chen, Z., Yang, J. et al. (2014). Removal of volatile fatty acid in landfill leachate by the microwave-hydrothermal method. *Desalin. Water Treat.* 52 (22–24): 4423–4429.

115 Zalat, O.A. and Elsayed, M.A. (2013). A study on microwave removal of pyridine from wastewater. *J. Environ. Chem. Eng.* 1 (3): 137–143.

116 Halasz, A., Thiboutot, S., Ampleman, G., and Hawari, J. (2010). Microwave-assisted hydrolysis of nitroglycerin (NG) under mild alkaline conditions: new insight into the degradation pathway. *Chemosphere* 79 (2): 228–232.

117 Remya, N. and Lin, J.-G. (2015). Microwave-granular activated carbon (MW-GAC) system for carbofuran degradation: analysis of characteristics and recyclability of the spent GAC. *Desalin. Water Treat.* 53 (6): 1621–1631.

118 Feng, Q.L., Wang, X.Q., Jia, Y., and Ning, P. (2013). Study on microwave combined with active carbon for the COD remove of amoxicillin wastewater. *Appl. Mech. Mater.* 295–298: 1348–1352.

119 Yuan, D., Fu, D.Y., and Luo, Z.W. (2013). Study on microwave combined with granular active carbon for treatment of *p*-nitrophenol wastewater. *Adv. Mater. Res.* 602–604: 2287–2290.

120 Xu, D., Cheng, F., Zhang, Y., and Song, Z. (2014). Degradation of methyl orange in aqueous solution by microwave irradiation in the presence of granular-activated carbon. *Water Air Soil Pollut.* 225: 1983.

121 Zhang, Z., Shan, Y., Wang, J. et al. (2007). Investigation on the rapid degradation of Congo red catalyzed by activated carbon powder under microwave irradiation. *J. Hazard. Mater.* 147 (1–2): 325–333.

122 Zhang, Z., Xu, Y., Shen, M. et al. (2013). Assisted activated carbon-microwave degradation of the sodium dodecyl benzene sulfonate by nano- or micro-Fe_3O_4

and comparison of their catalytic activity. *Environ. Prog. Sustainable Energy* 32 (2): 181–186.

123 Ahmed, A.B., Jibril, B., Danwittayakul, S., and Dutta, J. (2014). Microwave-enhanced degradation of phenol over Ni-loaded ZnO nanorods catalyst. *Appl. Catal., B* 156–157: 456–465.

124 Lee, C.-L., Lin, C., and Jou, C.-J.G. (2012). Microwave-induced nanoscale zero-valent iron degradation of perchloroethylene and pentachlorophenol. *J. Air Waste Manag. Assoc.* 62 (12): 1443–1448.

125 Bi, X.Y., Yang, H.Y., and Sun, P.S. (2012). Microwave-induced oxidation progress for treatment of imidacloprid pesticide wastewater. *Appl. Mech. Mater.* 229–231: 2489–2492.

126 Dias, A.P.S., Puna, J., Correia, M.J.N. et al. (2013). Effect of the oil acidity on the methanolysis performances of lime catalyst biodiesel from waste frying oils (WFO). *Fuel Process. Technol.* 116: 94–100.

127 Hong, I.K., Park, J.W., Kim, H., and Lee, S.B. (2014). Alcoholysis of canola oil using a short-chain (C1–C3) alcohols. *J. Ind. Eng. Chem.* 20 (5): 3689–3694.

128 Hong, I.K., Lee, J.R., and Lee, S.B. (2015). Fuel properties of canola oil and lard biodiesel blends: higher heating value, oxidative stability, and kinematic viscosity. *J. Ind. Eng. Chem.* 22: 335–340.

129 Kamath, H.V., Regupathi, I., and Saidutta, M.B. (2011). Optimization of two step karanja biodiesel synthesis under microwave irradiation. *Fuel Process. Technol.* 92 (1): 100–105.

130 Farooq, M., Ramli, A., and Subbarao, D. (2013). Biodiesel production from waste cooking oil using bifunctional heterogeneous solid catalysts. *J. Cleaner Prod.* 59: 131–140.

131 Li, M., Zheng, Y., Chen, Y., and Zhu, X. (2014). Biodiesel production from waste cooking oil using a heterogeneous catalyst from pyrolyzed rice husk. *Bioresour. Technol.* 154: 345–348.

132 Ghoreishi, S.M. and Moein, P. (2013). Biodiesel synthesis from waste vegetable oil via transesterification reaction in supercritical methanol. *J. Supercrit. Fluids* 76: 24–31.

133 Enciso, S.R.A., Fellmann, T., Dominguez, I.P., and Santini, F. (2016). Abolishing biofuel policies: possible impacts on agricultural price levels, price variability and global food security. *Food Policy* 61: 9–26.

134 Renzaho, A.M.N., Kamara, J.K., and Toole, M. (2017). Biofuel production and its impact on food security in low and middle income countries: implications for the post-2015 sustainable development goals. *Renewable Sustainable Energy Rev.* 78: 503–516.

135 Li, J. and Liang, X. (2017). Magnetic solid acid catalyst for biodiesel synthesis from waste oil. *Energy Convers. Manage.* 141: 126–132.

136 Sharma, Y.C., Singh, B., and Korstad, J. (2010). High yield and conversion of biodiesel from a nonedible feedstock (*Pongamia pinnata*). *J. Agric. Food. Chem.* 58 (1): 242–247.

137 Silitonga, A.S., Masjuki, H.H., Ong, H.C. et al. (2016). Pilot-scale production and the physicochemical properties of palm and *Calophyllum inophyllum* biodiesels and their blends. *J. Cleaner Prod.* 126: 654–666.

138 Mosarof, M.H., Kalam, M.A., Masjuki, H.H. et al. (2016). Optimization of performance, emission, friction and water characteristics of palm and *Calophyllum inophyllum* biodiesel blends. *Energy Convers. Manage.* 118: 119–134.

139 Sharma, Y.C. and Singh, B. (2010). An ideal feedstock, kusum (*Schleichera triguga*) for preparation of biodiesel: optimization of parameters. *Fuel* 89 (7): 1470–1474.

140 Khan, T.M.Y., Atabani, A.E., Badruddin, I.A. et al. (2015). *Ceiba pentandra, Nigella sativa* and their blend as prospective feedstocks for biodiesel. *Ind. Crops Prod.* 65: 367–373.

141 Rashtizadeh, E., Farzaneh, F., and Ghandi, M. (2010). A comparative study of KOH loaded on double aluminosilicate layers, microporous and mesoporous materials as catalyst for biodiesel production via transesterification of soybean oil. *Fuel* 89 (11): 3393–3398.

142 Ito, T., Sakurai, Y., Kakuta, Y. et al. (2012). Biodiesel production from waste animal fats using pyrolysis method. *Fuel Process. Technol.* 94 (1): 47–52.

143 Vahid, B.R. and Haghighi, M. (2017). Biodiesel production from sunflower oil over $MgO/MgAl_2O_4$ nanocatalyst: effect of fuel type on catalyst nanostructure and performance. *Energy Convers. Manage.* 134: 290–300.

144 Karmakar, A., Karmakar, S., and Mukherjee, S. (2010). Properties of various plants and animals feedstocks for biodiesel production. *Bioresour. Technol.* 101 (19): 7201–7210.

145 Rahmani, F., Haghighi, M., and Mohammadkhani, B. (2017). Enhanced dispersion of Cr nanoparticles over nanostructured ZrO_2-doped ZSM-5 used in CO_2-oxydehydrogenation of ethane. *Microporous Mesoporous Mater.* 242: 34–49.

146 Mendonça, I.M., Paes, O.A.R.L., Maia, P.J.S. et al. (2019). New heterogeneous catalyst for biodiesel production from waste tucumã peels (*Astrocaryum aculeatum* Meyer): parameters optimization study. *Renewable Energy* 130: 103–110.

147 Jeon, Y., Chi, W.S., Hwang, J. et al. (2019). Core-shell nanostructured heteropoly acid-functionalized metal-organic frameworks: bifunctional heterogeneous catalyst for efficient biodiesel production. *Appl. Catal., B* 242: 51–59.

148 Wang, S., Shan, R., Wang, Y. et al. (2019). Synthesis of calcium materials in biochar matrix as a highly stable catalyst for biodiesel production. *Renewable Energy* 130: 41–49.

149 Vahid, B.R. and Haghighi, M. (2016). Urea-nitrate combustion synthesis of $MgO/MgAl_2O_4$ nanocatalyst used in biodiesel production from sunflower oil: influence of fuel ratio on catalytic properties and performance. *Energy Convers. Manage.* 126: 362–372.

150 Dehghani, S. and Haghighi, M. (2017). Sono-sulfated zirconia nanocatalyst supported on MCM-41 for biodiesel production from sunflower oil: influence of ultrasound irradiation power on catalytic properties and performance. *Ultrason. Sonochem.* 35 (Part A): 142–151.

151 Nayebzadeh, H., Saghatoleslami, N., Haghighi, M., and Tabasizadeh, M. (2017). Influence of fuel type on microwave-enhanced fabrication of KOH/$Ca_{12}Al_{14}O_{33}$ nanocatalyst for biodiesel production via microwave heating. *J. Taiwan Inst. Chem. Eng.* 75: 148–155.

152 Helwani, Z., Othman, M.R., Aziz, N. et al. (2009). Solid heterogeneous catalysts for transesterification of triglycerides with methanol: a review. *Appl. Catal. A: General* 363 (1–2): 1–10.

153 Lozano, P., Bernal, J.M., and Vaultier, M. (2011). Towards continuous sustainable processes for enzymatic synthesis of biodiesel in hydrophobic ionic liquids/supercritical carbon dioxide in biphasic systems. *Fuel* 90 (11): 3461–3467.

154 Arzamendi, G., Campo, I., Arguiñarena, E. et al. (2007). Synthesis of biodiesel with heterogeneous NaOH/alumina catalysts: comparison with homogeneous NaOH. *Chem. Eng. J.* 134 (1–3): 123–130.

155 Cai, Z.-Z., Wang, Y., Teng, Y.-L. et al. (2015). A two-step biodiesel production process from waste cooking oil via recycling crude glycerol esterification catalyzed by alkali catalyst. *Fuel Process. Technol.* 137: 186–193.

156 Serio, M.D., Tesser, R., Pengmei, L., and Santacesaria, E. (2008). Heterogeneous catalysts for biodiesel production. *Energy Fuels* 22 (1): 207–217.

157 Sandouqa, A., Al-Hamamre, Z., and Asfar, J. (2019). Preparation and performance investigation of a lignin-based solid acid catalyst manufactured from olive cake for biodiesel production. *Renewable Energy* 132: 667–682.

158 Jeon, K.-W., Shim, J.-O., Cho, J.-W. et al. (2019). Synthesis and characterization of Pt-, Pd-, and Ru-promoted Ni-$Ce_{0.6}Zr_{0.4}O_2$ catalysts for efficient biodiesel production by deoxygenation of oleic acid. *Fuel* 236: 928–933.

159 Seffati, K., Honarvar, B., Esmaeili, H., and Esfandiari, N. (2019). Enhanced biodiesel production from chicken fat using $CaO/CuFe_2O_4$ nanocatalyst and its combination with diesel to improve fuel properties. *Fuel* 235: 1238–1244.

160 Liu, X., He, H., Wang, Y. et al. (2008). Transesterification of soybean oil to biodiesel using CaO as a solid base catalyst. *Fuel* 87 (2): 216–221.

161 Soltani, S., Khanian, N., Choong, T.S.Y. et al. (2021). Microwave-assisted hydrothermal synthesis of sulfonated TiO_2-GO core-shell solid spheres as heterogeneous esterification mesoporous catalyst for biodiesel production. *Energy Convers. Manage.* 238: 114165.

162 Bekhradinassab, E., Haghighi, M., Tavakoli, A., and Shabani, M. (2022). Mn-Fe catalyzed microwave combustion-plasma hybrid synthesis of 2D chips-like Mn-Fe boosted TiO_2 architecture self-assembled of nano-walled honeycomb-like super-macroporous: green fuel generation. *Energy Convers. Manage.* 270: 116178.

163 Yang, J., Cong, W.-J., Zhu, Z. et al. (2023). Microwave-assisted one-step production of biodiesel from waste cooking oil by magnetic bifunctional SrO-ZnO/MOF catalyst. *J. Cleaner Prod.* 395: 136182.

164 Qu, S., Chen, C., Guo, M. et al. (2021). Microwave-assisted in-situ transesterification of Spirulina platensis to biodiesel using PEG/MgO/ZSM-5 magnetic catalyst. *J. Cleaner Prod.* 311: 127490.

165 Amani, T., Haghighi, M., and Rahmanivahid, B. (2019). Microwave-assisted combustion design of magnetic Mg-Fe spinel for MgO-based nanocatalyst used in biodiesel production: influence of heating-approach and fuel ratio. *J. Ind. Eng. Chem.* 80: 43–52.

166 Dutta, S., Jaiswal, K.K., Verma, R. et al. (2019). Green synthesis of zinc oxide catalyst under microwave irradiation using banana (*Musa* spp.) corm (rhizome) extract for biodiesel synthesis from fish waste lipid. *Biocatal. Agric. Biotechnol.* 22: 101390.

167 Zhang, M., Ramya, G., Brindhadevi, K. et al. (2022). Microwave assisted biodiesel production from chicken feather meal oil using bio-nano calcium oxide derived from chicken egg shell. *Environ. Res.* 205: 112509.

168 Hsiao, M.-C., Kuo, J.-Y., Hsieh, S.-A. et al. (2020). Optimized conversion of waste cooking oil to biodiesel using modified calcium oxide as catalyst via a microwave heating system. *Fuel* 266: 117114.

169 Athar, M., Imdad, S., Zaidi, S. et al. (2022). Biodiesel production by single-step acid-catalysed transesterification of Jatropha oil under microwave heating with modelling and optimisation using response surface methodology. *Fuel* 322: 124205.

170 Li, H., Wang, Y., Han, Z. et al. (2022). Nanosheet like CaO/C derived from Ca-BTC for biodiesel production assisted with microwave. *Appl. Energy* 326: 120045.

171 Hong, I.K., Jeon, H., Kim, H., and Lee, S.B. (2016). Preparation of waste cooking oil based biodiesel using microwave irradiation energy. *J. Ind. Eng. Chem.* 42: 107–112.

172 Ali, M.A.M., Gimbun, J., Lau, K.L. et al. (2020). Biodiesel synthesized from waste cooking oil in a continuous microwave assisted reactor reduced PM and NO_x emissions. *Environ. Res.* 185: 109452.

173 Allami, H.A., Tabasizadeh, M., Rohani, A. et al. (2019). Precise evaluation the effect of microwave irradiation on the properties of palm kernel oil biodiesel used in a diesel engine. *J. Cleaner Prod.* 241: 117777.

174 Milano, J., Ong, H.C., Masjuki, H.H. et al. (2018). Optimization of biodiesel production by microwave irradiation-assisted transesterification for waste cooking oil-*Calophyllum inophyllum* oil via response surface methodology. *Energy Convers. Manage.* 158: 400–415.

175 Yadav, G., Yadav, N., and Ahmaruzzaman, M. (2023). Microwave-assisted sustainable synthesis of biodiesel on *Oryza sativa* catalyst derived from agricultural waste by esterification reaction. *Chem. Eng. Process.* 187: 109327.

176 Wahidin, S., Idris, A., Yusof, N.M. et al. (2018). Optimization of the ionic liquid-microwave assisted one-step biodiesel production process from wet microalgal biomass. *Energy Convers. Manage.* 171: 1397–1404.

177 Helmi, F., Helmi, M., and Hemmati, A. (2022). Phosphomolybdic acid/chitosan as acid solid catalyst using for biodiesel production from pomegranate seed oil via microwave heating system: RSM optimization and kinetic study. *Renewable Energy* 189: 881–898.

178 Gouda, S.P., Ngaosuwan, K., Assabumrungrat, S. et al. (2022). Microwave assisted biodiesel producing using sulfonic acid-functionalized metal-organic frameworks UiO-66 as a heterogeneous catalyst. *Renewable Energy* 197: 161–169.

179 Gouda, S.P. and Rokhum, S.L. (2023). Robust microwave-assisted sulfonic acid-functionalized metal-organic framework UiO-66 for biodiesel production. *Sci. Talks* 7: 100242.

180 Pham, E.C., Le, T.V.T., Le, K.C.T. et al. (2022). Optimization of microwave-assisted biodiesel production from waste catfish using response surface methodology. *Energy Rep.* 8: 5739–5752.

181 Lin, Y.-C., Amesho, K.T.T., Chen, C.-E. et al. (2020). A cleaner process for green biodiesel synthesis from waste cooking oil using recycled waste oyster shells as a sustainable base heterogeneous catalyst under the microwave heating system. *Sustainable Chem. Pharm.* 17: 100310.

182 Ergan, B.T., Yılmazer, G., and Bayramoğlu, M. (2022). Fast, high quality and low-cost biodiesel production using dolomite catalyst in an enhanced microwave system with simultaneous cooling. *Cleaner Chem. Eng.* 3: 100051.

183 Ye, W., Gao, Y., Ding, H. et al. (2016). Kinetics of transesterification of palm oil under conventional heating and microwave irradiation, using CaO as heterogeneous catalyst. *Fuel* 180: 574–579.

184 Nazir, M.H., Ayoub, M., Zahid, I. et al. (2021). Development of lignin based heterogeneous solid acid catalyst derived from sugarcane bagasse for microwave assisted-transesterification of waste cooking oil. *Biomass Bioenergy* 146: 105978.

185 Hsiao, M.-C., Liao, P.-H., Lan, N.V., and Hou, S.-S. (2021). Enhancement of biodiesel production from high-acid-value waste cooking oil via a microwave reactor using a homogeneous alkaline catalyst. *Energies* 14 (2): 437–447.

186 Amesho, K.T.T., Lin, Y.-C., Chen, C.-E. et al. (2022). Kinetic studies of sustainable biodiesel synthesis from *Jatropha curcas* oil by exploiting bio-waste derived CaO-based heterogeneous catalyst via microwave heating system as a green chemistry technique. *Fuel* 323: 123876.

187 Silitonga, A.S., Shamsuddin, A.H., Mahlia, T.M.I. et al. (2020). Biodiesel synthesis from *Ceiba pentandra* oil by microwave irradiation-assisted transesterification: ELM modeling and optimization. *Renewable Energy* 146: 1278–1291.

188 Binnal, P., Amruth, A., Basawaraj, M.P. et al. (2021). Microwave-assisted esterification and transesterification of dairy scum oil for biodiesel production: kinetics and optimization studies. *Indian Chem. Eng.* 63 (4): 374–386.

189 Nayak, M.G. and Vyas, A.P. (2022). Parametric study and optimization of microwave assisted biodiesel synthesis from Argemone Mexicana oil using response surface methodology. *Chem. Eng. Process.* 170: 108665.

190 Lawan, I., Garba, Z.N., Zhou, W. et al. (2020). Synergies between the microwave reactor and CaO/zeolite catalyst in waste lard biodiesel production. *Renewable Energy* 145: 2550–2560.

191 Kumar, A., Vashistha, V.K., and Das, D.K. (2021). Recent development on metal phthalocyanines based materials for energy conversion and storage applications. *Coord. Chem. Rev.* 431: 213678.

192 Kumar, A., Ibraheem, S., Nguyen, T.A. et al. (2021). Molecular-MN$_4$ vs atomically dispersed M-N$_4$-C electrocatalysts for oxygen reduction reaction. *Coord. Chem. Rev.* 446: 214122.

193 Yasin, G., Ibrahim, S., Ajmal, S. et al. (2022). Tailoring of electrocatalyst interactions at interfacial level to benchmark the oxygen reduction reaction. *Coord. Chem. Rev.* 469: 214669.

194 Li, Y., Gerdes, K., Horita, T., and Liu, X. (2013). Surface exchange and bulk diffusivity of LSCF as SOFC cathode: electrical conductivity relaxation and isotope exchange characterizations. *J. Electrochem. Soc.* 160 (4): F343–F350.

195 Zhu, Y., Tahini, H.A., Hu, Z. et al. (2020). Boosting oxygen evolution reaction by creating both metal ion and lattice-oxygen active sites in a complex oxide. *Adv. Mater.* 32 (1): e1905025.

196 Kim, J.H., Yoo, S., Murphy, R. et al. (2021). Promotion of oxygen reduction reaction on a double perovskite electrode by a water-induced surface modification. *Energy Environ. Sci.* 14 (3): 1506–1516.

197 Ziegelbauer, J.M., Olson, T.S., Pylypenko, S. et al. (2008). Direct spectroscopic observation of the structural origin of peroxide generation from Co-based pyrolyzed porphyrins for ORR applications. *J. Phys. Chem. C* 112 (24): 8839–8849.

198 Rao, C.V. and Viswanathan, B. (2009). ORR activity and direct ethanol fuel cell performance of carbon-supported Pt-M (M = Fe, Co, and Cr) alloys prepared by polyol reduction method. *J. Phys. Chem. C* 113 (43): 18907–18913.

199 Yang, C.-J. (2009). An impending platinum crisis and its implications for the future of the automobile. *Energy Policy* 37 (5): 1805–1808.

200 Dai, Y., Ou, L., Liang, W. et al. (2011). Efficient and superiorly durable Pt-lean electrocatalysts of Pt-W alloys for the oxygen reduction reaction. *J. Phys. Chem. C* 115 (5): 2162–2168.

201 Huang, L., Han, Y., and Dong, S. (2016). Highly-branched mesoporous Au-Pd-Pt trimetallic nanoflowers blooming on reduced graphene oxide as an oxygen reduction electrocatalyst. *Chem. Commun.* 52 (56): 8659–8662.

202 Rahsepar, M., Pakshir, M., and Kim, H. (2013). Synthesis of multiwall carbon nanotubes with a high loading of Pt by a microwave-assisted impregnation method for use in the oxygen reduction reaction. *Electrochim. Acta* 108: 769–775.

203 Yin, S., Shen, P.K., Song, S., and Jiang, S.P. (2009). Functionalization of carbon nanotubes by an effective intermittent microwave heating-assisted HF/H$_2$O$_2$ treatment for electrocatalyst support of fuel cells. *Electrochim. Acta* 54 (27): 6954–6958.

204 Sahin, N.E., Napporn, T.W., Dubau, L. et al. (2017). Temperature-dependence of oxygen reduction activity on Pt/C and PtCr/C electrocataysts synthesized from microwave-heated diethylene glycol method. *Appl. Catal., B* 203: 72–84.

205 Sandström, R., Gracia-Espino, E., Hu, G. et al. (2018). Yttria stabilized and surface activated platinum (Pt$_x$YO$_y$) nanoparticles through rapid microwave assisted synthesis for oxygen reduction reaction. *Nano Energy* 46: 141–149.

206 Cui, S.-K. and Guo, D.-J. (2021). Microwave-assisted preparation of PtCu/C nanoalloys and their catalytic properties for oxygen reduction reaction. *J. Alloys Compd.* 874: 159869.

207 El-Deeb, H. and Bron, M. (2015). Microwave-assisted polyol synthesis of PtCu/carbon nanotube catalysts for electrocatalytic oxygen reduction. *J. Power Sources* 275: 893–900.

208 Li, Y., Li, W., Ke, T. et al. (2016). Microwave-assisted synthesis of sulfur-doped graphene supported PdW nanoparticles as a high performance electrocatalyst for the oxygen reduction reaction. *Electrochem. Commun.* 69: 68–71.

209 Li, Y., Yang, J., Zhao, N. et al. (2017). Facile fabrication of N-doped three-dimensional reduced graphene oxide as a superior electrocatalyst for oxygen reduction reaction. *Appl. Catal. A: General* 534: 30–39.

210 Li, Y., Yang, J., Huang, J. et al. (2017). Soft template-assisted method for synthesis of nitrogen and sulfur co-doped three-dimensional reduced graphene oxide as an efficient metal free catalyst for oxygen reduction reaction. *Carbon* 122: 237–246.

211 Zhao, D., Li, L., Xie, L. et al. (2018). Sulfur codoping enables efficient oxygen electroreduction on FeCo alloy encapsulated in N-doped carbon nanotubes. *J. Alloys Compd.* 741: 368–376.

212 Ren, G., Li, Y., Chen, Q. et al. (2018). Sepia-derived N, P co-doped porous carbon spheres as oxygen reduction reaction electrocatalyst and supercapacitor. *ACS Sustainable Chem. Eng.* 6 (12): 16032–16038.

213 Yan, Z., Gao, L., Dai, C. et al. (2018). Metal-free mesoporous carbon with higher contents of active N and S codoping by template method for superior ORR efficiency to Pt/C. *Int. J. Hydrogen Energy* 43 (7): 3705–3715.

214 Wang, X., Xu, J., Zhi, M. et al. (2019). Synthesis of Co_2P nanoparticles decorated nitrogen, phosphorus Co-doped carbon-CeO_2 composites for highly efficient oxygen reduction. *J. Alloys Compd.* 801: 192–198.

215 Fan, T., Zhang, G., Jian, L. et al. (2019). Facile synthesis of defect-rich nitrogen and sulfur Co-doped graphene quantum dots as metal-free electrocatalyst for the oxygen reduction reaction. *J. Alloys Compd.* 792: 844–850.

216 Wang, Y., Tao, L., Xiao, Z. et al. (2018). 3D carbon electrocatalysts in situ constructed by defect-rich nanosheets and polyhedrons from NaCl-sealed zeolitic imidazolate frameworks. *Adv. Funct. Mater.* 28 (11): 1705356.

217 Zhang, J., Sun, Y., Zhu, J. et al. (2018). Defect and pyridinic nitrogen engineering of carbon-based metal-free nanomaterial toward oxygen reduction. *Nano Energy* 52: 307–314.

218 Xie, G., Yang, R., Chen, P. et al. (2014). A general route towards defect and pore engineering in graphene. *Small* 10 (11): 2280–2284.

219 Muthuswamy, N., Buan, M.E.M., Walmsley, J.C., and Rønning, M. (2018). Evaluation of ORR active sites in nitrogen-doped carbon nanofibers by KOH post treatment. *Catal. Today* 301: 11–16.

220 Yan, Z., Meng, H., Shen, P.K. et al. (2012). Effect of the templates on the synthesis of hollow carbon materials as electrocatalyst supports for direct alcohol fuel cells. *Int. J. Hydrogen Energy* 37 (5): 4728–4736.

221 She, W., Wang, J., Zhang, X. et al. (2018). Structural engineering of S-doped Co/N/C mesoporous nanorods via the Ostwald ripening-assisted template method for oxygen reduction reaction and Li-ion batteries. *J. Power Sources* 401: 55–64.

222 Rajendiran, R., Nallal, M., Park, K.H. et al. (2019). Mechanochemical assisted synthesis of heteroatoms inherited highly porous carbon from biomass for electrochemical capacitor and oxygen reduction reaction electrocatalysis. *Electrochim. Acta* 317: 1–9.

223 Yan, Z., Dai, C., Zhang, M. et al. (2019). Nitrogen doped porous carbon with iron promotion for oxygen reduction reaction in alkaline and acidic media. *Int. J. Hydrogen Energy* 44 (8): 4090–4101.

224 Liu, Z., Li, Z., Tian, S. et al. (2019). Conversion of peanut biomass into electrocatalysts with vitamin B12 for oxygen reduction reaction in Zn-air battery. *Int. J. Hydrogen Energy* 44 (23): 11788–11796.

225 Tang, J., Wang, Y., Zhao, W. et al. (2019). Biomass-derived hierarchical honeycomb-like porous carbon tube catalyst for the metal-free oxygen reduction reaction. *J. Electroanal. Chem.* 847: 113230.

226 Kim, I.T., Song, M.J., Kim, Y.B., and Shin, M.W. (2016). Microwave-hydrothermal synthesis of boron/nitrogen co-doped graphene as an efficient metal free electrocatalyst for oxygen reduction reaction. *Int. J. Hydrogen Energy* 41 (47): 22026–22033.

227 Dai, C., Gao, Z., Yan, Z. et al. (2020). Microwave-assisted synthesis of mesoporous hemispherical graphite promoted with iron and nitrogen doping for reduction of oxygen. *J. Alloys Compd.* 838: 155608.

228 Xu, X., Zhang, X., Xia, Z. et al. (2021). Solid phase microwave-assisted fabrication of Fe-doped ZIF-8 for single-atom Fe-N-C electrocatalysts on oxygen reduction. *J. Energy Chem.* 54: 579–586.

229 Kuo, H.-C., Lin, Y.-G., Chiang, C.-L., and Liu, S.-H. (2021). FeN@N-doped graphitic biochars derived from hydrothermal-microwave pyrolysis of cellulose biomass for fuel cell catalysts. *J. Anal. Appl. Pyrolysis* 153: 104991.

230 Zhang, J., Wu, S., Chen, X. et al. (2014). Egg derived nitrogen-self-doped carbon/carbon nanotube hybrids as noble-metal-free catalysts for oxygen reduction. *J. Power Sources* 271: 522–529.

231 Li, Y., Cheng, F., Zhang, J. et al. (2016). Cobalt-carbon core-shell nanoparticles aligned on wrinkle of N-doped carbon nanosheets with Pt-like activity for oxygen reduction. *Small* 12 (21): 2839–2845.

232 Jiao, L., Hu, Y., Ju, H. et al. (2017). From covalent triazine-based frameworks to N-doped porous carbon/reduced graphene oxide nanosheets: efficient electrocatalysts for oxygen reduction. *J. Mater. Chem. A* 5 (44): 23170–23178.

233 Wang, Z., Jin, H., Meng, T. et al. (2018). Fe, Cu-coordinated ZIF-derived carbon framework for efficient oxygen reduction reaction and zinc-air batteries. *Adv. Funct. Mater.* 28 (39): 1802596–1802604.

234 Liu, Z.-Q., Cheng, H., Li, N. et al. (2016). $ZnCo_2O_4$ quantum dots anchored on nitrogen-doped carbon nanotubes as reversible oxygen reduction/evolution electrocatalysts. *Adv. Mater.* 28 (19): 3777–3784.

235 Hu, Z., Guo, Z., Zhang, Z. et al. (2018). Bimetal zeolitic imidazolite framework-derived iron-, cobalt- and nitrogen-codoped carbon nanopolyhedra electrocatalyst for efficient oxygen reduction. *ACS Appl. Mater. Interfaces* 10 (15): 12651–12658.

236 Meng, Z., Cai, S., Wang, R. et al. (2019). Bimetallic-organic framework-derived hierarchically porous Co-Zn-N-C as efficient catalyst for acidic oxygen reduction reaction. *Appl. Catal., B* 244: 120–127.

237 Palaniselvam, T., Kashyap, V., Bhange, S.N. et al. (2016). Nanoporous graphene enriched with Fe/Co-N active sites as a promising oxygen reduction electrocatalyst for anion exchange membrane fuel cells. *Adv. Funct. Mater.* 26 (13): 2150–2162.

238 Kuang, M., Wang, Q., Han, P., and Zheng, G. (2017). Cu, Co-embedded N-enriched mesoporous carbon for efficient oxygen reduction and hydrogen evolution reactions. *Adv. Energy Mater.* 7 (17): 1700193–1700200.

239 Guo, Y., Yuan, P., Zhang, J. et al. (2018). Carbon nanosheets containing discrete Co-N_x-B_y-C active sites for efficient oxygen electrocatalysis and rechargeable Zn-air batteries. *ACS Nano* 12 (2): 1894–1901.

240 Yin, M., Zhang, Y., Bian, Z. et al. (2019). Efficient and stable nanoporous functional composited electrocatalyst derived from Zn/Co-bimetallic zeolitic imidazolate frameworks for oxygen reduction reaction in alkaline media. *Electrochim. Acta* 299: 610–617.

241 Li, H., Gao, D., and Cheng, X. (2014). Simple microwave preparation of high activity Se-rich $CoSe_2$/C for oxygen reduction reaction. *Electrochim. Acta* 138: 232–239.

242 Kang, X., Fu, G., Song, Z. et al. (2019). Microwave-assisted hydrothermal synthesis of MOFs-derived bimetallic CuCo-N/C electrocatalyst for efficient oxygen reduction reaction. *J. Alloys Compd.* 795: 462–470.

243 González, A.S., Delgado, F.P., Sebastian, P.J., and Arco, E.B. (2014). Microwave synthesis of an electrocatalyst based on CoFeRu for the oxygen reduction reaction in the absence and presence of methanol. *J. Power Sources* 267: 793–798.

244 Gnanaprakasam, P., Mangalaraja, R.V., and Salvo, C. (2022). Microwave driven synthesis of tungsten sulfide nanosheets: an efficient electrocatalyst for oxygen reduction reaction. *Mater. Sci. Semicond. Process.* 137: 106213.

245 Deng, Z., Zheng, X., Deng, M. et al. (2021). Catalytic activity of V_2CO_2 MXene supported transition metal single atoms for oxygen reduction and hydrogen oxidation reactions: a density functional theory calculation study. *Chin. J. Catal.* 42 (10): 1659–1666.

246 Geng, C., Yuan, J., Hong, T. et al. (2022). Improved oxygen reduction reaction activity by in-situ synthesizing $Sr_3Fe_{1.8}Nb_{0.2}O_{7-\delta}$ coating on $SrFe_{0.9}Nb_{0.1}O_{3-\delta}$ cathode via the microwave water bath heating method. *J. Eur. Ceram. Soc.* 42: 6557–6565.

247 Sandhiran, N., Ganapathy, S., Manoharan, Y. et al. (2022). CuO-NiO binary transition metal oxide nanoparticle anchored on rGO nanosheets as high-performance electrocatalyst for the oxygen reduction reaction. *Environ. Res.* 211: 112992.

248 Mahato, D., Gurusamy, T., Ramanujam, K. et al. (2022). Unravelling the role of interface of CuO_x-TiO_2 hybrid metal oxide in enhancement of oxygen reduction reaction performance. *Int. J. Hydrogen Energy* 47 (80): 34048–34065.

249 Zhang, X., Li, B., Liu, C. et al. (2013). Rapid microwave-assisted hydrothermal synthesis of morphology-tuned MnO_2 nanocrystals and their electrocatalytic activities for oxygen reduction. *Mater. Res. Bull.* 48 (7): 2696–2701.

250 Cabello, G., Davoglio, R.A., and Pereira, E.C. (2017). Microwave-assisted synthesis of anatase-TiO_2 nanoparticles with catalytic activity in oxygen reduction. *J. Electroanal. Chem.* 794: 36–42.

251 Kiran, G.K., Sreekanth, T.V.M., Nagajyothi, P.C. et al. (2022). Microwave induced synthesis of mandarin-flower shaped $ZnCo_2O_4$/GO nanocomposites as efficient electrocatalyst for oxygen reduction reaction. *Mater. Lett.* 316: 132026.

252 Zhang, Z., Zhou, D., Zou, S. et al. (2019). One-pot synthesis of $MnFe_2O_4$/C by microwave sintering as an efficient bifunctional electrocatalyst for oxygen reduction and oxygen evolution reactions. *J. Alloys Compd.* 786: 565–569.

253 Nissinen, T., Valo, T., Gasik, M. et al. (2002). Microwave synthesis of catalyst spinel $MnCo_2O_4$ for alkaline fuel cell. *J. Power Sources* 106: 109–115.

254 Zhang, Z., Zhou, D., Liao, J. et al. (2019). One-pot synthesis of Fe_3O_4/Fe/C by microwave sintering as an efficient bifunctional electrocatalyst for oxygen reduction and oxygen evolution reactions. *J. Alloys Compd.* 786: 134–138.

255 Liu, M., Yin, X., Guo, X. et al. (2019). High efficient oxygen reduction performance of Fe/Fe_3C nanoparticles in situ encapsulated in nitrogen-doped carbon via a novel microwave-assisted carbon bath method. *Nano Mater. Sci.* 1 (2): 131–136.

256 Ajmal, S., Kumar, A., Yasin, G. et al. (2023). A microwave-assisted decoration of carbon nanotubes with Fe_3O_4 nanoparticles for efficient electrocatalytic oxygen reduction reaction. *J. Alloys Compd.* 943: 169067.

257 Mohamed, M.J.S. (2021). High bifunctional electrocatalytic activity of $FeWO_4$/Fe_3O_4@NrGO nanocomposites towards electrolyzer and fuel cell technologies. *J. Electroanal. Chem.* 897: 115587.

258 Ellis, M.W., Von Spakovsky, M.R., and Nelson, D.J. (2001). Fuel cell systems: efficient, flexible energy conversion for the 21st century. *Proc. IEEE* 89 (12): 1808–1818.

259 Stambouli, A.B. and Traversa, E. (2002). Fuel cells, an alternative to standard sources of energy. *Renewable Sustainable Energy Rev.* 6 (3): 295–304.

260 Wang, C. and Nehrir, M. (2006). Distributed generation applications of fuel cells. *2006 Power Systems Conference: Advanced Metering, Protection, Control, Communication, and Distributed Resources*. Clemson, SC, USA: IEEE, pp. 244–248.

261 Bonesi, A., Garaventa, G., Triaca, W.E., and Luna, A.M.C. (2008). Synthesis and characterization of new electrocatalysts for ethanol oxidation. *Int. J. Hydrogen Energy* 33 (13): 3499–3501.

262 Wang, Z.-H., Li, J., Dong, X. et al. (2008). Ethanol oxidation on a nichrome-supported spherical platinum microparticle electrocatalyst prepared by electrodeposition. *Int. J. Hydrogen Energy* 33 (21): 6143–6149.

263 Kadirgan, F., Beyhan, S., and Atilan, T. (2009). Preparation and characterization of nano-sized Pt–Pd/C catalysts and comparison of their electro-activity toward methanol and ethanol oxidation. *Int. J. Hydrogen Energy* 34 (10): 4312–4320.

264 Xin, Y., Liu, J.-G., Jie, X. et al. (2012). Preparation and electrochemical characterization of nitrogen doped graphene by microwave as supporting materials for fuel cell catalysts. *Electrochim. Acta* 60: 354–358.

265 Chen, C.-C., Chen, C.-F., Chen, C.-M., and Chuang, F.-T. (2007). Modification of multi-walled carbon nanotubes by microwave digestion method as electrocatalyst supports for direct methanol fuel cell applications. *Electrochem. Commun.* 9 (1): 159–163.

266 Sakthivel, M., Schlange, A., Kunz, U., and Turek, T. (2010). Microwave assisted synthesis of surfactant stabilized platinum/carbon nanotube electrocatalysts for direct methanol fuel cell applications. *J. Power Sources* 195 (20): 7083–7089.

267 Kumar, L.V., Ntim, S.A., Sae-Khow, O. et al. (2012). Electro-catalytic activity of multiwall carbon nanotube-metal (Pt or Pd) nanohybrid materials synthesized using microwave-induced reactions and their possible use in fuel cells. *Electrochim. Acta* 83: 40–46.

268 Hu, X., Song, P., Yang, X. et al. (2020). One-step microwave-assisted synthesis of carbon-supported ternary Pt-Sn-Rh alloy nanoparticles for fuel cells. *J. Taiwan Inst. Chem. Eng.* 115: 272–278.

269 Burhan, H., Arikan, K., Alma, M.H. et al. (2023). Highly efficient carbon hybrid supported catalysts using nano-architecture as anode catalysts for direct methanol fuel cells. *Int. J. Hydrogen Energy* 48 (17): 6657–6665.

270 Jiang, L., Sun, G., Wang, S. et al. (2005). Electrode catalysts behavior during direct ethanol fuel cell life-time test. *Electrochem. Commun.* 7 (7): 663–668.

271 Bianchini, C. and Shen, P.K. (2009). Palladium-based electrocatalysts for alcohol oxidation in half cells and in direct alcohol fuel cells. *Chem. Rev.* 109 (9): 4183–4206.

272 Antolini, E. (2009). Palladium in fuel cell catalysis. *Energy Environ. Sci.* 2 (9): 915–931.

273 An, L. and Zhao, T.S. (2011). An alkaline direct ethanol fuel cell with a cation exchange membrane. *Energy Environ. Sci.* 4 (6): 2213–2217.

274 An, H., Pan, L., Cui, H. et al. (2013). Synthesis and performance of palladium-based catalysts for methanol and ethanol oxidation in alkaline fuel cells. *Electrochim. Acta* 102: 79–87.

275 He, Q., Chen, W., Mukerjee, S. et al. (2009). Carbon-supported PdM (M = Au and Sn) nanocatalysts for the electrooxidation of ethanol in high pH media. *J. Power Sources* 187: 298–304.

276 Qiu, C., Shang, R., Xie, Y. et al. (2010). Electrocatalytic activity of bimetallic Pd–Ni thin films towards the oxidation of methanol and ethanol. *Mater. Chem. Phys.* 120 (2–3): 323–330.

277 Wang, Y., Nguyen, T.S., Liu, X., and Wang, X. (2010). Novel palladium–lead (Pd-Pb/C) bimetallic catalysts for electrooxidation of ethanol in alkaline media. *J. Power Sources* 195 (9): 2619–2622.

278 Kannan, R., Karunakaran, K., and Vasanthkumar, S. (2012). PdNi-coated manganite nanorods as catalyst for electrooxidation of methanol in alkaline medium. *Appl. Nanosci.* 2 (2): 149–155.

279 Safavi, A., Kazemi, H., Momeni, S. et al. (2013). Facile electrocatalytic oxidation of ethanol using Ag/Pd nanoalloys modified carbon ionic liquid electrode. *Int. J. Hydrogen Energy* 38 (8): 3380–3386.

280 Wang, Y., Zou, S., and Cai, W.-B. (2015). Recent advances on electro-oxidation of ethanol on Pt and Pd-based catalysts: from reaction mechanisms to catalytic materials. *Catalysis* 5 (3): 1507–1534.

281 Xu, C., Tian, Z., Shen, P., and Jiang, S.P. (2008). Oxide (CeO_2, NiO, Co_3O_4 and Mn_3O_4)-promoted Pd/C electrocatalysts for alcohol electrooxidation in alkaline media. *Electrochim. Acta* 53 (5): 2610–2618.

282 Mahendiran, C., Maiyalagan, T., Scott, K., and Gedanken, A. (2011). Synthesis of a carbon-coated NiO/MgO core/shell nanocomposite as a Pd electro-catalyst support for ethanol oxidation. *Mater. Chem. Phys.* 128 (3): 341–347.

283 Lim, E.J., Kim, H.J., and Kim, W.B. (2012). Efficient electrooxidation of methanol and ethanol using MoO_x-decorated Pd catalysts in alkaline media. *Catal. Commun.* 25: 74–77.

284 Liang, R., Hu, A., Persic, J., and Zhou, Y.N. (2013). Palladium nanoparticles loaded on carbon modified TiO_2 nanobelts for enhanced methanol electrooxidation. *Nano-Micro Lett.* 5 (3): 202–212.

285 Kumar, R., da Silva, E.T.S.G., Singh, R.K. et al. (2018). Microwave-assisted synthesis of palladium nanoparticles intercalated nitrogen doped reduced graphene oxide and their electrocatalytic activity for direct-ethanol fuel cells. *J. Colloid Interface Sci.* 515: 160–171.

286 Sikeyi, L.L., Ntuli, T.D., Mongwe, T.H. et al. (2021). Microwave assisted synthesis of nitrogen doped and oxygen functionalized carbon nano onions supported palladium nanoparticles as hybrid anodic electrocatalysts for direct alkaline ethanol fuel cells. *Int. J. Hydrogen Energy* 46 (18): 10862–10875.

287 Mphahlele, N.E., Ipadeola, A.K., Haruna, A.B. et al. (2022). Microwave-induced defective PdFe/C nano-electrocatalyst for highly efficient alkaline glycerol oxidation reactions. *Electrochim. Acta* 409: 139977.

288 Abdel Hameed, R.M. (2017). Enhanced ethanol electro-oxidation reaction on carbon supported Pd-metal oxide electrocatalysts. *J. Colloid Interface Sci.* 505: 230–240.

289 Abdel Hameed, R.M., Fahim, A.E., and Allam, N.K. (2020). Tin oxide as a promoter for copper@palladium nanoparticles on graphene sheets during ethanol electro-oxidation in NaOH solution. *J. Mol. Liq.* 297: 111816.

290 Li, P., Gu, Y., Yu, Z. et al. (2019). TiO_2-SnO_2/SO_4^{2-} mesoporous solid superacid decorated nickel-based material as efficient electrocatalysts for methanol oxidation reaction. *Electrochim. Acta* 297: 864–871.

291 Rahmani, K. and Habibi, B. (2020). Excellent electro-oxidation of methanol and ethanol in alkaline media: electrodeposition of the NiMoP metallic nano-particles on/in the ERGO layers/CE. *Int. J. Hydrogen Energy* 45 (51): 27263–27278.

292 Yagizatli, Y., Ulas, B., Cali, A. et al. (2020). Improved fuel cell properties of nano-TiO_2 doped poly(vinylidene fluoride) and phosphonated poly(vinyl alcohol) composite blend membranes for PEM fuel cells. *Int. J. Hydrogen Energy* 45 (60): 35130–35138.

293 Eisa, T., Mohamed, H.O., Choi, Y.-J. et al. (2020). Nickel nanorods over nickel foam as standalone anode for direct alkaline methanol and ethanol fuel cell. *Int. J. Hydrogen Energy* 45 (10): 5948–5959.

294 Li, L., Gao, W., Tang, K. et al. (2021). Structure engineering of Ni_2P by Mo doping for robust electrocatalytic water and methanol oxidation reactions. *Electrochim. Acta* 369: 137692.

295 Jin, D., Li, Z., and Wang, Z. (2021). Hierarchical $NiCo_2O_4$ and $NiCo_2S_4$ nanomaterials as electrocatalysts for methanol oxidation reaction. *Int. J. Hydrogen Energy* 46 (63): 32069–32080.

296 Abbasi, M., Noor, T., Iqbal, N., and Zaman, N. (2022). Electrocatalytic study of Cu/Ni MOF and its g-C_3N_4 composites for methanol oxidation reaction. *Int. J. Energy Res.* 46 (10): 13915–13930.

297 Lv, X., Lam, F.L.-Y., and Hu, X. (2022). Developing $SrTiO_3/TiO_2$ heterostructure nanotube array for photocatalytic fuel cells with improved efficiency and elucidating the effects of organic substrates. *Chem. Eng. J.* 427: 131602.

298 Pezeshkvar, T., Norouzi, B., Moradian, M., and Mirabi, A. (2022). Fabrication of new nanocomposites based on NiO-MWCNT-sodium dodecyl sulfate in the presence of *Gundelia tournefortii* extract: application for methanol electrooxidation in alkaline solution. *J. Solid State Electrochem.* 26 (6–7): 1479–1492.

299 Boostani, N., Vardak, S., Amini, R., and Mohammadifard, Z. (2023). Optimization of Ni-Co-metallic-glass powder ($Ni_{60}Cr_{10}Ta_{10}P_{16}B_4$) (MGP) nanocomposite coatings for direct methanol fuel cell (DMFC) applications. *Int. J. Hydrogen Energy* 48 (27): 10002–10015.

300 Ghalkhani, M., Mirzaie, R.A., Banimostafa, A. et al. (2023). Electrosynthesis of ternary nonprecious Ni, Cu, Fe oxide nanostructure as efficient electrocatalyst for ethanol electro-oxidation: design strategy and electrochemical performance. *Int. J. Hydrogen Energy* 48 (55): 21214–21223.

301 Mahalingam, S., Ayyaru, S., and Ahn, Y.-H. (2021). Facile one-pot microwave assisted synthesis of rGO-CuS-ZnS hybrid nanocomposite cathode catalysts for microbial fuel cell application. *Chemosphere* 278: 130426.

302 Mehdinia, A., Ziaei, E., and Jabbari, A. (2014). Facile microwave-assisted synthesized reduced graphene oxide/tin oxide nanocomposite and using as anode material of microbial fuel cell to improve power generation. *Int. J. Hydrogen Energy* 39 (20): 10724–10730.

303 Chen, H., Duan, J., Zhang, X. et al. (2014). One step synthesis of Pt/CeO_2-graphene catalyst microwave-assisted ethylene glycol process for direct methanol fuel cell. *Mater. Lett.* 126: 9–12.

304 Abdel Hameed, R.M. (2017). Facile preparation of Pd-metal oxide/C electrocatalysts and their application in the electrocatalytic oxidation of ethanol. *Appl. Surf. Sci.* 411: 91–104.

305 El-Khatib, K.M., Abdel Hameed, R.M., Amin, R.S., and Fetohi, A.E. (2019). Core-shell structured Pt-transition metals nanoparticles supported on activated carbon for direct methanol fuel cells. *Microchem. J.* 145: 566–577.

306 Fahim, A.E., Abdel Hameed, R.M., and Allam, N.K. (2018). Synthesis and characterization of core-shell structured M@Pd/SnO$_2$-graphene [M = Co, Ni or Cu] electrocatalysts for ethanol oxidation in alkaline solution. *New J. Chem.* 42 (8): 6144–6160.

12

Comparative Studies on Thermal, Microwave-Assisted, and Ultrasound-Promoted Preparations

Tri P. Adhi[1], Aqsha Aqsha[1,2], and Antonius Indarto[2]

[1] Institut Teknologi Bandung, Department of Chemical Engineering, Jalan Ganesha 10, Bandung 40132, Indonesia
[2] Institut Teknologi Bandung, Department of Bioenergy Engineering and Chemurgy, Jl. Let. Jen. Purn. Dr. (HC) Mashudi No. 1, Jatinangor 45363, Indonesia

12.1 Introduction

12.1.1 Background on Preparative Techniques in Chemistry

Preparative techniques are used in various areas of chemistry, including solid-state chemistry, organic chemistry, and polymer chemistry [1, 2]. The preparation of compounds will always be an important part of chemistry, whether it be industrial production or the synthesis of new compounds on the research scale [3]. Preparative techniques can involve a variety of methods, including heating, cooling, mixing, and separating [1, 2]. The choice of preparative technique depends on the specific application and the desired outcome [1, 3].

In recent decades, there has been a growing emphasis on green chemistry, as stated by the U.S. EPA (United States Environmental Protection Agency), which focuses on describing the design of chemical products and processes that reduce or eliminate the use or generation of substances hazardous to human health, and minimizing environmental impact [4, 5]. This goal can be achieved by use of 12 principles of green chemistry which are as follows [6]:

1. **Prevention**: It is better to prevent waste than to treat or clean up waste after it has been created.
2. **Atom economy**: Synthetic methods should be designed to maximize the incorporation of all materials used in the process, into the final product.
3. **Less hazardous chemical syntheses**: Synthetic methods should be designed to use and generate less hazardous/toxic chemicals.
4. **Designing safer chemicals**: Chemical products should be designed to affect their desired function while minimizing their toxicity.
5. **Benign solvent and auxiliaries**: The use of solvents and auxiliary substances should be made unnecessary wherever possible and innocuous when used.

Green Chemical Synthesis with Microwaves and Ultrasound, First Edition.
Edited by Dakeshwar Kumar Verma, Chandrabhan Verma, and Paz Otero Fuertes.
© 2024 WILEY-VCH GmbH. Published 2024 by WILEY-VCH GmbH.

6. **Design for energy efficiency**: Energy requirements of chemical processes should be minimized, and synthetic methods should be conducted at ambient temperature and pressure if possible.
7. **Use of renewable feedstocks**: A raw material should be renewable rather than depleting whenever practicable.
8. **Reduce derivatives**: Unnecessary derivatization should be minimized or avoided if possible.
9. **Catalysis**: Catalytic reagents are superior to stoichiometric reagents.
10. **Design for degradation**: Chemical products should be designed so that at the end of their function, they break down into innocuous degradation products that do not persist in the environment.
11. **Real-time analysis for pollution prevention**: Analytical methodologies need to be further developed to allow for real-time, in-process monitoring and control prior to the formation of hazardous substances.
12. **Inherently safer chemistry for accident prevention**: Substances and the form of a substance used in a chemical process should be chosen to minimize the potential for chemical accidents.

Researchers have developed innovative techniques such as microwave-assisted and ultrasound-promoted preparations to contribute to the principles of green chemistry by offering more sustainable and efficient alternatives to conventional methods [7]. By harnessing the power of microwave and ultrasound energy, these techniques not only accelerate reaction kinetics but also substantially reduce reaction times and energy consumption, thereby minimizing the environmental footprint of chemical processes [8, 9].

12.1.2 Overview of Thermal, Microwave-Assisted, and Ultrasound-Promoted Preparations

Thermal, microwave-assisted, and ultrasound-promoted are techniques used in chemical synthesis and preparation, these techniques are commonly employed to accelerate chemical reactions and enhance product yields, improve efficiency, and enhance reaction rates, leading to reduced environmental impact and the facilitation of synthesizing various compounds [7, 10–12]. Here is an overview of each method.

Thermal preparation is a widely used technique and conventional method of heating a reaction mixture to promote a chemical reaction [3, 13]. It involves applying heat to the reaction mixture, typically through a heating source such as a Bunsen burner or a hot plate [3]. Thermal reactions involve the application of heat to increase the kinetic energy of molecules, leading to enhanced reaction rates; according to studies, the effect of temperature on reaction rates is typically an increase in rate with increasing temperature [14]. When the temperature is raised, the average kinetic energy of the reactant molecules increases, causing them to move more quickly and collide with greater energy. This results in a higher frequency and force of collisions between the reactant molecules, leading to an increased rate of reaction [15]. Heat energy speeds up the motion of molecules and lowers the activation energy barrier, making it easier for the reaction to occur [16]. However, it is important to note that thermal reactions can sometimes lead to side reactions or the degradation of sensitive compounds due to the high temperatures involved [17]. Heating reactions with conventional methods, such as Bunsen burners, oil baths, sand baths, and heating mantles, is not only slow but also creates a hot surface on the reaction vessel where products, substrates, and reagents often decompose over time [18]. These methods are not only time-consuming and tedious, but also energy inefficient and wasteful [19]. Conventional methods of organic synthesis usually need longer heating time, and tedious apparatus setup, which result in higher cost of process and the excessive use of solvents/reagents [20]. Mitigating these issues and optimizing the efficiency of thermal reactions can be overcome by the use of alternate methods, the use of microwaves being one such method [7, 21].

Microwave-assisted techniques utilize microwave radiation to heat reaction mixtures or samples, leading to faster and more efficient reactions compared to traditional heating methods [22]. Microwave-assisted preparation utilizes microwave heating to accelerate chemical reactions; the heating process is efficient and selective for specific molecules, allowing for rapid and uniform heating [10, 23]. Microwaves interact with the molecules in the reaction mixture, causing them to rotate and generate heat [23]. The use of microwaves can lead to a reduction in processing time and increase the production yield. This technique has been studied in the context of heterogeneous catalysis, where microwave heating can provide rapid and efficient heating to the reaction system, resulting in improved reaction rates [24]. The reactions can be carried out in solvent-free conditions as well, either neat or with solid-support or phase transfer catalysts, making them less toxic and less wasteful [12]. Microwave-assisted synthesis has several advantages over the conventional methods – high temperature is easily and quickly attained and cooling is also fast, microwave heating does not heat the whole surface inside the appliance, as is the case with conventional heating that heat up the reaction mixture by conduction, but only uses the waves to heat the reaction mixture kept in it. This reduces the formation of unwanted side products, so the yield is enhanced and the synthesis is cleaner. Furthermore, the use of harmful organic solvents in large amounts is avoided, which is usually not possible in conventional synthesis methods, all these features make microwave-assisted synthesis an example of green chemistry, that is, it is energy efficient, atom efficient, faster, uses fewer solvents, and is cleaner [21]. This technique has found applications in the synthesis of nanomaterials, organics, and pharmaceutical compounds [25]. The advantages of this enabling technology have, more recently, been exploited in the context of multistep total synthesis and medicinal chemistry/drug discovery and have additionally penetrated related fields such as polymer synthesis, material sciences, nanotechnology, and biochemical processes [7].

All these features make microwave-assisted synthesis an example of green chemistry, that is, it is energy efficient, atom efficient, faster, uses fewer solvents, and is cleaner [12]. Microwave-assisted synthesis is carried out in special microwave

Table 12.1 The comparison of a conventional heating and microwave heating [12].

	Conventional heating	Microwave
Heating method	Uses fans to circulate hot air, hence creating a uniform heating environment	Heats by polarizing effect [26]
Heating times	Slow heating, may take several hours	Fast heating, takes few minutes [27]
Heating characteristics	Heats the material from outside and then toward the inside of the material	Heats the material inside out

reactors that generally comprise five main components, such as a high voltage transformer, magnetron, waveguide, cooling fan, and cavity [22]. The comparison between conventional heating and microwave heating for synthesis is shown in Table 12.1 [12]. Conventional synthesis usually involves the use of a furnace or oil bath which heats the walls of the reactors by convection or conduction, whereas, in microwave-assisted synthesis, the microwave penetrates inside the material, and heat is generated through direct microwave-material interaction (Figure 12.1) [28].

Ultrasound-promoted techniques involve the application of high-frequency sound waves to a reaction mixture or sample, enhancing mixing, accelerating reactions, and improving mass transfer [29]. Ultrasound waves create small bubbles in the reaction mixture through a phenomenon called cavitation, the collapse of these bubbles generates heat and pressure, which can promote the reaction [25]. The application of ultrasound can improve reaction rates, reduce reaction times, and enhance product yields [30]. Reaction conditions, such as temperature and reactant concentration, were optimized to achieve the best results [31]. The method offers simplicity, mild reaction conditions, and high efficiency in short reaction times without the need for a complicated synthesis pathway [32]. This technique has been applied in the synthesis of various compounds, such as pharmaceuticals and organic compounds, showing higher yields compared to classical heating

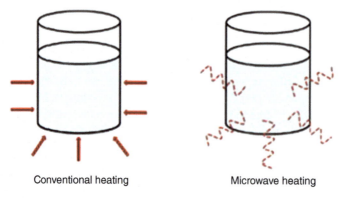

Figure 12.1 Comparison between microwave heating and conventional heating. Source: Adapted from Collins [28].

Table 12.2 Overview of conventional, microwave and ultrasound heating.

Conventional heating	Microwave heating	Ultrasonic heating
Thermal gradient (outside to inside)	Inverse thermal gradient (inside to outside)	Limited thermal gradient due to mixing
Conduction and convection currents	Molecular lever hot spots	Microbubble formation and collapse (compression and rarefaction cycles)
Longer processing time	Very short and instant heating	Relatively very short reaction times, not as quick as microwave
No or low solvent savings	No or low solvent reactions possible	Solvent savings possible
Product quality and quantity can be affected	Higher product quality and quantity possible	Same as conventional heating
High energy consumption	Moderate to low consumption	Moderate to low consumption
Simple process configuration	Very simple process	Moderate complexity

Source: Veera and Edith [34]/Springer Nature.

methods [32]. Both microwave-assisted and ultrasound-promoted preparations have been employed in various fields, including organic synthesis, catalysis, and material science. These techniques provide advantages such as shorter reaction times, decreased energy consumption, and improved selectivity [10]. The combination of microwave and ultrasound techniques has also been explored, leveraging the unique heating properties of microwaves and the cavitation phenomenon of ultrasound for even more effective and economical synthetic procedures [25]. Ultrasound-assisted synthesis was found to be the most efficient method, with high productivity and low energy consumption [33]. Factors such as process productivity, product purity, energy consumption, and overall cost need to be considered when selecting a synthesis method [33]. Table 12.2 summarizes the differences between conventional, microwave, and ultrasound heating [34].

12.1.3 Significance of Comparative Studies in Enhancing Synthetic Methodologies

Comparative studies contribute significantly to the progress of chemical synthesis and its applications across various fields, leading to optimized conditions, increased efficiency, enhanced selectivity, and the discovery of innovative techniques. The significance of comparative studies in enhancing synthetic methodologies can be highlighted in several ways.

12.1.3.1 Optimization of Conditions

The purpose of comparative studies is to identify the most optimal reaction conditions for a specific synthesis, to achieve higher yields, improved selectivity,

Table 12.3 Comparative study of conversions reaction to 5-HMF and LA.

Catalyst	Method	Substrate	Reaction condition	Result	References
Mn/ZSM-5	Conventional heating	Cellulose	5 h, 130 °C	1.17% 5-HMF	[36]
MnO_x/Hi_ZSM-5	Conventional heating	Cellulose	4 h, 100 °C	4.51% LA	[37]
			8 h, 100 °C	15.83% LA	
Hierarchical Mn_3O_4/ZSM-5	Conventional heating	Cellulose	2 h, 130 °C	8.77% LA	[38]
			4 h, 130 °C	16.23% LA	
			8 h, 130 °C	39.75% LA	
MOF MIL-88B(Fe)	Conventional heating	Glucose	3 h, 140 °C	3.8% 5-HMF	[39]
HCl	Conventional heating	Corn Starch	1 h, 165 °C	54% LA	[40]
SO_3H-functionalized ionic liquids (SFILs)	Laboratory microwave heating	Cellulose	5 min, 160 °C	15.7% LA	[41]

Source: Mazizah et al. [35]/with permission from Elsevier.

and reduced waste. Table 12.3 is the comparative study of the conversion reaction to 5-hydroxylmethylfurfural (5-HMF) and levulinic acid (LA) [35].

Table 12.3, the hierarchical Mn_3O_4/ZSM-5 catalyst was successfully synthesized and used in the conversion of biomass to LA. It is confirmed that the conventional heating method at 130 °C for eight hours of reaction time gives the highest conversion and LA yield. From the study of microwave-assisted reactions of biomass conversion into LA over hierarchical Mn_3O_4/ZSM-5 zeolite catalysts, the reaction with a 600 W microwave for 180 seconds shows results that are comparable with the four hours reaction using the conventional heating methods, with better purity of LA. In addition, the used catalyst, after reactivation, could be applied for three cycles of reaction without losing too much of its activity. This shows that the conversion using a microwave has the potential to be explored in the future to achieve cleaner reaction conditions in a very short reaction time [35].

12.1.3.2 Efficiency Improvement

Comparative analysis can help pinpoint the most efficient synthetic route among different options. This leads to shortened reaction times, reduced energy consumption, and increased overall efficiency of the synthesis process. Table 12.4 has a few reactions that were carried out using microwave heating and is compared with conventional heating, indicating the time and energy efficiency of the technique in Table 12.4 [42]. Based on experimental data in Tables 12.4 and 12.5, it has been found that microwave-enhanced chemical reaction rates can be faster than those of conventional heating methods, can use higher temperatures than conventional heating systems, and consequently, the reactions are completed in

Table 12.4 Comparison of reaction times using microwave versus conventional heating.

Compound synthesized	Reaction time: microwave	Reaction time: conventional
Methyl benzoate	5 min	8 hours
4-nitrobenzyl ester	2 min	1.5 h
Zeolite synthesis	30 s	60 min
Cubanite	3 min	3 d
$NaAlH_4$	2 h	8 h
$CuBi_2O_4$	5 min	18 h
Ag_3In	2 min	48 h

Source: Grewal et al. [7]/scienztech: International Journal of Dermatopathology and Surgery/Licensed under CC BY 4.0.

a few minutes instead of hours [7]. Table 12.5 shows that the other advantage of microwave-assisted organic synthesis is energy saving. Heating by means of microwave radiation is a highly efficient process and results in significant energy savings, because microwaves heat up just the sample and not the apparatus, and therefore energy consumption is less [7].

Table 12.5 compares the energy requirements for conventional, microwave-mediated catalytic, and microwave-mediated non-catalytic transesterification reactions [34].

12.1.3.3 Methodological Advances

Comparative studies drive methodological innovation to develop hybrid or combined approaches. These approaches can leverage the strengths of multiple techniques, leading to the creation of novel synthetic methodologies that offer enhanced efficiency and selectivity. The combination of microwave heating and ultrasound irradiation has been successfully exploited in applied chemistry, besides saving energy, these green techniques promote faster and more selective transformations [46]. Table 12.6 shows a combination of ultrasound and microwaves from the Peng and Gonghua experiment [47].

Table 12.6 reports the yield improvements between conventional reflux, ultrasound reaction, microwave-assisted reaction, and combination thereof in 73%, 79%, 80%, and 84%, respectively. More than an improvement in yield, this fusion of technologies allowed a drastic reduction of the reaction time from 9 hours to 40 seconds under optimum conditions. This effect was attributed to a combination of enforced heat transfer due to microwave irradiation and intensive mass transfer at phase interfaces due to ultrasound activation [9].

12.1.3.4 Sustainability and Green Chemistry

Comparative studies contribute to the principles of green chemistry by enabling the evaluation of synthetic methodologies in terms of their environmental impact. Thermal preparation is a widely used technique for the conventional method of

Table 12.5 Comparison of energy consumption per unit biodiesel production under different heating methods.

Type of heating	Conditions	Energy consumption (kJ/l)	References
Conventional	Continuous (industrial scale)	94.3	[43]
Microwave	Continuous, 7.2 l/min	26	[44]
	Continuous, 2 l/min (a power consumption of 1700 W and microwave input of 1045 W)	60.3	
	Continuous, 2 l/min (a power consumption of 2600 W and microwave input of 1600 W)	92.3	
	Batch 4.6 l (a power consumption of 1300 W, a microwave input of 800 W, a time to reach 50 °C of 3.5 min and a hold time at 50 °C of 1 min)	90.1	
Microwave (non-catalytic)	Supercritical, 10 ethanol/oleic acid molar ratio, 150 °C, 3.6 min (ml scale)	265	[45]
	Supercritical, 10 ethanol/oleic acid molar ratio, 200 °C, 5.7 min (ml scale)	762	
	Supercritical, 10 methanol/oleic acid molar ratio, 150 °C, 3.7 min (ml scale)	251	
	Supercritical, 20 methanol/oleic acid molar ratio, 200 °C, 3.7 min (ml scale)	609	
	Supercritical, 10 methanol/oleic acid molar ratio, 200 °C, 5.5 min	753	
	Supercritical, 5 methanol/oleic acid molar ratio, 200 °C, 5.1 min	804	
Ultrasound		137.5	[43]

heating a reaction mixture to promote a chemical reaction [3, 13]. Conventional methods of organic synthesis usually need longer heating time, and tedious apparatus setup, which results in higher cost of process and the excessive use of solvents/reagents [20]. To mitigate these issues and optimize the efficiency of thermal reactions, the use of alternate methods, the use of microwaves being one such method, is recommended [7, 21]. To mitigate these issues and optimize the efficiency of thermal reactions, various strategies have been developed, the use of ultrasound and/or microwave is in complete agreement with the principles of green chemistry/engineering through their numerous advantages – change of reactivity, improvement of yields and selectivity, reduction of reaction time, limitation of energy consumption and waste production, use of water/PEG as solvent instead

Table 12.6 Hydrazinolysis of methyl salicylate using different methods.

methyl salicylate + N$_2$H$_4$·H$_2$O (75%), US + MW (40 s) → salicylhydrazide

Methods	Reaction time	Isolated yield (%)
Conventional reflux	9 h	73
Ultrasound (20 kHz, 50 W) + reflux	1.5 h	79
Microwave (2.45 GHz, 200 W)	18 min	80
Simultaneous microwave (200 W) and ultrasound (50 W)	40 s	84

Source: Adapted from Yanqing and Gonghua [47].

of volatile organic solvents or solventless reactions, and activation of catalysts, to name a few [9].

12.2 Fundamentals of Thermal, Microwave-Assisted, and Ultrasound-Assisted Reactions

12.2.1 Explanation of Thermal Reactions and Their Advantages and Limitations

Thermal reactions refer to chemical reactions that occur as a result of changes in temperature. These reactions involve the conversion of reactants into products, accompanied by the release or absorption of heat energy. The study of thermal reactions is important for understanding the behavior and kinetics of energetic materials, such as all-metal energetic structural materials [48]. One advantage of thermal reactions is that they are relatively simple and widely applicable. They can be carried out in various types of reactors, such as batch reactors or continuous flow reactors, making them versatile for different reaction systems and scales. Additionally, thermal reactions can often be easily controlled by adjusting the temperature, allowing for precise control of reaction rates and selectivity. The range of thermal reaction temperatures can vary depending on the specific reaction and the nature of the reactants. In general, thermal reactions involving solids can occur over a wide temperature range, from relatively low temperatures to very high temperatures. For example, in the study of Ni/Al energetic structural materials, the thermal reactions were observed to occur at temperatures ranging from around 600 to 1200 K [49]. Another study on the thermal reactions of Ni/Al samples reported reaction temperatures ranging from 400 to 1000 K [49]. The schematic figure of the reaction mechanism of Ni/Al materials is shown in Figure 12.2.

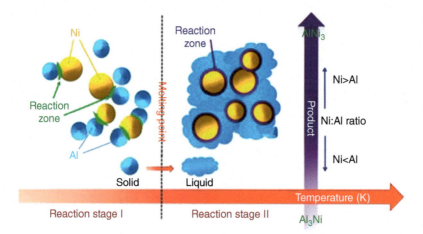

Figure 12.2 Schematic diagram of the reaction mechanism of Ni/Al materials (thermal reaction). Source: Zhang et al. [49]/MDPI/Public Domain.

Another advantage of thermal reactions is that they can be easily scaled up for industrial production. The equipment and infrastructure required for thermal reactions are generally well-established and readily available, making it easier to transition from laboratory-scale to large-scale production. However, there are also limitations to thermal reactions. One limitation is that they can sometimes require high temperatures, which can lead to energy inefficiencies and potential safety hazards. Additionally, thermal reactions may not be suitable for reactions that are sensitive to high temperatures or require specific reaction conditions.

In recent years, alternative reaction methods such as microwave-assisted and ultrasound-assisted reactions have gained attention due to their advantages over traditional thermal reactions. These methods offer faster reaction rates, milder reaction conditions, improved selectivity, and higher yields [50]. However, the choice of reaction method depends on the specific reaction and the desired outcome, and factors such as reaction kinetics, reactant properties, and scalability should be considered [51]. In conclusion, thermal reactions are widely used in chemical processes and offer simplicity and scalability. However, alternative reaction methods such as microwave-assisted and ultrasound-assisted reactions provide additional advantages and should be considered depending on the specific requirements of the reaction.

12.2.2 Introduction to Microwave-Assisted Reactions and How They Differ from Traditional Method

Microwave-assisted reactions involve the use of microwave irradiation to heat the reaction mixture, leading to faster reaction rates and improved yields compared to traditional thermal reactions. In microwave-assisted reactions, microwaves are a form of electromagnetic radiation that can rapidly and selectively heat polar molecules or ions in the reaction mixture [22]. One key difference between microwave-assisted reactions and traditional methods is the heating mechanism.

In traditional thermal reactions, heat is transferred through conduction, convection, or radiation from an external heat source to the reaction mixture. This can result in slower and less efficient heating, as the heat must penetrate the reaction mixture from the outside. The range of microwave-assisted temperatures can vary depending on the specific reaction and the nature of the reactants. Microwave heating is known for its ability to rapidly and selectively heat materials, including solids, liquids, and gases. In general, microwave-assisted reactions can occur at temperatures ranging from room temperature to several hundred degrees Celsius. For example, in a study on the microwave-assisted synthesis of metal–organic frameworks, the reactions were carried out at temperatures ranging from 60 to 180 °C [52].

In microwave-assisted reactions, however, the microwaves directly interact with the polar molecules or ions in the reaction mixture, causing them to rapidly rotate and generate heat through molecular friction. This localized heating leads to faster and more uniform heating of the reaction mixture, resulting in accelerated reaction rates. Another difference is the selectivity of microwave-assisted reactions. The rapid and efficient heating provided by microwaves can lead to specific activation of certain reaction pathways, allowing for improved selectivity and control over the desired products. This can be particularly advantageous in organic synthesis, where microwave-assisted reactions have been shown to enhance yields and reduce unwanted side reactions [53]. Additionally, microwave-assisted reactions often require shorter reaction times compared to traditional methods. The efficient heating provided by microwaves can accelerate reaction kinetics, allowing for faster completion of reactions. This can be beneficial in terms of time and energy savings, as well as increased productivity in industrial applications. It is important to note that the choice between microwave-assisted reactions and traditional methods depends on the specific reaction and desired outcome. Factors such as reaction kinetics, reactant properties, and scalability should be considered when selecting the appropriate method [54]. In summary, microwave-assisted reactions offer several advantages over traditional methods, including faster reaction rates, improved selectivity, and shorter reaction times. The direct interaction of microwaves with the reaction mixture allows for efficient and uniform heating, leading to enhanced yields and reduced side reactions. However, the suitability of microwave-assisted reactions depends on the specific reaction system and should be carefully evaluated.

12.2.3 Understanding the Principles and Mechanisms of Ultrasound-Promoted Reactions

Ultrasound-promoted reactions involve the use of high-frequency sound waves to induce and enhance chemical reactions. The principles and mechanisms of ultrasound-promoted reactions can be attributed to a phenomenon called acoustic cavitation (see Figure 12.3). Acoustic cavitation occurs when sound waves create alternating high and low-pressure regions in a liquid medium. This leads to the formation and rapid collapse of small bubbles, known as cavitation bubbles. During bubble collapse, high temperatures, and pressures are generated locally, creating

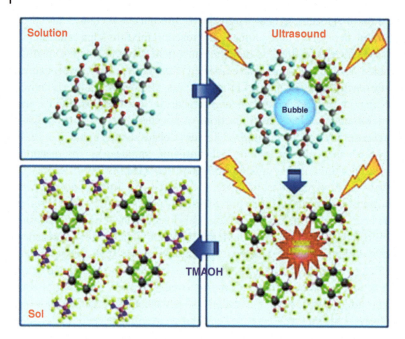

Figure 12.3 Reaction mechanism of ultrasound assisted synthesis method. Source: Yan et al. [55]/with permission of Frontiers Media S.A.

highly reactive conditions. The range of ultrasound reaction waves can vary depending on the specific reaction and the experimental conditions. Ultrasound waves are high-frequency sound waves that can be used to induce and enhance chemical reactions. In general, ultrasound-assisted reactions can occur at frequencies ranging from a few kilohertz to several megahertz. The intensity of the ultrasound wave, measured in terms of power density or acoustic pressure, can also vary depending on the desired reaction conditions. For example, in a study on ultrasound-assisted synthesis of nanoparticles (NPs), the ultrasound waves were applied at frequencies ranging from 20 kHz to 1 MHz [29]. Another study on ultrasound-assisted organic reactions reported the use of ultrasound waves at frequencies ranging from 20 to 100 kHz [56].

The collapse of cavitation bubbles generates intense localized energy, including shockwaves, microjets, and high temperatures. These energetic events can break chemical bonds, enhance mass transfer, and promote reaction kinetics. The high temperatures and pressures generated during bubble collapse can also facilitate reactions that would otherwise require high temperatures or harsh conditions.

The mechanisms by which ultrasound promotes reactions can vary depending on the specific reaction and reactants involved. Some possible mechanisms include [57]:

1. **Enhanced mass transfer**: The rapid formation and collapse of cavitation bubbles create microstreaming and turbulence in the reaction mixture, improving the mixing and transport of reactants to reaction sites.

2. **Sonoluminescence**: During bubble collapse, the release of energy can result in the emission of light, known as sonoluminescence. This phenomenon can generate reactive species, such as free radicals, which can participate in chemical reactions.
3. **Sonochemistry**: The high temperatures and pressures generated during bubble collapse can induce chemical reactions through thermal effects, radical formation, and the generation of reactive species.
4. **Mechanical effects**: The mechanical forces generated by cavitation bubbles, such as shockwaves and microjets, can break chemical bonds and enhance reaction rates.

It is important to note that the specific mechanisms and effects of ultrasound-promoted reactions can vary depending on factors such as ultrasound frequency, power, and reaction conditions. Optimization of these parameters is crucial for achieving desired outcomes in ultrasound-assisted reactions [57].

12.3 Case Studies in Organic Synthesis

12.3.1 Examining Examples of Organic Reactions Performed Under Thermal Conditions

The application of heat in organic reactions helps facilitate bond-breaking and bond-forming processes to achieve the desired chemical transformations. However, many organic compounds have low heat resistance characteristics. These compounds burn or carbonize when heated over a few 100 °C. C—C and C—H bonds have a low polar character that affects the thermal stability of organic compounds [58, 59].

12.3.1.1 Esterification Reaction Under Thermal Conditions
Some studies have proven that the reaction temperature had a favorable influence on the degree of esterification. With an increase in temperature, the conversion rose at a faster rate [60–62]. One of the applications of thermal conditions in esterification reaction is the esterification of FFA (Free Fatty Acid) to methyl ester. FFA conversion was increased by increasing temperature. From the study results presented in Figure 12.4, it can be seen that the conversion was 98% at 60 °C (optimum temperature). As the temperature rises more, the FFA conversion remains constant [62].

12.3.1.2 Dehydration of Alcohols
As shown in Figure 12.5, the dehydration reaction of alcohols to generate alkene proceeds by heating the alcohols in the presence of a strong acid (such as sulfuric or phosphoric acid) at high temperatures. With increasing substitution of the hydroxy-containing carbon, the required range of reaction temperature lowers as described below [63]:

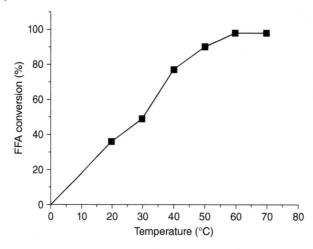

Figure 12.4 Effect of temperature on esterification of FFA. Source: Ferdous et al. [62]/Science and Education Publishing.

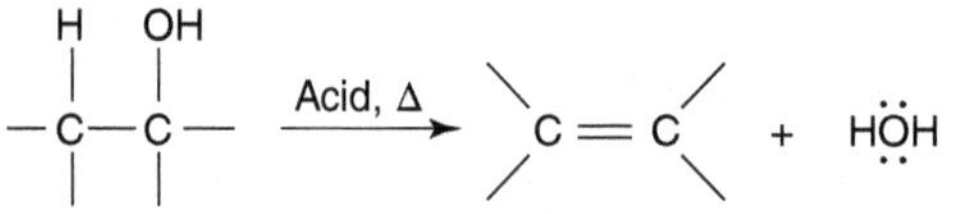

Figure 12.5 Dehydration of alcohols. Source: Jeffrey [63].

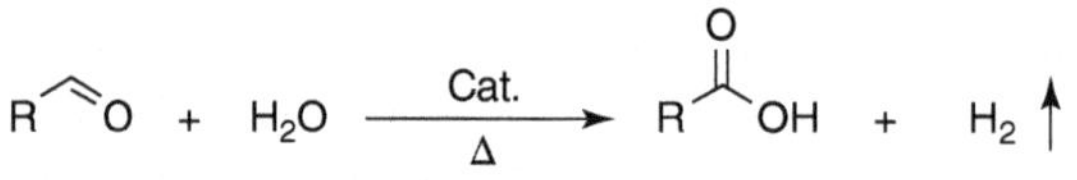

Figure 12.6 Aqueous aldehyde reforming under neutral conditions.

- 1° alcohols: 170–180 °C
- 2° alcohols: 100–140 °C
- 3° alcohols: 25°–80 °C

If the reaction is not sufficiently heated, the alcohols do not dehydrate to form alkenes but react with one another to form ethers [63].

12.3.1.3 Oxidation of Aldehydes to Carboxylic Acids Using Water

The oxidation of aldehydes to carboxylic acids in water releasing H_2 (or aldehyde-water shift reaction, as shown in Figure 12.6) is usually carried out using heterogeneous catalysts at extremely high temperatures (>200 °C). In a journal written by Kar and Milstein [64], Cundari et al. reported that aldehyde oxidation to carboxylic acids can be conducted under neutral conditions at 0.4–0.8 mol% catalyst loading, and a 105 °C reaction temperature.

12.3.2 Case Studies Showcasing the Application of Microwave-Assisted Reactions

Due to its capacity to deliver quick and effective heating, microwave-assisted irradiation has drawn a lot of interest as a method for quickening chemical reactions. By lowering reaction times and energy consumption, microwave irradiation can

Table 12.7 Cross-coupling reactions meaning.

Cross-coupling reactions	Meaning
Heck coupling	Palladium-catalyzed coupling reaction that involves the cross-coupling of aryl or vinyl halides with alkenes (usually styrene) to form substituted alkenes. This reaction is typically performed under mild conditions in the presence of a base and a suitable solvent [71].
Stille coupling	Palladium-catalyzed cross-coupling of aryl or alkenyl halides with aryl, alkenyl, or alkynyl tin compounds results in C—C bond formation [69].
Suzuki coupling	In this reaction, boronic acids or boronic esters are coupled with aryl or alkenyl halides and catalyzed by various palladium species and require a stoichiometric base [69].
Sonogashira coupling	This reaction uses a palladium catalyst to couple aryl or alkenyl halides to terminal alkynes to give disubstituted alkynes [69].

contribute to the promotion of sustainable and environment-friendly synthesis techniques [65]. The principles of microwave heating involve generating heat directly within reactants through molecular-level interaction with electromagnetic radiation [66, 67]. The main applications of microwave heating are in the attenuation of environmental pollution, medical uses, food processing, agriculture, the ink and paint industry, and wood treatments. Regarding material processing and microwave chemistry, microwave heating is applied to organic and analytical chemistry, biochemistry, catalysis, photochemistry, inorganic materials, and metal chemistry (Horikoshi and Sepone, 2013 in Palma et al. [24]). Microwave heating is applicable to various synthetic applications across diverse organic transformations, including coupling reactions, cyclization, and heterocyclic synthesis [67].

12.3.2.1 Microwave-Assisted C—C Bond Formation

Microwave-assisted C—C bond formation refers to the use of microwave irradiation to facilitate and accelerate reactions that involve the creation of carbon–carbon (C—C) bonds [68]. Microwave heating has proven to be a valuable technique in organic synthesis, offering several advantages over conventional heating methods, such as faster reaction rates, improved yields, and enhanced selectivity [19]. One of the reactions that involve changing C—C bond is a cross-coupling reaction. Cross-coupling reactions are a class of organic reactions that involve the formation of a new C—C bond between two different organic fragments, often using a metal catalyst. In cross-coupling reactions, two different types of organic substrates, often referred to as "partners," are combined to form a single product [69]. These are some cross-coupling reactions: Heck coupling, Stille coupling, Suzuki coupling, and Sonogashira coupling [70], as described in Table 12.7.

Many studies about cross-coupling applications have been conducted and summarized in Table 12.8.

Table 12.8 Cross-coupling reactions applications from literature study.

Cross-coupling reactions	Application	Study results	References
Heck coupling	Palladium-catalyzed coupling reaction of aryl iodides with styrene in water	The synthesis under microwave irradiation was 18–42 times faster than under conventional heating. Microwave irradiation provides the same selectivity, with respect to halide displacement as compared to the original reaction.	[72]
Stille coupling	Transforming 1-alkynes into 1,3-dienes or styrene	This method provides a high-yielding and robust method for the stereo-defined synthesis of 1,3-dienes. The conventional methods for these transformations typically require more time, but this microwave-assisted approach allows for faster reactions	[73]
Suzuki coupling	Cross-coupling of aryl bromides and boronic acids	Microwave irradiation of aryl bromides and boronic acids using polyurea microencapsulated palladium catalyst (Pd EnCat 30) was carried out at 120 °C, 250 W. This method gave the coupling adducts in excellent yields in just 20 min (compared to ~24 h under thermal conditions). That means microwave irradiation increases the coupling reaction rate.	[74]
Sonogashira coupling	Synthesis of poly-substituted aromatic acetylene compounds	The temperature reaction was maintained at 100–115 °C. High product yields are produced as a result of the improved reaction conditions, which also lessen reliance on anaerobic conditions. This technique has a lot of potential uses, particularly for the synthesis of porous aromatic frameworks.	[75]

12.3.2.2 Microwave-Assisted Cyclization

Microwave-assisted cyclizations have been used to synthesize various heterocyclic compounds. Microwave irradiation offers an eco-friendly, solvent-free approach for rapid heterocyclic synthesis, reducing reaction time and improving yields [76]. The study conducted by Ochoa-Terán et al. [77] uses microwave in cyclization–methylation reaction of amino alcohols and acetylated derivatives with dimethyl carbonate and TBAC. This reaction applied microwave irradiation at

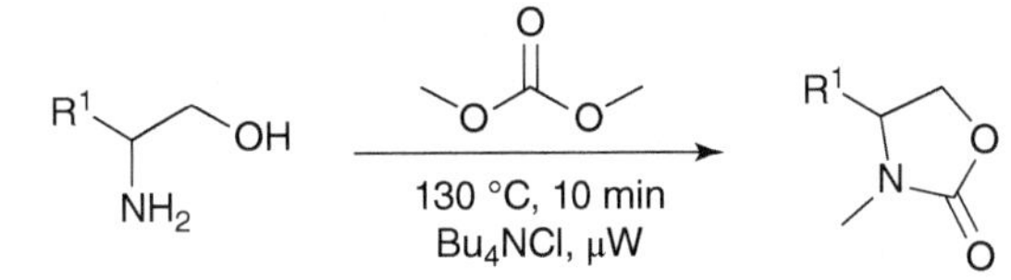

Figure 12.7
Cyclization-methylation reaction.
Source: Ochoa-Terán et al. [77].

130 °C (see Figure 12.7). The writer stated that this method has several noteworthy benefits, including a one-step protocol, reproducibility, quick and easy product purification, low amounts of TBAC with extremely high yields and purity, and beneficial environmental and synthetic benefits from the use of nontoxic compounds and catalysts.

12.3.2.3 Microwave-Assisted Dehydrogenation Reactions

Microwave-assisted processes can be advantageous for dehydrogenation reactions. The study conducted by Kim et al. (2001), summarized in a journal written by Palma et al. [24], stated that the interaction of microwaves with the active metal in catalysts generates localized hot spots, preserving the catalyst's selectivity. By using this method, an 87% yield of ethylene can be obtained, approximately 10–15% higher than the conventional steam cracking of ethane. The effectiveness of microwave-assisted approaches is demonstrated in the case of ethane dehydrogenation to ethylene, where α-SiC catalyst is reported to enhance the reaction's yield.

12.3.2.4 Microwave-Assisted Organic Synthesis

The study conducted by Caliman et al. [78] about the synthesis of amine-functionalized graphene oxide (amine-GO) using microwave-assisted reactions. Amine-GO can be used as a promising option for supporting materials in supercapacitors. Supercapacitors are energy storage devices with high power density and fast charge/discharge rates, making them valuable for various applications. By using graphene-based materials, supercapacitors' performance can be enhanced due to their excellent conductivity and large surface area. The results of the experiment show that all synthesized amine-GO materials presented a good electrochemical behavior, with long life cycle stability and reaching specific capacitance values, confirming their potential application as supporting materials in supercapacitors.

12.3.3 Highlighting Successful Instances of Ultrasound-Promoted Organic Synthesis

Ultrasound can be an effective method for promoting complex radical reactions, offering potential applications in the synthesis of diverse organic compounds [79]. The principle of ultrasound-assisted organic synthesis by ultrasound waves generates mechanical vibrations that result in the formation of cavitation bubbles. The collapse of these bubbles generates localized heating and pressure changes, accelerating reaction rates and facilitating reactions that might otherwise require harsh conditions [80]. In addition, this method is used in many phase transfer-catalyzed reactions, such as the N-alkylation of amines, the production of ethers and esters, and the hydrolysis of esters; ultrasound also facilitates effective mixing in solid–liquid and liquid–liquid mixtures [81].

12.3.3.1 Ultrasound-Promoted in Organic Synthesis

From the review by Bonarth [82], Rinderspacher and Prijs were the first to synthesize 4-methyl-oxazole-5-carbonitrile by dehydrating 4-methyl-oxazole-5-carboxylic acid amide. This compound is a building block in the vitamin B6 synthesis. The reaction yields 61% in a shorter time. The dehydration used phosphorus pentoxide.

Nandurkar et al. [83] conducted a study of synthesizing metal-1,3-diketonates by reacting metal nitrates/halides with the corresponding 1,3-diketone ligands under ultrasound. Metal-1,3-diketonates can be used as MOCVD precursors for the deposition of metal-oxide films, which are used as insulating layers in microelectronics, as protective or optical coatings, and as fuel cells. This ultrasound-assisted method showed a significant rate enhancement for metal complex formation and provided higher yields.

12.3.3.2 Ultrasound-Promoted Oxidations

The study conducted by Khatri et al. [84] used hydrogen peroxide as the oxidant in the presence of poly(ethylene glycol) dimethyl ether as a recyclable reaction medium under ultrasonic irradiation to oxidize sulfides to sulfoxides with good selectivities, without the use of a catalyst. A number of benefits, including clean, cost-effective synthesis, excellent selectivity, quicker reaction times, recyclability of the reaction medium, and moderate reaction conditions are provided by the discovered approach. This technology is more appealing, both from an environmental and a financial standpoint, thanks to the special effects of ultrasound, recycling, and cost-effectiveness of the reaction media PEGDME500, and mild reaction conditions. They believe that due to the effective mixing of the substrates under ultrasonic irradiation and the electrostatic attraction of the hydrogen peroxide with PEGDME500, the activation of hydrogen peroxide in the described technique may be feasible.

12.3.3.3 Ultrasound-Promoted Esterification

One of the applications of this method on esterification is methyl esterification of carboxylic acids catalyzed by polymer-supported triphenylphosphine. This reaction is conducted under certain conditions, the triphenylphosphine catalyst supported by polymer is essential in promoting the reaction, and ultrasonic irradiation speeds up the reaction's kinetics. The research shows that when a solid catalyst is used along with ultrasonic irradiation, the reaction time is greatly sped up and the yields of methyl esters from carboxylic acids are improved. Methyl esters are produced in high purities of 95% [85].

12.3.3.4 Ultrasound-Promoted Cyclization

One of the applications of ultrasound-promoted cyclization reaction is the synthesis of 5-methylselanyl-4,5-dihydroisoxazoles through the radical cyclization of unsaturated oximes with selenium reagents. In this study, ultrasound is used as a tool to facilitate the formation of complex organic molecules through radical pathways. The positive effect of this application resulted in high selectivity, short reaction times, and moderate to excellent yields of the expected products [79].

12.4 Scope and Limitations

12.4.1 Discussing the Applicability of Each Method to Different Reaction Types

Microwave and ultrasound processing are two techniques used in chemical engineering for energy transfer and reaction control. Chemical reactions of many kinds, including organic synthesis, extraction, catalysis, polymerization, and crystallization, can be processed using microwave and ultrasonic technology. Broadly applicable in a variety of chemical reactions, microwave, and ultrasound processing methods both provide benefits such as quick heating, increased reaction speeds, increased selectivity, and improved product quality [86]. Microwave processing has been widely used in organic synthesis since it provides quick and effective heating for reactions including esterification, oxidation, reduction, and cyclization. Here are a few examples:

1. In extraction processes, such as the extraction of essential oils from plants [86].
2. In catalysis, microwave irradiation has been utilized for the synthesis of heterogeneous catalysts and the activation of catalytic reactions [87]. Microwave heating can speed up reaction rates and increase selectivity, resulting in higher yields and reduced reaction times [86]. For example, the synthesis of composite catalysts $Cu-CeO_2/C$ for selective hydrogenolysis [87].
3. In crystallization processes, to control crystal size and morphology to develop desirable crystals [86].
4. Material synthesis, including polymers, polyesters, polyamides, and polyurethanes [86], NPs, metal oxides, carbon-based materials [88], carbon nanotubes [46]. Hydrogenation of various compounds, such as furfural, nitrobenzene, LA, jatropha oil biodiesel, and polyunsaturated fatty acid methyl esters [87]. Microwave heating is quick and precise, which can speed up the polymerization processes.
5. **Thermal treatment and drying procedures**: Microwave heating has been successfully used in commercial thermal treatment and drying procedures [88], for instance in the food industry [89]. Microwaves also have potential applications in biodiesel production. Microwaves may be used in the transesterification reaction, which is a crucial step in turning vegetable oils or animal fats into biodiesel. The reaction mixture can be heated using microwaves, which speeds up the reaction process and increases yields. The application of microwaves in biodiesel production offers advantages such as faster reaction rates, improved energy efficiency, and enhanced reaction control [90].

Microwave processing is a useful tool in many processes because it has a lot of advantages over traditional techniques. These advantages make microwave processing a valuable technique in various reaction types and applications. Here are a few advantages:

1. **Faster reaction times**: When compared to traditional heating techniques, microwave heating produces much faster response times by allowing for fast and effective heating of reaction mixtures. This can speed up work and cut down on processing time overall [88].
2. **Improved product quality**: By improving control over reaction parameters like temperature and pressure, microwave processing can increase the selectivity and yield of desired products [88]. This can lead to improved product quality and reduced side reactions [86].
3. **Energy efficiency**: Microwaves used for direct heating of the reaction mixture reduce energy loss through conduction and convection, which lowers energy use. This makes microwave processing a more environmentally responsible and sustainable choice [88].
4. **Enhanced safety**: Microwave processing can offer improved safety compared to conventional methods. Microwaves' quick and focused heating diminishes the possibility of mishaps and lowers the likelihood of thermal runaway reactions [88].
5. **Versatility**: A wide variety of reactions, including the creation of NPs, drying, chemical processes, and the creation of pharmaceuticals, can be handled by microwave processing. It is a useful tool in many sectors and research fields because of its adaptability [88].
6. **Provides a clean environment**: Microwave processing provides a clean environment at the point of use, as it does not require the use of combustion or external heating sources. This can reduce the emission of pollutants and improve the sustainability of chemical processes [86].

Ultrasound processing is commonly used in organic synthesis, for instance in sonochemical reactions. Numerous fields, including organic synthesis, oxidation, reduction, and degradation of contaminants, have used sonochemical reactions. Another crucial application is ultrasonic-assisted extraction, in which ultrasound waves increase the extraction efficiency by encouraging mass transfer and rupturing cell walls. Similar to microwave processing, ultrasound has also been employed in the catalysis, polymerization, and crystallization processes [86]. Numerous reactions in numerous disciplines have been shown to be accessible to ultrasound processing. Here are a few instances:

1. **Homogenization and emulsification**: In a variety of industries, such as food, pharmaceuticals, and cosmetics, ultrasound can be used to disintegrate particles and generate stable emulsions.
2. **Extraction**: The extraction of bioactive components from plant materials, such as flavonoids, phenolics, and essential oils, is frequently improved by the application of ultrasound-assisted extraction.
3. **Degradation and degradation control**: When treating organic contaminants in water and wastewater, ultrasound can help them break down. Heavy metals from wastewater have been removed using ultrasound, using techniques like

ultrasonic-assisted adsorption and ultrasonic-assisted precipitation [91]. It can also be applied to stop materials and polymer deterioration [88].
4. **Synthesis and catalysis**: Sonochemistry, which uses ultrasound to speed up chemical reactions and catalytic processes, can make use of ultrasound as a tool. It has been used in a variety of organic synthesis processes, including those that produce NPs and activate catalysts.
5. **Crystallization**: Using ultrasound to regulate and accelerate crystallization can produce desirable crystal shapes with enhanced characteristics. These examples demonstrate the versatility and applicability of ultrasound processing in various reaction types.

Utilizing ultrasound can yield substantial advantages, including enhanced processing, minimized waste production, intrinsically secure operations, conservation of materials and energy, and heightened productivity [92].

12.4.2 Identifying the Limitations and Challenges Faced by Each Technique

Microwave processing, while offering numerous advantages, also faces certain limitations and challenges. The limitations and challenges faced by microwave methods include scalability, limited application, health risks, uneven heating, and expensive manufacturing costs. Due to the limited wavelength of microwave technology, reactor scale-up is an important factor. Low product yields for non-polar reactants might also come from the reliance of microwave heating on the dielectric characteristics of the material. Long-term exposure to high-frequency microwave radiation can harm DNA strands, bodily tissues, and cells, among other health problems [93].

1. **Scalability**: While microwave processing has been successfully applied in laboratory-scale reactions, scaling up to industrial production can be challenging. The design and engineering of large-scale microwave reactors that maintain efficient and uniform heating throughout the reaction volume is a complex task [3]. Scalability presents an additional obstacle for microwave techniques, especially in the context of sizable batch reactors and tubular flow reactors, owing to the extent of microwave penetration and inherent temperature gradients [93].
2. **Limited penetration depth**: Microwaves have a limited penetration depth into materials, which can result in uneven heating and the formation of hotspots. This can lead to localized overheating and inhomogeneous dissipation of microwave energy within a sample, affecting reaction kinetics and product quality [88].
3. **Reactor design limitations**: The dimensions and design of microwave reactors can pose limitations on the scale and volume of reactions that can be performed [88]. The distribution of microwave energy within the reactor can vary, leading to nonuniform heating and temperature gradients [88].
4. **Temperature and pressure control**: Precise control of temperature and pressure is crucial for many reactions. However, achieving accurate and uniform temperature distribution throughout the reaction mixture can be challenging

Table 12.9 Comparison between microwave and ultrasound differences.

Application	Microwave	Ultrasound
Reaction media	MW-absorbing liquids; solvent-free protocols	aqueous and organic solvents
Use of bulk metals	Forbidden practice	Favorite domain
Acceleration	Large (min, even seconds!)	Variable (from min to h)
Scaling up	Possible	Possible but still a challenge
Chemical effects	Selectivity changes, waste reductions	Selectivity changes, mechanistic switching, waste reductions
Other effects	Heating above boiling points, change in solvent properties	Light emission, cleaning, microstreaming

Source: Adapted from Cravoto and Cintas [94].

in microwave processing. Pressure control can also be difficult, especially for reactions that require elevated pressures [88]. Stand-alone microwave reactors have limitations such as arching, overheating and explosion, expensive fiber optic thermocouples for temperature measurement, and nonuniform heating [93]. The nonuniformity in temperature dispersion and the creation of areas with elevated temperatures can result in thermal loss and influence the final result of the reaction [93].

5. **Material compatibility**: Microwave processing may not be suitable for all types of materials. Some materials may be sensitive to microwave radiation or may not absorb microwaves efficiently, limiting their applicability in microwave reactions [88].
6. **Safety considerations**: Microwave processing can pose safety risks, such as the potential for thermal runaway reactions and the generation of high pressures. Proper safety measures, including the use of appropriate reaction vessels and monitoring systems, are necessary to mitigate these risks.

Table 12.9 shows the comparison between microwave and ultrasound differences.

12.4.3 Opportunities for Combining Approaches to Overcome Specific Limitations

The combined use of microwaves and ultrasound is a solution. Microwaves expedite reaction rates by employing selective or volumetric heating (dependent on the nature of the reaction mixture, homogeneous or heterogeneous). Meanwhile, ultrasound enhances phase-to-phase mass transfer, improves mixing, and facilitates more even temperature distribution within homogeneous mediums [95]. Combining microwave and ultrasound approaches offers opportunities to overcome the limitations of microwave methods. For e.g. in biodiesel and biolubricant production. Combining the two methods can improve heat and mass transfer, shorten the duration reaction, and reduce energy consumption [93]. The study conducted by [96] a simultaneous ultrasound–microwave irradiation to produce

Table 12.10 Comparison of methods for synthesis Pb(OH)Br nanowires [98].

	Reaction time	Temperature (°C)	Yield (%)	Diameter (nm)	Length (μm)
Conventional heating	24 h	70	23.0	20 000–30 000	2 000–3 000
Microwave (50 W)	10 min	89	35.1	1 800–2 000	60–80
Ultrasound (50 W)	10 min	34	32.9	700–1500	20–40
Microwave–ultrasound (50–50 W)	10 min	66	45.0	80–800	50–100
Microwave–ultrasound (250 W to 50 W)	80 s	95	48.2	100–500	10–30

biodiesel from palm oil. This combined approach significantly accelerated the transesterification reaction, resulting in a high biodiesel yield. Through concurrent irradiation, the transesterification time can be decreased by a factor of 26.4 compared to the conventional approach. Another study by [97], with a two-step method combining microwave sequential effect followed by ultrasound for the intensification of biodiesel synthesis, resulted in shorter reaction durations and a decreased requirement for excess methanol, leading to notable energy conservation. The combination of micro-emulsification's physical effects, ultrasound's acoustic streaming, swift heating, and microwaves' dipolar rotation collectively contributed to boosting the efficiency of the transesterification process [86].

Another study of combining microwave and ultrasound methods is applied in the synthesis of Pb(OH)Br nanowires [98]. The conclusion of this study suggested a method (combining microwave and ultrasound) that offers substantial yields, rapid reaction times, heightened efficiency, and a notable aspect ratio. It is anticipated that this approach will have diverse practical applications in the fields of nanoscience and nanotechnology.

A comparison was made between conventional heating, microwave, or ultrasound irradiation alone, and microwave–ultrasound combined irradiation, tabulated in Table 12.10.

This combining method review compiles these advancements and offers insight into potential future possibilities, especially concerning the creation of NPs and heterogeneous catalysis. Additionally, it sheds light on the adaptation of processes to pilot-scale operations, with a note that industrial applications will require specialized engineering designs [46].

12.5 Future Directions and Emerging Trends

12.5.1 Overview of Recent Advancements and Ongoing Research in Thermal, Microwave, and Ultrasound-Assisted Preparations

There have been several advancements and ongoing research in the field of thermal preparations. Here is an overview of some recent advancements and ongoing research.

12.5.1.1 Food Processing Technologies

Thermal food processing, developed by Nicholas Appert, preserves food safety and quality by canning glass bottles; Louis Pasteur studied spoilage reduction for improved shelf life [99]. Thermal processing improves food safety by inactivating pathogens, extending shelf life, and destroying anti-nutritional factors. It also enhances product quality by denaturing proteins, making them more accessible to digestive enzymes, breaking down plant cell walls, and promoting the bioavailability of beneficial compounds. It initiates caramelization and Maillard reactions, enhancing flavor, aroma, color, texture, and functional properties [100]. Negative effects of thermal processing can cause vitamin loss, lipid oxidation, and Maillard browning, reducing nutritional value and potentially producing toxicants [101]. The flavor, texture, and color of food products can also lead to undesirable changes as a result of high-temperature heat treatment [102]. Food thermal processes are complex and require mathematical models to guide design and operation. Models describe transport phenomena and can be developed at different length scales, depending on the purpose. Research is increasing on more complex transient dynamics. One-, two-, and three-dimensional models are used to model food thermal processing at smaller scales. More detailed computational fluid dynamics (CFD) simulation techniques and finite-element-based models are being developed to describe food structural changes during thermal processing. Kinetic models are also being developed, but specialized experimental techniques may be required. AI and computer-based learning models are being developed to describe complex food thermal processes [103].

12.5.1.2 Chemical Routes to Materials: Thermal Oxidation of Graphite for Graphene Preparation

Oxidation of graphite is the first important step toward the industrial scale preparation of graphene by a top-down method using graphite as a starting material [104]. Graphene, introduced by Geim and Novoselov in 2004, is a wonder material with remarkable properties and potential applications [105–107]. Graphene preparation involves oxidation, separation/exfoliation, and scalability, but requires strong chemicals, expensive equipment, and long production times [104]. Thermal oxidation is a quicker, less expensive, and environment-friendly method for oxidizing graphite. This method is economically viable and environmentally friendly but requires understanding the appropriate conditions, temperature, and time for efficient oxidation [104]. From the research by Nair et al., graphite was oxidized using a furnace under atmospheric conditions between 200 and 1000 °C for specific time periods (5–20 minutes); significant oxidation occurred when heated between 200 °C and 600 °C for a shorter time (<5 minutes), while partial reduction occurred when heated for longer periods (>10 minutes). The maximum oxidation was achieved at 400 °C for five minutes, while reduction occurred at 800 and 1000 °C due to the lack of oxygen and moisture, lesser Van der Walls force of attraction, and decomposition of functional groups [104].

12.5.1.3 Environmental and Sustainable Applications: Waste to Energy

Global waste generation is expected to continue to grow due to economic development and population growth [108]. Waste-to-energy processes could constitute a way to recover energy from waste, helping access renewable energy to the world population, in addition to a waste management system [109]. The present review describes different wastes that can be employed in waste-to-energy processes, using thermochemical, biochemical, and chemical processes [110]. The energy produced can be in the form of heat, electricity, or biofuels [111]. The trend toward biorefinery concepts in thermochemical processes is driven by waste as a renewable carbon source [111]. This can be achieved through recovery or recycling, resulting in higher-value products [111]. Research on waste oil management and biodiesel yield is crucial for effective waste oil collection and improved conversion [111]. Proper waste characterization is essential for energy recovery, promoting the selection of suitable conversion pathways, and final energy products [111]. Operational strategies should focus on feedstock requirements and energy conversion processes for a technically and economically feasible waste biorefinery [111].

12.5.2 Recent Findings in Microwave-Assisted Preparation

Recent advancements in microwave-assisted preparations have been made in various fields. Microwave-assisted synthesis has become one of the latest achievements in green chemistry, which is environmentally friendly solvents and the product produced in a clean state that does not require additional purification steps [112]. There is also other research that focuses on nanostructured materials. The following below are the findings from preparation using a microwave.

12.5.2.1 Catalyst

There are many different applications of microwave radiation. One of the applications is a technology that assists a process or works independently to synthesize catalysts. The processes are chemical vapor deposition, chemical reduction, and even biomediated processes, among others. In addition to warming up the object, microwave exposure to catalysts causes atoms to migrate, changing the angle at which the catalyst's atoms are viewed, so revealing the active sites. Additionally, it has been noted that after being exposed to microwave radiation, the materials become porous and their specific pore volume or surface area rises. In comparison to non-microwave irradiation catalysts, the gas or liquid flow passing by the catalysts produces more pores, which leads to increased reaction rates with the catalysts. As a result, it is determined that increasing the catalytic characteristics of a catalyst by the use of microwave technology or other technologies supported by microwaves is beneficial [113].

Some parties have already scaled up production using microwave assistance because of its advantages. As an example in the field of microwave-assisted carbohydrate chemistry, glycosylation reactions are summarized in the journal by Richel et al. [114]. The study conducted by Richel et al. [114], developed a batch microwave platform for the glycosylation of decanol with peracetyl derivatives. The method allowed for the synthesis of several hundred grams of glycosides per run, with yields comparable to laboratory-scale experiments.

Another study about glycosylation reactions achieved the scaling-up of microwave-assisted Fischer glycosylations up to the kilogram scale, improving economic efficiency. In batch reactions, various monosaccharides reacted with alcohols in the presence of a catalytic amount of homogeneous acid. Yields were nearly quantitative after a short microwave exposure time.

12.5.2.2 Nanotechnology

By using microwave irradiation, homogeneous heating can be achieved in the reaction mixture. That makes microwave-assisted a reliable method because of its capability of making uniform particle size distribution. Nucleation is a crucial part of the formation of molecular arrangements in NPs. This product characteristic is an advantage to some applications, such as [115]:

1. Fuel cell
   Microwave can be used in metal oxide or metal-doped production. This method can produce a uniform material structure. Shih et al. reported that PtNiCo/rGO nanocomposites produced by this method have high electrochemical performance compared to PtNiCo catalysts. Also, because of the smaller particle size, nanocomposites have a larger electrochemical active surface area. Those nanocomposites can be used for direct methanol fuel cells [116].
2. Batteries technology
   In battery technology, it has been studied that nanocomposite has the potential as an anode to replace Na^+ anode. This nanocomposite has a spherical structure that can give maximum reversible specific capacities and excellent rate performance [115].
3. Solar cell
   Microwave is reported to enhance the efficiency of produced solar cells with low-cost processes. Shome et al. conducted a study about microwave-assisted synthesis of non-fullerene electron acceptors for solution-processed organic solar cells [117]. This study used a cross-coupling method to introduce two molecules: benzo (2, 1, 3) thiadiazole (BT) unit with a fuse aromatic ring. By using a palladium catalyst, this direct arylation yields 80%. Also, impurities can be decreased by using microwave irradiation.

12.5.3 Food Processing Technologies

Microwaves can enhance food processing techniques with high efficiency and minimal changes to food quality attributes. This approach can produce desirable food products and increase commercial adoption of microwave technology. Microwave-assisted extraction techniques provide high extraction efficiency and greener alternatives, while samples pre-treated with microwaves can generate extracts with enhanced nutraceutical compounds. Other microwave-assisted applications, such as tempering, frying, roasting, and freezing, retain nutritional and sensory profiles. However, high assembly costs may be a concern. There is potential for improvement in performance and applications in MW-assisted food processing technologies, particularly in thawing, blanching, and roasting. Most technologies are in experimental phases, and future studies should investigate the feasibility and effects of combining microwave energy with other emerging technologies. Robust theoretical models for process optimization are crucial [118].

12.5.4 Ultrasound-Assisted Preparations

12.5.4.1 Biomedical

Recent advancements in ultrasound transducer technologies have focused on enhancing their performance for various biomedical applications. These include material strategies and device designs based on both piezoelectric and photoacoustic mechanisms. Some of the practical applications of these technologies include ultrasound imaging (obtaining information through echoes is a common practice), ultrasound therapy (tumor ablation using high-intensity ultrasound), particle/cell manipulation, drug delivery, and nerve stimulation. Advancements in ultrasound transducer technologies have also focused on improving stability and functionality. For example, research has been conducted on enhancing the stability of piezoelectric and optoacoustic transducers. These advancements have led to the development of novel transducer designs and materials, such as carbon nanotubes and polymer-based composites [119].

12.5.4.2 Artificial Intelligence (AI)

Recent advancements and ongoing research in ultrasound preparation include the application of artificial intelligence (AI) techniques to improve image quality, standardize medical practice, and reduce training and examination times. AI-based methods have been developed to enhance ultrasound image formation, such as beamforming, super-resolution, and image enhancement, which can provide less blurry and more significant images. These approaches require modifications to hardware elements and overcome traditional reconstruction algorithms [120].

12.6 Identification of Potential Areas for Further Exploration and Improvement

Some potential areas for further exploration and improvement in thermal, microwave, and ultrasound-assisted preparations include the following.

12.6.1 Reaction Mechanisms and Kinetics

Thermal reactions, microwave-assisted preparations, and ultrasound-assisted preparations can all affect reaction mechanisms and kinetics in different ways. Here are some specific functions of each:

1. **Thermal reactions** [22]: Thermal reactions typically involve heating a reaction mixture to a specific temperature to initiate or speed up a reaction. The rate of a thermal reaction is typically proportional to the temperature of the reaction mixture, according to the Arrhenius equation. Thermal reactions can also cause changes in the reaction mechanism, such as the formation of different products or intermediates.
2. **Microwave-assisted preparations** [22]: Microwave irradiation can rapidly heat reaction mixtures, resulting in faster reaction times and higher yields. Microwave heating can also selectively heat certain components of a reaction mixture, such as catalysts or reagents, leading to different reaction mechanisms and products. The unique heating mechanism of microwaves can also lead to specific microwave effects, such as the superheating of solvents or the formation of molecular radiators.
3. **Ultrasound-assisted preparations** [22]: Ultrasound waves can create cavitation bubbles in a reaction mixture, which can cause physical and chemical changes in the mixture. The high energy of the cavitation bubbles can break chemical bonds and initiate or speed up reactions. Ultrasound can also increase the surface area of reactants, leading to faster reaction times and higher yields.

Overall, each of these methods can have specific advantages and disadvantages depending on the reaction being performed. Researchers often choose a specific method based on factors such as reaction time, yield, selectivity, and scalability [22].

12.6.2 Synergistic Effects

Exploring the combined effects of microwave and ultrasound, or thermal and ultrasound treatments on reaction outcomes for enhanced efficiency [121]. NPs are important building blocks for a variety of applications. Research spans a wide range of fields, such as polymer science, pharmaceutical manufacturing, nanotechnology, chemistry, and physics. The dimensions, morphology, and size of NPs have a significant impact on the characteristics of nanostructured materials. Due to their innovative applications, unusual characteristics, and quantum size effects, nanostructured metallic and semiconducting materials have recently attracted a lot of attention [122]. Thus, a range of synthetic processes has been used to create nanostructured materials, such as gas phase techniques, liquid phase techniques (such as the reduction of metal salts), and mixed-phase approaches (such as the production of typical heterogeneous catalysts on oxide substrates). However, in the creation of metallic NPs, the intensity of the reducing agent as well as interactions with stabilizing agents and solvents can impact the size and structure of metal NPs. Due to the significant metal dilution during the ab initio production of NPs, the use of MW irradiation in this context is unaffected by safety concerns. The employment of microwave radiation as the heat source in this case is of special relevance since it may quickly manufacture high-quality NPs [123]. Additionally, ultrasound irradiation has been well-studied in this area for many years [124, 125], and is now regarded as one of the most effective methods for creating nanostructured materials. It was only logical to take into account the interaction between MW and US irradiation, which has, in fact, recently been used to synthesize NPs in a short time and was also able to tune particle properties and size [124, 125].

A typical photocatalyst that is active when exposed to UV light is ZnO. Due to its special optical and elective properties, nanoscaled ZnO is actually of tremendous interest. ZnO microstructures outperform NPs, nanorods, and nanosheets in terms of performance. They must be prepared using a time-consuming process and with the addition of a surfactant or structure-directing reagent. In the absence of surfactants, well-defined flower-like nanostructures can be created thanks to the synergistic impact of MW and US [126]. The recommended method involves sonicating a zinc acetate solution in water with a power of 1000 W for five minutes, followed by 30 minutes of MW heating combined with intermittent US irradiation (one second of sonication and two second of interruption). Compared to ZnO microrods, the flower-like ZnO nanostructures demonstrated better catalytic activity for the degradation of methylene blue.

Goethite and akageneite are ferric oxyhydroxides (– and –FeOOH), which are used in the hydrogen peroxide-catalyzed breakdown of organic molecules under comparatively modest UV irradiation. An article by Z. Xu et al. describes how to make –FeOOH by SMUIing a solution of $FeCl_3 \cdot 6H_2O$ in urea and deionized water. In this procedure, a high-intensity ultrasonic probe (model JY92-2D from Xinzhi Co., China) with a titanium horn of 10 mm in diameter operating in pulsed mode was used [127]. It was submerged directly into the solution at a depth of 1 cm and had a duty cycle of one second. For the purpose of comparison, the akageneite was also created using standard procedures, using just MW and US irradiation, as well as the combination approach. XRD and TEM were used to describe each sample. It was established that the heating method had a significant impact on morphologies. The crystals are largest under conventional circumstances, whereas the smallest are formed during SMUI, as was previously found for the creation of Pb(OH)Br nanowires. While MW irradiation alone caused a higher agglomeration, US alone reduced crystal size and aggregation. The -FeOOH sample showed the highest catalytic activity under visible light irradiation and has potential for usage in practical photo-Fenton-like processes that degrade organic pollutants. It was generated using SMUI (MW 400 W, US 200 W) at 70 °C for three hours as shown in Table 12.11.

Table 12.11 Comparison of conventional heating, MW, US, and SMUI in the synthesis of β-FeOOH.

Preparation technique	Width (nm)	Length (nm)
Convention condition	75	250
MW	25	80
US	50	175
MW-US combines	35	35

12.6.3 Green Chemistry and Sustainability

Thermal reactions, microwave-assisted preparations, and ultrasound-assisted preparations are techniques that can be used in green chemistry and sustainability to extract bioactive compounds from natural sources. In particular, microwave-assisted extraction (MAE) has been evaluated as a potential green technology for the isolation of bioactive compounds from saffron floral by-products [128]. MAE is a technique that combines microwave heating with traditional solid–liquid extraction, and it has several advantages over conventional techniques, such as an increase in extraction kinetics, a shorter extraction time, and a higher efficiency and extraction yield, as well as lower energy consumption and cost [128]. The use of MAE in saffron floral by-products resulted in high-value-added compounds with potential applications in the food, pharmaceutical, and cosmetic industries [128]. The specific functions of these techniques are to increase the efficiency of extraction, reduce the use of toxic solvents, and minimize the environmental impact of the extraction process. By using these techniques, it is possible to obtain bioactive compounds from natural sources in a sustainable and environmentally friendly way.

12.6.4 Scale-Up and Industrial Application

Studying the scalability and practicality of these techniques for industrial applications should consider equipment design and cost-effectiveness [129]. The need for high-quality convenient meals with natural taste and flavor, and without chemical additives or preservatives, has led to an increase in the use of non-thermal processing processes. In comparison to other non-thermal processing techniques, ultrasonic technology has proven to be extremely useful. Ultrasound processing has been deemed effective since it improves food quality significantly, whether used alone or in conjunction with other processing techniques. It is thought that the cavitation phenomenon and enhanced mass transfer have an impact on food processing when ultrasound is used. It is regarded as a new and promising technology that has been successfully used in the food processing sector for a number of operations, including freezing, filtration, drying, separation, emulsion, sterilization, and extraction.

US waves, as previously mentioned, are sound waves with a frequency range of 20 kHz to 10 MHz. Additional subdivisions within this range have been found, and depending on the frequency at which they are produced and the energy level of the acoustic field, these subdivisions may really differ significantly from one another in terms of features. Particularly, US power affects chemical reactivity and can be divided into two subfamilies: (i) high-energy US processes with low frequencies (20–100 kHz), which are used in some food technologies, and intermediate-power US processes with medium frequencies (100 kHz–1 MHz); and (ii) low-energy diagnostic US with high frequencies (5–10 MHz), which is used in physical measurements, primarily for medical and diagnostic purposes.

The low-frequency (30 Hz) emulsion was split into its water and oil components using high-energy US waves, which is a novel particle separation technique that forms the basis of this technology [130]. Before it can be commercialized, this approach must be further refined because high-strength US waves can easily have the opposite effect by causing the production of a more stable emulsion or dispersion. A crucial step in the dairy industry is cleaning the membranes used for protein concentration and separation in order to preserve the permeability and selectivity of the membrane and restore the plant's ability to reduce the risk of bacterial contamination and produce products that are acceptable. US effectively cleans membranes, and its efficiency rises at lower frequencies, as demonstrated in a study by Luján-Facundo et al. [131].

12.6.5 Catalysis and Selectivity

Thermal reactions, microwave, and ultrasound-assisted preparations can all be used to enhance catalysis and selectivity in the valorization of lignocellulosic biomass [132]. Specifically, ultrasound-assisted catalysis has been shown to be a promising method for upgrading biomass into value-added chemical feedstocks. The combination of catalysis with sonication provides new strategies for the valorization of lignocellulosic compounds into value-added chemical feedstocks. Ultrasound-assisted reactions offer opportunities to develop environmentally friendly and cost-effective processes for biomass upgrading. The use of ultrasound in organic synthesis has been shown to accelerate chemical reactions, increase conversion, improve yield, and enhance selectivity in both homogeneous and heterogeneous systems. In heterogeneous systems, the use of ultrasound improves mass transfer from turbulent mixing and acoustic streaming, generates cavitation erosion at liquid–solid interfaces, and is responsible for the deformation of solid surfaces, which can increase the accessibility of the internal surface for the reagents. In the case of gas–liquid–solid systems, sonication increases the interphase surface and favors the removal of outer oxide or other passivating layers from the catalyst surface. Overall, the use of thermal reactions, microwave, and ultrasound-assisted preparations can all enhance catalysis and selectivity in the valorization of lignocellulosic biomass, with ultrasound-assisted catalysis showing particular promise.

12.6.6 In Situ Monitoring and Control

Thermal reactions, microwave-assisted preparations, and ultrasound-assisted preparations can all affect in situ monitoring and control in different ways. Here are some specific functions of each:

- **Thermal reaction**: Thermal reactions involve the application of heat to accelerate chemical reactions. The specific function of thermal reaction in in situ monitoring and control is to provide controlled heating conditions to regulate the reaction rate and optimize the reaction conditions.
- **Microwave-assisted preparations**: Microwave-assisted preparations utilize microwave radiation to heat the reaction mixture. The specific function of microwave-assisted preparations in in situ monitoring and control is to provide

rapid and uniform heating, which can enhance reaction rates, improve product yields, and enable real-time monitoring of reaction progress.
- **Ultrasound-assisted preparations**: Ultrasound-assisted preparations involve the use of ultrasonic waves to enhance chemical reactions. The specific function of ultrasound-assisted preparations in in situ monitoring and control are to provide mechanical agitation and cavitation effects, which can improve mass transfer, increase reaction rates, and enable real-time monitoring of reaction progress.

Overall, these techniques offer unique advantages in terms of reaction control, efficiency, and real-time monitoring, making them valuable tools for in situ monitoring and control of chemical reactions.

12.6.7 Mechanistic Studies

Thermal reactions, microwave, and ultrasound-assisted preparations are used for the mechanistic study of chemical reactions [133]. Here are the specific functions of each method:

- **Thermal reactions**: These reactions use heat to increase the rate of a chemical reaction. The heat provides the activation energy required for the reaction to occur. Thermal reactions can be used to study the kinetics of a reaction, as well as the thermodynamics of the reaction. They are also useful for studying the effect of temperature on a reaction.
- **Microwave-assisted preparations**: These preparations use microwave radiation to heat a reaction mixture. Microwave radiation can penetrate the reaction mixture and heat it uniformly, which can lead to faster and more efficient reactions. Microwave-assisted preparations are useful for studying the effect of microwave radiation on a reaction, as well as for synthesizing compounds in a shorter amount of time.
- **Ultrasound-assisted preparations**: These preparations use ultrasound waves to create high-energy chemistry through the process of acoustic cavitation. During the cavitational collapse, intense heating of the bubbles occurs, creating localized hot spots with temperatures of roughly 5000 °C, pressures of about 500 atm, and lifetimes of a few microseconds. Ultrasound-assisted preparations are useful for studying the effect of acoustic cavitation on a reaction, as well as for creating clean, highly reactive surfaces on metals, and for initiation or enhancement of catalytic reactions.

12.6.8 Temperature and Energy Management

Investigating temperature distribution and energy deposition for efficient and uniform heating of reaction mixtures [134], the optimum drying uses are between 20% and 25% of the total energy used applied in the food sector [135]. When choosing a drying method, energy usage, and product quality are also important considerations [136]. To increase energy efficiency, alternative food drying methods including atmospheric freeze drying (AFD) have been suggested. The low drying rate of the AFD process is the process's limiting phase, hence many ways for accelerating it have been suggested [137]. It is suggested incorporating an absorbent substance into the final product to constantly dry the process air and utilize the heat produced by water adsorption to increase the sublimation rate. However, it may be difficult to separate the desiccant from the dried product, and it is always advisable to make sure the absorbent is compatible with the food product.

Power ultrasound (US), i.e. acoustic waves with frequencies between 20 and 100 kHz and a power of over 1 W/cm, appears to be particularly successful at increasing drying kinetics [138, 139], only slightly affecting product quality [140], mainly because of the moderate thermal effect compared to other techniques, e.g. microwaves, infrared radiation, or superheated steam. Power ultrasonography can considerably quicken the atmospheric freeze-drying process. The purpose of this article is to examine the implications of this technology on the process's overall energy usage and environmental impact. Because of the differences in their internal structures and water contents, the representative products chosen were the apple, the carrot, and the eggplant. To replicate the atmospheric ultrasound-assisted freeze-drying process in silico, a mathematical model of an industrial-scale plant was created. Model parameters were adjusted based on the outcomes of a pilot-scale unit. The life cycle assessment (LCA) method was used to determine how the procedure would affect the environment. According to the findings, applying ultrasound can lower the total energy consumption of the entire process by up to 70%. The LCA research also demonstrated that each impact category saw savings between 58% and 82% depending on the product. The most crucial step has been identified as the moisture removal device (dehumidifier). The internal design of the product has a significant impact on the process's energy requirements and, consequently, on the environmental impact.

Thus, even though the use of ultrasound increased the energy consumption per hour, the total energy consumption of the entire process (kWh) was lower since the total operation time was shorter in the case of ultrasonically aided drying. Specifically, a 70% energy saving was attained, independent of the product processed, in those drying studies with an ultrasonic power of 10.3 kW/m³. The amount of energy used overall, and the drying time, are both decreased when ultrasound is used. The increase in energy use per hour of operation does not, however, make up for the decrease in drying time when the ultrasound power is above a particular threshold.

12.6.9 Materials Processing

Thermal reaction, microwave, and ultrasound-assisted preparations are methods used to process materials. The specific functions of each method are as follows [141]:

- **Thermal reaction**: This method involves the use of heat to initiate chemical reactions in materials. The heat causes the molecules in the material to vibrate, which increases their energy and allows them to react with other molecules. Thermal reactions are commonly used in the production of polymers, ceramics, and metals.

- **Microwave-assisted preparations**: This method uses electromagnetic waves to heat materials. The microwaves cause the molecules in the material to rotate, which generates heat. Microwave heating is often used in the food industry to cook or thaw food quickly, and it is also used in materials processing to dry, cure, or sinter materials.
- **Ultrasound-assisted preparations**: This method uses high-frequency sound waves to extract bioactive compounds from materials. The sound waves cause the material to vibrate, which creates microscopic channels that facilitate solvent diffusion into the material. Ultrasound-assisted extraction is environmentally friendly, time-saving, and operated at a lower temperature, leading to an increase in mass transfer. This method is commonly used in the food industry to extract polyphenols, carotenoids, aromatic compounds, and polysaccharides from plant matrices.

In summary, thermal reaction, microwave, and ultrasound-assisted preparations are methods used to process materials. Thermal reaction uses heat to initiate chemical reactions, microwave uses electromagnetic waves to heat materials, and ultrasound-assisted uses high-frequency sound waves to extract bioactive compounds from materials [141].

12.6.10 Biomedical Applications

Thermal reaction, microwave, and ultrasound-assisted preparations have specific functions in biomedical applications. Here are some of their functions [142]:

- **Thermal reaction**: Thermal reactions can be used to sterilize medical equipment and supplies, such as surgical instruments and bandages [142].
- **Microwave-assisted preparations**: Microwaves can be used to heat and thaw frozen biological samples, such as tissues and cells, for research purposes, new drugs, and drug delivery systems [142].
- **Ultrasound-assisted preparations**: Ultrasound's applications in pharmaceutics encompass drug delivery, skin permeability enhancement, microfeeding, solid-free forming, contaminant destruction, nanoparticle production, and sonocrystallization [142].

Overall, thermal reaction, microwave, and ultrasound-assisted preparations have various functions in biomedical applications, ranging from sterilization to drug delivery and synthesis.

12.7 The Role of Artificial Intelligence and Computational Approaches in Optimizing Preparative Techniques

Artificial intelligence and computational methodologies have a crucial function in optimizing microwave techniques. These methodologies, collectively referred to as soft computing methods, have been widely used in microwave design applications. They provide effective modeling and optimization capabilities, allowing microwave designers to improve the speed and accuracy of their design processes. Various computing methods have seen widespread use in microwave design [143], such as:

1. **Artificial neural networks (ANNs)**: For microwave modeling and prediction tasks.
2. **Genetic algorithms (GAs)**: For optimizing design parameters.
3. **Particle swarm optimization (PSO)**: For optimizing design parameters.
4. **Support vector machines**: For microwave modeling, offering efficient and accurate predictions.
5. Ant colony optimization (ACO).
6. Bacterial foraging optimization (BFO).

AI and computational approaches have revolutionized microwave design by providing efficient modeling and optimization capabilities. These techniques have significantly improved the speed and accuracy of microwave designs, allowing designers to generate more data and obtain the desired accuracy [143]. In a study by Jing Liu et al., flavonoids from Rosa sterilis were extracted using the ultrasonic technique, and the extraction parameters were modeled and fine-tuned through a combination of response surface methodology and artificial intelligence approaches [144]. In the ultrasonic welding field, different AI models have been developed to predict and optimize joint quality such as the conventional ANN model, ANN integrated with FEM, and hybrid ANN/GA [145]. These optimization tools effectively enhanced and maximized the performance of AI models across various engineering applications [143]. Microwave design using evolutionary algorithms [143].

1. GA
2. PSO in microwave design
3. ANN-based microwave modeling
4. Steps in modeling microwave components using SVR

A study by Elsheikh et al., suggested a hybrid AI methodology that combines an ensemble random vector functional link (RVFL) model with a gradient-based optimizer (GBO), to simulate the ultrasonic welding process of a polymeric material blend. The model was trained using experimental data on acrylonitrile butadiene styrene (ABS) and polycarbonate (PC) blends. It demonstrated high accuracy in predicting joint strength and average temperature [145].

AI algorithms have been applied to measurement, quantification, and computer-aided detection in ultrasound imaging. Deep learning-based approaches have replaced traditional feature engineering and have shown competitive solutions in ultrasound imaging. Computer-assisted diagnosis, triage, detection, and quantification have also attracted attention in reducing physicians' workload. AI algorithms have been used to assist in ultrasound scanning by reducing the learning curve for physicians, nurses, and technicians. Computer-assisted tools can help in standard plane recognition and organ identification, extracting clinical planes from 3D ultrasound volumes, and guiding ultrasound acquisitions performed by humans or robots. These AI-assisted techniques aim to improve the accuracy and repeatability of ultrasound outcomes [146].

References

1 Hagenmuller, P. (1972). *Preparative Methods in Solid State Chemistry*. Academic Press.
2 Sabel, W. (1967). *Basic Techniques of Preparative Organic Chemistry*. Pergamon.
3 Nicholls, D. (1974). General preparative methods. In: *Complexes and First-Row Transitions Elements. A Macmillan Chemistry Text*, 128–138. London: Palgrave.
4 Sheldon, R.A., Arends, I., and Hanefed, U. (2007). *Green Chemistry and Catalysis*, 278–285. Germany: Wiley-VCH.
5 EPA Green Chemistry, *Basics of Green Chemistry*, US EPA, 2023. https://www.epa.gov/greenchemistry/basics-green-chemistry (accessed 7 January 2024).
6 Anastas, P.T. and Warner, J.C. (2000). *Green Chemistry: Theory and Practice*. Boston: Oxford University Press.
7 Grewal, A.S., Kumar, K., Redhu, S., and Bhardwaj, S. (2013). Microwave assisted synthesis: a green chemistry approach. *Int. Res. J. Pharm. Appl. Sci.* 5: 278–285.
8 Baig, R.B.N. and Varma, R.S. (2012). Alternative energy input: mechanochemical, microwave and ultrasound-assisted organic synthesis. *Chem. Soc. Rev.* 41 (4): 1559–1584.
9 Chatel, G. and Varma, R.S. (2019). Ultrasound and microwave irradiation: contributions of alternative physicochemical activation methods to green chemistry. *Green Chem.* 21 (22): 6043–6050.
10 Frecentese, F., Sodano, F., Corvino, A. et al. (2023). The application of microwaves, ultrasounds, and their combination in the synthesis of nitrogen-containing bicyclic heterocycles. *Int. J. Mol. Sci.* 24 (13): 1–43.
11 Karkare, Y.Y., Rathod, W.R., Sathe, V.S., and Chavan, A.R. (2023). Combined ultrasound-microwave assisted synthesis of aripiprazole: process optimization using RSM-ANN. *Chem. Eng. Process. Process Intensif.* 183: 109250.
12 Abraham, A., Bano, A., and Raza, S. (2021). Microwave-assisted organic synthesis: a green chemistry strategy. *Indian J. Adv. Chem. Sci.* 4: 288–292.
13 Skoog, D.A., Holler, F.J., and Crouch, S.R. (2017). *Principal of Instrumental Analysis*. New York: Sunder College Publisher.
14 Petrucci, R.H., Harwood, W.S., Herring, F.G., and Madura, J.D. (1972). The effect of temperature on reaction rates. In: *General Chemistry Principles and Modern Application* (ed. R.H. Petrucci, W.S. Harwood, and F.G. Herring), 14.9. New Jersey: Pearson Education International.
15 Chemical Education Division Groups (2004). The activation energy of chemical reactions. In: *General Chemistry Topic Review*, 22. West Lafayette: Purdue University.
16 Lumen Learning Activation energy and temperature dependence. In: *Introductory Chemistry*, 61. UEN Pressbooks.

17 Reeves, R.V., Mukasyam, A.S., and Son, S. (2010). Thermal and impact reaction initiation in Ni/Al heterogeneous reactive systems. *J. Phys. Chem. C* 114 (35): 14772–14780.

18 Khair, M. and van Rantwijk, F. (2010). Green chemistry perspective of hydroxy-methylfurfural microwave assisted organic syntheses. *Jurnal Sainstek* 2: 105–109.

19 Nain, S., Singh, R., and Ravichandran, S. (2019). Importance of microwave heating in organic synthesis. *Adv. J. Chem.* 2: 94–104.

20 Yadav, A.R., Mohite, S.K., and Magdum, C.S. (2020). Comparative study of conventional and microwave assisted synthesis of some organic reactions. *Asian J. Pharm. Res.* 10 (3): 217–220.

21 A. K. Nagariya, A. K. Meena, K. Kiran, A. K. Yadav, U. S. Niranjan, A. K. Pathak, B. Singh and M. M. Rao, "Microwave assisted organic reaction as new tool in organic synthesis," *J. Pharm. Res.*, vol. 3, pp. 575–580, 2010.

22 Kappe, C.O. (2004). Controlled microwave heating in modern organic synthesis. *Angew. Chem. Int. Ed.* 46: 6250–6284.

23 Gawande, M.B., Shelke, S.N., Zboril, R., and Varma, R.S. (2014). Microwave-assisted chemistry: synthetic applications for rapid assembly of nanomaterials and organics. *Acc. Chem. Res.* 4: 1338–1348.

24 Palma, V., Barba, D., Cortese, M. et al. (2020). Microwaves and heterogeneous catalysis: a review on selected catalytic processes. *Catalysts* 10 (2): 1–58.

25 Pawelczyk, A., Sowa-Kasprzak, K., Olender, D., and Zaprutko, L. (2018). Microwave (MW), ultrasound (US) and combined synergic MW-US strategies for rapid functionalization of pharmaceutical use phenols. *Molecules* 23 (9): 2360.

26 De la Hoz, A., Diaz-Ortiz, Á., and Moreno, A. (2005). Microwaves in organic synthesis: thermal and non-thermal microwave effects. *Chem. Soc. Rev.* 34: 164–178.

27 Bogdal, D. (2005). *Microwave Assisted Organic Synthesis*. Elsevier Publications.

28 Collins, M.J. (2010). Future trends in microwave synthesis. *Future Med. Chem.* 2: 151–155.

29 Mason, T.J. and Lorimer, J.P. (2003). *Applied Sonochemistry: The Uses of Power Ultrasound in Chemistry and Processing*. Weinheim: Wiley-VCH.

30 Macías-Benitez, P., Sierra-Padilla, A., Yeste, M.P. et al. (2022). Ultrasound-promoted synthesis of a copper–iron-based catalyst for the microwave-assisted acyloxylation of 1,4-dioxane and cyclohexene. *Org. Biomol. Chem.* 21 (3): 590–599.

31 Sadjadi, S., Sadjadi, S., and Hekmatshoar, R. (2010). Ultrasound-promoted greener synthesis of benzoheterocycle derivatives catalyzed by nanocrystalline copper(II) oxide. *Ultrason. Sonochem.* 17: 764–767.

32 Shakib, P., Dekamin, M.G., Valiey, E. et al. (2023). Ultrasound-promoted preparation and application of novel bifunctional core/shell Fe_3O_4@SiO_2@PTS-APG as a robust catalyst in the expeditious synthesis of Hantzsch esters. *Sci. Rep.* 13: 1–17.

33 Bubalo, M.C., Sabotin, I., Rados, I. et al. (2013). A comparative study of ultrasound-, microwave-, and microreactor-assisted imidazolium-based ionic liquid synthesis. *Green Process. Synth.* 2: 579–590.

34 Gude, V.G. and Martinez-Guerra, E. (2015). Green chemistry of microwave-enhanced biodiesel production. *Biofuels Biorefin.* 3: 224–250.

35 Helmi, M.R., Rahayu, D.U.C., Pratama, A.P. et al. (2023). Comparative study of microwave-assisted versus conventional heated reactions of biomass conversion into levulinic acid over hierarchical Mn_3O_4/ZSM-5 zeolite catalysts. *Carbon Resour. Convers.* 6 (3): 245–252.

36 Chen, Y., Li, G., Yang, F., and Zhang, S.M. (2011). Mn/ZSM-5 participation in the degradation of cellulose under phosphoric acid media. *Polym. Degrad. Stab.* 5: 863–869.

37 Krisnandi, Y.K., Nurani, D.A., Agnes, A. et al. (2019). Hierarchical MnO_x/ZSM-5 as heterogeneous catalysts in conversion of Delignified. *Indones. J. Chem.* 1: 115–123.

38 Pratama, A.P., Rahayu, D.U.C., and Krisnandi, Y.K. (2020). Levulinic acid production from Delignified Rice husk waste over manganese catalysts: heterogeneous versus homogeneous. *Catalyst* 3 (10): 327.

39 Pertiwi, R., Oozeerally, R., Burnett, D.L. et al. (2019). Replacement of chromium by non-toxic metals in Lewis-acid MOFs: assessment of stability as glucose conversion catalysts. *Catalyst* 5 (9): 437.

40 Mukherjee, A. and Dumont, M.J. (2016). Levulinic acid production from starch using microwave and oil bath heating: a kinetic modeling approach. *Ind. Eng. Chem. Res.* 33 (55): 8941–8949.

41 Ren, H., Zhou, Y., and Liu, L. (2013). Selective conversion of cellulose to levulinic acid via microwave-assisted synthesis in ionic liquids. *Bioresour. Technol.* 129: 616–619.

42 Saxena, V.K. (2011). Microwave synthesis: a physical concept. In: *Microwave Heating* (ed. U. Chandra), 3–22. IntechOpen.

43 Chand, P., Chintareddy, V.R., Verkade, J.G., and Grewell, D. (2010). Enhancing biodiesel production from soybean oil using ultrasonics. *Energy Fuels* 24: 2010–2015.

44 Barnard, T., Leadbeater, N., Boucher, M. et al. (2007). Continuous-flow preparation of biodiesel using microwave heating. *Energy Fuels* 21 (3): 1777–1781.

45 Melo-Junior, C.A., Albuquerque, C.E., Fortuny, M. et al. (2009). Use of microwave irradiation in the noncatalytic esterification of C18 fatty acids. *Energy Fuels* 1 (23): 580–585.

46 Martina, K., Tagliapietra, S., Barge, A., and Cravotto, G. (2016). Combined microwaves/ultrasound, a hybrid technology. *Top. Curr. Chem.* 79: 1–27.

47 Peng, Y. and Song, G. (2001). Simultaneous microwave and ultrasound irradiation: a rapid synthesis of hydrazides. *Green Chem.* 3: 302–304.

48 Wang, K., Deng, P., Liu, R. et al. (2022). A novel understanding of the thermal reaction behavior and mechanism of Ni/Al energetic structural materials. *Crystals* 12: 1632.

49 Zhang, Y. et al. (2019). Thermal reaction kinetics of Ni/Al energetic structural materials. *J. Therm. Anal. Calorim.* 135 (2): 1091–1100.

50 Szterner, P., Legendre, B., and Sghaier, M. (2010). Thermodynamic properties of polymorphic forms of theophylline. *J. Therm. Anal. Calorim.* 520: 201.

51 Vyazovkin, S., Burnham, A.K., Criado, J.M. et al. ICTAC kinetics committee recommendations for performing kinetic computations on thermal analysis data. *Thermochim. Acta* 520: 1–19.

52 Wang, Z. (2019). Microwave-assisted synthesis of metal-organic frameworks: a mini-review. *Microporous Mesoporous Mater.* 274: 1–9.

53 Loupy, A. (2002). *Microwaves in Organic Synthesis.* Whiley-VCH.

54 Galwey, A.K. (2020). Thermal reactions involving solids: a personal view of selected features of decompositions, thermal analysis and heterogeneous catalysis. *J. Therm. Anal. Calorim.* 142: 1123–1144.

55 Yan, Q., Qiu, M., Chen, X., and Fan, Y. (2019). Ultrasound assisted synthesis of size-controlled aqueous colloids for the fabrication of nanoporous zirconia membrane. *Front Chem.* 7: 337.

56 Bogdal, D. (2006). *Microwave-Assisted Organic Synthesis: One Hundred Reaction Procedures.* Elsevier.

57 Suslick, K.S. *Kirk-Orthmer Encyclopedia of Chemical Technology: Ultrasound.* Wiley.

58 Temperature Behaviour of Organic Compounds, Phywe.

59 Moldoveanu, S. (2010). *Techniques and Instrumentation in Analytical Chemistry*, vol. 28, 3–6. Elsevier.

60 Ding, J., Xia, Z., and Lu, J. (2012). Esterification and deacidification of a waste cooking oil (TAN 68.81 mg KOH/g) for biodiesel production. *Energies* 5: 2683–2691.

61 Zeki, N.S.A., Al-Hassani, M.H., and Al-Jendeel, H.A. (2010). Kinetic study of esterification reaction. *Al-Khwarizmi Eng. J.* 6 (2): 33–42.

62 Ferdous, K., Uddin, M., Uddin, M.R. et al. (2013). Preparation and optimization of biodiesel production from mixed feedstock oil. *Chem. Eng. Sci.* 1: 62–66.

63 Jeffrey, M. (2020). Dehydration reactions of alcohols. In: *Organic Chemistry* (ed. M. Jeffrey), 14.4. California: LibreTexts Chemistry.

64 Kar, S. and Milstein, D. (2022). Oxidation of organic compounds using water as the oxidant with H2 liberation catalyzed by molecular metal complexes. *Acc. Chem. Res.* 55 (16): 2304–2315.

65 Strauss, C.R. and Trainor, R.W. (1995). Developments in microwave- assisted organic chemistry. *Aust. J. Chem.* 48: 1665–1692.

66 Caddick, S. (1995). Microwave assisted organic reactions. *Terrahedron* 51 (38): 10403–10432.

67 Kappe, C.O. (2004). Synthetic methods controlled microwave heating in modern organic synthesis C. *Angew. Chem. Int. Ed* 43: 6250–6284.

68 Mehta, V.P. and Van der Eycken, E.V. (2011). Microwave-assisted C–C bond forming cross-coupling reactions: an overview. *Chem. Soc. Rev.* 40: 4925–4936.

69 Varnado, C. and Bielawski, C. (2012). Condensation polymers via metal-catalyzed coupling reactions. *Polym. Sci.* 5: 175–194.

70 Salih, K.S. and Baqi, Y. (2019). Microwave-assisted palladium-catalyzed cross-coupling reactions: generation of carbon–carbon bond. *Catalysts* 10 (1): 4.

71 Omae, I. (2007). Three types of reactions with intramolecular five-membered ring compounds in organic synthesis. *J. Organomet. Chem.* 692 (13): 608–2632.

72 Wang, J.X., Yulai Hu, Z.L., Wei, B., and Bai, L. (2000). Microwave-promoted palladium catalysed heck cross coupling reaction in water. *J. Chem. Res.* 484–485.

73 Maleczka, R.E., Lavis, J.M., Clark, D.H., and Gallagher, W.P. (2000). Microwave-assisted one-pot hydrostannylation/stille couplings. *Org. Lett.* 2 (23): 3655–3658.

74 Sharma, A.K., Gowdahalli, K., Krzeminski, J., and Amin, S. (2007). Microwave-assisted Suzuki cross-coupling reaction, a key step in the synthesis of polycyclic aromatic hydrocarbons and their metabolites. *J. Organomet. Chem.* 72 (23): 8987–8989.

75 Tian, Y., Wang, J., Cheng, X. et al. (2020). Microwave-assisted unprotected Sonogashira reaction in water for the synthesis of polysubstituted aromatic acetylene compounds. *Green Chem.* 22 (4): 1338–1344.

76 Bougrin, K., Loupy, A., and Soufiaoui, M. (2005). Microwave-assisted solvent-free heterocyclic synthesis. *J. Photochem. Photobiol., C* 6 (2–3): 139–167.

77 Ochoa-Terán, A., Guerrero, L., and Rivero, I.A. (2014). A novel one-pot and one-step microwave-assisted cyclization-methylation reaction of amino alcohols and acetylated derivatives with dimethyl carbonate and TBAC. *Sci. World J.* 2014.

78 Caliman, C.C., Mesquita, A.F., Cipriano, D.F. et al. (2018). One-pot synthesis of amine-functionalized graphene oxide by microwave-assisted reactions: an outstanding alternative for supporting material in supercapacitors. *RSC Adv.* 8: 6136.

79 Araujo, D.R., Lima, Y.R., Barcellos, A.M. et al. (2019). Ultrasound-promoted radical synthesis of 5-methylselanyl-4,5-dihydroisoxazoles. *Eur. J. Org. Chem.* 2020 (5): 586–592.

80 Bremner, D.H. (1994). Recent advances in organic synthesis utilizing ultrasound. *Ultrason. Sonochem.* 1 (2): S119–S124.

81 Davidson, R.S., Safdar, A., Spencer, J.D., and Robinson, B. (1987). Applications of ultrasound to organic chemistry. *Ultrasonics* 25 (1): 35–39.

82 Bonrath, W. (2004). Chemical reactions under "non-classical conditions", microwaves and ultrasound in the synthesis of vitamins. *Ultrason. Sonochem.* 11: 1–4.

83 Nandurkar, N.S., Patil, D.S., and Bhanage, B.M. (2008). Ultrasound assisted synthesis of metal-1,3-diketonates. *Inorg. Chem. Commun.* 11: 733–736.

84 Khatri, P.K., Jain, S.L., and Sain, B. (2011). Ultrasound-promoted oxidation of sulfides with hydrogen peroxide under catalyst-free conditions. *Ind. Eng. Chem. Res.* 50: 701–704.

85 Jaita, S., Phakhodee, W., and Pattarawarapan, M. (2015). Ultrasound-assisted methyl esterification of carboxylic acids catalyzed by polymer-supported triphenylphosphine. *Synlett* 26 (14): 2006–2008.

86 Leonelli, C. and Mason, T.J. (2010). Microwave and ultrasonic processing: now a realistic option for industry. *Chem. Eng. Process.* 49: 885–900.

87 Strekalova, A.A., Shesterkina, A.A., Kustov, A.L., and Kustov, L.M. (2023). Recent studies on the application of microwave-assisted method for the preparation of heterogeneous catalysts and catalytic hydrogenation processes. *Mol. Sci.* 24: 8272.

88 Dabrowska, S., Chudoba, T., Wojnarowicz, J., and Lojkowski, W. (2018). Current trends in the development of microwave reactors for the synthesis of nanomaterials in laboratories and industries: a review. *Crystals* 141: 02–507.

89 Punathil, L. and Basak, T. (2016). *Microwave Processing of Frozen and Packaged Food Materials: Experimental*. Chennai: Department of Chemical Engineering, Indian Institute of Technology Madras.

90 Calinescu, I., Vinatoru, M., Ghimpeteanu, D. et al. (2021). A new reactor for process intensification involving the simultaneous application of adjustable ultrasound and microwave radiation. *Ultrason. Sonochem.* 77: 105701.

91 Stankiewicz, A.I. and Moulijn, J.A. (2000). Process intensification: transforming chemical engineering. *Process Des. Trends* 96 (1): 22–34.

92 Sancheti, S.V. and Gogate, P.R. (2016). A review of engineering aspects of intensification of chemical synthesis using ultrasound. *Ultrason. Sonochem.* 36: 527–543.

93 Mohamad Aziz, N.A., Yunus, R., Kania, D., and Abd Hamid, H. (2021). Prospects and challenges of microwave-combined technology for biodiesel and biolubricant production through a transesterification: a review. *Molecules* 26: 788.

94 Cravoto, G. and Cintas, P. (2007). The combined use of microwaves and ultrasound: improved tools in process chemistry and organic synthesis. *Chemistry* 13: 1902–1909.

95 Vinatoru, M. and Calinescu, I. (2019). Microwave and ultrasounds together – a challenge. In: *17th International Conference on Microwave and High Frequency Heating*. Bucharest, Romania.

96 Safieddin Ardebili, S.M., Hashjin, T.T., Ghobadian, B. et al. (2015). Optimization of biodiesel synthesis under simultaneous ultrasound-microwave irradiation using response surface methodology (RSM). *Green Porcess. Synth.* 4: 259–267.

97 Malani, R.S., Moholkar, V.S., Elbashir, N.O., and Choudhury, H.A. (2021). Advancements of cavitation technology in biodiesel production – from fundamental concept to commercial scale up. In: *Liquid Biofuels: Fundamentals, Characterization, and Applications*, 39–76. Wiley.

98 Shen, X.F. (2009). Combining microwave and ultrasound irradiation for rapid synthesis of nanowires: a case study on Pb(OH)Br. *J. Chem. Technol. Biotechnol.* 84: 1811–1817.

99 van Boekel, M., Fogliano, V., Pellegrini, N. et al. (2010). A review on the beneficial aspects of food processing. *Mol. Nutr. Food Res.* 54: 1215–1247.

100 Zhang, Q., Liu, R., Geirsdóttir, M. et al. (2022). Thermal-induced autolysis enzymes inactivation, protein degradation and physical properties of sea cucumber *Cucumaria Frondosa*. *Processes* 10: 847.

101 Demirok, N.T. and Yikmis, S. (2022). Combined effect of ultrasound and microwave power in tangerine juice processing: bioactive compounds, amino acids, minerals and pathogens. *Processes* 10: 2100.

102 Tsai, C.F. and Jioe, I.P.J. (2021). The analysis of Chlorogenic acid and caffeine content and its correlation with coffee bean color under different roasting degree and sources of coffee (*Coffea Arabica* Typica). *Processes* 9: 2040.

103 Hii, C.L., Tan, C.H., and Woo, M.W. (2023). Special issue "recent advances in thermal food". *Processes* 11: 288.

104 Nair, S.S., Saha, T., Dey, P., and Bhadra, S. (2021). Thermal oxidation of graphite as the first step for graphene preparation: effect of heating temperature and time. *J. Mater. Sci.* 56: 1–17.

105 Novoselov, K.S., Geim, A.K., Morozov, S.V. et al. (2004). Electric field effect in atomically thin carbon films. *Science* 5696: 666–669.

106 Aristov, V.Y., Urbanik, G., Kummer, K. et al. (2010). Graphene synthesis on cubic SiC/Si wafers: perspectives for mass production of graphene-based electronic devices. *Nano Lett.* 3: 992–995.

107 Michon, A., Vezian, S., Ouerghi, A. et al. (2010). Direct growth of few-layer graphene on 6H-SiC and 3C-SiC/Si via propane chemical vapor deposition. *Appl. Phys. Lett.* 17 (97): 171909.

108 Wienchol, P., Szlęk, A., and Ditaranto, M. (2020). Waste-to-energy technology integrated with carbon capture-challenges and opportunities. *Energy* 198: 117352.

109 dos Santos, R.E., dos Santos, I.F.S., Barros, R.M. et al. (2019). Generating electrical energy through urban solid waste in Brazil: an economic and energy comparative analysis. *J. Environ. Manage.* 231: 198–206.

110 Gabbar, A. and Aboughaly, M. (2021). Conceptual process design, energy and economic analysis of solid waste to hydrocarbon fuels via thermochemical processes. *Processes* 12 (9): 2149.

111 Lisbona, P., Pascual, S., and Pérez, V. (2023). Waste to energy: trends and perspectives. *Chem. Eng. J. Adv.* 14: 100494.

112 Strekalova, A.A., Shesterkina, A.A., Kustov, A.L., and Kustov, L.M. (2023). Recent studies on the application of microwave-assisted method for the preparation of heterogeneous catalysts and catalytic hydrogenation processes. *Int. J. Mol. Sci.* 9: 24.

113 Chia, S.R., Nomanbhay, S., Milano, J. et al. (2022). Microwave-absorbing catalysts in catalytic reactions of biofuel production. *Energies* 21: 15.

114 Richel, A., Laurent, P., Wathelet, B. et al. (2011). Microwave-assisted conversion of carbohydrates. State of the art and outlook. *C.R. Chim.* 14: 224–234.

115 Kumar, A., Kuang, Y., Liang, Z., and Sun, X. (2020). Microwave chemistry, recent advancements, and eco-friendly microwave-assisted synthesis of nanoarchitectures and their applications: a review. *Mater. Today Nano* 11: 100076.

116 Shih, K.Y., Wei, J.J., and Tsai, M.C. (2021). One-step microwave-assisted synthesis of PtNiCo/rGO Electrocatalysts with high electrochemical performance for direct methanol fuel cells. *Nanomaterials (Basel)* 9 (2206): 11.

117 Shome, S., Shin, H.J., Yang, J. et al. (2021). Microwave-assisted synthesis of non-fullerene acceptors and their photovoltaic studies for high-performance organic solar cells. *ACS Appl. Energy Mater.* 9 (4): 9816–9826.

118 Ekezie, F.G.C., Sun, D.W., Han, Z., and Cheng, J.H. (2017). Microwave-assisted food processing technologies for enhancing product quality and process efficiency: a review of recent developments. *Trends Food Sci. Technol.* 67: 58–69.

119 Li, J., Ma, Y., Zhang, T. et al. (2022). Recent advancements in ultrasound transducer: from material strategies to biomedical applications. *BME Front.* 2022: 9764501.

120 Tenajas, R., Miraut, D., Illana, C.I. et al. (2023). Recent advances in artificial intelligence-assisted ultrasound scanning. *Appl. Sci.* 13: 3693.

121 Martina, K., Tagliapietra, S., Barge, A., and Cravotto, G. (2016). Combined microwaves/ultrasound, a hybrid technology. *Top. Curr. Chem. (Cham)* 374: 175–201.

122 Cui, Y. and Lieber, C.M. (2001). Functional nanoscale electronic devices assembled using silicon nanowire building blocks. *Science* 291 (5505): 851–853.

123 Boffa, L., Tagliapietra, S., and Cravotto, G. (2013). Combined energy sources in the synthesis of nanomaterials in microwaves. In: *Nanoparticle Synthesis – Fundamentals and Applications*, 55–74. CRC Press.

124 Bang, J.H. and Suslick, K.S. (2010). Application of ultrasound to the nanostructured materials. *Adv. Mater.* 22: 1039–1059.

125 Colmenares, J.C. (2014). Sonication-induced pathways in the synthesis of light-active catalysts for photocatalytic oxidation of organic contaminants. *ChemSusChem* 7: 1512–1527.

126 Li, H., Liu, E., Chan, F.Y.F. et al. (2011). Fabrication of ordered flower-like ZnO nanostructures by a microwave and ultrasonic combined technique and their enhanced photocatalytic activity. *Mater. Lett.* 65: 3440–3443.

127 Xu, Z., Yu, Y., Fang, D. et al. (2015). Microwave–ultrasound assisted synthesis of β-FeOOH and its catalytic property in a photo-Fenton-like process. *Ultrason. Sonochem.* 27: 287–295.

128 Cerdá-Bernad, D., Baixinho, J.P., Fernández, N., and Frutos, M.J. (2022). Evaluation of microwave-assisted extraction as a potential green technology for the Isolation of bioactive compounds from saffron (*Crocus sativus* L.) floral by-product. *Foods* 11: 2335.

129 Gallo, M., Ferrara, L., and Naviglio, D. (2018). Application of ultrasound in food science and technology: a perspective. *Foods* 10: 164.

130 Pangu, G.D. and Feke, D.L. (2004). Acoustically aided separation of oil droplets from aqueous emulsions. *Chem. Eng. Sci.* 59: 3183–3193.

131 Luján-Facundo, M.J., Mendoza-Roca, J.A., Cuartas-Uribe, B., and Álvarez-Blanco, S. (2016). Cleaning efficiency enhancement by ultrasounds for membranes used in dairy industries. *Ultrason. Sonochem.* 33: 18–25.

132 Kuna, E., Behling, R., Valange, S. et al. (2017). Sonocatalysis: a potential sustainable pathway for the valorization of lignocellulosic biomass and derivatives. *Top. Curr. Chem.* 41.

133 Suslick, K.S. (1998). Sonochemistry. *Science* 4949 (247): 1439–1445.

134 Merone, D., Colucci, D., Fissore, D. et al. (2020). Energy and environmental analysis of ultrasound-assisted atmospheric freeze-drying of food. *J. Food Eng.* 283: 110031.

135 Kumar, C., Karim, M.A., and Joardder (Omar), M.U. (2013). Intermittent drying of food products: a critical review. *J. Food Eng.* 121: 48–57.

136 Sagar, V.R. and Suresh Kumar, P. (2010). Recent advances in drying and dehydration of fruits and vegetables: a review. *J. Food Sci. Technol.* 47: 15–26.

137 Rahman, S. and Mujumdar, A. (2008). A novel atmospheric freeze-drying system using a vibro-fluidized bed with adsorbent. *Drying Technol.* 26: 393–403.

138 Garcia-Perez, J.V., Carcel, J.A., Rossello, C. et al. (2012). Intensification of low-temperature drying by using ultrasound. *Drying Technol.* 30: 1199–1208.

139 Colucci, D., Fissore, D., Mulet, A., and Carcel, J.A. (2017). On the investigation into the kinetics of the ultrasound assisted atmospheric freeze drying of eggplant. *Drying Technol.* 35: 1818–1831.

140 Colucci, D., Fissore, D., Rossello, C., and Carcel, J.A. (2018). On the effect of ultrasound assisted atmospheric freeze-drying on the antioxidant properties of eggplant. *Food Res. Int.* 106: 580–588.

141 Sanjaya, Y.A., Tola, P.S., and Rahmawati, R. (2022). Ultrasound-assisted extraction as a potential method to enhanced extraction of bioactive compound. *3rd International Conference Eco-Innovation in Science, Engineering, and Technology, Surabaya, Indonesia.* pp. 191–198.

142 Ishtiaq, F., Farooq, R., Farooq, U. et al. (2009). Application of ultrasound in pharmaceutics. *World Appl. Sci. J.* 7 (6): 886–893.

143 Chauhan, N., Karikeyan, M.V., and Mittal, A. (2009). A review on the use of soft computing methods for microwave design applications. *Frequenz* 63: 24–31.

144 Liu, J., Li, C., Ding, G., and Quan, W. (2021). Artificial intelligence assisted ultrasonic extraction of Total flavonoids from Rosa sterilis. *Molecules* 13: 26.

145 Elsheikh, A.H., Abd Elaziz, M., and Vendan, A. (2022). Modeling ultrasonic welding of polymers using an optimized artificial intelligence model using a gradient-based optimizer. *Weld. World* 66: 27–44.

146 Tenajas, R., Miraut, D., Illana, C.I. et al. (2023). Recent advances in artificial intelligence-assisted ultrasound scanning. *Appl. Sci.* 6: 13.

Index

a

acoustic cavitation   1, 3–7, 142, 150, 190, 191, 220, 262, 347, 368
acridine derivatives, sonicated synthesis of   92
activated carbon   249, 287, 307
Ag-NPs, green chemistry-inspired synthesis of   33
Ag$_2$S nanoparticles   285
akageneite   365
alcoholic fuel cells   306–313
2-aminopyrimidine synthesis   139
anthocyanin extraction   193
Arrhenius law   97
artificial intelligence (AI)   47, 363
  algorithms   49
  and computational methodologies   370
2-aryl benzimidazoles   148
azabicyclo-[2.2.2]octan-5-ones   125
azide-alkyne Huisgen (3+2) cycloaddition   136

b

Baeyer–Villiger oxidations   112
BaTiO$_3$ nanoparticles   226
batteries technology   362
benzimidazoloquinazolines, sonochemical preparation of   87–88
1,5-benzodiazepines   101
benzoimidazo-pyrimidine derivatives   87
benzylamine coupling reaction   146

Biginelli reaction   104, 122
biodegradable, edible polymer   220
biodiesel production   199, 289
  enzyme-catalysed ultrasound-assisted transesterification   207
  heterogenous acid-catalysed ultrasound-assisted transesterification   201–205
  heterogenous base-catalysed ultrasound-assisted transesterification   205–207
  homogenous acid-catalysed ultrasound-assisted transesterification   199
  homogenous base-catalysed ultrasound-assisted transesterification   199–201
bioremediation   186
Bischler–Möhlau indole synthesis   162
Boger pyridine synthesis   174
Boger reaction   174
Bohlmann–Rahtz pyridine synthesis   173–174
3-Br-indazoles   147, 148
Buchwald–Hartwig coupling   120
Buchwald–Hartwig reactions   142

c

Ca-based catalyst assisted transformation reaction   108
Ca-Benzenetricarboxylic acid (Ca-BTC)   293, 294

Green Chemical Synthesis with Microwaves and Ultrasound, First Edition.
Edited by Dakeshwar Kumar Verma, Chandrabhan Verma, and Paz Otero Fuertes.
© 2024 WILEY-VCH GmbH. Published 2024 by WILEY-VCH GmbH.

Cadogan–Sundberg indole synthesis 163–164
carbon-based nanofillers 39
carbon dots, synthesis of 239
carbon fiber-reinforced bioresin (CFRB) 34
carbonylation/carboxylation 121
cardiology 7
catalysis and selectivity 367
cavitation 2, 3
   growth and collapse mechanism 4–5
   history 3–7
   types of 5–7
cavity collapse 3
cellulose-supported copper nanoparticles 111
chemical reactions, energy sources 42
chitosan-based hydrogels 29, 30
chitosan-based nanoparticles 29
chitosan beads assisted condensation reaction 99
chitosan/cellulose-Pd(II) catalyzed Suzuki–Miyaura coupling 116
chitosan/Pd(II) catalyst assisted coupling reaction 114
chitosan-SO_3H assisted MCR reaction 105
chloroauric acid, chemical and ultrasonic reduction of 236–237
ciprofloxacin (CIP) 10
circular economy 22
Clauson–Kaas pyrrole synthesis 158–159
Click Chemistry 29, 30, 125–127, 169
$CoFe_2O_4$/GO-SO_3H assisted MCR reaction 105
Co/SBA-15 catalyst assisted multi-component reaction 124
comparative studies in enhancing synthetic methodologies 341–345
condensation reactions 98–101
conductive polymer composites (CPCs) 39, 40
conventional heating vs. microwave heating 340

copper based catalysts 89
   dihydropyrimidinones by Cu based catalysts 91–92
   dihydroquinazolinones by Cu based catalysts 92–94
copper oxide nanoparticles 85, 240
coumarins-containing pyrrole, one pot synthesis of 135
coumarin-3-yl-1,2,4-triazolin-3-ones 141
coupling reactions 113
   Buchwald–Hartwig coupling 120
   carbonylation/carboxylation 121
   click chemistry 125–127
   Heck coupling 118–119
   micelliances reactions 121–125
   Sonogashira coupling 113–114
   Suzuki–Miyaura-coupling 114–118
   Ullmann coupling 120
covalent organic frameworks (COF)
   applications 267–268
   factors affecting 266–267
   microwave-assisted synthesis 262–263
   structure of 263–266
   ultrasound-assisted synthesis 262
cradle-to-cradle 29
cresol reaction 106
C–S cross-coupling reaction 142
Cu based catalysts
   dihydropyrimidinones by 91–92
   dihydroquinazolinones by 92–94
Cu/cellulose (20%) catalyst assisted oxidation reaction 111
Cu-HAP catalyst assisted Click reaction 126
CuI@amine-functionalized halloysite 89
CuLDH-3 catalyst assisted Ullmann coupling 120
CuO/Al_2O_3 catalyst 113
$Cu(OCH_3)_2$/porous glass catalyzed Click reaction 127
Cu-Pd catalyzed Suzuki–Miyaura coupling 117
Cu/TiO_2 catalyst assisted reduction reaction 110
cyclization reactions 100–104, 354

d

Debus–Radziszewski reaction   164–166
dehydration reaction of alcohols   349
1,2-diaryl azaindoles   148
3,5-dibenzyl-4-amino1,2,4-triazole
   derivative design pathway   137
dichloromethane (DCM)   90, 134
dielectric energy transfer   157
dihydropyridine derivatives synthesis
   138
1,4-dihydropyridine (1,4-DHP)
   analogues, synthesis of   87
   derivatives   86
dihydropyrimidinones
   Cu based catalysts   91–92
   derivatives, sonicated synthesis of   91
2,3-dihydroquinazolin-4(1H)-one
   synthesis   93
dihydroquinazolinones, Cu based
   catalysts   92–94
dimethylformamide (DMF)   134, 138,
   163, 166, 258
dimethylsulfoxide (DMSO)   134, 146, 258
1,8-dioxodecahydroacridine (DODHAs)
   derivatives   149, 150
direct alcohol fuel cells   284
Doppler effect   2
DPPH free radical scavenging activity
   194

e

Einhorn–Brunner reaction   168–169
electromagnetic interference (EMI)   39,
   40
electromagnetic radiation   21, 35–37,
   43–44, 134, 185, 220, 250, 252, 346,
   351
   based devices   43
electromagnetic spectrum theory
   fundamental   35–38
   microwave heating and equipment
      46–53
   microwave-promoted synthesis   56
electro-stimulated drug release devices
   (EDRDs)   33

enhanced mass transfer   10, 348, 366
enzymes   185
   catalysed ultrasound-assisted
      transesterification   207
   stability   186–190
   stability and activity, microwave
      radiation effect on   186–190
esterification reaction under thermal
   conditions   349

f

Fe nanopowders   288
Fe_3O_4@CNF@Cu catalyzed Ullmann
   reaction   144
Fe_3O_4@CNF@Cu nanocomposites   144
Fe_3O_4@CS-Starch/Pd nanocomposite
   146
Fe_3O_4/Fe/C nanocatalyst   304
Fe_2O_3 nanoparticles   240, 285
$FeWO_4$/Fe_3O_4 nanoparticles   305
Finnegan tetrazole synthesis   164,
   171–172
Fischer indole synthesis   162
fluorescent carbon nanoparticles (FCNS)
   226
food processing technologies   360,
   362–363
formamidine-based ligand synthesis   141
four-component Ugi-azide reaction   172
Fourier-transform infrared spectroscopy
   (FTIR)   88, 93
free radical polymerization   222
frequency, ultrasound waves   2
Friedel–Crafts reactions   106–107
Friedländer quinoline synthesis   176
fuel cell, microwave   362

g

gellan gum (GG)   33
Gewald reaction   161
glass/polystyrene-supported
   Pd(II)-catalyst   119
   assisted heck coupling reaction   119
global waste generation   361
goethite   365

gold nanoparticles 236
Gould–Jacobs reaction 175–176
green chemicals and solvents 77–78
green chemistry (GrC) 69
   Ag-NPs 33
   definition 69–70
   green chemicals and solvents 77–78
   microwave induced synthesis 74–75
   multicomponent reactions 71–73
   12 principles of 22–23, 337, 343
   significance 70–71
   solid phase reactions 73
   sustainable system 28–35, 366
   through microwave heating 53–57
   ultrasound induced synthesis 75–76
green solvents 69–78, 90
Gunn effect 48

h

HA-based hydrogels 30
Hantzsch and Biginelli reactions 104
Hantzsch thiazole synthesis 167–168
H-beta-zeolites assited Friedel–Crafts reaction 107
Heck coupling 118–119, 351
Heck reactions 142, 145
Hemetsberger–Knittel indole synthesis 163
3-heteroarylmethylene substituted isoindolin-1-one derivatives 146
heterocyclic compound synthesis
   Bischler–Möhlau indole synthesis 162
   Boger pyridine synthesis 174
   Boger reaction 174
   Bohlmann–Rahtz pyridine synthesis 173–174
   Cadogan–Sundberg indole synthesis 163–164
   classical Piloty–Robinson pyrrole synthesis 158
   Clauson–Kaas pyrrole synthesis 158–159
   Debus–Radziszewski reaction 164–166
   Einhorn–Brunner reaction 168–169
   Finnegan tetrazole synthesis 171–172
   Fischer indole synthesis 162
   four-component Ugi-azide reaction 172
   Friedländer quinoline synthesis 176
   Gewald reaction 161
   Gould–Jacobs reaction 175–176
   Hantzsch thiazole synthesis 167–168
   Hemetsberger–Knittel indole synthesis 163
   Huisgen reaction 169–171
   Kröhnke pyridine synthesis 172–173
   Leimgruber–Batcho indole synthesis 163
   Paal–Knorr furan synthesis 160
   Paal–Knorr pyrrole synthesis 159–160
   Paal–Knorr thiophene synthesis 160–161
   Pechmann pyrazole synthesis 164
   Pellizzari reaction 169
   Povarov reaction 176–177
   Robinson–Gabriel reaction 167
   Skraup reaction 174–175
   van Leusen imidazole synthesis 166
   van Leusen oxazole synthesis 166–167
heterogeneous catalysis 84, 93, 99, 122, 124, 127, 128, 201, 260, 339, 359
heterogeneous sonochemistry 83
heterogenous acid-catalysed ultrasound-assisted transesterification 201–205
heterogenous base-catalysed ultrasound-assisted transesterification 205–207
hierarchical Mn_3O_4/ZSM-5 catalyst 342
high-intensity focused ultrasound (HIFU) 8
homogeneous sonochemistry 83
homogenous acid-catalysed ultrasound-assisted transesterification 199
homogenous base-catalysed ultrasound-assisted transesterification 199–201

$H_4[SiW_{12}O_{40}]$ catalyst assisted
  multi-component reaction   125
Huisgen reaction   169–171
  of benzyl azide and vinyl acetate   170
hyaluronic acid (HA)
  based hydrogels   30
  hydrogels   30
hydrodynamic cavitation   3, 6, 7
4-hydroxy-3[aryl(piperidin-1-yl/
    morpholino/pyrrolidin-1-l)
    methyl]-2H-2-chromenones   141
hydroxylated β-carboline derivatives   140
5-hydroxymethylfurfural (HMF)   107, 342
hydroxy methyl xanthine derivatives   140
HY-zeolite assisted cyclisation reaction   101

i

IEDDA click reaction-based hyaluronic
    acid (HA) hydrogels   30, 31
Indeno[2′,1′:5,6]pyrido[2,3-d]pyrimidines   150
inertial cavitation   5, 6
in situ monitoring and control   367–368
in-situ ultrasound-assisted emulsion
    polymerization process   226
in-situ ultrasound-assisted hydrolysis
    process   226
intensity   1, 2, 4–8, 12, 51, 92, 194, 198, 207, 209, 348, 363–365
Internet of Underwater Things (IoUT)   12
intramolecular Huisgen reaction   170
ionic liquids (ILs)   28, 29, 70, 76–78, 97, 148
ionizing radiation   2, 219
iron based catalysts   86–89
iron oxide nanoparticles   240

k

Klystrons   48
Kröhnke pyridine ring annelation   173
Kröhnke pyridine synthesis   172–173

l

large-scale sonochemical reactors   4
Leimgruber–Batcho indole synthesis   163
Lepidiline B   165
  microwave-assisted synthesis of   165
liquid medium   1–3, 5–7, 9, 133, 347
low-frequency (30 Hz) emulsion   366

m

macromolecular biological catalysts   185
magnetic nanoparticles (MNPs)   85, 87, 88, 96
magnetic permeability   39, 40
Maillard reaction   193, 194, 360
mechanistic study of chemical reactions   368
mesoporous bi-functionalized
    organosilica (MPBOS) catalyst   107
metal-catalyzed reactions
  microwave-assisted reactions   97–127
  ultrasonic irradiation   83–96
metal-1,3-diketonates   354
metal nanoparticles, synthesis of   236–239
metal-organic frameworks (MOF)
  application of   260–61
  factors that affect
    solvent system   257–258
    temperature and pH   258–260
  microwave-assisted synthesis   253–255
  ultrasound-assisted synthesis   256–257
metal oxides, synthesis of   240–243
4-methyl-oxazole-5-carbonitrile   354
5-methylselanyl-4,5-dihydroisoxazoles   354
$MgAl_2O_4$ nano-crystals   105
$MgAl_2O_4$/NPs nanocomposite catalyst   96
micelliances reactions   121–125
microwave (MW)   134
  assisted coupling reactions   135–142
  assisted reactions   142, 143

microwave (MW) (*contd.*)
  conventional *vs.* microwave heating 44–46
  reactor 74
  and ultrasonication methods 25
  ultrasound assisted coupling reactions 142–150
microwave-assisted C–C bond formation 351–352
microwave-assisted continuous-flow of organic synthesis (MACOS) 118
microwave-assisted coupling reactions 135–142
microwave-assisted cyclizations 352–353
microwave-assisted dehydrogenation reactions 353
microwave-assisted drying (MAD) 50
microwave-assisted heterogeneous catalysis 122
microwave-assisted organic synthesis 134, 343, 353
microwave-assisted preparation
  catalysts 361–362
  food processing techniques 362–363
  nanotechnology 362
microwave-assisted reactions 97, 142, 346
  advantages 53
  solid acid and base catalysts 98–127
microwave-assisted synthesis of polymers
  polymer modification 223
  radical polymerization 222
  ring-opening polymerization 222–223
  step-growth polymerizations 222
microwave-assisted synthesis, advantages and disadvantages of 252
microwave-assisted techniques, limitation of 219
microwave-based energy harvesting devices 43
microwave heating 220
  and equipment 46–53
  principle of 250
microwave induced synthesis 74–75

microwave irradiated MWCNTs-nickel foam nanocomposite 287
microwave irradiation 39, 133, 157
  conventional *vs.* microwave heating 44–46
  electrical, dielectric, and magnetic properties 38–40
  electromagnetic radiations and microwave 43–44
  electromagnetic theory 42–46
  method 283
  molecular rotation 41–42
  thiazolidin-4-one 54–55
  technology 26
microwave radiation (MW) 33, 54, 56, 186–190, 252, 262, 283, 287, 305, 339, 343, 357, 358, 361, 364, 367, 368
  facilitated synthesis 74
microwave reaction, continuous flow of 135
microwave *vs.* ultrasound processing 355
minimally invasive surgeries 8
misalliances metal based catalysts 94–96
Montmorillonite K-10 assisted condensation reaction 98–100
Montmorillonite K-10 assisted cyclisation reaction 102, 103
Montmorillonite K-10 assisted MCR reaction 104
Montmorillonite KSF catalysed cross-aldol condensation reaction 99
MPBOS assisted transformation reaction 107
multicomponent assembly processes (MCAPs) 71
multi component reactions (MCR) 69–72, 78, 87, 93, 94, 104–106, 121, 122, 124–126, 150
  synthesis, of tetrahydrobenzo-[b]-pyran derivatives 87

multi-walled carbon nanotubes (MWCNTs)   39, 93, 124, 287, 298, 306, 307

n

nanofibrillar cellulose/polystyrene (NFC/PS) composites   227
nano MgAl$_2$O$_4$ assisted MCR reaction   106
nano-NiO catalyst assisted reduction reaction   110
nanotechnology   34, 48, 51, 339, 359, 362, 364
N-arylisatins   101
natural fiber-reinforced polymer (NFRP)   34
Na-zeolite assited cyclisation reaction   101
N-based heterocycles   56
NHC-Pd catalyst assisted Suzuki–Miyaura coupling   115
n-heterocycles   56
N-heterocyclic carbene   114
Ni/Cg catalyst assisted reduction reaction   110
NiCl$_2$.6H$_2$O catalysed Buchwald–Hartwig coupling   120
NiFe$_2$O$_4$/FMNPs geopolymer   96
NiO catalyst assisted multicomponent reaction   124
non-destructive testing (NDT)   13
non-ionizing electromagnetic radiation   134
non-polar solvents   97, 134, 135, 190
Novozym 435   186, 190, 207

o

optic cavitation   7
organic synthesis, purification procedures   70
oxidation of aldehydes to carboxylic acids in water   350
oxygen reduction reaction (ORR)   297–306

p

Paal–Knorr furan synthesis   160
Paal–Knorr pyrrole synthesis   159–160
Paal–Knorr thiophene synthesis   160–161
palladium nanoparticles   113, 114, 121, 308
Palm kernel oil (PKO)   295
PANI/ZnMoO$_4$ nanocomposite   223
particle cavitation   6, 7
Pd catalyszed Suzuki–Miyaura coupling reaction   115
Pd/C catalyst assisted oxidation reaction   112
Pd@PS catalyst assisted carbonylation/carboxylation reaction   121
Pd/SiO$_2$ catalyst Heck coupling   118
Pechmann pyrazole synthesis   164
PECVD   52
Pellizzari reaction   169
peroxymonosulfate (PMS)   112
1-phenylpyrrole, Clauson-Kaas synthesis of   158
5-phenyltetrazole, one-pot synthesis of   172
pH-sensitive superabsorbent hydrogel composite   220
Piloty–Robinson pyrrole synthesis   158, 162
platinum carbon nanotube/NPs hybrid catalyzed one-pot US-assisted synthesis, of dihydroquinazolinones   93
PMMA/ZnO nanocomposites   226
polyamino acids   77
poly(GMA) nanocomposites   226
poly(lactic-co-glycolic acid) (PLGA)   29
polymer grafting   30
polymer modification   222, 223
poly(N-isopropylacrylamide) (PNIPAm) nanogel   288
polyvinyl alcohol/phytic acid (PVA/PA) polymer   226, 227
porous silica   84, 243

potassium peroxodisulfate (KPS)   220
Povarov reaction   176–177
power ultrasound (US)   228, 369
PPy-DBSNa/ZnO (polypyrrole-sodium dodecylbenzenesulfonate/zinc oxide) nanocomposite   226
Preparative techniques   337–338, 370–372
PTSA-Montmorillonite assisted Friedel–Crafts reaction   106
pyridine-linked hydrazinylimidazoles   148
11H-pyrido[2,1-b]quinazolin-11-one derivatives   147
pyrimidine-2,4-diamine (PDA)   85, 86
pyrrole-imidazole synthesis   140

r

radical polymerization   222, 226, 228
random vector functional link (RVFL) model   371
reaction mechanisms and kinetics   363–364
resin assisted condensation reaction   100
ring-opening polymerization   222–223
Robinson–Gabriel reaction   167
rosuvastatin based azaindole derivatives   147

s

Sapare cross-linked polymers   221
SBA-15/[AubpyCl$_2$]Cl nanocatalyst   95
scanning electron microscopy (SEM)   88, 91, 93–95, 257, 285, 286, 291, 293, 294, 308, 309
semiconductor transition metal oxides   303
silicoaluminophosphate (SAPO-n) zeolites   91
silicon dioxide, synthesis of   243
silver nanoparticles   236, 238, 239
SiO$_2$ catalyst Pd(0) assisted Suzuki–Miyaura coupling   117
SiO$_2$/CuCl$_2$ catalyst assisted multicomponent reaction   123
SiO$_2$.Lm Cu(I) catalyst assisted Click reaction   127
Skraup reaction   174, 175
  of aniline   175
  of nitrobenzene   175
SMC reaction for biaryls formation   137
SnO$_2$ NPs   240, 241
Sn-polymer catalyst assisted oxidation reaction   112
sodium alginate-g-poly(acrylic acid-co-acrylamide)   220
solar cells   267, 362
solid acid and base catalysts
  condensation reactions   98–100
  coupling reactions   113–121
    Buchwald–Hartwig coupling   120
    carbonylation/carboxylation   121
    click chemistry   125–127
    Heck coupling   118–119
    micelliances reactions   121–125
    Sonogashira coupling   113–114
    Suzuki–Miyaura-coupling   114–118
    Ullmann coupling   120
  cyclisation reactions   100–104
  Friedel–Crafts reactions   106–107
  multi-component reactions   104–106
  oxidation   110–113
  reduction   109–110
solid-phase diazotation method   104
solid phase reactions   69–78
solvent-free synthesis   73
sonication or ultrasonic treatment   1
sonocatalysis   83
sonochemical reactions   83, 84, 144, 356
sonochemistry   2, 4–6, 12, 83, 191, 220, 228, 349, 357
  and catalysis   84
sonoelectrochemical technique   14
Sonogashira coupling   113, 114, 351
Sonogashira reactions   137
sonography-guided therapies   8
sonoluminescence   9–10, 349
sonolysis   83, 84, 227
sonophotocatalysis   9–10
spinel MgO/MgFe$_2$O$_4$ nanocatalysts   291

stable cavitation 5
starch-based materials (SHs) 31
step-growth polymerizations 222
Strecker-type reactions 124
substituted imidazoles
   one-pot synthesis of 91
   synthesis of 96
sulfur-modified ruthenium nanoparticles
   on gold 137
sustainable system
   green chemistry principles for 28
   sustainable system 28–35
Suzuki–Miyaura-coupling 114–118
Suzuki–Miyaura cross-coupling reaction
   (SMCR) 136–137
Suzuki–Miyaura reactions 116, 117, 142
Suzuki reaction of
   4-methylbromobenzene with
   phenyl boronic acid 146
synergistic effects 98, 144, 364–365
synthetic polymers 99, 219

t

temperature and energy management
   368–369
tert-butyl hydroperoxide (TBHP) 140
tetrahydrobenzo[b]pyrans 88
tetrahydropyranquinolines
   ultrasound-mediated MCR synthesis
   94
thermal food processing 360
thermal, microwave-assisted, and
   ultrasound-promoted preparations
   338–341
thermal oxidation of graphite for
   graphene preparation 360
thermal reactions, advantages and
   limitations 345–346
thiazolidin-4-one 54, 55
thiazoloquinolines preparation 95
3D cobalt (II) MOF 256
TiO_2/graphite nanocomposite 304
TiO_2 nanomaterials 240
TiO_2 nanoparticles 290, 304
tosylmethyl isocyanide 166

transient cavitation 5, 6
transmission electron microscopy (TEM)
   11, 93, 298–300, 307, 365
1,2,4-triazole formation 168
triazolothiadizepinylquinolines 138, 139
triboelectrification-induced
   electroluminescence (TIEL) 13

u

Ugi reaction 104
Ullmann coupling 120
Ullmann–Goldberg coupling 147
ultrapure nanomaterials 235
ultrasonic-assisted acceleration of
   hydrolysis time 195–196
ultrasonic-assisted enzymolysis
   applications of 186–187
   biological activity enhancement
   194–195
   principle of 190–191
   protein functionality modification
   193–194
   proteins and plant components
   extraction 192–193
   ultrasonic-assisted acceleration of
   hydrolysis time 195–196
ultrasonic-assisted tryptic protein
   digestion 198
ultrasonication assisted reaction, of
   2,3-dichloroquinoxaline with
   terminal alkyne 149
ultrasonic cleaning systems 11
ultrasonic irradiation 220
   copper based catalysts 89–94
   iron based catalysts 86–89
   misalliances metal based catalysts
   94–96
   in synthesis 13–14
ultrasonic treatment 1, 93, 193, 196, 226
ultrasound-aided enzymatic process 196
   lipase 196
   polysaccharide enzymes 198
   protease 196–198
ultrasound-assisted coupling reactions
   133, 134, 142–150

ultrasound assisted enzymatic reaction
   applications   12, 185–209
ultrasound-assisted preparations
   artificial intelligence (AI)   363
   biomedical   363
ultrasound-assisted synthesis   220
   advantages and disadvantages of   252
   of polymers   223–228
   principle of   250–252
ultrasound induced synthesis   75–76
ultrasound intensity   2
ultrasound irradiation   1, 242
   cardiology   7–8
   cavitation history   3–7
   chemical and biological reactions   12–14
   dental application   9
   diagnostic imaging   7
   high-frequency sound waves   2
   industrial cleaning   10–11
   interventional procedure   8
   material processing   11–12
   medical sciences   7
   non-destructive testing   13
   obstetrics and gynecology   7
   physiotherapy and rehabilitation   9
   sonography-guided therapies   8
   sonoluminescence and sonophotocatalysis   9–10
   underwater communication and sensing   12–13
   vascular imaging   8
ultrasound-promoted cyclization   354–355
ultrasound-promoted esterification   354
ultrasound-promoted organic synthesis   353–355
ultrasound-promoted oxidations   354
ultrasound-promoted reactions   347–349
underwater communication and sensing   12–13
underwater self-powered all-optical wireless ultrasonic sensor   13
US-facilitated synthesis   76

v

van Leusen imidazole synthesis   166
van Leusen oxazole synthesis   166–167
vibrating sample magnetometry (VSM)   91

w

waste cooking oil (WCO)   196, 293–296
waste-to-energy processes   361
wastewater treatment   1, 28, 34, 284–289
water pollution   283
wavelength   2, 35, 37, 38, 42, 48, 75, 134, 185, 220, 239, 357
Wire and Arc Additive Manufacturing (WAAM)   13

x

X-ray diffraction (XRD)   12, 88, 94, 365
   analysis   88

z

zeolite beta assisted transformation reaction   108
Zeolite Socony Mobil-5 (ZSM-5)   90, 291
zig-zag one dimensional polymer [Pb(l-2-pinh)$N_3H_2O]_n$ 223
ZnO nanorods   241
ZnO nanotubes   241